# Hypergeometry, Integrability and Lie Theory

CONTEMPORARY MATHEMATICS

780

# Hypergeometry, Integrability and Lie Theory

Virtual Conference
Hypergeometry, Integrability and Lie Theory
December 7–11, 2020
Lorentz Center
Leiden, The Netherlands

Erik Koelink
Stefan Kolb
Nicolai Reshetikhin
Bart Vlaar
Editors

AMS AMERICAN MATHEMATICAL SOCIETY

2020 *Mathematics Subject Classification.* Primary 13A35, 16S38, 17B37, 17B67, 17B80, 33C60, 33C67, 33D45, 43A90, 60J60.

For additional information and updates on this book, visit
**www.ams.org/bookpages/conm-780**

**Library of Congress Cataloging-in-Publication Data**

Cataloging-in-Publication Data has been applied for by the AMS.
See http://www.loc.gov/publish/cip/.
DOI: https://doi.org/10.1090/conm/780



∞ The paper used in this book is acid-free and falls within the guidelines
established to ensure permanence and durability.
Visit the AMS home page at `https://www.ams.org/`

10 9 8 7 6 5 4 3 2 1 27 26 25 24 23 22

# Contents

# Preface

The workshop *Hypergeometry, Integrability and Lie Theory* was planned to take place live at the Lorentz Center, Leiden, but due to COVID-19 measures it took place online in the week December 7-11, 2020. This volume contains the proceedings of this workshop.

## 1. Background

Hypergeometric functions have many spectacular appearances in mathematics and its applications. Ever since their introduction by Euler, Riemann, and Gauss, hypergeometric functions have played an important role. Initially they appear as solutions to classical differential equations that emerged from the analysis of specific applied problems. Their fundamental role in geometry and representation theory gradually became clear over many years. The workshop was focused on advances at the interface of the theory of hypergeometric type functions, quantum integrable systems and representation theory.

The study of quantum integrable systems, in particular the study of the spectrum of their Hamiltonians, is one of the most important applications of the theory of hypergeometric type functions and of the accompanying representation theory. These systems provide solvable models for quantum many-body problems, for quantum field theory and statistical mechanics. Together with topological and conformal field theories, quantum integrable systems provide an important class of models of quantum matter where the dynamics is almost completely determined by the symmetry. An important characteristic property of integrable systems is that they have sufficiently many commuting operators (quantum conservation laws) and one of the main challenges here is the description of the joint spectrum of these operators.

In all known examples the symmetries defining these models are given either by an action of a Lie group, a Lie algebra or by their deformations, such as quantum groups. This makes representation theory particularly important for the study of integrable systems. Hypergeometric functions appear in representation theory in many ways. Typically, they appear as spherical functions. For Lie groups and Lie algebras spherical functions are eigenfunctions of differential operators acting on functions on symmetric spaces and representing the radial part of Casimir elements. Generalizations of hypergeometric functions such as multivariable hypergeometric functions, basic hypergeometric functions, and elliptic hypergeometric functions appear in a similar way in representation theory of quantum groups, elliptic quantum groups, and related algebras such as Hecke algebras and double affine Hecke algebras. Some of the best known integrable systems that are related to spherical functions and similar objects in representation theory include Calogero-Moser-Sutherland (CMS) models, Toda lattices, spin chains and elliptic Painlevé equations.

Vector-valued spherical functions also appear as eigenfunctions of radial parts of Casimir operators acting on more complicated spaces, also associated with symmetric pairs. New important developments in the theory of vector-valued spherical type functions, their asymptotic expansions in terms of Harish-Chandra series, and their applications in harmonic analysis emerged recently.

The corresponding vector-valued analogue of the (single-variable) hypergeometric differential operator was introduced in works of Tirao, Grünbaum and others. It is naturally related to matrix-valued spherical functions for symmetric spaces of rank one. More recently, the role of matrix-valued spherical functions and matrix-valued orthogonal polynomials in the integrability of the non-abelian Toda lattice has become well understood. Another application of the vector-valued spherical functions is to solutions of boundary Khnizhnik-Zamolodchikov equations and boundary fusion rules. Matrix-valued spherical functions arise as eigenfunctions of explicit realizations of the matrix-valued partial differential operators which are expected to have a close relation to quantum commuting Hamiltonians for super-integrable quantum spin CMS systems. These new connections have the potential for establishing important breakthroughs in both fields.

The quantum inverse scattering method in the study of quantum integrable systems with periodic boundary conditions led to the quantum Yang-Baxter equation and to the discovery of quantum groups such as $q$-deformed enveloping algebras in the 1980s by Drinfeld and Jimbo. Similarly, work by Cherednik, Sklyanin, Kulish and others in the 1980s and 1990s suggested that systems with open boundary conditions lead to the quantum reflection equation and to comodule algebras over quantum groups. However, in the early days these comodule algebras remained largely formal objects. A parallel development in the 1990s aimed to identify Macdonald polynomials as zonal spherical functions on quantum group analogues of symmetric spaces. The construction of quantum symmetric spaces is based on coideal subalgebras in quantum universal enveloping algebras. The resulting theory of quantum symmetric pairs has seen a lot of progress within the last decade, and is presently emerging as a vast generalization of the theory of Drinfeld-Jimbo quantum groups. Their representation theory is a natural extension of the representation theory of real reductive Lie groups and is a natural place where many $q$-hypergeometric functions appear. From the perspective of quantum integrable systems, the coideal subalgebras in the theory of quantum symmetric pairs are the symmetry objects associated to the boundary in the same way as the quantum universal enveloping algebras are associated to the bulk. Various versions of boundary Knizhnik-Zamolodchikov type equations appear in this context.

From the above it is clear that one should expect more important connections between representation theory and quantum integrable systems. The main goal of this workshop was to develop stronger interactions between these fields.

## 2. Structure of workshop

Each day was centred on a designated topic, which was briefly introduced in a general and accessible way by a leading expert (a moderator), after which presentations followed. Even though the workshop was online, the discussions around the talks were very lively. Going online meant that the number of participants of the workshop was almost double the originally planned number.

To give a better impression of the workshop, we summarize each day below.

**Monday: Hypergeometric functions and representation theory.** The moderator Hjalmar Rosengren gave an introduction to the connection of hypergeometric functions to Lie groups, Dunkl operators and non-symmetric special functions by showing how these ideas work for single-variable Jacobi polynomials.

Margit Rösler discussed the limit transition from Heckman-Opdam functions of type BC to type A. In this limit transition she paid particular attention to the interpretation on symmetric spaces, and related integral representations, and as spherical functions on Gelfand pairs of infinite-dimensional groups as introduced by G. Olshanski. These proceedings include joint work of M. Rösler and M. Voit on the construction of martingales of Heckman-Opdam diffusion processes for root systems of type A and B.

Catharina Stroppel discussed an approach to Verlinde rings using double affine Hecke algebras (DAHA) at roots of unity. A Verlinde ring is a category consisting of isomorphism classes of certain representations of the loop group associated to a Lie group, and it comes with a tensor product. She showed how properties of Verlinde rings can be obtained from representations of DAHA.

Ivan Cherednik discussed a super-version, or spinor version, of the Dunkl operators in relation to degenerate DAHA in the rank one case. Apart from several results, such as the Cherednik-Matsuo theorem, he discussed how the Bessel functions can be used to model the spread of COVID-19, especially in the first wave.

Pavel Etingof discussed quantized Kleinian singularities of type $A_{n-1}$, also known as generalized Weyl algebras, and twisted traces on them. These twisted traces can be used to discuss particular classes of $*$-products, and connect to 3-dimensional superconformal field theory. In the case $n = 2$ this connects to unitary spherical representations of $SL(2, \mathbb{C})$, and moreover it connects to orthogonal polynomials, which for $n = 2, 3$ can be identified as polynomials from the Askey scheme, such as Meixner-Pollaczek and continuous Hahn polynomials. These proceedings also contain joint work by P. Etingof and D. Kazhdan on characteristic functions of certain $p$-adic integral operators, which restrict to the associated eigenspace as deformed Wronskians of solution sets of basic hypergeometric equations.

In the spirit of the theme of the first day, we also highlight a contribution to this volume by Tom Koornwinder on a natural classification of polynomials from the $q$-Askey scheme, developing an approach by Verde-Star.

**Tuesday: Symmetric spaces and coideal subalgebras.** The moderator Tom Koornwinder gave an introduction to harmonic analysis on symmetric spaces, interpolation polynomials and quantum symmetric pairs to prepare for the talks by Eric Opdam, Siddhartha Sahi, Gail Letzter and Stefan Kolb.

E. Opdam discussed some open problems in harmonic analysis related to hypergeometric functions for root systems and recent progress by T. Honda, H. Oda and N. Shimeno. He explained how this contributes to the classification of the exceptional discrete series. Interpolation polynomials are inhomogeneous symmetric polynomials indexed by partitions whose leading terms are given by Jack polynomials. It was conjectured by F. Knop and S. Sahi in 1996 that suitable normalized interpolations polynomials have positive coefficients when expressed in terms of monomials symmetric functions. In his talk, S. Sahi presented a proof of this conjecture, see also `arXiv:2104.08598`.

The remaining two talks of the day were devoted to the theory of quantum symmetric pairs (QSP). G. Letzter focused on the interplay between the Faddeev-Reshetikhin-Takhtajan type construction of QSP (Noumi, Sugitani, Dijkhuizen) and her own Drinfeld-Jimbo type approach. She pointed out similarities between the structure of the center and the construction of Cartan subalgebras for QSP. S. Kolb presented recent joint work with M. Yakimov on the interpretation of QSP as star-product deformations of partial quantum parabolic algebras. This perspective allows a conceptual, bar-involution free description of the quasi $K$-matrix. This suggests constructing the bar-involution via the quasi $K$-matrix, as explained in S. Kolb's contribution to this volume.

**Wednesday: Coideal subalgebras and integrable spin chains.** On the third day of the conference the focus moved to the applications of quantum group theory to integrable quantum models and exactly solvable models from statistical mechanics (Heisenberg spin chains, vertex models, asymmetric simple exclusion processes (ASEPs), etc.). The moderator Christian Korff introduced the topic by explaining the role of representations of quantum groups and coideal subalgebras in such models.

Bernard Nienhuis, presenting joint work with O. Huygen, discussed how the affine Temperley-Lieb algebra can be used to construct new explicit expressions of quasi-local conserved quantities of the closed XXZ chain.

Robert Weston talked about ongoing work with A. Cooper and B. Vlaar on Baxter's Q-operators for the open XXZ chain, via an isomorphism of infinite-dimensional Borel subalgebra tensor product representations. Although the additional coideal subalgebras make the setting rather complex, the compatibility of the isomorphism with the boundaries can be shown.

Finally, Jan de Gier discussed joint work with Z. Chen and M. Wheeler on an application of Hecke algebra representation theory, expressed in terms of non-symmetric Macdonald polynomials, to multi-species ASEPs. Dualities between such models arise as particular cases of quantum KZ equations and allow for the exact computations of certain observables.

**Thursday: Special functions and integrable systems.** On the penultimate day we continued the theme of applications to integrability, with a stronger emphasis on the role of special functions. The moderator Simon Ruijsenaars presented a bird's eye view on Calogero-Moser-Sutherland type and Toda type models and important classes of hypergeometric and other special functions.

Marta Mazzocco discussed recent joint work with L. Chekhov and V. Rubtsov on the generalized Sklyanin-Painlevé algebra, a Calabi-Yau algebra with desirable properties whose rational degenerations include many known algebras with various geometric (del Pezzo surfaces) and special functions (Askey-Wilson polynomials) interpretations, which at the same time directly relate to the double affine Hecke algebra of type $\widetilde{C}_1$. This was followed by a talk by Masatoshi Noumi on collaborative work with E. Langmann and J. Shiraishi, which showcased two types of joint eigenfunctions for elliptic Ruijsenaars difference operators: symmetric deformations of Macdonald polynomials defined in a neighbourhood of the torus, and asymptotically free eigenfunctions.

Volker Schomerus talked about conformal field theories and the role of objects closely related to multivariate $BC_2$-symmetric Heckman-Opdam hypergeometric functions as partial waves in these theories. His contribution to these proceedings is a survey of recent works and works in progress connecting conformal field theories and integrable systems of Calogero-Moser type.

In the last talk of the day, Paul Zinn-Justin discussed the connecting role of solutions of quantum KZ equations in Brauer loop models, generalized "pipe dreams" and, through Maulik-Okounkov stable envelopes, Gröbner geometry. His joint submission with A. Garbali to this volume is on a related topic, namely the description of an isomorphism between the commutative trigonometric Feigin-Odesskii shuffle algebra and the center of the Hecke algebra, with an application in enumerative combinatorics.

**Friday: Interactions.** On the last day of the workshop connections between the various themes were considered. The moderator Milen Yakimov gave an introduction to the four talks of the day, which had wide-ranging topics.

In Bart Vlaar's talk on joint work with A. Appel, a formalism of cylindrical quasitriangularity for quantum symmetric pair coideal subalgebras of Kac-Moody type was discussed which generalizes results by M. Balagović and S. Kolb and produces universal solutions of twisted reflection equations. A classification of such cylindrical structures was conjectured in terms of pseudo-involutions and pseudo-fixed-point subalgebras; the Kac-Moody theory of this is featured in the joint submission of V. Regelskis and B. Vlaar.

This was followed by Yuri Berest, who talked about on algebras of quasi-invariants, which first appeared in work by Chalykh and Veselov on Calogero-Moser type systems. Y. Berest explained how his cohomological realization of this algebra generalizes a classical theorem of A. Borel on Weyl group invariants and pointed out the homotopy-theoretic nature of the construction, allowing for extensions to $p$-adic reflection groups.

Alexander Varchenko's presented recent joint work with R. Rimányi on a Selberg integral formula for finite fields $\mathbb{F}_p$ for an odd prime $p$, which is related to hypergeometric solutions of (differential) KZ equations modulo $p$. Two works of his have been included in this volume which are on closely related topics: one in which a determinantal formula is proven for solutions of KZ equations over $\mathbb{F}_p$ in the case the dimension of the solution space is maximal, and one which constructs polynomial solutions of these equations modulo $p^s$, with special attention paid to the $p$-adic limit $s \to \infty$.

In the workshop's last talk, given by Eric Rains, the double affine Hecke algebra featured once more, or rather its elliptic version. He explained how this new algebraic structure yields results on biorthogonal functions and the integrability of the $BC_n$-symmetric Hamiltonian due to J. F. van Diejen. This volume contains related work by E. Rains on families of filtered deformations of elliptic algebras (twisted homogeneous endomorphism rings of vector bundles on elliptic curves), which computes various examples and constructs deformations from noncommutative del Pezzo surfaces.

## 3. A special occasion

On Wednesday December 9, 2020 we held an online celebration of the 50th birthday of Jasper Stokman, who holds these topics dear and who made fundamental contributions to them. Jasper was a PhD student at the Universiteit van Amsterdam under the supervision of Tom Koornwinder and during this time visited Kobe for 10 weeks in Spring 1997. After graduation in 1998, he was a postdoctoral researcher in Université Paris 6 and Université de Strasbourg. In 2000 he took up a position as KNAW fellow at the Universiteit van Amsterdam, which included a 3-month stay at MIT in 2002. At the UvA Jasper was appointed as an assistant professor in 2005 and an associate professor in 2007. In 2014 he was appointed professor of Lie Theory. He was also a professor by special appointment at Radboud Universiteit in the period 2012-2015.

Initially, Jasper worked on multivariable orthogonal polynomials and special functions, and their relation to quantum group structures and quantum Grassmannians. Later his attention shifted to noncompact quantum groups and related spherical functions, as well as fixed-point Lie subalgebras and the associated coideal subalgebras (reflection equation algebras). This in turn led him to topics in mathematical physics such as his work on (quantum) Knizhnik-Zamolodchikov equations and their application in the study of spin chains and loop models.

As part of this work he has also supervised many PhD students and post-docs, and has collaborated with researchers representing this wide array of topics. Jasper has also been very active in undergraduate and postgraduate teaching, and the organization of conferences, seminars and working groups in the Netherlands and beyond.

It is our pleasure to be able to include in this volume Jasper's recent work with Nicolai Reshetikhin on boundary Knizhnik-Zamolodchikov-Bernard operators and quantum Calogero-Moser spin chains, a new class of superintegrable systems. It reflects many of his interests and fits the themes of the workshop very well.

Erik Koelink<br>
Stefan Kolb<br>
Nicolai Reshetikhin<br>
Bart Vlaar

Contemporary Mathematics
Volume **780**, 2022
https://doi.org/10.1090/conm/780/15685

# Characteristic functions of $p$-adic integral operators

Pavel Etingof and David Kazhdan

*To Jasper Stokman on his 50th birthday with admiration*

Abstract. Let $P \in \mathbb{Q}_p[x,y]$, $s \in \mathbb{C}$ with sufficiently large real part, and consider the integral operator

$$(A_{P,s}f)(y) := \frac{1}{1-p^{-1}} \int_{\mathbb{Z}_p} |P(x,y)|^s f(x)|dx|$$

on $L^2(\mathbb{Z}_p)$. We show that if $P$ is homogeneous of degree $d$ then for each character $\chi$ of $\mathbb{Z}_p^\times$ the characteristic function $\det(1-uA_{P,s,\chi})$ of the restriction $A_{P,s,\chi}$ of $A_{P,s}$ to the eigenspace $L^2(\mathbb{Z}_p)_\chi$ is the $q$-Wronskian of a set of solutions of a (possibly confluent) $q$-hypergeometric equation, where $q = p^{-1-ds}$. In particular, the nonzero eigenvalues of $A_{P,s,\chi}$ are the reciprocals of the zeros of such $q$-Wronskian.

## Contents



## 1. Introduction

Let $P \in \mathbb{Q}_p[x,y]$ be a $p$-adic polynomial and $s \in \mathbb{C}$. We consider the operator

$$(A_{P,s}f)(y) = \frac{1}{1-p^{-1}} \int_{\mathbb{Z}_p} |P(x,y)|^s f(x)|dx| \tag{1.1}$$

on the Hilbert space $H := L^2(\mathbb{Z}_p)$ when $\mathrm{Re}(s)$ is not too negative (meaning that $\mathrm{Re}(s) > s_0$ for some $s_0 \in \mathbb{R}$). In this case $A_{P,s}$ is a well-defined Hilbert-Schmidt (in particular, compact) operator. Therefore, its spectrum consists of eigenvalues $\lambda_i \neq 0$ (if present) and also 0, which may or may not be an eigenvalue. Moreover, if $|P(x,y)| = |P(y,x)|$ and $s \in \mathbb{R}$ then $A_{P,s}$ is self-adjoint, hence $\lambda_i \in \mathbb{R}$ and we have a

The first author's work was partially supported by the NSF grant DMS - 1916120.
The project was supported by the second author's ERC grant N° 669655.

spectral decomposition $H = \mathrm{Ker}\, A_{P,s} \oplus \bigoplus_i H(\lambda_i)$, where $H(\lambda_i)$ is the $\lambda_i$-eigenspace of $A_{P,s}$ ([**L**]). The goal of this paper is to compute $\lambda_i$.

We mostly focus on the case when $P$ is homogeneous of degree $d$. Then $A_{P,s}$ commutes with the group of units $\mathbb{Z}_p^\times$, hence preserves each eigenspace $H_\chi \subset H$ of $\mathbb{Z}_p^\times$ corresponding to a character $\chi : \mathbb{Z}_p^\times \to \mathbb{C}^\times$. Denote the restriction of $A_{P,s}$ to $H_\chi$ by $A_{P,s,\chi}$. If $\mathrm{Re}(s)$ is sufficiently large then $A_{P,s,\chi}$ is trace class with exponential decay of eigenvalues, so we can define its characteristic function – the Fredholm determinant

$$h_{P,s,\chi}(u) := \det(1 - uA_{P,s,\chi}) = \prod_i (1 - u\lambda_i)^{\mathrm{mult}(\lambda_i)},$$

where $\mathrm{mult}(\lambda_i)$ is the algebraic multiplicity of $\lambda_i$. Then $h_{P,s,\chi}(u)$ is an entire function of Hadamard order 0 whose zeros are the reciprocals of $\lambda_i$ ([**L**]).

Our main result is the following theorem. Let $P(x,y) := x^{d-l}y^l Q(\frac{x}{y})$, where $Q(0) = 1$. Let $k := k_0 + \delta_{1,\chi}$, where $k_0$ is the order of the pole at 0 of the 2-variable zeta-function $Z(Q, \chi, s, z)$ defined by (2.2). Define **the $q$-Wronski matrix** of a collection of functions $f_1, \dots, f_k$ of a complex variable $u$ to be the $k$-by-$k$ matrix with entries

$$W(f_1, \dots, f_k)(u)_{ji} := f_i(q^{j-1}u).$$

Finally, let $\Phi_i$ be the solutions of the $q$-hypergeometric equation defined by (4.6), and

$$f_i(u) := (u; q)_\infty \Phi_i(u),$$

where

$$(u; q)_\infty := \prod_{n=0}^{\infty} (1 - uq^n);$$

then $u^{-\nu_i} f_i(u)$ are entire functions taking value 1 at the origin.

THEOREM 1.1. *The characteristic function $h_{P,s,\chi}$ is a limit of functions of the form*

$$h(u) = \frac{(\beta u)^{-\sum_{i=1}^k \nu_i}}{\prod_{1 \le i < j \le k} (q^{\nu_j} - q^{\nu_i})} \det W(f_1, \dots, f_k)(\beta u).$$

*as $\nu_1, \dots, \nu_{k_0} \to +\infty$ and $\beta \to \infty$ if $k_0 > 0$, where $q := p^{-1-ds}$.*

We note that such a limit can itself be written as a Wronskian of a set $f_1, \dots, f_k$ of solutions of a confluent $q$-hypergeometric difference equation. Also note that the order of this difference equation is typically bigger than $k$, so that $f_1, \dots, f_k$ don't span the space of its meromorphic solutions over the field of $q$-elliptic functions, so that the $q$-Wronskian does not factor explicitly. As a result, the eigenvalues of $A_{P,s}$ are typically given by *transcendental* functions of $p^s$, in contrast with Igusa's theorem [**I**] that integrals $\int_{\mathbb{Z}_p^n} |P(\mathbf{x})|^s |d\mathbf{x}|$ for a $p$-adic polynomial $P$ are always *rational* functions of $p^s$. We show, however, that for special values of parameters these eigenvalues may be algebraic.

Theorem 1.1 is proved in Section 4. Our proof of Theorem 1.1 provides a method of computing the function $h_{P,s,\chi}$. This method is based on realizing the operator $A_{P,s,\chi}$ as a first order $q$-difference operator on a space of analytic functions.

By the same method, analogous results can be obtained over a general non-archimedian local field $\mathbf{F}$, and for more general integral operators in which the character $w \mapsto |w|^s$ is replaced by any multiplicative character of $\mathbf{F}$. Further, they can be extended to integral operators whose kernels are of the form $\prod_{i=1}^{m} \chi_i(P_i(x,y))$ where $P_i$ are homogeneous polynomials and $\chi_i$ are multiplicative characters, and to linear combinations of such kernels. Since these extensions are straightforward, we will not discuss them in detail.

REMARK 1.2. One of our motivations for writing this paper was the desire to understand the eigenvalues of the $p$-**adic Hecke operators** defined in [**BK**], in particular the operators $T_x$ discussed in [**K**], Theorem 2, which are the Hecke operators of [**BK**] for $\mathbb{P}^1$ with 4 parabolic points and $G = PGL_2$. More precisely, in this paper we study what happens for $p$-adic integral operators"generically", which helps to recognize, by contrast, the special features of the situation of [**K**],[**BK**]. Namely, as explained in [**K**], Subsection 2.4, this situation should belong to the realm of "$p$-adic integrable systems" (a notion which still somewhat vague; see [**KO**] for recent developments), and as a result the eigenvalues of $T_x$ are algebraic numbers,[1] in contrast with the present paper, where the eigenvalues are (almost certainly) transcendental.

The organization of the paper is as follows. In Section 2 we discuss zeta-functions of univariate $p$-adic polynomials, and define a 2-variable zeta-function of such a polynomial, which is a rational function of variables $s$ and $z$ depending on a character of $\mathbb{Z}_p^\times$. In Section 3 we realize the operator $A_{P,s,\chi}$ for homogeneous $P$ as a first order $q$-difference operator on the space of analytic functions on a disk written in terms of the 2-variable zeta-function of the univariate polynomial corresponding to $P$. This allows us to write the equation for an eigenvector of $A_{P,s}$ as a first order $q$-difference equation. In Section 4 we recall the definition and properties of $q$-hypergeometric functions and then use them to solve the $q$-difference equation for an eigenvector of $A_{P,s,\chi}$. This allows us to express the characteristic function $h_{P,s,\chi}$ as a $q$-Wronskian of a set of entire solutions of a $q$-hypergeometric equation, thereby proving Theorem 1.1. In Section 5 we give examples of explicit computation of $h_{P,s,\chi}$ for specific classes of homogeneous polynomials $P$. Finally, in Section 6 we consider the non-homogeneous case. We first show that the operator $A_{P,s}$ is trace class for sufficiently large $\mathrm{Re}(s)$, and then consider an example of a non-homogeneous $P$ and show that in this example the problem of finding eigenvalues of $A_{P,s}$ leads to more complicated functional equations, namely Mahler's recursions ([**AS**, **BCCD**]).

## 2. Zeta-functions

Let $\chi$ be a character of $\mathbb{Z}_p^\times$. Let $Q \in \mathbb{Q}_p[x]$ be a polynomial with constant term 1 of degree $r$. Let $s \in \mathbb{C}$ with $\mathrm{Re}(s) > -\frac{1}{r}$ if $r \geq 1$. Define the zeta-function

$$\zeta(Q,\chi,s) := \frac{1}{1-p^{-1}} \int_{\mathbb{Z}_p^\times} |Q(x)|^s \chi(x) |dx|,$$

where $|dx|$ is the additive Lebesgue measure in which the volume of $\mathbb{Z}_p$ is 1. For example, if $|Q(x)| = a$ when $|x| = 1$ then $\zeta(Q,\chi,s) = a^s \delta_{1,\chi}$. The function $\zeta(Q,\chi,s)$ is a special case of the Igusa zeta-function of $Q$ and therefore is rational in $p^s$ by the result of [**I**].

[1] This was checked in 2007 by M. Vlasenko in examples ([**V**]) but the general proof has not been written down and will be given in our forthcoming joint paper with E. Frenkel.

Write $Q_m$ for the polynomial $Q_m(x) := Q(xp^{-m})$. Then

$$\frac{1}{1-p^{-1}}\int_{p^{-m}\mathbb{Z}_p^\times} |Q(x)|^s\chi(xp^m)|dx| = p^m\zeta(Q_m,\chi,s).$$

Let $c$ be the leading coefficient of $Q$. If $m \ll 0$ then $|Q_m(x)| = 1$ when $|x| = 1$, so $\zeta(Q_m,\chi,s) = \delta_{1,\chi}$. Likewise, if $m \gg 0$, we have $|Q_m(x)| = |c|p^{mr}$ when $|x| = 1$, so $\zeta(Q_m,\chi,s) = |c|^s p^{mrs}\delta_{1,\chi}$. So if $|c| = 1$ and $m$ is far enough from zero then

$$\zeta(Q_m,\chi,s) = p^{\max(m,0)rs}\delta_{1,\chi}. \tag{2.1}$$

EXAMPLE 2.1. Assume that $Q \in \mathbb{Z}_p[x]$ and its reduction $\overline{Q}$ to $\mathbb{F}_p[x]$ also has degree exactly $r$ and no roots in $\mathbb{F}_p$ (so $r \geq 2$). Then $|c| = 1$, so $|Q(x)| = 1$ when $|x| \leq 1$ and $|Q(x)| = |x|^r$ when $|x| > 1$. So

$$\zeta(Q_m,1,s) = p^{\max(m,0)rs}\delta_{1,\chi}.$$

for all $m$.

EXAMPLE 2.2. Let $Q(x) = 1 - x$, so $r = 1$. If $|x| = 1$ then $|1 - p^{-m}x| = 1$ if $m < 0$ and $|1 - p^{-m}x| = p^m$ if $m > 0$. Thus formula (2.1) holds for $m \neq 0$. For $m = 0$, we compute:

$$(1-p^{-1})\zeta(1-x,\chi,s) = \int_{\mathbb{Z}_p^\times} |1-x|\chi(x)|dx| =$$

$$\sum_{k=2}^{p-1}\int_{k+p\mathbb{Z}_p}\chi(x)|dx| + \sum_{n\geq 1} p^{-ns}\sum_{k=1}^{p-1}\int_{1+kp^n+p^{n+1}\mathbb{Z}_p}\chi(x)|dx|.$$

Thus for $\chi = 1$ we get

$$\zeta(1-x,1,s) = 1 - \frac{p^{-1}(1-p^{-s})}{(1-p^{-1})(1-p^{-s-1})}.$$

If $\chi \neq 1$ then let $\ell$ be the smallest integer such that $\chi$ is trivial on $1 + p^\ell\mathbb{Z}_p$. Then we get

$$\zeta(1-x,\chi,s) = -\frac{p^{-1}(1-p^{-\ell(s+1)+1})}{(1-p^{-1})(1-p^{-s-1})}.$$

So for all $\chi$ we get

$$\zeta(1-x,\chi,s) = \delta_{1,\chi} - \frac{p^{-1}(1-p^{-\ell(s+1)+1})}{(1-p^{-1})(1-p^{-s-1})},$$

where for $\chi = 1$ we set $\ell = 0$.

Define the **2-variable zeta function** of $Q, \chi$ by

$$Z(Q,\chi,s,z) := \sum_{m\in\mathbb{Z}}(\zeta(Q_m,\chi,s) - \delta_{1,\chi})z^m. \tag{2.2}$$

It is easy to see that this series is finite in the negative direction and converges for $|z| < \min(1, p^{-r\operatorname{Re}(s)})$ to a rational function of $p^s, z$ which has the form

$$Z(Q,\chi,s,z) = Z_0(Q,\chi,s,z) + \delta_{1,\chi}\left(\frac{|c|^s}{1-p^{rs}z} - \frac{1}{1-z}\right), \tag{2.3}$$

where $Z_0 \in \mathbb{C}(p^s)[z,z^{-1}]$ is a Laurent polynomial. For instance, in Example 2.1 we have $Z_0 = 0$, while in Example 2.2 $Z_0 = -\frac{p^{-1}(1-p^{-\ell(s+1)+1})}{(1-p^{-1})(1-p^{-s-1})}$ is independent of $z$. However, it is clear that in general $Z_0$ may contain any integer power of $z$.

## 3. Realization of $A_{P,s,\chi}$ on analytic functions

Let $P \in \mathbb{Q}_p[x,y]$ be a homogeneous polynomial of degree $d$. Without loss of generality we may (and will) assume that

$$P(x,y) = x^{d-l}y^l Q(\tfrac{x}{y}),$$

where $Q$ is a polynomial with constant term 1, of some degree $r \le l$. Let $s \in \mathbb{C}$, and consider the operator $A_{P,s,\chi}$ given by (1.1) acting on $H_\chi$. We are interested in nonzero eigenvalues of $A_{P,s,\chi}$.

Recall that for two linear endomorphisms $B, C$ of a vector space, nonzero eigenvalues of $BC$ and $CB$ are the same. Thus we may assume without loss of generality that $l = d$.

The space $H_\chi$ can be identified with the space $\ell_2$ of sequences $\{f_n\}$ such that $\sum_{n\ge 0} |f_n|^2 < \infty$ by the assignment

$$f_n = p^{-\frac{n}{2}} f(p^n),\ n \ge 0.$$

In terms of this presentation, we have

$$(A_{P,s,\chi}f)(y) = \frac{|y|^{ds}}{1-p^{-1}} \sum_{n\ge 0} p^{\frac{n}{2}} f_n \int_{|x|=p^{-n}} |Q(\tfrac{x}{y})|^s \chi(xp^{-n})|dx|.$$

Therefore, replacing $\frac{x}{y}$ with $x$ in the integral, we obtain

$$(A_{P,s,\chi}f)_m = p^{-\frac{m}{2}}(A_{P,s,\chi}f)(p^m) =$$

$$\frac{p^{-\frac{3m}{2}-dms}}{1-p^{-1}} \sum_{n\ge 0} p^{\frac{n}{2}} f_n \int_{|x|=p^{m-n}} |Q(x)|^s \chi(xp^{m-n})\,dx|.$$

So we get that

$$(A_{P,s,\chi}f)_m = \sum_{n\ge 0} a_{mn} f_n \text{ where } a_{mn} := p^{-\frac{m+n}{2}-dms}\zeta(Q_{m-n},\chi,s). \tag{3.1}$$

EXAMPLE 3.1. Assume that we are in the situation of Example 2.1. Then

$$a_{mn} = p^{-\frac{m+n}{2}-dms} p^{\max(m-n,0)rs} = p^{-\frac{m+n}{2}-(d-r)ms-\min(m,n)rs}.$$

Let us represent the sequence $f_n$ by the analytic function

$$F(z) = \sum_{n\ge 0} f_n p^{\frac{n}{2}} z^n$$

in the disc $|z| < p^{-\frac{1}{2}}$. This identifies $H_\chi$ with the space $\mathcal{H}$ of such analytic functions whose boundary values are in $L^2$, with the norm being the usual $L^2$ norm on the boundary. Multiplying (3.1) by $p^{\frac{m}{2}} z^m$ and adding up over $m \ge 0$, we get

$$(A_{P,s,\chi}F)(z) = \sum_{m\ge 0, n\ge 0} p^{-\frac{n}{2}-dms}\zeta(Q_{m-n},\chi,s) z^{m-n} f_n z^n =$$

$$\delta_{1,\chi}\frac{F(p^{-1})}{1-p^{-ds}z} + \sum_{m\ge 0, n\ge 0} p^{-\frac{n}{2}-dms}(\zeta(Q_{m-n},\chi,s) - \delta_{1,\chi}) z^{m-n} f_n z^n =$$

$$\delta_{1,\chi}\frac{F(p^{-1})}{1-p^{-ds}z} + (Z(Q,\chi,s,p^{-ds}z)F(p^{-1-ds}z))_+$$

where $G_+$ is the regular part of a Laurent series $G$.

Thus we obtain the following proposition.[2]

PROPOSITION 3.2. *If $l = d$ then the operator $A_{P,s}$ on $H_\chi$ is equivalent to the operator on $\mathcal{H}$ given by the formula*

$$(A_{P,s,\chi}F)(z) =$$

$$\delta_{1,\chi}\left(\frac{F(p^{-1}) - F(p^{-1-ds}z)}{1-p^{-ds}z} + \frac{|c|^s F(p^{-1-ds}z)}{1-p^{-(d-r)s}z}\right) + (Z_0(Q,\chi,s,p^{-ds}z)F(p^{-1-ds}z))_+.$$

Conjugating this by $|x|^{d-l}$, we obtain a similar formula for general $l \in [r,d]$:

$$\begin{aligned}(A_{P,s,\chi}F)(z) =&\delta_{1,\chi}\left(\frac{F(p^{-1-(d-l)s}) - F(p^{-1-ds}z)}{1-p^{-(d-l)s}z} + \frac{|c|^s F(p^{-1-ds}z)}{1-p^{-(l-r)s}z}\right)\\ &+ (Z_0(Q,\chi,s,p^{-ls}z)F(p^{-1-ds}z))_+.\end{aligned}$$

This immediately implies

COROLLARY 3.3. *If $\mathrm{Re}(s) > -\frac{1}{\max(2(d-l),2(l-r),d)}$ then the operator $A_{P,s,\chi}$ is well defined and trace class, with exponential decay of eigenvalues. Moreover, if $\chi \neq 1$ then this is so for $\mathrm{Re}(s) > -\frac{1}{d}$.*

So we will assume from now on that $\mathrm{Re}(s)$ varies in this range.

## 4. $q$-hypergeometric functions and proof of Theorem 1.1

**4.1. Basic hypergeometric series.** The general **basic hypergeometric series** is defined by the formula

$$\begin{aligned}&{}_\ell\phi_{r-1}(a_1,\ldots,a_\ell;b_1,\ldots,b_{r-1};q,u)\\ &=\sum_{n=0}^\infty \frac{(a_1;q)_n\ldots(a_\ell;q)_n}{(b_1;q)_n\ldots(b_{r-1},q)_n(q;q)_n}\left((-1)^n q^{\frac{n(n-1)}{2}}\right)^{r-\ell} u^n,\end{aligned}$$

where $\ell \le r$ and $(a,q)_n$ is the $q$-Pochhammer symbol:

$$(a,q)_n := (1-a)(1-aq)\ldots(1-aq^{n-1}),$$

see [**GR**]. We assume that $|q| < 1$ and $b_j$ are not equal to a nonpositive integer power of $q$. Then this series converges for $|u| < 1$ if $r = \ell$ and for all $u$ if $r > \ell$.

There is also Bailey's variant of the basic hypergeometric series

$${}_\ell\varphi_{r-1}(a_1,\ldots,a_\ell;b_1,\ldots,b_{r-1};q,u) = \sum_{n=0}^\infty \frac{(a_1;q)_n\ldots(a_\ell;q)_n}{(b_1;q)_n\ldots(b_{r-1},q)_n(q;q)_n}u^n,$$

which converges in the disc $|u| < 1$ for all $\ell, r$. The two versions agree if $r = \ell$, and ${}_\ell\varphi_{r-1}$ can be obtained from ${}_m\phi_{m-1}$ with $m = \max(\ell, r)$ by specializing some parameters to 0. Also, ${}_\ell\phi_{r-1}$ can be obtained from ${}_r\varphi_{r-1}$ by sending $a_{\ell+1},\ldots,a_r$ to $\infty$.

[2]Note that Proposition 3.2 continues to hold when $d$ is not necessarily an integer. Indeed, the function $|y|^d|Q(x/y)|$ makes sense for non-integer $d$.

**4.2. The $q$-hypergeometric difference equation.** It is easy to see that ${}_\ell\varphi_{r-1}$ satisfies the difference equation

$$(4.1)\quad u(1-a_1T)\dots(1-a_\ell T)\Phi(u) = (1-b_1q^{-1}T)\dots(1-b_{r-1}q^{-1}T)(1-T)\Phi(u),$$

where $(T\Phi)(u) := \Phi(qu)$. This equation is called the **$q$-hypergeometric difference equation**.

We will mostly use the special case $r=\ell$, when $\phi=\varphi$ and

$${}_\ell\phi_{\ell-1}(a_1,\dots,a_\ell;b_1,\dots,b_{\ell-1};q,u) = \sum_{n=0}^{\infty}\frac{(a_1;q)_n\dots(a_\ell;q)_n}{(b_1;q)_n\dots(b_{\ell-1},q)_n(q;q)_n}u^n.$$

Equation (4.1) immediately implies that ${}_\ell\phi_{\ell-1}$ is meromorphic in the whole complex plane with at most simple poles at the points $1, q^{-1}, q^{-2}, \dots,$ present for generic parameter values.

EXAMPLE 4.1. The **Heine hypergeometric function** ${}_2\phi_1(a,b;c;q,u)$ satisfies the equation

$$u(1-aT)(1-bT)\Phi(u) = (1-cq^{-1}T)(1-T)\Phi(u).$$

One may also consider a more general $q$-hypergeometric equation

$$(4.2)\qquad u(1-a_1T)\dots(1-a_\ell T)\Phi(u) = (1-b_1T)\dots(1-b_rT)\Phi(u).$$

If $a_i, b_j \neq 0$, this equation is regular at 0 iff $r\ge \ell$ and at $\infty$ iff $r\le l$. If $b_j = q^{-\nu_j}$ are distinct, equation (4.2) has $r$ solutions with power behavior at 0:

$$(4.3)\quad \Phi_{\ell,r,i}(a_1,\dots,a_\ell;b_1,\dots,b_r;q,u) := u^{\nu_i}{}_\ell\varphi_{r-1}(\tfrac{a_1}{b_i},\dots,\tfrac{a_\ell}{b_i};\tfrac{qb_1}{b_i},\dots,\tfrac{qb_r}{b_i};q,u),$$

where the $i$-th term in the $b$-list is omitted. These solutions form a basis of solutions if and only if $r\ge\ell$ (or, equivalently, the equation is regular at 0).

**4.3. Confluent limits.** The $q$-hypergeometric difference equation has many interesting limits when various parameters go to 0 and $\infty$ in various coordinated ways, which are called **confluent limits**. Under these limits, the $q$-hypergeometric equation turns into **confluent hypergeometric equations** which may be irregular at 0 and $\infty$, and $q$-hypergeometric functions turn into **confluent $q$-hypergeometric functions**. For example, the $q$-hypergeometric functions ${}_\ell\varphi_{r-1}$ and ${}_\ell\phi_{r-1}$ are confluent unless $\ell=r$.

A basic example of a confluent $q$-hypergeometric function is the entire function

$$(4.4)\qquad J(a,q,u) := \sum_{n=0}^{\infty}\frac{(-1)^nq^{\frac{n(n-1)}{2}}u^n}{(a;q)_n(q;q)_n} = {}_1\phi_1(0;a;q,u).$$

which will arise in the next section. This function satisfies the **$q$-Bessel difference equation**

$$-uT\Phi(u) = (1-aq^{-1}T)(1-T)\Phi(u)$$

and expresses in terms of the **Hahn-Exton $q$-Bessel function**

$$J_\nu(x;q) := x^\nu\frac{(q^{\nu+1};q)_\infty}{(q,q)_\infty}J(q^{\nu+1},q,qx^2)$$

also known as **the third Jackson $q$-Bessel function**, see [**KSw, KS, GR**].

**4.4. The spectrum of a first order difference operator and proof of Theorem 1.1.** Let $\mathcal{H}$ be the Hilbert space of analytic functions in the disk $|z| < 1$ with $L^2$ boundary values, with norm

$$\|F\|^2 = \frac{1}{2\pi}\int_{|z|=1} |F(z)|^2 d\theta.$$

For a meromorphic function $G$ at a point $a \in \mathbb{C}$, let $\mathrm{pr}(G, a)$ be the principal part of $G$ at $a$ (a polynomial in $\frac{1}{z-a}$ with zero constant term). Let $q \in \mathbb{C}, |q| < 1$. Let $R(z) \in \mathbb{C}(z)$ be a nonzero rational function equipped with a subset $\{z_1, \dots, z_k\}$ of its poles in the disk $|z| < |q|^{-1}$ that we call **marked**, which includes all its poles in the smaller disk $|z| \le 1$.

Define an operator on $\mathcal{H}$ by the formula

$$(B_R F)(z) = R(z)F(qz) - \sum_{j=1}^{k} \mathrm{pr}(R(z)F(qz), z_j).$$

It is easy to see that this operator is well defined and trace class. The goal of this subsection is to compute the characteristic function $h_R(u) = \det(1 - uB_R)$ of $B_R$. This will imply Theorem 1.1, as the operator $A_{P,s,\chi}$ is a special case of $B_R$.

We start with the case when $z_1, \dots, z_k$ are simple poles and $z_i \ne 0$ for all $i$. Let $b_i = z_i^{-1}$. Then

$$R(z) = \beta\frac{(1-a_1z)\dots(1-a_\ell z)}{(1-b_1z)\dots(1-b_rz)}$$

for some $r \ge k$, $\beta \ne 0$. Let $b_j^{-1}R_j$ be the residues of $R$ at the poles $z_j$, $j \in [1,k]$; so we have

$$R_j = -\beta\frac{(1-\frac{a_1}{b_j})\dots(1-\frac{a_\ell}{b_j})}{(1-\frac{b_1}{b_j})\dots(1-\frac{b_r}{b_j})},$$

where the $j$-th factor in the denominator is omitted. Now the operator $B_R$ takes the form

$$(B_RF)(z) = \beta\frac{(1-a_1z)\dots(1-a_\ell z)}{(1-b_1z)\dots(1-b_rz)}F(qz) + \sum_{j=1}^{k}\frac{R_jF(qb_j^{-1})}{1-b_jz}.$$

So the eigenvalue equation for $B_R$ looks like

$$\lambda F(z) = \beta\frac{(1-a_1z)\dots(1-a_\ell z)}{(1-b_1z)\dots(1-b_rz)}F(qz) + \sum_{j=1}^{k}\frac{C_j}{1-b_jz},$$

where $C_j := R_jF(qb_j^{-1})$.

Consider the difference equation

$$F(z) = u\frac{(1-a_1z)\dots(1-a_\ell z)}{(1-b_1z)\dots(1-b_rz)}F(qz) + \frac{1}{1-b_jz}.$$

For small $u$ it has a unique power series solution, obtained by iterating the equation:

$$F_j(z,u) = \frac{1}{1-b_jz} + u\frac{(1-a_1z)\dots(1-a_\ell z)}{(1-b_1z)\dots(1-b_rz)}\frac{1}{1-b_jqz} + \dots =$$

$$\frac{1}{1-b_jz}\cdot {}_{\ell+1}\varphi_r(a_1z,\dots,a_\ell z, q; b_1z,\dots,qb_jz,\dots,b_rz; q, u).$$

So for the eigenvector we have

$$F(z,\lambda^{-1}) = \sum_{j=1}^{k} \lambda^{-1} C_j F_j(z,\beta\lambda^{-1}) =$$

$$\sum_{j=1}^{k} \frac{\lambda^{-1} C_j}{1-b_j z} \cdot {}_{\ell+1}\varphi_r(a_1 z,\dots,a_\ell z, q; b_1 z,\dots, q b_j z,\dots, b_r z; q, \beta\lambda^{-1}). \tag{4.5}$$

So we get

$$C_i = R_i F(q b_i^{-1}) =$$

$$\sum_{j=1}^{k} \frac{\lambda^{-1} R_i C_j}{1 - q b_i^{-1} b_j} \cdot {}_{\ell+1}\varphi_r(\tfrac{q a_1}{b_i},\dots,\tfrac{q a_\ell}{b_i}, q; \tfrac{q b_1}{b_i},\dots,\tfrac{q^2 b_j}{b_i},\dots,\tfrac{q b_r}{b_i}; q, \beta\lambda^{-1}).$$

This can be written as

$$\sum_{j=1}^{k} C_j M_{ji}(\beta\lambda^{-1}) = 0,$$

where

$$M_{ii}(u) := {}_{\ell}\varphi_{r-1}(\tfrac{a_1}{b_i},\dots,\tfrac{a_\ell}{b_i}; \tfrac{b_1}{b_i},\dots,\tfrac{b_r}{b_i}; q, u),$$

and for $j \neq i$

$$M_{ji}(u) := \frac{{}_{\ell}\varphi_{r-1}(\tfrac{a_1}{b_i},\dots,\tfrac{a_\ell}{b_i}; \tfrac{b_1}{b_i},\dots,\tfrac{q b_j}{b_i},\dots,\tfrac{b_r}{b_i}; q, u) - {}_{\ell}\varphi_{r-1}(\tfrac{a_1}{b_i},\dots,\tfrac{a_\ell}{b_i}; \tfrac{b_1}{b_i},\dots \tfrac{q b_j}{b_i},\dots,\tfrac{b_r}{b_i}; q, qu)}{1 - \tfrac{b_j}{b_i}},$$

where in both cases the $i$-th term in the $b$-list is omitted. Hence

$$M_{ji}(u) = \frac{u^{-\nu_i} \prod_{m\neq j}(1 - b_m T)}{\prod_{m\neq i}(1 - \frac{b_m}{b_i})} \Phi_i(u),$$

where $m$ varies in $[1,k]$ and

$$\Phi_i(u) := \Phi_{\ell,r,i}(a_1,\dots,a_\ell; b_1,\dots,b_k, q^{-1} b_{k+1},\dots,q^{-1} b_r; q, u) \tag{4.6}$$

are the solutions of the $q$-hypergeometric difference equation defined by formula (4.3) (with shifted parameters). Thus we have a $k$-by-$k$ matrix $M(u)$, and the eigenvalues of $B_R$ are solutions of the equation

$$\det M(\beta\lambda^{-1}) = 0.$$

The matrix $M(u)$ expresses in terms of the $q$-Wronski matrix of the functions $\Phi_1,\dots,\Phi_k$. Indeed, the $q$-Wronski matrix $W = W(\Phi_1,\dots,\Phi_k)$ has entries

$$W_{ni}(u) = T^{n-1}\Phi_i(u).$$

But by the Newton interpolation formula we have

$$T^{n-1} = \sum_{j=1}^{k} b_j^{1-n} \frac{\prod_{m\neq j}(1 - b_m T)}{\prod_{m\neq j}(1 - \frac{b_m}{b_j})}.$$

Thus, setting $D := \operatorname{diag}(\prod_{m\neq i}(1 - \frac{b_m}{b_i}))$ and $N(u) := \operatorname{diag}(u^{-\nu_i})$, we have

$$M(u) = V^{-1} D W(u) N(u) D^{-1},$$

where $V$ is the Vandermonde matrix with entries

$$V_{nj} = b_j^{1-n}.$$

In particular, we find that the eigenvalues of $B_R$ are solutions of the equation

$$\det W(\beta\lambda^{-1}) = 0.$$

If $k = r = \ell$, the matrix $W(u)$ is the $q$-Wronski matrix of a **basis** of solutions of the $q$-hypergeometric difference equation, so its determinant factorizes, resulting in the eigenvalues being of the form $\frac{\prod_i a_i}{\prod_i b_i} q^n$, $n \geq 0$. This is not surprising since in this case the matrix of the operator $B_R$ is triangular, so the eigenvalues are just the diagonal entries.

Now to obtain the characteristic function of $B_R$ we need to eliminate the poles of the function $\det W(\beta\lambda^{-1})$ (i.e., "clear denominators"). Since $q$-hypergeometric functions have first order poles at $1, q^{-1}, q^{-2}$ etc., this is achieved by replacing $\Phi_i(u)$ by

$$f_i(u) = (u;q)_\infty \Phi_i(u).$$

The functions $f_i(u)$ satisfy a simple modification of the $q$-hypergeometric equation, and we have

PROPOSITION 4.2.

$$\det(1 - u\beta^{-1}B_R) = \frac{u^{-\sum_{i=1}^k \nu_i}}{\prod_{i<j}(q^{\nu_j} - q^{\nu_i})} \det W(f_1, \ldots, f_k)(u). \tag{4.7}$$

PROOF. Since $\det V = \prod_{i<j}(q^{\nu_j} - q^{\nu_i})$, the two sides of (4.7) are entire functions of Hadamard order 0 with the same zeros, both having value 1 at the origin. The Proposition follows. □

REMARK 4.3. This method also gives explicit q-hypergeometric formulas for eigenvectors of the operator $B_R$, which are given by (4.5), where $\mathbf{C} := (C_1, \ldots, C_k)$ is a null row-vector of the matrix $V^{-1}DW(\beta\lambda^{-1})$. Moreover, if the operator $B_R$ is self-adjoint, the orthogonality relations for the eigenvectors yield interesting identities with $q$-hypergeometric functions.

Now, if the simple poles $z_j = b_j^{-1}$, $j \in [1, k]$ of $R$ are allowed collide and produce multiple poles, including one at the origin (which would give a fully general function $R(z)$) then the characteristic function of $B_R$ is given by a confluent limit of the formula of Proposition 4.2. But by Proposition 3.2, the operator $A_{P,s,\chi}$ is a special case of $B_R$, with $k = k_0 + \delta_{1,\chi}$, where $k_0$ is the order of pole of $Z(Q, \chi, s, z)$ at the origin, i.e., is obtained from the generic situation by sending $\nu_1, \ldots, \nu_{k_0}$ to $+\infty$ and $\beta \to \infty$ if $k_0 > 0$. Therefore Proposition 4.2 and Proposition 3.2 imply Theorem 1.1.

REMARK 4.4. The same results with obvious changes extend over any non-archimedian local field $\mathbf{F}$ instead of $\mathbb{Q}_p$. Namely, $\mathbb{Z}_p$ should be replaced by the ring of integers $\mathcal{O}_\mathbf{F}$, $p \in \mathbb{Z}_p$ by a uniformizer $\pi \in \mathcal{O}_\mathbf{F}$ and $p \in \mathbb{C}$ by the order of the residue field $\mathcal{O}_\mathbf{F}/\pi\mathcal{O}_\mathbf{F}$ of $\mathbf{F}$. Furthermore, the character $w \mapsto |w|^s$ in the integral can be replaced with any multiplicative character $w \mapsto \eta(w) := |w|^s \eta_0(w)$, where $\eta_0$ is a character of $\mathcal{O}_\mathbf{F}^\times$, and it is easy to see that the corresponding function $Z_0(P, \chi, s, \eta_0, z)$ is still a Laurent polynomial of $z$. Finally, we can further extend this theory to kernels of the form $\prod_{i=1}^m \eta_i(P_i(x, y))$, and in fact to any sufficiently nice homogeneous kernels, as well as linear combinations of such kernels. Since these extensions are straightforward, we will not discuss them in detail.

**4.5. The case when $R$ is a Laurent polynomial.** To illustrate the confluent limit of Proposition 4.2, let us consider the case when $R$ is a Laurent polynomial:

$$R(z) = z^{-k}(1 - a_1 z)\dots(1 - a_\ell z), \ k \geq 1, \ell \geq k.$$

This is an important special case, as it occurs for $A_{P,s,\chi}$ with $\chi \neq 1$. The simplest interesting example $k = 1, \ell = 2$ is worked out even more explicitly in Section 5.5.

For simplicity assume that $a_i$ are distinct; the general case is similar. Then we have

$$(B_R F)(z) = (R(z)F(qz))_+.$$

To find the spectrum of $B_R$, let us introduce a parameter $t$ and let

$$R_t(z) = (-t)^{-k} b_1 \cdots_k \frac{(1 - a_1 z)\dots(1 - a_\ell z)}{(1 - t^{-1}b_1 z)\dots(1 - t^{-1}b_k z)},$$

for some distinct $b_i \neq 0$. Then $R_t(z) \to R(z)$ as $t \to 0$, so $\lim_{t\to 0} B_{R_t} \to B_R$. Thus we can find the characteristic function of $B_R$ by finding the characteristic function of $B_{R_t}$ and taking the limit $t \to 0$.

In the limit $t \to 0$, the difference equation for the functions $\Phi_i$ takes the form

$$(1 - a_1 T)\dots(1 - a_\ell T)\Psi = u^{-1}T^k \Psi. \tag{4.8}$$

which can be written as

$$(-1)^\ell a_1 \dots a_\ell (1 - a_1^{-1}T^{-1})\dots(1 - a_\ell^{-1}T^{-1})\Psi = q^\ell u^{-1} T^{k-\ell}\Psi. \tag{4.9}$$

This equation is regular at $\infty$, so it has a basis of (confluent) $q$-hypergeometric solutions with power behavior at infinity, $\Psi_1, \dots, \Psi_\ell$, such that $\Psi_i(u) \in u^{\mu_i}\mathbb{C}[[u^{-1}]]$, where $\mu_i := -\log_q a_i$. Namely,

$$\Psi_i(u) = u^{\mu_i}\Psi_i^0(u),$$

and

$$\Psi_i^0(u) = {}_{\ell-k}\phi_{\ell-1}\left(0, \dots, 0; \tfrac{qa_i}{a_1}, \dots, \tfrac{qa_i}{a_\ell}; q, \frac{(-1)^k q^\ell a_i^{\ell-k}}{a_1 \dots a_\ell u}\right).$$

is an entire function in $u^{-1}$ equal to 1 at $\infty$.

Let

$$\theta(u) = \theta(u, q) := (q;q)_\infty(-u;q)_\infty(-qu^{-1};q)_\infty$$

be the Jacobi theta function. We have $\theta(qu) = u^{-1}\theta(u)$. Thus conjugating (4.8) by $\theta(u)^{\frac{1}{k}}$, we get for every solution $\Psi$ of this equation:

$$u(1 - a_1 u^{\frac{1}{k}}T)\dots(1 - a_\ell u^{\frac{1}{k}}T)\theta^{\frac{1}{k}}\Psi = (u^{\frac{1}{k}}T)^k \theta^{\frac{1}{k}}\Psi,$$

i.e.,

$$(1 - a_1 u^{\frac{1}{k}}T)\dots(1 - a_\ell u^{\frac{1}{k}}T)\Psi_* = q^{\frac{k-1}{2}}T^k\Psi_*.$$

where $\Psi_* := \theta^{\frac{1}{k}}\Psi$ (using appropriate branches of the function $z^{\frac{1}{k}}$). This equation has $k$ independent solutions with power behavior near zero (in terms of the variable $u^{\frac{1}{k}}$), $E_1, \dots, E_k$, which can be found using the power series method. Namely,

$$E_m(u) = u^{\frac{2\pi i m}{k \log q}} E(e^{\frac{2\pi i m}{k}} u^{\frac{1}{k}})$$

where $E$ is an entire function with $E(0) = 1$.

We have

$$E(v) = \sum_{j=1}^{\ell} \xi_j(v)\theta(v^k)^{\frac{1}{k}}\Psi_j(v^k),$$

where $\xi_j(q^{\frac{1}{k}}v)=\xi_j(v)$. Also the functions

$$\widetilde{\xi}_j(v):=v^{k\mu_j}\theta(v^k)^{\frac{1}{k}}\xi_j(v^k)$$

must be meromorphic in $v$ away from the origin. We have

$$\widetilde{\xi}_j(q^{\frac{1}{k}}v)=q^{\mu_j}v^{-1}\widetilde{\xi}_j(v).$$

Thus

$$\widetilde{\xi}_j(v)=\theta(q^{-\mu_j}v,q^{\frac{1}{k}})\eta_j(v),$$

where $\eta_j$ is an elliptic function with period $q^{\frac{1}{k}}$. We conclude that

$$E(v)=\sum_{j=1}^{\ell}\eta_j(v)\theta(a_jv,q^{\frac{1}{k}})\Psi_j^0(v^k).$$

LEMMA 4.5. *The elliptic functions $\eta_j$ are constant.*

PROOF. Assume the contrary. Let $p$ be a pole of $\eta_l$ for some $l$, and assume that its order $r$ is maximal among all $\eta_j$ at $p$. Let $\lim_{v\to p}(v-p)^r\eta_j(v)=c_j$; so $c_l\neq 0$. Thus

$$(v-p)^rE(vq^{\frac{n}{k}})=q^{-\frac{n(n-1)}{2k}}v^{-n}\sum_{j=1}^{\ell}(v-p)^r\eta_j(v)a_j^{-n}\theta(a_jv,q^{\frac{1}{k}})\Phi_j^0(q^nv^k).$$

So setting $v=p$, we obtain (since $E$ is entire):

$$\sum_{j=1}^{\ell}d_ja_j^{-n}\Psi_j^0(q^np^k)=0,$$

$d_j:=c_j\theta(a_jp,q^{\frac{1}{k}})$. Consider these equations for $n=-N,-N+1,\dots,-N+\ell-1$ as a linear system with respect to the unknowns $x_j:=d_ja_j^{-N}$. The matrix of this system has entries $\alpha_{ij}=a_j^{-i+1}\Psi_j^0(q^{-N+i-1}p^k)$. When $N\to+\infty$, this matrix approaches a Vandermonde matrix (as $\Psi_j^0(\infty)=1$). Thus it is nondegenerate for $N\gg 0$, which yields that $d_j=0$, i.e., $c_j\theta(a_jp,q^{\frac{1}{k}})=0$ for all $j$. We conclude that $\theta(a_lp,q^{\frac{1}{k}})=0$, so the only pole of $\eta_l$ on $\mathbb{C}^\times/q^{\frac{1}{k}\mathbb{Z}}$ is $a_l^{-1}$. Thus $r\geq 2$ (as there are no elliptic functions with just one simple pole). We also see that for all $j$ with $\theta(a_jp,q^{\frac{1}{k}})\neq 0$ we have $c_j=0$, i.e., the order of the pole of $\eta_j$ at $p$ is $\leq r-1$. So for such $j$ set $c_j':=\lim_{v\to p}(v-p)^{r-1}\eta_j(v)$. Then, evaluating the identity

$$(v-p)^{r-1}E(vq^{\frac{n}{k}})=q^{-\frac{n(n-1)}{2k}}v^{-n}\sum_{j=1}^{\ell}(v-p)^{r-1}\eta_j(v)a_j^{-n}\theta(a_jv,q^{\frac{1}{k}})\Psi_j^0(q^nv^k)$$

at $p$, we obtain

$$\sum_{j:c_j\neq 0}c_ja_j^{-n+1}\theta'(a_jp,q^{\frac{1}{k}})\Psi_j^0(q^nv^k)+\sum_{j:c_j=0}c_j'a_j^{-n}\theta(a_jp,q^{\frac{1}{k}})\Psi_j^0(q^nv^k)=0,$$

and the same argument with the Vandermonde matrix gives $c_l\theta'(a_lp,q^{\frac{1}{k}})=0$, hence $c_l=0$, a contradiction. $\square$

Thus, setting $\zeta := e^{\frac{2\pi i}{k}}$ we obtain

PROPOSITION 4.6.

$$E(v) = \sum_{j=1}^{\ell} \eta_j \theta(a_j v, q^{\frac{1}{k}}) \Psi_j^0(v^k),$$

*$\eta_j \in \mathbb{C}$, and the constants $\eta_j$ are uniquely determined by the condition that $E$ is holomorphic at 0 and $E(0) = 1$.*

PROOF. We just have to justify uniqueness, which follows since the difference of two such expressions will have an inadmissible power asymptotics at 0 for a solution of the corresponding difference equation. □

REMARK 4.7. The constants $\eta_j$ admit product formulas which can be derived by taking the confluent limit of the connection matrix of the $q$-hypergeometric equation. We will not give these formulas here except in the simplest example of the $q$-Bessel equation, see Subsection 5.5.

Recall that $\det(\zeta^{m(n-1)})_{1\le m,n\le k} = i^{\frac{(k-1)(k-2)}{2}} k^{\frac{k}{2}}$. So we arrive at the following result.

PROPOSITION 4.8. *We have*

$$\det(1 - uB_R) = i^{-\frac{(k-1)(k-2)}{2}} k^{-\frac{k}{2}} \det(\zeta^{m(n-1)} T^{n-1} E(\zeta^m u^{\frac{1}{k}}))_{1\le m,n\le k},$$

*where $E$ is the entire function given by the formula of Proposition 4.6.*

Note that while the entries of this matrix are functions of $u^{\frac{1}{k}}$, the determinant is a function of $u$.

## 5. Examples

In this section we consider a number of examples of computation of the characteristic function $h_{P,s,\chi}$. In all of these examples, the parameter $k$ (the size of the Wronski matrix) equals 1, so the characteristic function turns out to be (confluent) $q$-hypergeometric. While these examples are all instances of Theorem 1.1, to make the discussion more concrete, we do all computations explicitly from scratch and then compare the answers to existing literature on $q$-special functions.

**5.1. The situation of Example 2.1.** Our first example is the setting of Example 2.1. A typical polynomial $P$ of this kind is $P(x,y) = y^{d-r}(x^2+y^2)^{\frac{r}{2}}$ for even $r$, where $p$ has remainder 3 modulo 4. In this case $Z_0 = 0$, so $A_{P,s,\chi} = 0$ when $\chi \neq 1$, so it remains to consider the case $\chi = 1$. Then, by Proposition 3.2, we have

$$(A_{P,s,1}F)(z) = \frac{F(p^{-1})}{1-pqz} + \frac{(p^{rs}-1)pqzF(qz)}{(1-pqz)(1-p^{rs+1}qz)}.$$

So the eigenvalue equation has the form

$$\lambda F(z) = \frac{F(p^{-1})}{1-pqz} + \frac{(p^{rs}-1)pqzF(qz)}{(1-pqz)(1-p^{rs+1}qz)}.$$

If $r = 0$ then $P = y^d$, so $A_{P,s,1}$ has rank 1 with the only nonzero eigenvalue $\lambda = \frac{1}{1-q}$. The same thing happens if $s = 0$. So let us assume that $r > 0$, $s \neq 0$. Then we have

$$\lambda F(0) = F(p^{-1}).$$

This implies that $\lambda \neq 0$. Now, if $F(p^{-1}) = 0$ then there are no nonzero solutions (looking at the order of $F$ at zero), so $F(p^{-1}) \neq 0$. Then we may set $F(0) = 1$, so that $\lambda = F(p^{-1})$. So the difference equation for the eigenvector looks like

$$F(z) = \frac{1}{1-pqz} + \lambda^{-1}\frac{(p^{rs}-1)pqzF(qz)}{(1-pqz)(1-p^{rs+1}qz)}.$$

Iterating this, we get the expression for $F$:

$$F(z) = \frac{1}{1-pqz} + \lambda^{-1}\frac{(p^{rs}-1)pqz}{(1-pqz)(1-pq^2z)(1-p^{rs+1}qz)} +$$

$$\lambda^{-2}\frac{(p^{rs}-1)^2p^2q^3z^2}{(1-pqz)(1-pq^2z)(1-pq^3z)(1-p^{rs+1}qz)(1-p^{rs+1}q^2z)} + \dots$$

So we have

$$\lambda = F(p^{-1}) = \frac{1}{1-q} + \lambda^{-1}\frac{(p^{rs}-1)q}{(1-q)(1-q^2)(1-p^{rs}q)} +$$

$$\lambda^{-2}\frac{(p^{rs}-1)^2q^3}{(1-q)(1-q^2)(1-q^3)(1-p^{rs}q)(1-p^{rs}q^2)} + \dots$$

Multiplying the numerators and denominators on the right hand side of this equation by $1-p^{rs}$ and multiplying both sides by $\lambda^{-1}$, we obtain

PROPOSITION 5.1. *We have*

$$h_{P,s,1}(u) = J(p^{rs}, q, (1-p^{rs})u),$$

*where* $J(a,q,u) = {}_1\phi_1(0,a;q,u)$ *is the entire function given by the formula (4.4). In particular, the eigenvalues of the operator* $A = A_{P,s,1}$ *are the solutions of the equation*

$$J(p^{rs}, q, (1-p^{rs})\lambda^{-1}) = 0$$

*and are simple.*

Thus the characteristic function of $A_{P,s,1}$ expresses via the Hahn-Exton $q$-Bessel function with parameter

$$\nu = -1 - \frac{rs}{1+ds}.$$

REMARK 5.2. We obtain the following formula for the eigenvector $F(z)$ with eigenvalue $\lambda$:

$$F(z) = \frac{1}{1-pqz}\cdot {}_2\phi_2(0,q;pq^2z, apqz; q, (1-a)pqz\lambda^{-1}).$$

where $a := p^{rs}$. So setting $p^{\frac{1}{2}}q = b$ and replacing $z$ with $p^{-\frac{1}{2}}z$, we get that the functions

$$F_i(z) := \frac{1}{1-bz}\cdot {}_2\phi_2(0,q;bqz,abz;q,b^2q^{-1}zu_i),$$

for $u_i$ roots of the equation $J(a,q,u)=0$, are orthogonal on the unit circle $|z|=1$.

REMARK 5.3. This calculation implies that the unbounded self-adjoint operator $(1-p^{rs})A^{-1}_{P,s,1}$ has a limit $B_0$ as $s \to 0$, which has 1-dimensional kernel. Moreover, the inverse $\widehat{A}_0$ of $B_0$ on the orthogonal complement of $\operatorname{Ker} B_0$ is trace class and has characteristic function

$$\det(1-u\widehat{A}_0) = \lim_{a\to 1}(a-1)\frac{J(a,q,u)}{u} = \sum_{n=0}^{\infty}\frac{(-1)^nq^{\frac{n(n+1)}{2}}u^n}{(q;q)_n(q;q)_{n+1}}.$$

Consider now the case $l = \frac{r+d}{2}$ and $s > -\frac{1}{d}$. Then the operator $A = A_{P,s,1}$ has matrix

$$a_{mn} = q^{\frac{m+n}{2}} \mathbf{q}^{\frac{m+n}{2} - \min(m,n)}.$$

where $\mathbf{q} := p^{-1}q^{-1} = q^{\nu+1}$. Note that this operator is trace class whenever $|q| < 1, |\mathbf{q}| < |q|^{-1}$ and self-adjoint if in addition $q, \mathbf{q} \in \mathbb{R}$.

PROPOSITION 5.4. *If $0 < \mathbf{q}, q < 1$ then the operator $A$ is positive.*

PROOF. It suffices to show that for any $0 < \mathbf{q} < 1$, the matrix with entries

$$b_{mn} := \mathbf{q}^{-\min(m,n)}, 0 \le m, n \le N$$

is positive definite. By Sylvester's criterion, for this it suffices to show that its determinant $D_N$ is positive (as we know that 0 is not an eigenvalue). Subtracting from the first row of this matrix $\mathbf{q}$ times the second row, we get

$$D_N = \mathbf{q}^{-N}(1 - \mathbf{q})D_{N-1},$$

with $D_0 = 1$. So $D_N = \mathbf{q}^{\frac{-N(N+1)}{2}}(1 - \mathbf{q})^N > 0$. □

COROLLARY 5.5. *If $|q| < 1, |\mathbf{q}| < |q|^{-1}$ then the zeros of the function $(\mathbf{q}, q)_\infty J(\mathbf{q}, q, u)$ are all simple. Moreover, for $0 < q < 1$, they are real if in addition $0 < \mathbf{q} \le q^{-1}$, and nonnegative if $0 < \mathbf{q} \le 1$.*

PROOF. For $\mathbf{q} = 1$ the result follows from Remark 5.3, so assume $\mathbf{q} \neq 1$. Then the first statement follows from Proposition 5.1 (interpolated to non-integer values of $d$). For the second statement it suffices to assume that $\mathbf{q} \neq q^{-1}$. So the operator $A$ is a self-adjoint trace class operator. Hence its eigenvalues $\lambda_i$ are real, and by Proposition 5.4 they are positive if $\mathbf{q} < 1$. But $(1 - \mathbf{q})\lambda_i^{-1} = u_i$, where $u_i$ are the zeros of $J(\mathbf{q}, q, u)$. This implies the second statement. □

REMARK 5.6. 1. We note that Corollary 5.5 is not new and is given just for illustration purposes. Namely, the second statement of Corollary 5.5 is proved in [**KS**] using a different method. Also a proof similar to ours is given in [**SS**].

2. The assumptions of Corollary 5.5 cannot be dropped. For example, if $\mathbf{q} = q^{-m+1}$ for $m \ge 0$ (i.e., $\nu = -m$) then $(\mathbf{q}, q)_\infty J(\mathbf{q}, q, u)$ has a zero of order $m$ at $u = 0$ (so a multiple zero for $m \ge 2$). This also shows that when $q, \mathbf{q} \in (0, 1)$ and $\mathbf{q}$ is close to $q^{-m+1}$ then there have to be zeros of $(\mathbf{q}, q)_\infty J(\mathbf{q}, q, u)$ with argument close to $\frac{2\pi k}{m}$ or $\frac{2\pi(k+\frac{1}{2})}{m}$ (depending on the sign of $1 - q^{m-1}\mathbf{q}$). So for $m \ge 2$ they cannot be all real.

A more complete analysis of the zeros of the function $J$ is carried out in [**AM1**]. In particular, it is shown there that for $q, \mathbf{q} \in (0, 1)$ the number of non-real zeros is always finite.

**5.2. A rank 1 perturbation.** The operator $A$ from Subsection 5.1 for $r = d$ is a rank 1 perturbation of the inverse to a second order difference operator. Namely, consider the self-adjoint trace class operator $\widetilde{A}$ with kernel $\frac{|P(x,y)|^s - 1}{1-p^{-ds}}$; i.e., $\widetilde{A} = \frac{1}{1-pq}(A - \frac{1}{1-p^{-1}}\Pi)$, where $\Pi$ is the projector to constant functions. This operator has a limit at $s = 0$ where the kernel becomes $\frac{1}{d}\log|P(x,y)|$. As before, $\widetilde{A}$ vanishes on $H_\chi$ with $\chi \neq 1$, so let us consider it on $H_1 \cong \ell^2$. Then the matrix of $\widetilde{A}$ in the standard basis is given by

$$\widetilde{a}_{mn} = p^{-\frac{m+n}{2}} \frac{p^{-\min(m,n)ds} - 1}{1 - p^{-ds}}.$$

There is an obvious eigenvector $(1,0,\dots)$ with eigenvalue 0 (the indicator function of $\mathbb{Z}_p^\times$), so let's consider the action of $\widetilde{A}$ on its orthogonal complement, i.e., on the space $\ell_{2,0}$ of square summable sequences $\{f_n\}$ such that $f_0=0$. This space is identified with $\mathcal{H}$ by the assignment $F(z)=\sum_{n\ge 1}p^{\frac{n}{2}}f_nz^{n-1}$. In terms of this realization we have

$$(\widetilde{A}F)(z)=\frac{qzF(qz)-p^{-1}F(p^{-1})}{(1-z)(1-pqz)}.$$

Thus the eigenvalue equation has the form

$$\lambda F(z)=\frac{qzF(qz)-p^{-1}F(p^{-1})}{(1-z)(1-pqz)}.$$

If $\lambda=0$ then we get $qzF(qz)=p^{-1}F(p^{-1})$, i.e. $F=0$. Thus $\lambda\ne 0$. Now, if $F(p^{-1})=0$ then there are no nonzero solutions (looking at the order of $F$ at zero), so $F(p^{-1})\ne 0$. Then we may set $F(0)=1$, so that

$$\lambda=-p^{-1}F(p^{-1}).$$

Thus we have

$$F(z)=\frac{1}{(1-z)(1-pqz)}+\lambda^{-1}\frac{qz}{(1-z)(1-pqz)}F(qz).$$

Iterating this, we get the expression for $F$:

$$F(z)=\frac{1}{(1-z)(1-pqz)}+\lambda^{-1}\frac{qz}{(1-z)(1-qz)(1-pqz)(1-pq^2z)}+$$
$$+\lambda^{-2}\frac{q^3z^2}{(1-z)(1-qz)(1-q^2z)(1-pqz)(1-pq^2z)(1-pq^3z)}+\dots$$

So setting $z=p^{-1}$, we have

$$1+\lambda^{-1}\frac{p^{-1}}{(1-p^{-1})(1-q)}+\lambda^{-2}\frac{qp^{-2}}{(1-p^{-1})(1-p^{-1}q)(1-q)(1-q^2)}$$
$$+\lambda^{-3}\frac{q^3p^{-3}}{(1-p^{-1})(1-p^{-1}q)(1-p^{-1}q^2)(1-q)(1-q^2)(1-q^3)}+\dots=0.$$

Thus we get

PROPOSITION 5.7. *The characteristic function of $\widetilde{A}$ is given by the formula*

$$\det(1-u\widetilde{A})=J(p^{-1},q,-p^{-1}u),$$

*so the eigenvalues of $\widetilde{A}$ are the solutions of the equation*

$$J(p^{-1},q,-p^{-1}\lambda^{-1})=0 \tag{5.1}$$

*and are simple.*

Note that the Hahn-Exton parameter $\nu$ now equals $-\frac{ds}{ds+1}$.

Consider now the inverse operator $\widetilde{A}^{-1}$. Let $F=\widetilde{A}^{-1}G$. Then we have

$$G(z)=\frac{qzF(qz)-p^{-1}F(p^{-1})}{(1-z)(1-pqz)}.$$

Thus $G(0)=-p^{-1}F(p^{-1})$. So we have

$$(\widetilde{A}^{-1}G)(z)=(1-q^{-1}z)(z^{-1}-p)G(q^{-1}z)-z^{-1}G(0)=((1-q^{-1}z)(z^{-1}-p)G(q^{-1}z))_+.$$

This shows that $\widetilde{A}^{-1}$ is an unbounded self-adjoint operator, which is a self-adjoint extension of the symmetric operator $L = L_\nu$ given by

$$(LG)(z) := ((1-q^{-1}z)(z^{-1}-p)G(q^{-1}z))_+$$

with initial domain being the space of polynomials $\mathbb{C}[z]$ with norm

$$\|G\|^2 = \frac{1}{2\pi}\int_{|z|=p^{-\frac{1}{2}}} |G(z)|^2 d\theta.$$

Writing $G(z)$ as $\sum_{n\geq 0} p^{\frac{n}{2}} g_n z^n$, we obtain an expression for $L$ as a difference operator on the space of sequences whose matrix in the standard basis is a Jacobi matrix:

$$(Lg)_n = q^{-n-1}(p^{\frac{1}{2}}g_{n+1} - (1+pq)g_n + p^{\frac{1}{2}}qg_{n-1}),$$

where we agree that $g_{-1} = 0$. It follows that the self-adjoint extension of $L$ given by $\widetilde{A}^{-1}$ has discrete spectrum, and its eigenvalues are the solutions $u_i$ of the equation

$$J(p^{-1}, q, -p^{-1}u) = 0.$$

This analysis is carried out in [**SS**]. Let us explain its results in more detail. Consider the homogeneous difference equation $Lg = 0$. Its basic solutions (generically) are $g_n = \beta^n$ where $\beta$ is a root of the characteristic equation

$$p^{\frac{1}{2}}\beta^2 - (1+pq)\beta + p^{\frac{1}{2}}q = 0.$$

The roots of this equation are $\beta_1 = p^{-\frac{1}{2}}$ and $\beta_2 = p^{\frac{1}{2}}q = p^{-\frac{1}{2}-ds}$. So in the range $-\frac{1}{d} < s \leq -\frac{1}{2d}$ (i.e., $\nu \geq 1$) we have $p^{-\frac{1}{2}-ds} \geq 1$, so the equation $Lg = 0$ has a unique, up to scaling, solution in $\ell_2$ (namely, $p^{-\frac{n}{2}}$). This implies (after some work, done in [**SS**]) that the operator $L$ is essentially self-adjoint on $\mathbb{C}[z]$. On the other hand, if $s > -\frac{1}{2d}$ (i.e., $|\nu| < 1$) then $p^{-\frac{1}{2}-ds} < 1$, so both basic solutions are in $\ell_2$. This implies (again after some work done in [**SS**]) that the operator $L$ is not essentially self-adjoint but rather has von Neumann deficiency indices $(1,1)$. So $L$ has a 1-parameter family of self-adjoint extensions parametrized by $(\alpha_1,\alpha_2) \in \mathbb{RP}^1$ obtained by adding the function $\psi_{\alpha_1,\alpha_2} := \frac{\alpha_1}{1-z} + \frac{\alpha_2}{1-pqz}$ to the initial domain of $L$, which makes it essentially self-adjoint (if $s = 0$, i.e., $q = p^{-1}$, we should add $\psi_{\alpha_1,\alpha_2} = \frac{\alpha_1}{1-z} + \frac{\alpha_2}{(1-z)^2}$). Moreover, the self-adjoint extension of $L$ given by $\widetilde{A}^{-1}$ corresponds to the point $(1,0) \in \mathbb{RP}^1$. Finally, if $-\frac{1}{2d} < s \leq 0$ (i.e., $0 \leq \nu < 1$) then $L$ is positive and $\widetilde{A}^{-1}$ is its Friedrichs extension, as it has the smallest domain, see [**L**]. This comes from the fact that the decay of Taylor coefficients of the function $\psi_{\alpha_1,\alpha_2}$ is fastest if and only if $\alpha_2 = 0$.

For example, in the case $s = 0$, i.e., $q = p^{-1}$, we get the difference operator

$$(Lg)_n = p^n(p^{\frac{1}{2}}g_{n+1} - 2x_n + p^{-\frac{1}{2}}g_{n-1}),$$

i.e. $L = DD^*$ where

$$(Dg)_n = p^{\frac{n}{2}}(g_n - g_{n-1}).$$

The eigenvalues of the Friedrichs extension of $L$ are, therefore, solutions of the equation $J(p^{-1}, p^{-1}, -p^{-1}u) = 0$. This is shown in [**KRS**].

REMARK 5.8. A more complete account of the spectral theory of difference operators (also called $q$-Sturm-Liouville theory) can be found in the book [**AM2**].

REMARK 5.9. We obtain the following formula for the eigenvector $F(z)$ with eigenvalue $\lambda$:

$$F(z) = \frac{1}{(1-z)(1-pqz)} \cdot {}_2\phi_2(0, q; pq^2z, pqz; q, -z\lambda^{-1}).$$

Thus the eigenvectors are

$$F_i(z) = \frac{1}{(1-z)(1-pqz)} \cdot {}_2\phi_2(0, q; pq^2z, pqz; q, pzu_i).$$

where $u_i$ are the roots of the equation $J(p^{-1}, q, u) = 0$.

The fact that the inverse of $\widetilde{A}$ is a difference operator $L$ implies that we also have

$$F_i(z) = \sum_{n=0}^{\infty} J(p^{-1}, q, q^{n+1}u_i)z^n.$$

Thus we obtain the orthogonality relations

$$\sum_{n\geq 0} p^{-n} J(p^{-1}, q, q^{n+1}u_i)J(p^{-1}, q, q^{n+1}u_j) = 0, \ i \neq j$$

which are the Hahn-Exton $q$-analogs of the Fourier-Bessel orthogonality relations with $p^{-1} = q^{\alpha+1}$ (see [**KSw**], p.2).

**5.3. Continuous limit.** We can also consider the "archimedian" limit of the operator $\widetilde{A}|_{H_1}$ from the previous subsection as $p \to 1$ (this does not have a $p$-adic interpretation but makes sense analytically). For this pick $a > 0$ and think of $q^{\frac{n\nu}{2a}} f_n$ as values of some function $\mathbf{f}$ at $q^{-\frac{n\nu}{a}}$. Then the limit $p \to 1$ will be a continuous limit which transforms Riemann sums into integrals over $[0, 1]$. The condition $\sum_n |f_n|^2 < \infty$ translates into the condition that

$$\int_0^1 |\mathbf{f}(x)|^2 dx < \infty,$$

i.e., $\mathbf{f} \in L^2[0, 1]$. Also the operator $\widetilde{A}$ after appropriate renormalization converges to

$$(\widetilde{A}_\infty(a)\mathbf{f})(x) = \frac{1}{a}\int_0^1 (\max(x, y)^{-a} - 1)(xy)^{-\frac{a}{2ds}-\frac{1}{2}}\mathbf{f}(y)dy. \tag{5.2}$$

Note that the operator $\widetilde{A}_\infty(a)$ is unitarily equivalent to $\widetilde{A}_\infty(1)$ via the change of variable $x \to x^a$. Since $q \to 1$ as $p \to 1$ (for fixed $s$), and since the $q$-Bessel function $J_\nu(x; q)$ degenerates in this limit to the classical Bessel function $J_\nu(x)$, the eigenvalues of $\widetilde{A}_\infty(a)$ are proportional to inverse squared zeros of $J_\nu$, where $\nu = -\frac{ds}{ds+1}$.

Furthermore, $\widetilde{A}_\infty(a)$ is inverse to a second order differential operator. Namely, consider the differential operator

$$\mathcal{L}_\nu := -\partial^2 + \frac{\nu^2 - \frac{1}{4}}{x^2}.$$

Assume that $\nu > -1$ and consider the action of $\mathcal{L}_\nu$ on the space of functions $g$ on $[0, 1]$ such that $gx^{-\nu-\frac{1}{2}}$ is a polynomial and $g(1) = 0$. This defines an essentially self-adjoint operator. Its eigenvalues and eigenvectors are found in the standard way (for $\nu = 0$ this is the classical problem of vibrations of a circular membrane, i.e., the Dirichlet problem for the Laplacian on the disk). Namely, consider the solution of the equation $\mathcal{L}_\nu\psi = \psi$ which behaves as $x^{\nu+\frac{1}{2}}$ at 0. This solution is $x^{\frac{1}{2}}J_\nu(x)$. Thus the solution of $\mathcal{L}_\nu\psi = \lambda^2\psi$ with such behavior at 0, up to scaling, is $x^{\frac{1}{2}}J_\nu(\lambda x)$. So the eigenvalues of $\mathcal{L}_\nu$ are squared zeros of the Bessel function $J_\nu(x)$.

On the other hand, let us compute the Green function of $\mathcal{L}_\nu$. This is also standard. We need to find the solution of the equation $\mathcal{L}_\nu^x g(x,y) = \delta(x-y)$ with the above boundary conditions. So $g(x,y) = b(y)(1-y^{-2\nu})x^{\nu-\frac{1}{2}}$ when $x<y$ and $g(x,y) = b(y)(x^{\nu+\frac{1}{2}} - x^{-\nu+\frac{1}{2}})$ when $x>y$ with gluing condition $b(y) = -\frac{y^{\nu+\frac{1}{2}}}{2\nu}$. Thus

$$g(x,y) = \frac{1}{2\nu}(\max(x,y)^{-2\nu}-1)(xy)^{\nu+\frac{1}{2}}.$$

So we have

$$\widetilde{A}_\infty(2\nu)^{-1} = \mathcal{L}_\nu$$

where $\nu = -\frac{ds}{ds+1}$. This gives another way to see that eigenvalues of $\mathcal{L}_\nu$ are squared zeros of the Bessel function $J_\nu(x)$.

REMARK 5.10. The differential operator $\mathcal{L}_\nu$ is the continuous limit of the difference operator $L_\nu$ considered in the previous subsection.

**5.4. The case when $Z_0$ is a constant.** Assume now that $Q \in \mathbb{Z}_p[x]$ and the reduction $\overline{Q}$ of $Q$ modulo $p$ also has degree $d$. In this case the function $Z_0(Q,\chi,s,z)$ defined by (2.2),(2.3) is constant as a function of $z$, i.e.,

$$Z_0(Q,\chi,s,z) = \zeta(Q,\chi,s) - \delta_{1,\chi}.$$

The case $\chi \neq 1$ is trivial, so we now consider the case $\chi = 1$.

For brevity we will denote $\zeta(Q,1,s)-1$ by $\beta = \beta(s)$. The case when $\beta(s) = 0$ has already been discussed in Subsection 5.1, so we assume that $\beta(s) \neq 0$. For instance, if $\overline{Q}$ has $m$ zeros in $\mathbb{F}_p$ all of which are simple then by the computation in Example 2.2 we have

$$\beta(s) = -m\frac{p^{-1}(1-p^{-s})}{(1-p^{-1})(1-p^{-s-1})}.$$

We have

$$R(z) = \frac{1}{1-z} - \frac{1}{1-pqz} + \beta = \beta\frac{(1-az)(1-a^{-1}pqz)}{(1-z)(1-pqz)}.$$

for $a$ found from the quadratic equation

$$\beta(1-a)(1-a^{-1}pq) = 1-pq.$$

Thus eigenvectors of $A_{P,s,1}$ are solutions of the $q$-difference equation

$$\lambda F(z) = \frac{\beta(1-az)(1-a^{-1}pqz)}{(1-z)(1-pqz)}F(qz) + \frac{F(p^{-1})}{1-pqz},$$

Setting $z=0$, we have

$$(\lambda-\beta)F(0) = F(p^{-1}).$$

First consider the case when $F(p^{-1}) \neq 0$. Then $F(0)\neq 0$, so setting $F(0)=1$, we get $\lambda - \beta = F(p^{-1})$. Thus, using that $\lambda \neq \beta$, we have

$$F(z) = \frac{1-\beta\lambda^{-1}}{1-pqz} + \beta\lambda^{-1}\frac{(1-az)(1-a^{-1}pqz)}{(1-z)(1-pqz)}F(qz).$$

Hence, iterating, we obtain

$$F(z) = \frac{1-\beta\lambda^{-1}}{1-pqz} + \beta\lambda^{-1}\frac{(1-az)(1-a^{-1}pqz)}{(1-z)(1-pqz)}\frac{1-\beta\lambda^{-1}}{1-pq^2z}$$

$$+\beta^2\lambda^{-2}\frac{(1-az)(1-aqz)(1-a^{-1}pqz)(1-a^{-1}pq^2z)}{(1-z)(1-qz)(1-pqz)(1-pq^2z)}\frac{1-\beta\lambda^{-1}}{1-pq^3z}+\ldots$$

Substituting $z = p^{-1}$, we have

$$1 = \lambda^{-1}\frac{1}{1-q} + \beta\lambda^{-2}\frac{(1-ap^{-1})(1-a^{-1}q)}{(1-p^{-1})(1-q)}\frac{1}{1-q^2}$$

$$+\beta^2\lambda^{-3}\frac{(1-ap^{-1})(1-ap^{-1}q)(1-a^{-1}q)(1-a^{-1}q^2)}{(1-p^{-1})(1-p^{-1}q)(1-q)(1-q^2)}\frac{1}{1-q^3}+\ldots$$

Since $\beta = \frac{1-pq}{(1-a)(1-a^{-1}pq)} = -\frac{1-p^{-1}q^{-1}}{(1-a^{-1})(1-ap^{-1}q^{-1})}$, this can be written as

$$ {}_2\phi_1(ap^{-1}q^{-1}, a^{-1}; p^{-1}q^{-1}; q, \beta\lambda^{-1}a^{-1}) = 0. \tag{5.3}$$

So we see that the eigenvalues of $A$ are the solutions of the equation (5.3).

It remains to consider the case $F(p^{-1}) = 0$. In this case we get

$$\lambda F(z) = \frac{\beta(1-az)(1-a^{-1}pqz)}{(1-z)(1-pqz)}F(qz).$$

Setting $z = p^{-1}q^{k-1}$, $k \geq 1$, we see that $F(p^{-1}q^{k-1}) = 0$ implies $F(p^{-1}q^k) = 0$ unless $ap^{-1}q^{k-1} = 1$ or $a^{-1}q^k = 1$. If $F(p^{-1}q^k) = 0$ for all $k \geq 0$ then $F = 0$, so we see that we must have $a = pq^{-k+1}$ or $a = q^k$ for some $k \geq 1$. In this case

$$\beta = \frac{1-pq}{(1-q^k)(1-pq^{-k+1})},$$

and we have eigenvalues $\beta q^j$, $j \geq 0$, whose eigenvectors have the form $z^j + O(z^{j+1})$, in addition to $k$ eigenvalues which solve equation (5.3) (as the $q$-hypergeometric function on the left hand side of this equation specializes to a polynomial of degree $k$).

So in this case the spectrum consists of two different parts, finite and infinite. The infinite part is fairly trivial ($\beta q^j$, $j \geq 0$), while the finite part is more interesting and consists of the numbers $\beta\gamma_i^{-1}$, where $\gamma_1, \ldots, \gamma_k$ are the algebraic functions of $q$ which are zeros of the $k$-th little $q$-Jacobi polynomial $p_k(q^{-1}x; p^{-1}q^{-2}, 1; q)$ ([KoS]). If $s \in \mathbb{Q}$, these are algebraic numbers.

We note that as $k \to \infty$ (i.e., $\beta \to 0$), we find ourselves in the situation of Proposition 5.1 for $r = d$. This recovers [**KSw**], Proposition A1, stating that the little $q$-Jacobi polynomials can be degenerated to the $q$-Bessel function $J(a; q, u)$.

These two cases ($F(p^{-1}) \neq 0$ and $F(p^{-1}) = 0$) can be nicely unified as follows. Recall that the function ${}_2\phi_1$ for generic parameters has at most simple poles at $1, q^{-1}, q^{-2}, \ldots$ and no other singularities. Thus the function

$${}_2\widetilde{\phi}_1(a, b, c, q, u) := {}_2\phi_1(a, b, c, q, u)(u; q)_\infty$$

is entire (it is obtained from ${}_2\phi_1$ by "clearing denominators").

PROPOSITION 5.11. *The characteristic function of $A_{P,s,1}$ is*

$$h_{P,s,1}(u) = {}_2\widetilde{\phi}_1(ap^{-1}q^{-1}, a^{-1}; p^{-1}q^{-1}; q, \beta u).$$

*Thus the eigenvalues of $A_{P,s,1}$ are the solutions of the equation*

$$ {}_2\widetilde{\phi}_1(ap^{-1}q^{-1}, a^{-1}; p^{-1}q^{-1}; q, \beta\lambda^{-1}) = 0. $$

EXAMPLE 5.12. Let $P(x,y) = x - y$. Then $d = m = 1$ and

$$ \beta = \frac{1-pq}{(1-p)(1-q)}. $$

So the equation for $a$ takes the form

$$ (1-a)(1-a^{-1}pq) = (1-p)(1-q), $$

which gives $a = p$ or $a = q$. We have

$$ {}_2\phi_1(q^{-1}, p^{-1}; p^{-1}q^{-1}; q, u) = 1 - \frac{1-p}{1-pq}u. $$

(the first little $q$-Jacobi polynomial $p_1$). Thus for the eigenvalues we get the equation

$$ \left(1 - \frac{1-p}{1-pq}\beta\lambda^{-1}\right)(\beta\lambda^{-1}; q)_\infty. $$

So the eigenvalues are

$$ \lambda_0 = \frac{1}{1-q}, \quad \lambda_{k+1} = \frac{1-pq}{(1-p)(1-q)}q^k, \quad k \geq 0. \tag{5.4} $$

In fact, this is easy to see in a different way: the operator $A$ is the convolution with the function $\frac{|x|^s}{1-p^{-1}}$ on $\mathbb{Z}_p$, so its eigenvalues are the values of the Fourier transform of this function, which are easily shown to be exactly the list (5.4).

**5.5. The case $\chi \neq 1$ with $Z_0$ having a first order pole at 0 and $\infty$.** Consider now the case $\chi \neq 1$ with $Z_0$ having the form $z^{-1}(1-az)(1-bz)$. This happens, for instance, in the case

$$ P(x,y) = x^d - p^{-1}xyP_*(x,y) + y^d, $$

where $P_*(x,y) \in \mathbb{Z}_p[x,y]$ is a homogeneous polynomial of degree $d-2$ whose coefficients of $x^{d-2}$ and $y^{d-2}$ have norm 1. In this case

$$ Q(x) = 1 - p^{-1}xQ_*(x) + x^d, $$

where $Q_*(x) = P_*(x,1)$. Then we have

$$ Z_0(P,\chi,s,z) = \zeta(1-x,\chi,s)(p^{-1}q^{-1}z + z^{-1}) + p^s\zeta(Q_*,\chi,s). $$

Let

$$ \gamma = \gamma(s) := \frac{p^s\zeta(Q_*,\chi,s)}{\zeta(1-x,\chi,s)} $$

and let

$$ z^{-1} + \gamma + p^{-1}q^{-1}z = z^{-1}(1-az)(1-bz). $$

Then after rescaling $z$ the eigenvalue equation for $A_{P,s,\chi}$ takes the form

$$ \lambda F(z) = (z^{-1}(1-az)(1-bz)F(qz))_+. \tag{5.5} $$

So let us solve (5.5) for general $a, b$.

We can rewrite (5.5) in the form

$$ \lambda F(z) = z^{-1}(1-az)(1-bz)F(qz) - z^{-1}F(0). \tag{5.6} $$

To solve (5.6), consider the deformed equation

$$\lambda F(z) = -\frac{t^{-1}(1-az)(1-bz)F(qz)}{1-t^{-1}z} + \frac{t^{-1}(1-at)(1-bt)F(qt)}{1-t^{-1}z} \tag{5.7}$$

which degenerates to (5.5) as $t \to 0$, and let us solve (5.7). We have

$$(1-at)(1-bt)F(qt) = (1+t\lambda)F(0).$$

Assume first that $F(qt) \neq 0$. Then $F(0) \neq 0$, and setting $F(0) = 1$, we get

$$F(z) = -\lambda^{-1}t^{-1}\frac{(1-az)(1-bz)}{1-t^{-1}z}F(qz) + \frac{1+\lambda^{-1}t^{-1}}{1-t^{-1}z}.$$

Iterating this, we obtain

$$F(z) = \frac{1+\lambda^{-1}t^{-1}}{1-t^{-1}z} - \lambda^{-1}t^{-1}\frac{(1-az)(1-bz)}{1-t^{-1}z}\frac{1+\lambda^{-1}t^{-1}}{1-t^{-1}qz}$$
$$+\lambda^{-2}t^{-2}\frac{(1-az)(1-aqz)(1-bz)(1-bqz)}{(1-t^{-1}z)(1-t^{-1}qz)}\frac{1+\lambda^{-1}t^{-1}}{1-t^{-1}q^2z} - \dots$$

Thus, setting $z = qt$, we have

$$\frac{1+t\lambda}{(1-at)(1-bt)} = F(qt) = \frac{1+\lambda^{-1}t^{-1}}{1-q} - \lambda^{-1}t^{-1}\frac{(1-aqt)(1-bqt)}{1-q}\frac{1+\lambda^{-1}t^{-1}}{1-q^2}$$
$$+\lambda^{-2}t^{-2}\frac{(1-aqt)(1-aq^2t)(1-bqt)(1-bq^2t)}{(1-q)(1-q^2)}\frac{1+\lambda^{-1}t^{-1}}{1-q^3} - \dots$$

So we get that the eigenvalues $\lambda$ are the solutions of the equation

$${}_2\phi_1(at, bt;\ 0;\ -\lambda^{-1}t^{-1}) = 0.$$

Now we have to determine what happens to this equation as $t \to 0$. This is done in [**Mo**], but we give the details for the reader's convenience. Recall the Watson connection formula for the Heine $q$-hypergeometric function ${}_2\phi_1$ ([**GR**]):

$${}_2\phi_1(a,b;c;q,x) =$$
$$\frac{(b,\frac{c}{a};q)_\infty}{(c,\frac{b}{a};q)_\infty}\frac{\theta(-ax;q)}{\theta(-x,q)}{}_2\phi_1(a,\tfrac{aq}{c};\tfrac{aq}{b};q,\tfrac{cq}{abx}) + \frac{(a,\frac{c}{b};q)_\infty}{(c,\frac{a}{b};q)_\infty}\frac{\theta(-bx;q)}{\theta(-x;q)}{}_2\phi_1(b,\tfrac{bq}{c};\tfrac{bq}{a};q,\tfrac{cq}{abx}).$$

So replacing $a$ with $at$, $b$ with $bt$, and $x$ with $-ut^{-1}$, we get

$${}_2\phi_1(at,bt;c;q,-ut^{-1}) = \frac{(bt,\frac{c}{at};q)_\infty}{(c,\frac{b}{a};q)_\infty}\frac{\theta(au;q)}{\theta(ut^{-1},q)}{}_2\phi_1(at,\tfrac{atq}{c};\tfrac{aq}{b};q,-\tfrac{cq}{abtu})$$
$$+\frac{(at,\frac{c}{bt};q)_\infty}{(c,\frac{a}{b};q)_\infty}\frac{\theta(bu;q)}{\theta(ut^{-1};q)}{}_2\phi_1(bt,\tfrac{btq}{c};\tfrac{bq}{a};q,-\tfrac{cq}{abtu}).$$

Now let us send $c$ to 0 and use that

$$\lim_{c\to 0}{}_2\phi_1(\alpha,\tfrac{\beta}{c};\gamma;q,cz) = {}_1\phi_1(\alpha;\gamma;q,\beta z)$$

(see [**GR**]). This yields

$${}_2\phi_1(at,bt;0;q,-ut^{-1}) =$$
$$\frac{(bt;q)_\infty}{(\frac{b}{a};q)_\infty}\frac{\theta(au;q)}{\theta(ut^{-1},q)}{}_1\phi_1(at;\tfrac{aq}{b};q,-\tfrac{q^2}{bu}) + \frac{(at;q)_\infty}{(\frac{a}{b};q)_\infty}\frac{\theta(bu;q)}{\theta(ut^{-1};q)}{}_1\phi_1(bt;\tfrac{bq}{a};q,-\tfrac{q^2}{au}).$$

Thus the equation ${}_2\phi_1(at,bt;0;q,-ut^{-1}) = 0$ yields

$$\frac{(bt;q)_\infty}{(\frac{b}{a};q)_\infty}\theta(au;q){}_1\phi_1(at;\tfrac{aq}{b};q,-\tfrac{q^2}{bu}) + \frac{(at;q)_\infty}{(\frac{a}{b};q)_\infty}\theta(bu;q){}_1\phi_1(bt;\tfrac{bq}{a};q,-\tfrac{q^2}{au}) = 0$$

Now sending $t$ to 0, we get

PROPOSITION 5.13. *The eigenvalues of $A_{P,s,\chi}$ are the solutions of the equation*

$$E(a,b;q,\lambda^{-1})=0,$$

*where*

$$E(a,b;q,u):=\frac{\theta(au;q)J(\frac{aq}{b};q,-\frac{q^2}{bu})}{(\frac{b}{a};q)_\infty(q;q)_\infty}+\frac{\theta(bu;q)J(\frac{bq}{a};q,-\frac{q^2}{au})}{(\frac{a}{b};q)_\infty(q,q)_\infty}. \tag{5.8}$$

Indeed, the case $F(qt)=0$ is irrelevant for us since then we must have $\lambda=t^{-1}q^m$ for some $m$, and these eigenvalues, if present, go to $\infty$ as $t\to 0$, since $m$ has to remain bounded.

As before, by growth considerations Proposition 5.13 implies that

$$\det(1-uA_{P,\chi,s})=Cu^mE(a,b;q;u)$$

for some $C\neq 0$, $m\in\mathbb{Z}$. In particular, this means that $E(a,b;q;u)$ is meromorphic at 0. Note also that the two summands in $E$ form a basis in the space of solutions of the $q$-Bessel difference equation transformed by $u\mapsto u^{-1}$ over the field of elliptic functions:

$$abqu^2K(q^2u)-(1+(a+b)u)K(qu)+K(u)=0. \tag{5.9}$$

This equation is irregular at 0, but it has a single, up to scaling, solution $\widetilde{E}(u)$ meromorphic at 0, which can be found by the power series method. Moreover, it is easy to see that $\widetilde{E}(u)$ is in fact holomorphic at 0 and does not vanish there (as equation (5.9) degenerates to the equation $K(qu)=K(u)$ near $u=0$). So up to scaling we must have $\widetilde{E}(u)=E(u)$, hence $m=0$. Thus we get

COROLLARY 5.14. *The function $E(a,b;q,u)$ extends holomorphically to the origin with $E(a,b;q,0)\neq 0$, and*

$$\det(1-uA_{P,s,\chi})=\frac{E(a,b;q,u)}{E(a,b;q,0)}.$$

In fact, it is shown in [**Mo**], Theorem 1 that $E(a,b;q,0)=1$. Thus one has

$$\det(1-uA_{P,s,\chi})=E(a,b;q,u).$$

## 6. The non-homogeneous case

**6.1. The trace class property.** Now let $P\in\mathbb{Q}_p[x,y]$ be any polynomial (not necessarily homogeneous). Let us first show that if $\mathrm{Re}(s)$ is sufficiently large then the operator $A_{P,s}$ is trace class. First recall the following well known lemma.

LEMMA 6.1. *Let $A$ be a bounded operator on a Hilbert space $H$ and let $A_k\to A$ be a sequence of bounded operators on $H$ of ranks $\le r_k$, $k\ge 0$, such that $r_{k-1}\le r_k$. If $M:=\sum_{k\ge 1}r_k\,\|A_{k-1}-A\|<\infty$ then $A$ is trace class.*

PROOF. Recall that for any bounded operator $B$ of rank $r$ we have $\mathrm{Tr}(B)\le r\,\|B\|$. Now, we have

$$A=\lim_{k\to\infty}A_k=A_0+(A_1-A_0)+\dots.+(A_k-A_{k-1})+\dots,$$

an absolutely convergent series with respect to the operator norm. So given a finite rank operator $C$, we have

$$\mathrm{Tr}(AC) = \mathrm{Tr}(A_0C) + \sum_{k\geq 1} \mathrm{Tr}((A_k - A_{k-1})C).$$

Now, the rank of $A_k - A_{k-1}$ is at most $r_k + r_{k-1} \leq 2r_k$, while its norm is at most $\|A_{k-1} - A\| + \|A_k - A\|$. Thus

$$|\mathrm{Tr}(AC)| \leq (\|A_0\| + \sum_{k\geq 1} 2r_k(\|A_{k-1} - A\| + \|A_k - A\|))\,\|C\|\,.$$

Hence

$$\mathrm{Tr}\,|A| \leq \|A_0\| + \sum_{k\geq 1} 2r_k(\|A_{k-1} - A\| + \|A_k - A\|) \leq \|A_0\| + 4M.$$

□

Now let $\gamma_P$ be the log-canonical threshold of $P$ ([**Mu**]), i.e.,

$$\gamma_P = \sup(\gamma : \mathrm{vol}(|P(x,y)| < t) = O(t^\gamma), t \to 0).$$

PROPOSITION 6.2. *If* $\mathrm{Re}(s) > 1 - \frac{1}{2}\gamma_P$ *then* $A_{P,s}$ *is trace class.*

PROOF. Let $\mathrm{Re}(s) = \rho > 0$. Let $A_k$ be the integral operator on $\mathbb{Z}_p$ with kernel $D_k(x,y) := \frac{1}{1-p^{-1}} \max(p^{-k}, |P(x,y)|)^s$. It is clear that $\|A - A_k\|$ is dominated by its Hilbert-Schmidt norm, hence

$$\|A - A_k\| \leq Cp^{-k(\rho + \frac{1}{2}\gamma_P)}.$$

for large $k$. On the other hand, we have

$$P(x,y) - P(x',y') = P_1(x,y)(x - x') + P_2(x',y)(y - y')$$

for some polynomials $P_1, P_2$, so

$$|P(x,y) - P(x',y')| < p^\ell \max(|x - x'|, |y - y'|)$$

for some $\ell$. This implies that if $|P(x,y)| = p^{-k}$ and $|x' - x|, |y' - y| \leq p^{-k-\ell}$ then $|P(x',y')| = |P(x,y)|$. Thus if $|x' - x|, |y' - y| \leq p^{-k-\ell}$ then $D_k(x,y) = D_k(x',y')$. We conclude that the rank of $A_k$ is $\leq p^{k+\ell}$. Thus by Lemma 6.1, $A$ is trace class if $\rho > 1 - \frac{1}{2}\gamma_P$. □

For example, if $P$ has zeros in $\mathbb{Z}_p^2$ but they are all smooth points of the curve $P(x,y) = 0$ then $\gamma_P = 1$ and $A_{P,s}$ is trace class whenever $\mathrm{Re}(s) > \frac{1}{2}$.

**6.2. An example.** Proposition 6.2 allows us to define the characteristic function $h_{P,s}(u)$ for a general polynomial $P$. Computing this function, however, seems difficult in general, as we no longer have a decomposition into the sectors $H_\chi$, and even when we do, the answer goes beyond the theory of $q$-hypergeometric functions.

To demonstrate what may happen in the non-homogeneous case, consider the polynomial

$$P(x,y) = x^{2d} + y^2,$$

where $d > 1$ and $p = 4k + 3$. We have $|P(x,y)| = \max(|x|^d, |y|)$. Thus the corresponding operator $A_{P,s}$ still commutes with $\mathbb{Z}_p^\times$ and $A_{P,s,\chi} = 0$ if $\chi \neq 1$, while the matrix of $A = A_{P,s,1}$ is

$$a_{mn} = p^{-\frac{m+n}{2} - 2s\min(m,dn)}.$$

For simplicity assume that $s \geq 0$. In terms of analytic functions $F(z)$ we get

$$(AF)(z) = \sum_{n\geq 0}\sum_{m>dn} z^m p^{-2dns-\frac{n}{2}} f_n + \sum_{n\geq 0}\sum_{m\leq dn} z^m p^{-2ns-\frac{n}{2}} f_n =$$

$$\frac{(1-p^{-2s})z}{(1-z)(1-p^{-2s}z)}F(p^{-2ds-1}z^d) + \frac{F(p^{-1})}{1-p^{-2s}z}.$$

So the eigenvalue equation looks like

$$\lambda F(z) = \frac{(1-p^{-2s})z}{(1-z)(1-p^{-2s}z)}F(p^{-2ds-1}z^d) + \frac{F(p^{-1})}{1-p^{-2s}z}.$$

Thus $\lambda F(0) = F(p^{-1})$. If $F(p^{-1}) = 0$ then $F = 0$, so $F(p^{-1}) \neq 0$, hence $F(0) \neq 0$, $\lambda \neq 0$. So we may set $F(0) = 1$, which gives $\lambda = F(p^{-1})$. Hence we get

$$F(z) = \frac{1}{1-p^{-2s}z} + \lambda^{-1}\frac{(1-p^{-2s})z}{(1-z)(1-p^{-2s}z)}F(p^{-2ds-1}z^d).$$

Let $s > -\frac{1}{2d}$ and $q = p^{-\frac{2s+1}{d-1}}$. Then, replacing $z$ by $p^{-1}q^{-d}z$, we have

$$F(p^{-1}q^{-d}z) = \frac{1}{1-q^{-1}z} + \lambda^{-1}\frac{(p^{-1}q^{-d}-q^{-1})z}{(1-p^{-1}q^{-d}z)(1-q^{-1}z)}F(p^{-1}q^{-d}z^d).$$

So setting $G(z) := F(p^{-1}q^{-d}z)$ and $u = \lambda^{-1}$, we get

$$G(z) = \frac{1}{1-q^{-1}z} + u\frac{(p^{-1}q^{-d}-q^{-1})z}{(1-p^{-1}q^{-d}z)(1-q^{-1}z)}G(z^d). \tag{6.1}$$

This shows that $G$ is a $d$-**Mahler series** (of degree 2) in the sense of [**BCCD**], which is a generalization of the notion of $d$-automatic and $d$-regular series (see [**AS**]).

Iterating (6.1), for any $u$ we get a solution

$$G(z,u) := \frac{1}{1-q^{-1}z} + u\frac{(p^{-1}q^{-d}-q^{-1})z}{(1-p^{-1}q^{-d}z)(1-q^{-1}z)(1-q^{-1}z^d)} +$$

$$u^2\frac{(p^{-1}q^{-d}-q^{-1})^2 z^{d+1}}{(1-p^{-1}q^{-d}z)(1-p^{-1}q^{-d}z^d)(1-q^{-1}z)(1-q^{-1}z^d)(1-q^{-1}z^{d^2})} + \ldots;$$

Namely, it is clear that this series converges (uniformly on every compact set of points $(z,u)$) in the region $|z| < \min(|q|, |pq^d|)$. Now set $z = q^d$, which belongs to this region. So setting $\beta := u(q^{-1} - p^{-1}q^{-d})$, we obtain

$$1 = \frac{\beta q}{(1-p^{-1}q^{1-d})(1-q^{d-1})} - \frac{\beta^2 q^{d+1}}{(1-p^{-1}q^{1-d})(1-p^{-1})(1-q^{d-1})(1-q^{d^2-1})}$$

$$+\frac{\beta^3 q^{d^2+d+1}}{(1-p^{-1}q^{1-d})(1-p^{-1})(1-p^{-1}q^{d^2-d})(1-q^{d-1})(1-q^{d^2-1})(1-q^{d^3-1})} + \ldots$$

Define the "automatic Pochhammer symbol"

$$[a;q]_{d,n} = (1-aq)(1-aq^d)(1-aq^{d^2})\ldots.(1-aq^{d^{n-1}}),\ [a;q]_{d,0} = 1.$$

Let

$$K(a;q,u) = \sum_{n=0}^{\infty}\frac{(-1)^n q^{\frac{d^n-1}{d-1}}u^n}{[a;q]_{d,n}[q^{-1};q^d]_{d,n}}$$

(an entire function). We obtain

PROPOSITION 6.3. *We have*

$$\det(1 - uA_{P,s,1}) = K(p^{-1}q^{-d}; q, (q^{-1} - p^{-1}q^{-d})u).$$

*Thus the eigenvalues $A_{P,s,1}$ are the solutions of the equation*

$$K(p^{-1}q^{-d}; q, (q^{-1} - p^{-1}q^{-d})\lambda^{-1}) = 0.$$

PROOF. Indeed, the right hand side is an entire function of order 0 since its Taylor coefficients are rapidly decaying. □

## Acknowledgments

We thank Maxim Kontsevich and Erik Koelink for useful comments.

## References

[AS] Jean-Paul Allouche and Jeffrey Shallit, *Automatic sequences: Theory, applications, generalizations*, Cambridge University Press, Cambridge, 2003, DOI 10.1017/CBO9780511546563. MR1997038

[AM1] M. H. Annaby and Z. S. Mansour, *On the zeros of the second and third Jackson q-Bessel functions and their associated q-Hankel transforms*, Math. Proc. Cambridge Philos. Soc. **147** (2009), no. 1, 47–67, DOI 10.1017/S0305004109002357. MR2507309

[AM2] Mahmoud H. Annaby and Zeinab S. Mansour, *q-Fractional calculus and equations*, Lecture Notes in Mathematics, vol. 2056, Springer, Heidelberg, 2012. With a foreword by Mourad Ismail, DOI 10.1007/978-3-642-30898-7. MR2963764

[BCCD] Jason P. Bell, Frédéric Chyzak, Michael Coons, and Philippe Dumas, *Becker's conjecture on Mahler functions*, Trans. Amer. Math. Soc. **372** (2019), no. 5, 3405–3423, DOI 10.1090/tran/7762. MR3988615

[BK] Alexander Braverman and David Kazhdan, *Some examples of Hecke algebras for two-dimensional local fields*, Nagoya Math. J. **184** (2006), 57–84, DOI 10.1017/S0027763000009314. MR2285231

[GR] George Gasper and Mizan Rahman, *Basic hypergeometric series*, Encyclopedia of Mathematics and its Applications, vol. 35, Cambridge University Press, Cambridge, 1990. With a foreword by Richard Askey. MR1052153

[I] Jun-ichi Igusa, *Complex powers and asymptotic expansions. I. Functions of certain types*, J. Reine Angew. Math. **268(269)** (1974), 110–130, DOI 10.1515/crll.1974.268-269.110. MR347753

[KSw] Tom H. Koornwinder and René F. Swarttouw, *On q-analogues of the Fourier and Hankel transforms*, Trans. Amer. Math. Soc. **333** (1992), no. 1, 445–461, DOI 10.2307/2154118. MR1069750

[KoS] R. Koekoek, R. Swarttouw, *The Askey-scheme of hypergeometric orthogonal polynomials and its q-analogue*, `http://homepage.tudelft.nl/11r49/askey/`

[KS] H. T. Koelink and R. F. Swarttouw, *On the zeros of the Hahn-Exton q-Bessel function and associated q-Lommel polynomials*, J. Math. Anal. Appl. **186** (1994), no. 3, 690–710, DOI 10.1006/jmaa.1994.1327. MR1293849

[KRS] Slawomir Klimek, Sumedha Rathnayake, and Kaoru Sakai, *A note on spectral properties of the p-adic tree*, J. Math. Phys. **57** (2016), no. 2, 023512, 14, DOI 10.1063/1.4940339. MR3456151

[K] Maxim Kontsevich, *Notes on motives in finite characteristic*, Algebra, arithmetic, and geometry: in honor of Yu. I. Manin. Vol. II, Progr. Math., vol. 270, Birkhäuser Boston, Boston, MA, 2009, pp. 213–247, DOI 10.1007/978-0-8176-4747-6_7. MR2641191

[KO] Maxim Kontsevich and Alexander Odesskii, *Multiplication kernels*, Lett. Math. Phys. **111** (2021), no. 6, Paper No. 152, 59, DOI 10.1007/s11005-021-01491-1. MR4353308

[L] Peter D. Lax, *Functional analysis*, Pure and Applied Mathematics (New York), Wiley-Interscience [John Wiley & Sons], New York, 2002. MR1892228

[Mo] Takeshi Morita, *A connection formula of the Hahn-Exton q-Bessel function*, SIGMA Symmetry Integrability Geom. Methods Appl. **7** (2011), Paper 115, 11, DOI 10.3842/SIGMA.2011.115. MR2861226

[Mu] Mircea Mustaţă, *IMPANGA lecture notes on log canonical thresholds*, Contributions to algebraic geometry, EMS Ser. Congr. Rep., Eur. Math. Soc., Zürich, 2012, pp. 407–442, DOI 10.4171/114-1/16. Notes by Tomasz Szemberg. MR2976952

[SS] F. Štampach and P. Šťovíček, *The Hahn-Exton q-Bessel function as the characteristic function of a Jacobi matrix*, Spec. Matrices **1** (2013), 131–147. MR3155378

[V] M. Vlasenko, private communication.



DEPARTMENT OF MATHEMATICS, MASSACHUSETTS INSTITUTE OF TECHNOLOGY, CAMBRIDGE, MASSACHUSETTS 02139

EINSTEIN INSTITUTE OF MATHEMATICS, EDMOND J. SAFRA CAMPUS, GIVAAT RAM THE HEBREW UNIVERSITY OF JERUSALEM, JERUSALEM, 91904, ISRAEL

Contemporary Mathematics
Volume **780**, 2022
https://doi.org/10.1090/conm/780/15686

# Shuffle algebras, lattice paths and the commuting scheme

Alexandr Garbali and Paul Zinn-Justin

*Dedicated to Jasper Stokman on the occasion of his 50th birthday*

Abstract. The commutative trigonometric shuffle algebra A is a space of symmetric rational functions satisfying certain wheel conditions. We describe a ring isomorphism between A and the center of the Hecke algebra using a realization of the elements of A as partition functions of coloured lattice paths associated to the $R$-matrix of $\mathcal{U}_{t^{1/2}}(\widehat{gl}_\infty)$. As an application, we compute under certain conditions the Hilbert series of the *commuting scheme* and identify it with a particular element of the shuffle algebra A, thus providing a combinatorial formula for it as a "domain wall" type partition function of coloured lattice paths.

## 1. Introduction

**1.1. Center of Hecke and shuffle products.** Consider the *Hecke algebra* $\mathcal{H}_n$ ($n \in \mathbb{Z}_{\geq 0}$), which we define for our purposes to be the algebra (over a field extension $\mathbb{F}$ of $\mathbb{Q}(t)$ to be specified below) with generators $T_1, \ldots, T_{n-1}$ and relations

$$T_i T_{i+1} T_i = T_{i+1} T_i T_{i+1} \qquad T_i^2 = (t-1)T_i + t \qquad T_i T_j = T_j T_i \ (|i-j| \geq 2)$$

The standard basis $(T_w)_{w \in \mathcal{S}_n}$ of $\mathcal{H}_n$ is obtained by writing $T_w := T_{i_1} \ldots T_{i_k}$ if $w = s_{i_1} \ldots s_{i_k}$ is a reduced word in the elementary transpositions $s_i$. Denote by $|w| = k$ the length of $w$.

Combine the centers $\mathrm{Z}(\mathcal{H}_n)$ into a graded algebra

$$\mathrm{Z} := \bigoplus_{n \geq 0} \mathrm{Z}(\mathcal{H}_n)$$

by defining a shuffle product

$$(1) \qquad \begin{aligned} * : \mathrm{Z}(\mathcal{H}_k) \otimes \mathrm{Z}(\mathcal{H}_\ell) &\to \mathrm{Z}(\mathcal{H}_{k+\ell}) \\ x &\mapsto \sum_{w \in \mathcal{S}^{k,\ell}} t^{-|w|} T_w x T_{w^{-1}} \end{aligned}$$

where $\mathcal{H}_k \otimes \mathcal{H}_\ell$ is embedded inside $\mathcal{H}_{k+\ell}$ in the obvious way, and $\mathcal{S}^{k,\ell}$ is the set of shortest representatives of cosets in $\mathcal{S}_{k+\ell}/(\mathcal{S}_k \times \mathcal{S}_\ell)$.

Let us also consider $\Lambda$, the algebra of *symmetric functions* with coefficients in $\mathbb{F}$. One can for example define it as a polynomial ring in countably many variables: $\Lambda := \mathbb{F}[p_1, p_2, \ldots]$ where $\deg p_r = r$.

The authors gratefully acknowledge financial support from the Australian Research Council.

There is a graded algebra isomorphism $\Phi$ between Z and $\Lambda$ [**WW15**], which will be reviewed in §2; it is a deformation of the classical Frobenius map in the context of the symmetric group.

In what follows we fix the base field to be

$$\mathbb{F} = \mathbb{Q}(q,t)$$

Define $\mathbb{F}[x_1^\pm,\dots,x_n^\pm]_0$ to be the space of Laurent polynomials with specific degree bounds:

(2)

$$\mathbb{F}[x_1^\pm,\dots,x_n^\pm]_0 = \mathbb{F}\text{-span of } \left\{ x_1^{i_1}\dots x_n^{i_n},\ i_1,\dots,i_n\in\mathbb{Z} : \left|\sum_{j=1}^r i_j\right| \le r(n-r),\ 0\le r\le n \right\}$$

(in particular, these are homogeneous polynomials of degree 0) and $\mathbb{F}[x_1^\pm,\dots,x_n^\pm]_0^{\mathcal{S}_n}$ to be its subspace of symmetric Laurent polynomials.

Consider[1]

$$\mathrm{A}_n = \left\{ P\in\mathbb{F}[x_1^\pm,\dots,x_n^\pm]_0^{\mathcal{S}_n} :\ P(x,qx,tx,x_4,\dots,x_n) = P(tqx,qx,tx,x_4,\dots,x_n) = 0 \right\}$$

One can again introduce a shuffle product

$$\begin{aligned} * : \mathrm{A}_k\otimes\mathrm{A}_\ell &\to \mathrm{A}_{k+\ell} \\ P\otimes Q & \\ &\mapsto \sum_{w\in\mathcal{S}^{k,\ell}} \left[ P(x_{w(1)},\dots,x_{w(k)})Q(x_{w(k+1)},\dots,x_{w(k+\ell)}) \prod_{\substack{1\le i\le k\\ k+1\le j\le k+\ell}} \tilde\omega(x_{w(i)}/x_{w(j)}) \right] \end{aligned}$$

where $\tilde\omega(x) = \frac{(1-q^{-1}x)(1-qt^{-1}x)(1-tx)}{-x(1-x)}$ making

$$\mathrm{A} = \bigoplus_{n\ge 0} \mathrm{A}_n$$

a graded algebra. There is another graded algebra isomorphism $\Upsilon$ from A to $\Lambda$ [**FHH$^+$09**], which will be reviewed in §3.3.

It is natural to ask if there is an interpretation of the isomorphism $\Upsilon^{-1}\circ\Phi$ from Z to A. Note however that $\Lambda$ possesses graded algebra automorphisms, and in particular we shall allow ourselves to work modulo the automorphisms that rescale the variables $p_r$.

### 1.2. Lattice paths.

Consider a $n\times n$ square grid, and a set of $n$ paths entering from the left external edges of the grid and exiting at the top. We label incoming paths on the left $1,\dots,n$ from top to bottom. We say that a set of paths $P$ has connectivity $\mathrm{conn}(P) = w\in\mathcal{S}_n$ if the path labelled $i$ (i.e., the $i$th incoming path on the left counted from the top) exits at the $w^{-1}(i)$th location at the top (counted from the left).

EXAMPLE 1. Here are all the paths with connectivity 2431:

[1] $\mathrm{A}_n$ is not quite the shuffle algebra as normally defined, but is more convenient for us; the standard shuffle algebra, denoted by $\mathcal{A}_n$, will be given in §3.

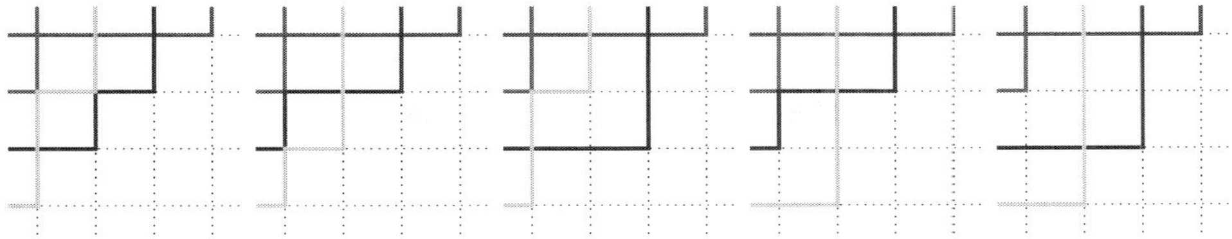

To each such set of lattice paths $P$ we associate a (Boltzmann) weight defined as the product over each vertex of the square grid of a factor that depends on the configuration around that vertex:

(3) $$\mathrm{wt}(P)=\prod_{i,j=1}^{n}\begin{cases} 1-t \\ q^{-1}x_ix_j^{-1}(1-t) \\ t(1-q^{-1}x_ix_j^{-1}) \\ 1-q^{-1}x_ix_j^{-1} \\ 1-tq^{-1}x_ix_j^{-1} \end{cases}$$

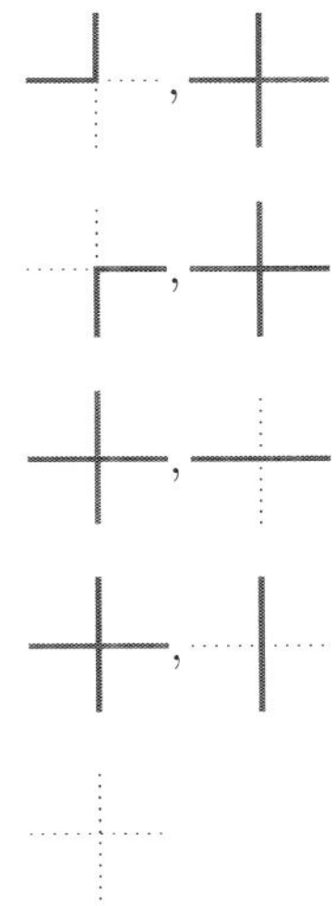

where the convention is that colours can be substituted as long as one preserves the ordering; that is, if two paths are present at a vertex, red (resp. green) stands for the one with the lower (resp. greater) label.

To any element $c=\sum_{v\in\mathcal{S}_n} c_vT_v\in \mathrm{Z}(\mathcal{H}_n)$ ($c_v\in\mathbb{F}$) of the center of the Hecke algebra we associate the *partition function*

(4) $$f(c):=\alpha_n\sum_{\substack{\text{lattice paths } P\\ \text{on the } n\times n \text{ grid}}}\mathrm{wt}(P)c_{\mathrm{conn}(P)}$$

where $\alpha_n=(q/t)^{\frac{n(n-1)}{2}}(1-t)^{-n}$ is a constant which we introduce for convenience.

We can now formulate our first theorem:

THEOREM 1. *There is a commuting square of graded algebra morphisms*

$$\begin{array}{ccc} \mathrm{Z} & \xrightarrow{\Phi} & \Lambda \\ \downarrow{\scriptstyle f} & & \downarrow{\scriptstyle \sigma_{q^{-1}}\sigma_t^{-1}} \\ \mathrm{A} & \xrightarrow{\Upsilon} & \Lambda \end{array}$$

*where $\sigma_{q^{-1}}\sigma_t^{-1}$ is the automorphism that sends $p_r$ to $(-1)^{r-1}\frac{(1-q)^r(1-t^r)}{(1-q^r)(1-t)^r}p_r$.*

**1.3. Partition function with connectivity** $12\ldots n$. We consider here the special case of the unit $1_n\in\mathrm{Z}(\mathcal{H}_n)$ of $\mathcal{H}_n$ which corresponds to the partition function with connectivity $12\dots n$

$$f(1_n) = \alpha_n \sum_{\substack{\text{lattice paths } P \\ \text{on the } n\times n \text{ grid} \\ \text{conn}(P)=1_n}} \text{wt}(P)$$

EXAMPLE 2. Here are the identity paths at $n = 3$:

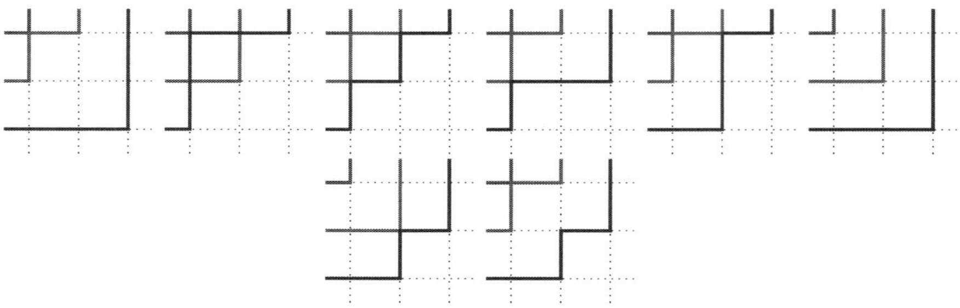

THEOREM 2. *The partition function* $f(1_n)$*, as an element of* $\mathbb{F}[x_1^\pm, \dots, x_n^\pm]_0^{\mathcal{S}_n}$*, is determined by* $f(1_0) = 1$ *and the recurrence relation*

(5)
$$(x_1+\dots+x_n)f(1_n)=\sum_{i=1}^n x_i f(1_{n-1})[\hat{x_i}]\prod_{\substack{j=1\\ j\neq i}}^n \frac{(1-q^{-1}x_j/x_i)(1-qt^{-1}x_j/x_i)(1-tx_i/x_j)}{-x_j/x_i(1-x_j/x_i)}$$

*where* $f(1_{n-1})[\hat{x_i}] \in \mathbb{F}[x_1^\pm, \dots, x_{i-1}, x_{i+1}, \dots, x_n^\pm]_0^{\mathcal{S}_n}$.

In §2.4 we will show that the partition functions $f(c)$ can be interpreted using the $R$-matrices of $\mathcal{U}_{t^{1/2}}(\widehat{gl}_\infty)$. This means that one can study them using the quantum inverse scattering method (QISM) [**KBI97**] (see also [**Lam18**]), and indeed the partition function $f(1_n)$ was already considered in [**BW18**]. We note, however, that the recurrence relation (5) has an unusual form from the point of view of the QISM due to the division by $x_1 + \dots + x_j$ at every $j$th step of the recursive computation of $f(1_n)$. It is not clear to us if one can prove (5) or find an alternative recursion relation for $f(1_n)$ using the QISM. Our proof of (5) is indirect; it follows as a corollary of our proof of Theorem 1 given in §4.

**1.4. The commuting scheme.** We now give a geometric interpretation of the partition function $f(1_n)$ of the previous section in terms of the commuting scheme. (The geometric interpretation of other $f(c)$ is left for future work.)

Given $n \in \mathbb{Z}_{\geq 0}$, define the *commuting scheme* $\mathfrak{C}_n$ to be the affine scheme in $\mathfrak{gl}_n(\mathbb{C})^2$ consisting of pairs of commuting $n \times n$ matrices:

$$\mathfrak{C}_n := \{(A, B) \in \mathfrak{gl}_n(\mathbb{C}) \times \mathfrak{gl}_n(\mathbb{C}) : [A, B] = 0\} \tag{6}$$

$\mathfrak{C}_n$ is known to be irreducible, of dimension $n(n+1)$ [**MT55**], smooth in codimension 1 [**Pop08**], and we have the conjecture, generally attributed to Artin and Hochster in 1982:

CONJECTURE 1. $\mathfrak{C}_n$ *is Cohen–Macaulay.*

which would imply that $\mathfrak{C}_n$ is reduced and normal. This conjecture will not be addressed in the present paper.

The group $GL_2(\mathbb{C}) \times GL_n(\mathbb{C})$ acts on $\mathfrak{gl}_n(\mathbb{C})^2$ by

$$(f,g) \in GL_2(\mathbb{C}) \times GL_n(\mathbb{C}) : \begin{pmatrix} A \\ B \end{pmatrix} \mapsto f \begin{pmatrix} gAg^{-1} \\ gBg^{-1} \end{pmatrix}$$

and leaves $\mathfrak{C}_n$ invariant. In particular, the maximal torus $T = (\mathbb{C}^\times)^{2+n}$ induces a $\mathbb{Z}^{2+n}$-grading of the corresponding coordinate rings; explicitly,

$$\deg A_{ij} = \varepsilon_1 + \varepsilon_{i+2} - \varepsilon_{j+2} \qquad \deg B_{ij} = \varepsilon_2 + \varepsilon_{i+2} - \varepsilon_{j+2} \qquad i,j = 1,\ldots,n$$

where $\varepsilon_i$ is the unit vector with nonzero coordinate $i$.

We consider the Hilbert series $\chi(\mathfrak{C}_n)$ of the commuting scheme with the multigrading above (i.e., the $T$-character of its coordinate ring); in fact, it is convenient to consider instead the *$K$-polynomial* (or Poincaré polynomial) of $\mathfrak{C}_n$, defined by

$$K_n = \frac{\chi(\mathfrak{C}_n)}{\chi(\mathfrak{gl}_n(\mathbb{C})^2)}$$

Explicitly, introducing formal variables $q_1, q_2, x_1, \ldots, x_n$ corresponding to $(\mathbb{C}^\times)^{2+n}$, one has

$$\chi(\mathfrak{gl}_n(\mathbb{C})^2) = \prod_{i,j=1}^{n} \frac{1}{(1 - q_1 x_i x_j^{-1})(1 - q_2 x_i x_j^{-1})}$$

$K_n$ is known to be a Laurent polynomial in $\mathbb{Z}[q_1^\pm, q_2^\pm, x_1^\pm, \ldots, x_n^\pm]$. Furthermore, because of the $GL_2(\mathbb{C}) \times GL_n(\mathbb{C})$-invariance of $\mathfrak{C}_n$, $K_n$ is invariant under the action of the corresponding Weyl group, i.e., it is a *symmetric* Laurent polynomial in the variables $q_1, q_2$ and $x_1, \ldots, x_n$ separately.

Finally, one can perform the substitution $q_i \mapsto 1 - q_i, x_i \mapsto 1 - x_i$ in $K_n$, expand in power series and keep only the lowest degree terms in those variables. We obtain this way a homogeneous polynomial of degree $\operatorname{codim} \mathfrak{C}_n = n(n-1)$ in $\mathbb{Z}[q_1, q_2, x_1, \ldots, x_n]$ (with its ordinary grading), called the *multidegree* of $\mathfrak{C}_n$ and denoted by

$$D_n := \operatorname{mdeg} \mathfrak{C}_n$$

Our first result on the commuting scheme is the following:

THEOREM 3. *Assuming Conjecture 1, the following formula holds for the $K$-polynomial of the commuting scheme:*

$$K_n = (q_1 q_2)^{\frac{n(n-1)}{2}} f(1_n) \tag{7}$$

*with the identification $q = q_1^{-1}$, $t = (q_1 q_2)^{-1}$.*

Performing the substitution of variables above, it is convenient to slightly rearrange the Boltzmann weights of §1.2 to:

$$\mathrm{wt}_K(P) = \prod_{i,j=1}^{n} \begin{cases} 1 - q_1 q_2 \\ q_1(1 - q_1 q_2) \\ 1 - q_1 x_i x_j^{-1} \\ q_1 q_2 (1 - q_1 x_i x_j^{-1}) \\ q_2 (1 - q_1 x_i x_j^{-1}) \\ x_i x_j^{-1} (1 - q_2 x_i^{-1} x_j) \end{cases} \tag{8}$$

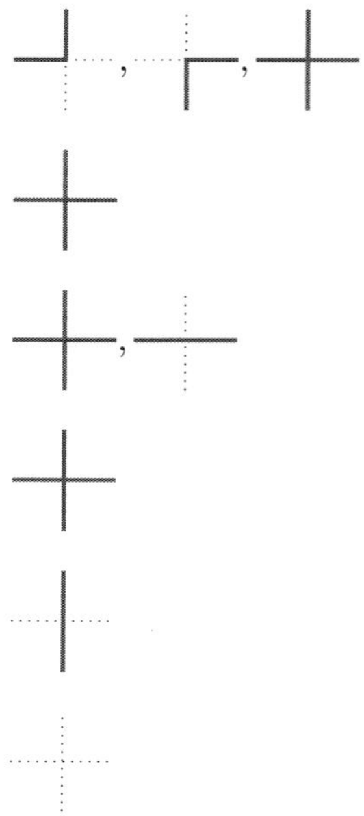

Then (7) can be formulated equivalently as

$$K_n = (1 - q_1 q_2)^{-n} \sum_{\substack{\text{lattice paths } P \\ \text{on the } n\times n \text{ grid} \\ \text{with identity connectivity}}} \mathrm{wt}_K(P)$$

EXAMPLE 3. At $n = 2$, the weights are

$\mathrm{wt}_K$ 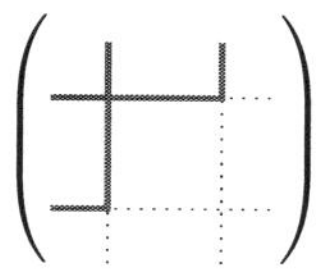 $= (1 - q_1 q_2)^3 (1 - q_2)$

$\mathrm{wt}_K$ 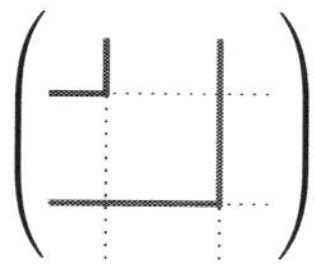 $= q_2 (1 - q_1 q_2)^2 (1 - q_1\, x_1 x_2^{-1})(1 - q_1\, x_2 x_1^{-1})$

and one can check that their sum divided by the common factor $(1 - q_1 q_2)^2$ indeed reproduces the $K$-polynomial of $\mathfrak{C}_2$

$$K_2 = 1 + q_1^2 q_2 - q_1 q_2 x_1 x_2^{-1} - q_1 q_2 - q_1 q_2 x_1^{-1} x_2 + q_1 q_2^2$$

Define another (Boltzmann) weight in a similar fashion:

$$\mathrm{wt}_H(P) = \prod_{i,j=1}^{n} \begin{cases} q_1 + q_2 \\ q_1 + x_i - x_j \\ q_2 - x_i + x_j \end{cases} \tag{9}$$

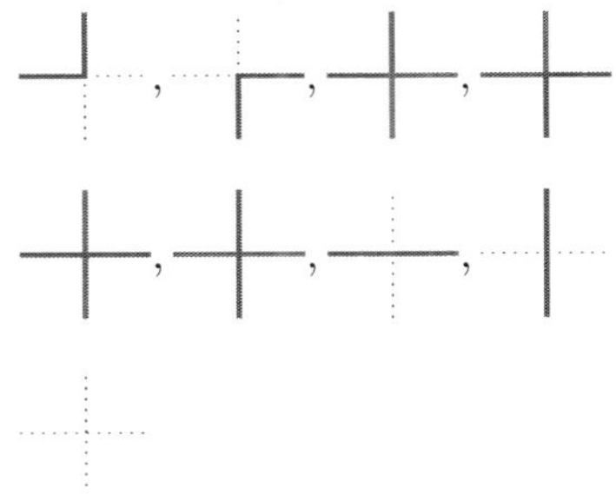

where colours of paths can be substituted freely.

We can state our second result:

THEOREM 4. *The following formula holds for the multidegree of the commuting scheme:*

$$D_n := (q_1 + q_2)^{-n} \sum_{\substack{\text{lattice paths } P \\ \text{on the } n\times n \text{ grid} \\ \text{with identity connectivity}}} \mathrm{wt}_H(P) \tag{10}$$

Note that Conjecture 1 and Theorem 3 imply Theorem 4, as can be easily checked. However Theorem 4 can be proven independently of Conjecture 1.

**1.5. Plan of the paper.** The rest of the paper is organized as follows. In §2, we provide details on the Hecke algebra construction of lattice paths partition functions, and on the map $\Phi$ from the center of Hecke to symmetric functions. In §3, we give details on the shuffle algebra, the map $\Upsilon$ and derive the recurrence relations which are required for proving Theorems 1-3. In §4, we prove our Theorem 1 by an explicit comparison of $\Phi$ and $\Upsilon \circ f$, and obtain Theorem 2 as a corollary of all that precedes. In §5, we turn to the geometry of the commuting scheme and prove Theorems 3 and 4, ending with some comments.

## 2. Hecke algebra and lattice paths

**2.1. The Hecke algebra and its center.** As in §1.1, we consider the Hecke algebra $\mathcal{H}_n$, an $\mathbb{F}$-algebra with generators $T_1, \ldots, T_{n-1}$ and relations

$$T_i T_{i+1} T_i = T_{i+1} T_i T_{i+1} \qquad T_i^2 = (t-1)T_i + t \qquad T_i T_j = T_j T_i \ (|i-j| \geq 2)$$

In this paper we choose $\mathbb{F}$ to be $\mathbb{Q}(q,t)$, though we shall not need the variable $q$ until §2.3. Note $\mathcal{H}_0 \cong \mathcal{H}_1 \cong \mathbb{F}$.

We have the simple lemma:

LEMMA 1.

$$T_i T_w = \begin{cases} T_{s_i w} & |s_i w| > |w| \\ (t-1)T_w + t T_{s_i w} & |s_i w| < |w| \end{cases}$$

$$T_w T_i = \begin{cases} T_{w s_i} & |w s_i| > |w| \\ (t-1)T_w + t T_{w s_i} & |w s_i| < |w| \end{cases}$$

expressing the action of $T_i$ on the standard basis $(T_w)_{w\in\mathcal{S}_n}$, where $T_w := T_{i_1}\dots T_{i_k}$ if $w = s_{i_1}\dots s_{i_k}$ is a reduced word in the elementary transpositions $s_i$, and $|w| = k$.

We shall often need two particular elements of the center $\mathrm{Z}(\mathcal{H}_n)$, namely the complete symmetrizer

$$S_n = \sum_{w\in\mathcal{S}_n} T_w \tag{11}$$

and the complete antisymmetrizer

$$A_n = \sum_{w\in\mathcal{S}_n} (-t)^{-|w|} T_w \tag{12}$$

Both are primitive idempotents up to normalization.

Next, we consider the shuffle product structure:

PROPOSITION 1. *The product $*$ in* (1) *is a well-defined map, making* $\mathrm{Z} = \bigoplus_{n\geq 0}\mathrm{Z}(\mathcal{H}_n)$ *a graded (associative) algebra.*

The nontrival part of the proof can be formulated by repeatedly applying Lemma 1; instead, we shall illustrate it by using the standard diagrammatic way to depict the Hecke algebra, as acting on $n$ strands:

$T_i =$ 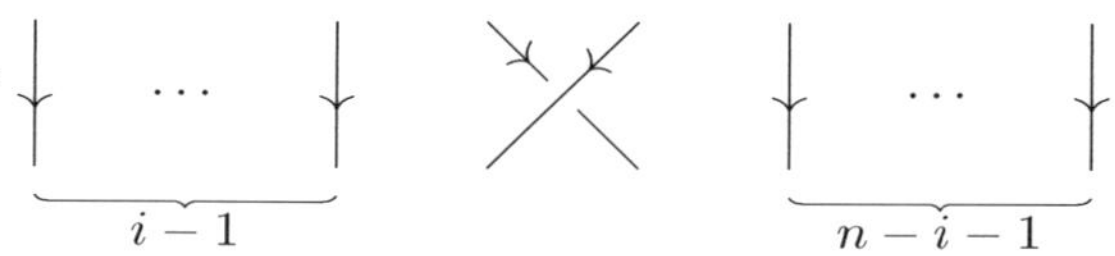

The arrows keep track of the ordering of operators: moving forward w.r.t. the orientation of a line is reading an expression right to left. Sometimes we will omit arrows; then all lines are implicitly oriented downwards. (The diagrammatic calculus that follows can be viewed as warming up for the considerably more difficult proof of our Theorem 1, which will also be diagrammatic.)

PROOF. Associativity of the product is obvious: an $m$-fold product can be expressed as

$$x_1 * \cdots * x_m = \sum_{w\in\mathcal{S}^{\lambda_1,\dots,\lambda_m}} t^{-|w|} T_w (x_1\otimes\cdots\otimes x_m) T_{w^{-1}} \tag{13}$$

where $x_i \in \mathcal{H}_{\lambda_i}$ and $\mathcal{S}^{\lambda_1,\dots,\lambda_m}$ is the set of shortest representatives in $\mathcal{S}_n/\mathcal{S}_{\lambda_1,\dots,\lambda_m}$, $\mathcal{S}_{\lambda_1,\dots,\lambda_m} = \mathcal{S}_{\lambda_1}\times\cdots\times\mathcal{S}_{\lambda_m}$.

The only nontrivial part is to show that $*$ is well-defined as a map from $\mathrm{Z}(\mathcal{H}_k)\otimes\mathrm{Z}(\mathcal{H}_\ell)$ to $\mathrm{Z}(\mathcal{H}_{k+\ell})$, i.e., we need to show that the right hand side of (1) is central, or equivalently that it commutes with all $T_i$s.

We recall that $\mathcal{S}^{k,\ell}$ is the set of shortest representatives in $\mathcal{S}_{k+\ell}/(\mathcal{S}_k\times\mathcal{S}_\ell)$; equivalently, it is the set of (Grassmannian) permutations of size $k+\ell$ with a unique possible descent at $(k, k+1)$.

Fixing $i = 1,\dots,n-1$, we distinguish four subsets of $\mathcal{S}^{k,\ell}$ according to the preimages of $i$ and $i+1$:

$$\mathcal{S}_{\lessgtr,\lessgtr} = \{w\in\mathcal{S}^{k,\ell} : \ w^{-1}(i)\lessgtr k+1/2,\ w^{-1}(i+1)\lessgtr k+1/2\}$$

Note that $w\mapsto s_i w$ induces a bijection between $\mathcal{S}_{<,>}$ and $\mathcal{S}_{>,<}$. We now discuss the corresponding terms in (1) separately:

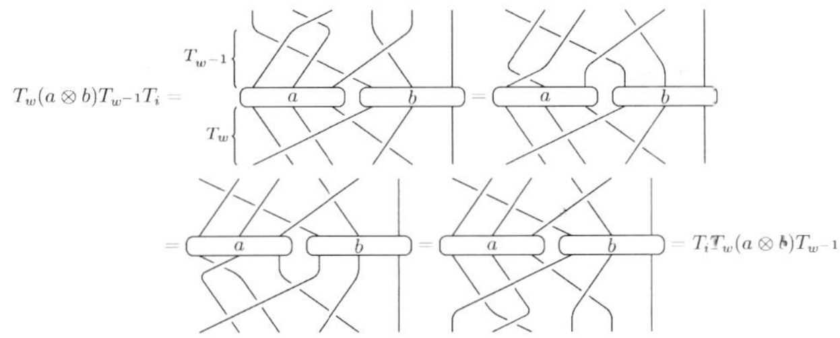


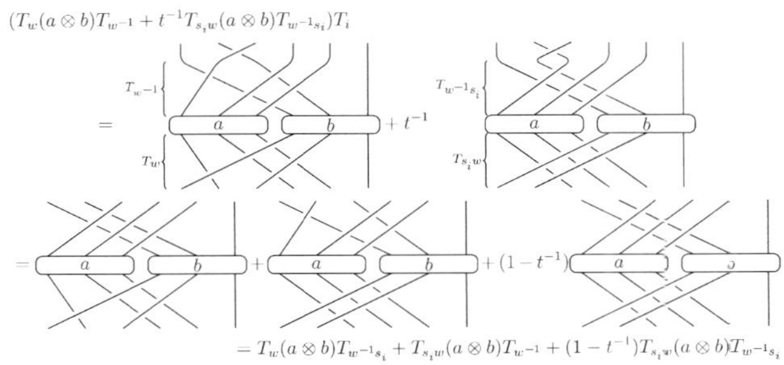


- If $w \in \mathcal{S}_{<,<}$, then $T_w(a \otimes b)T_{w^{-1}}$ commutes with $T_i$ provided $a \in Z(\mathcal{H}_k)$, as shown diagrammatically:
- Similarly, if $w \in \mathcal{S}_{>,>}$, then $T_w(a \otimes b)T_{w^{-1}}$ commutes with $T_i$ provided $b \in Z(\mathcal{H}_\ell)$.
- Finally, we group together the two terms corresponding to $w \in \mathcal{S}_{<,>}$ and $s_iw \in \mathcal{S}_{>,<}$ to obtain: Clearly, the same result is obtained by multiplying by $T_i$ on the left.

By multiplying by $t^{-|w|}$ and summing over $w$, we obtain the desired commutation property for the right hand side of (1). □

We define an algebra morphism $\Psi$ from $\Lambda = \mathbb{F}[p_1, p_2, \dots]$ to Z as follows. Define Jucys–Murphy elements in $\mathcal{H}_n$ (see e.g. [**Ram97**]) to be

$$J_{j,n} = \sum_{i=1}^{j-1} t^{i-j+1} T_i T_{i+1} \dots T_{j-1} T_{j-2} \dots T_i, \qquad j = 2, \dots, n$$

They form a commutative subalgebra of $\mathcal{H}_n$, and it is well-known that symmetric polynomials in the $J_{j,n}$ are central. Then $\Psi$ sends $p_n$ to the element

$$\Psi(p_n) = [n]_{t^{-1}} \prod_{j=2}^{n} J_{j,n}$$

where $[n]_{t^{-1}} = 1 + \cdots + t^{-(n-1)}$.

PROPOSITION 2 ([**WW15**]). *$\Psi$ is a graded algebra isomorphism from $\Lambda$ to Z.*

Its inverse is the $\Phi$ of §1.1. As a corollary, Z is commutative.

Denote by

$$\tilde{h}_n = \Phi(1_n) \tag{14}$$

$$\tilde{e}_n = \Phi(t^{-\frac{n(n-1)}{2}} T_{w_0}^2) \tag{15}$$

$$h_n = \Phi(S_n) \tag{16}$$

$$e_n = \Phi(A_n) \tag{17}$$

where $1_n$ is the identity of $\mathcal{H}_n$, and $w_0$ the longest element of $\mathcal{S}_n$. We can identify these symmetric functions. Given $u \in \mathbb{F}$, define $\sigma_u$ to be the automorphism of $\Lambda$

$$\sigma_u : \, p_r \mapsto \frac{(1-u)^r}{1-u^r} p_r \tag{18}$$

PROPOSITION 3 ([**Las06**]). *$h_n$ (resp. $e_n$) is the complete (resp. elementary) symmetric function of degree $n$.*

*Furthermore, $\tilde{h}_n = \sigma_t(h_n)$ and $\tilde{e}_n = \sigma_t(e_n)$.*

We also denote by $\sigma_\infty : \, p_r \mapsto (-1)^{r-1} p_r$. Note $\sigma_{u^{-1}} = \sigma_\infty \sigma_u$. It is well-known that $\sigma_\infty$ exchanges $h_n$ and $e_n$ (and similarly for $\tilde{h}_n$ and $\tilde{e}_n$).

**2.2. Partition functions from the Hecke algebra.** We now work inside the algebra $\mathcal{H}_{2n}$. There are two obvious embeddings of $\mathcal{H}_n$ into $\mathcal{H}_{2n}$, given by $T_i \mapsto T_i$ and $T_i \mapsto T_{i+n}$ respectively, and we denote the corresponding subalgebras by $\mathcal{H}_n^{(1)}$ and $\mathcal{H}_n^{(2)}$. Similarly, denote by $\mathcal{S}_n^{(1)}$ and $\mathcal{S}_n^{(2)}$ the two subgroups of $\mathcal{S}_{2n}$ (both isomorphic to $\mathcal{S}_n$) acting nontrivially on $\{1, \ldots, n\}$ (resp. $\{n+1, \ldots, 2n\}$) only.

Also define the *R-matrix*

$$\check{R}_i(u) = 1 - t + (1-u)T_i \qquad i = 1, \ldots, 2n-1 \tag{19}$$

as an element of $\mathcal{H}_{2n}[u^\pm]$.

Extending the diagrammatic calculus introduced in the proof of Proposition 1, we depict $R$-matrices as (flat) crossings of two lines carrying variables (so-called spectral parameters), the parameter of the $R$-matrix being the ratio of spectral parameters:

$\check{R}_i(u) =$ 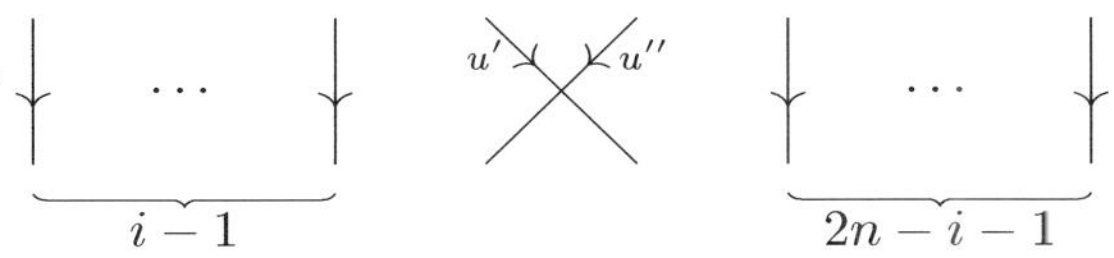

$u = u''/u'$

so that (19) can be expressed as:

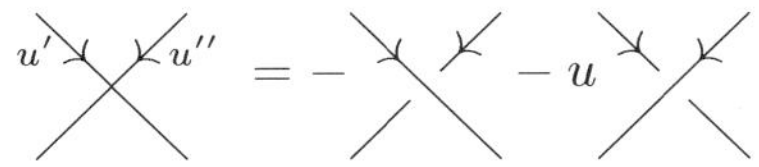

$u = u''/u'$

The *Yang–Baxter equation* is known to hold:

(20) $$\check{R}_i(u)\check{R}_{i+1}(uv)\check{R}_i(v) = \check{R}_{i+1}(v)\check{R}_i(uv)\check{R}_{i+1}(u) \qquad i = 1, \ldots, 2n-2$$

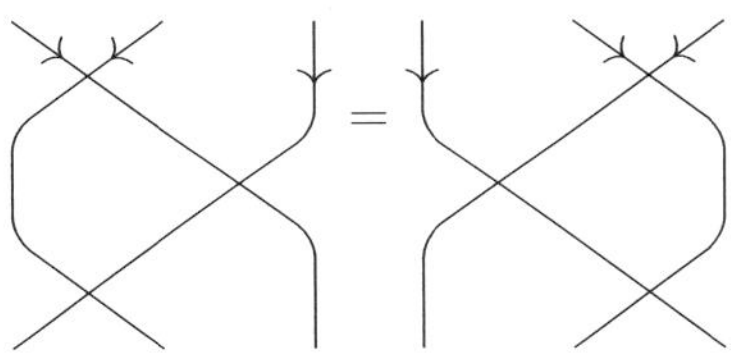


as an identity in $\mathcal{H}_{2n}[u^{\pm}, v^{\pm}]$, as well as the *unitarity equation*

(21) $$\check{R}_i(u^{-1})\check{R}_i(u) = (1-tu)(1-tu^{-1})$$

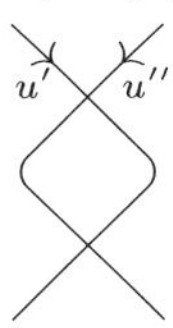

$= (1-tu)(1-tu^{-1})$ 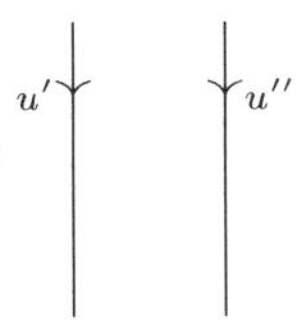

$u = u''/u'$

The central object of study of this section is the "formal partition function", which is the following element of $\mathcal{H}_{2n}[x_1^{\pm}, \ldots, x_n^{\pm}, y_1^{\pm}, \ldots, y_n^{\pm}]$:

(22)
$$Z(x_1, \ldots, x_n, y_1, \ldots, y_n) = \check{R}_n(x_n/y_1)\check{R}_{n-1}(x_{n-1}/y_1)\check{R}_{n+1}(x_n/y_2) \ldots \check{R}_n(x_1/y_n)$$

$=$ 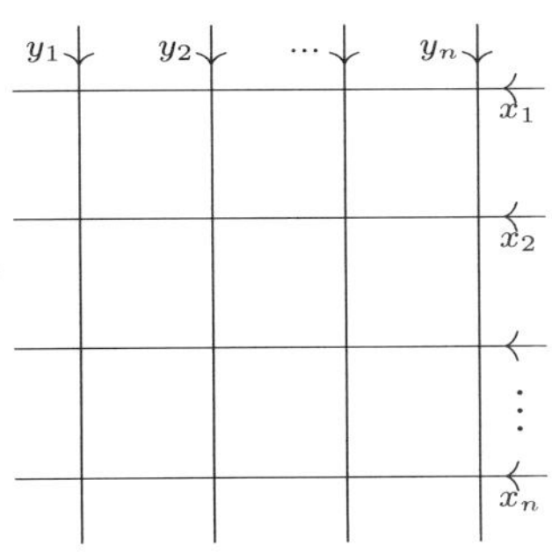


We also write $Z = Z(x_1, \ldots, x_n, y_1, \ldots, y_n)$ when there is no need to emphasize the dependence on the variables.

$Z$ has an expansion of the form

$$Z = \sum_{w \in \mathcal{S}_{2n}} Z_w T_w \tag{23}$$

where each $Z_w \in \mathbb{Z}[t^{\pm}, x_1^{\pm}, \ldots, x_n^{\pm}, y_1^{\pm}, \ldots, y_n^{\pm}]$.

**2.3. Exchange relation and symmetry.** Consider the operator $\tau_i$ (resp. $\tau'_j$) that permutes variables $x_i$ and $x_{i+1}$ (resp. $y_j$ and $y_{j+1}$), $i, j = 1, \ldots, n-1$.

We have the following fundamental lemma:

LEMMA 2. *$Z$ satisfies the* exchange relations*:*

$$(\tau_i Z)\check{R}_{n+i}(x_{i+1}/x_i) = \check{R}_i(x_{i+1}/x_i) Z \qquad i = 1, \ldots, n-1 \tag{24}$$

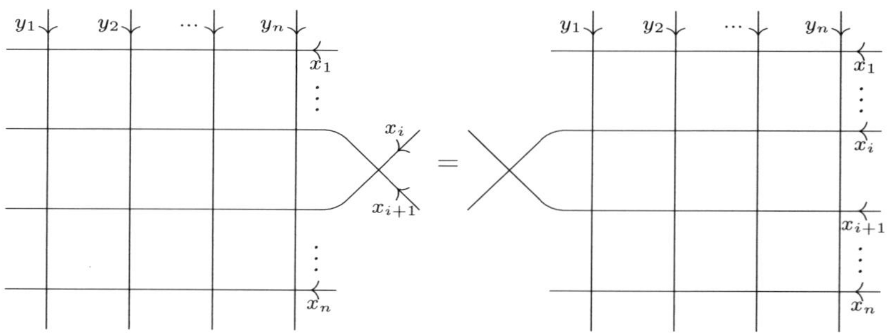


$$\check{R}_{n+j}(y_j/y_{j+1})\tau'_j Z = Z\check{R}_j(y_j/y_{j+1}) \qquad j = 1, \ldots, n-1 \tag{25}$$

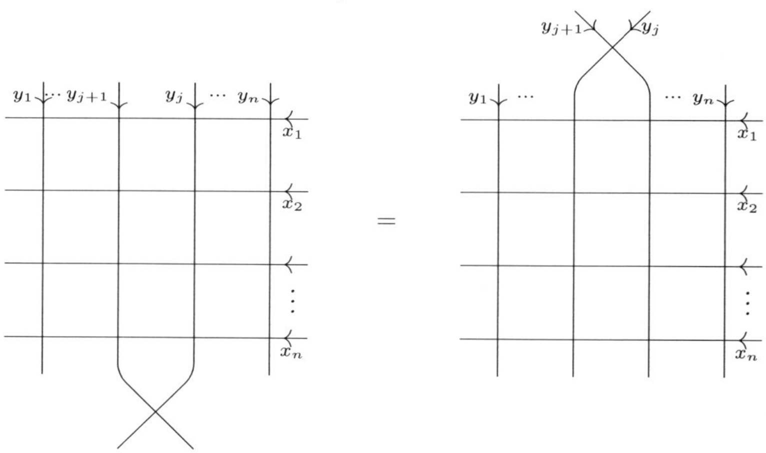


PROOF. Both equalities result from repeated application of the Yang–Baxter equation (20). $\square$

We now consider the specialization $y_i = qx_i$, $i = 1, \ldots, n$. We first note

LEMMA 3. $Z_w|_{y_i = qx_i} \in \mathbb{F}[x_1^{\pm}, \ldots, x_n^{\pm}]_0$.

PROOF. Recall that each of the $n^2$ factors of $Z$ corresponds to some vertex $(i, j)$ on the diagram of (22), and is a polynomial of degree 1 in the parameter $x_i/y_j$, so after specialization of the Lemma, in $x_i/(qx_j)$. Now fix $r \in \{0, \ldots, n\}$ and consider the degree $i_1 + \cdots + i_r$ in the first $r$ variables $x_1, \ldots, x_r$. Among the factors of $Z$, the ones corresponding in the diagram to the top-left $r \times r$ square have degree zero, and similarly for the bottom $(n-r)\times(n-r)$ square. The ones for the top-right $r\times(n-r)$ square can have monomials of degree 0 or 1, whereas for the bottom-left $(n-r)\times r$ square they can be of degree 0 or $-1$. Summing these degrees, we find the desired inequality $|i_1+\cdots+i_r|\le r(n-r)$. □

Of course the same proof would work with any $r$-subset of variables, suggesting that we should study the behaviour of $Z(x_1,\ldots,x_n,qx_1,\ldots,qx_n)$ under the interchange of the variables $x_i$. We have the following condition:

PROPOSITION 4. *Let $\chi$ be a linear form on $\mathcal{H}_{2n}$. $\chi(Z(x_1,\ldots,x_n,qx_1,\ldots,qx_n))$ is a symmetric Laurent polynomial in the $x_i$s provided $\chi$ satisfies*

$$\chi(ab)=\chi(ba)\qquad \forall a\in\mathcal{H}_{2n},\ b\in\mathcal{H}_n^{(1)}\otimes\mathcal{H}_n^{(2)} \tag{26}$$

PROOF. We assume $n\ge 2$, otherwise the Proposition is trivial. By definition exchanging $x_i$ and $x_{i+1}$ in $\chi(Z(x_1,\ldots,x_n,qx_1,\ldots,qx_n))$ amounts to considering $\chi(\tau_i\tau_i'Z)$ and then substituting $y_i=qx_i$. One has

$$\chi(\tau_i\tau_i'Z)=\chi(\check R_i(x_{i+1}/x_i)\check R_{n+i}(y_i/y_{i+1})^{-1}Z\check R_i(y_i/y_{i+1})\check R_{n+i}(x_{i+1}/x_i)^{-1})$$

Assuming the property (26) of the Proposition, one can rewrite this

$$\chi(\tau_i\tau_i'Z)=\chi(\check R_i(y_i/y_{i+1})\check R_i(x_{i+1}/x_i)Z\check R_{n+i}(y_i/y_{i+1})^{-1}\check R_{n+i}(x_{i+1}/x_i)^{-1})$$

Substituting $y_i=qx_i$ and simplifying using unitarity equation (21), we obtain the desired symmetry of $\chi(Z(x_1,\ldots,x_n,qx_1,\ldots,qx_n))$. □

In what follows, we restrict ourselves to the following class of linear forms. Denote by $\langle\bullet\rangle$ the linear form that extracts the coefficient of 1 in a linear combination of $T_w$, i.e., $\langle T_w\rangle:=\delta_{w,1}$. One checks that

$$\langle ab\rangle=\langle ba\rangle=\sum_{w\in\mathcal{S}_{2n}}t^{|w|}a_wb_{w^{-1}},\qquad a=\sum_{w\in\mathcal{S}_{2n}}a_wT_w,\ b=\sum_{w\in\mathcal{S}_{2n}}b_wT_w$$

Given $c\in\mathrm{Z}(\mathcal{H}_n)$, we then define

$$f(c):=\alpha_n\,\langle(c\otimes S_n)Z(x_1,\ldots,x_n,qx_1,\ldots,qx_n)\rangle \tag{27}$$

where $S_n$ was defined in (11), and $\alpha_n=(q/t)^{\frac{n(n-1)}{2}}(1-t)^{-n}$ is the same constant as in the introduction. Clearly, $\chi=\langle(c\otimes S_n)\bullet\rangle$ fulfills the condition of Proposition 4.

EXAMPLE 4. Three important examples are:

- the identity: $f(1_n)$ is the object of §1.3 and §5.
- the complete antisymmetrizer: $f(A_n)$ will be computed in §4.
- the complete symmetrizer: $f(S_n)$ is also interesting, since the symmetrization makes the model "color-blind", and thus equal to the famous Izergin determinant [**Ize87**] for the partition function of the six-vertex model with domain wall boundary conditions, as can be shown by the usual symmetry and recurrence arguments [**Kor82**]. We leave the proof to the interested reader.

We finally introduce a lattice path representation of $f(c)$, reconnecting it to the content of §1.2.

**2.4. Lattice path representation.** Consider the following representation of $\mathcal{H}_{2n}$. As a vector space, $\mathcal{M} = (\mathbb{F}^{n+1})^{\otimes 2n}$. Letting $(e_i)_{i=1,\dots,n+1}$ be the standard basis of $\mathbb{F}^{n+1}$, define the action of $\mathcal{H}_{2n}$ by

$$T_i(e_{k_1} \otimes \cdots \otimes e_{k_{2n}}) = \sum_{k'_i, k'_{i+1}=1}^{n+1} \begin{bmatrix} & k_i & \\ k'_i & & k_{i+1} \\ & k'_{i+1} & \end{bmatrix} e_{k_1} \otimes \cdots \otimes e_{k_{i-1}} \otimes e_{k'_i} \otimes e_{k'_{i+1}} \otimes e_{k_{i+2}} \otimes \cdots \otimes e_{k_{2n}}$$

where

$$\begin{bmatrix} & i & \\ j & & k \\ & \ell & \end{bmatrix} = \begin{cases} t-1 & i = j > k = \ell \\ 1 & j = k > i = \ell \\ t & j = k \le i = \ell \\ 0 & \text{else} \end{cases} \tag{28}$$

Given $v \in \mathcal{S}_{2n}$, define

$$|v\rangle = e_{\omega(v(1))} \otimes \cdots \otimes e_{\omega(v(2n))}$$

where $\omega(i) = i$ for $1 \le i \le n$ and $\omega(i) = n+1$ for $n+1 \le i \le 2n$. Note that $|v\rangle$ only depends on the class of $v$ inside $\mathcal{S}_n^{(2)} \backslash \mathcal{S}_{2n}$; we define this way a basis indexed by $\mathcal{S}_n^{(2)} \backslash \mathcal{S}_{2n}$ of the submodule of $\mathcal{M}$ of interest to us.

Similarly, define dual basis elements

$$\langle v| = e^*_{\omega(v(1))} \otimes \cdots \otimes e^*_{\omega(v(2n))}$$

in $\mathcal{M}^*$.

We have the easy lemma

LEMMA 4. *For all $w \in \mathcal{S}_{2n}$,*

(a) $\langle 1| T_w = t^{|w|} \langle w|$.

(b) $T_w |1\rangle = t^{|w_2|} |w^{-1}\rangle$ *where $w_2$ is defined by $w^{-1} = w_1 w_2$, $w_1 \in \mathcal{S}_n^{(2)}$ and $w_2$ a minimal representative in $\mathcal{S}_n^{(2)} \backslash \mathcal{S}_{2n}$.*

PROOF. (a) Write $w = w_1 w_2$ where $w_1 \in \mathcal{S}_n^{(2)}$ and $w_2$ minimal representative in $\mathcal{S}_n^{(2)} \backslash \mathcal{S}_{2n}$. One has directly $\langle 1| T_{w_1} = t^{|w_1|} \langle 1|$. Then use a reduced decomposition of $w_2$ and note that in (28), one only ever uses the case $j < \ell$ which leads to $\langle 1| T_{w_2} = t^{|w_2|} \langle w_2|$ inductively. At both stages, the only nonzero contributions come from the third line of (28). Case (b) is similar except the first (resp. second) stage involves the second (resp. third) line of (28).

□

Define, noting that $\mathcal{S}_n^{(1)}$ naturally sits inside $\mathcal{S}_n^{(2)} \backslash \mathcal{S}_{2n}$,

$$F_v(x_1, \dots, x_n, y_1, \dots, y_n) = \langle 1| Z(x_1, \dots, x_n, y_1, \dots, y_n) |v\rangle \qquad v \in \mathcal{S}_n^{(1)} \tag{29}$$

This is a slightly more general definition than we need for the interpretation of the right hand side of (4); indeed, we have:

PROPOSITION 5.

$$\sum_{\substack{\textit{lattice paths } P \\ \textit{on the } n \times n \textit{ grid} \\ \mathrm{conn}(P) = v}} \mathrm{wt}(P) = F_v(x_1, \dots, x_n, q x_1, \dots, q x_n)$$

PROOF. Consider the diagrammatic representation (22) of $Z$ and apply the representation described at the start of this section; this amounts to assigning to each edge a label in $\{1,\dots,n+1\}$. $F_v$ corresponds, in accordance with (29), to further imposing that labels of external edges be fixed: $1,\dots,n$ on the left side, $v(1),\dots,v(n)$ on the top side, and $n+1$ on the right and bottom sides. Now identify labels $1,\dots,n$ with the corresponding lattice paths, and $n+1$ with "empty". Then this produces precisely a lattice path with connectivity $P$. Finally, what needs to be checked is that the weight assigned to a lattice path coincides with the entry of $Z$. This can be done locally at the level of each individual crossing ($R$-matrix); and indeed, combining (28) with (19) and setting $y_i = qx_i$ leads to the weights in (3). □

Finally, we can show that the definition (27) given above is equivalent to the one (4) in the introduction:

LEMMA 5. *If* $c = \sum_{v\in\mathcal{S}_n} c_v T_v \in \mathrm{Z}(\mathcal{H}_n)$, *then* $c_v = c_{v^{-1}}$ *for all* $v \in \mathcal{S}_n$ *and*

$$f(c) = \alpha_n \sum_{v\in\mathcal{S}_n} c_v F_v(x_1,\dots,x_n,qx_1,\dots,qx_n)$$

PROOF. First we observe that

$$\langle x(1\otimes S_n)\rangle = \langle 1|\, x\, |1\rangle \qquad \forall x\in\mathcal{H}_{2n} \tag{30}$$

by testing this on $x = T_w$, $w\in\mathcal{S}_{2n}$, and noting that both sides of the equality are 0 unless $w\in\mathcal{S}_{2n}^{(2)}$, in which case they are equal to $t^{|w|}$ (using the definition (11) for the left hand side and Lemma 4 for the right hand side). We now start from the definition (27) of $f(c)$ and apply the equality above, as well as Lemma 4 (b):

$$f(c) = \alpha_n \langle 1|\, Zc\, |1\rangle = \alpha_n \sum_{v\in\mathcal{S}_n} c_{v^{-1}} F_v(x_1,\dots,x_n,qx_1,\dots,qx_n)$$

This is almost (4), except a small simplification was made: any central element satisfies $c_v = c_{v^{-1}}$ (short proof: it is obviously true for the generators $S_n$ of Z, and $*$ preserves this property). □

We conclude this section by reformulating more explicitly the exchange relations of Lemma 2 in terms of the $F_v$. The next proposition will not be used in what follows, but is included to reconnect to the existing literature [**BW18**].

PROPOSITION 6. *For* $v\in\mathcal{S}_n^{(1)}$ *and* $i,j=1,\dots,n-1$,

$$(1-tx_{i+1}/x_i)\tau_i F_v = \begin{cases} (1-t)F_v + (1-x_{i+1}/x_i)F_{s_iv} & s_iv > v \\ (1-t)(x_{i+1}/x_i)F_v + t(1-x_{i+1}/x_i)F_{s_iv} & s_iv < v \end{cases}$$

$$(1-ty_j/y_{j+1})\tau'_j F_v = \begin{cases} (1-t)F_v + (1-y_j/y_{j+1})F_{vs_j} & vs_j > v \\ (1-t)(y_j/y_{j+1})F_v + t(1-y_j/y_{j+1})F_{vs_j} & vs_j < v \end{cases}$$

PROOF. First, note that (23), (29), and Lemma 4 (a) imply

$$F_v = \sum_{w\in\mathcal{S}_n^{(2)}v} t^{|w|} Z_w \tag{31}$$

where the summation is over the coset of $v$ in $\mathcal{S}_n^{(2)}\backslash\mathcal{S}_{2n}$.

We then start from the identity (24) of Lemma 2.

Using (23) and Lemma 1, we obtain for the right hand side:

$$\check R_i(x_{i+1}/x_i)Z = \sum_{w\in\mathcal{S}_{2n}} Z_w \begin{cases}(1-t)T_w + (1-x_{i+1}/x_i)T_{s_i w} & s_i w > w\\ (1-t)(x_{i+1}/x_i)T_w + t(1-x_{i+1}/x_i)T_{s_i w} & s_i w < w\end{cases}$$

We now take the bra-ket $\langle 1|\cdot|v\rangle$ and note that only $w$ such that either $w\in\mathcal{S}_n^{(2)}v$ or $w\in\mathcal{S}_n^{(2)}s_i v$ contribute. We get for $s_i v > v$

$$\begin{aligned}\langle 1|\,\check R_i(x_{i+1}/x_i)Z\,|v\rangle &= (1-t)\sum_{w\in\mathcal{S}_n^{(2)}v} t^{|w|}Z_w + t(1-x_{i+1}/x_i)\sum_{w\in\mathcal{S}_n^{(2)}s_i v} t^{|s_i w|}Z_w\\ &= (1-t)F_v + (1-x_{i+1}/x_i)F_{s_i v}\end{aligned}$$

and for $s_i v < v$

$$\begin{aligned}\langle 1|\,\check R_i(x_{i+1}/x_i)Z\,|v\rangle &= (1-t)(x_{i+1}/x_i)\sum_{w\in\mathcal{S}_n^{(2)}v} t^{|w|}Z_w + (1-x_{i+1}/x_i)\sum_{w\in\mathcal{S}_n^{(2)}s_i v} t^{|s_i w|}Z_w\\ &= (1-t)(x_{i+1}/x_i)F_v + t(1-x_{i+1}/x_i)F_{s_i v}\end{aligned}$$

The bra-ket $\langle 1|\cdot|v\rangle$ of the left hand side is simple to evaluate: we note that $\check R_{n+i}(x_{i+1}/x_i)\,|v\rangle = (1-tx_{i+1}/x_i)\,|v\rangle$ because the $i$th and $(i+1)$th factors of $|v\rangle$ are $e_{n+1}\otimes e_{n+1}$; therefore

$$\langle 1|\,(\tau_i Z)\check R_{n+i}(x_{i+1}/x_i)\,|v\rangle = (1-tx_{i+1}/x_i)\tau_i F_v$$

This proves the first identity of the Proposition.

The second identity is actually simpler to prove (there isn't a full symmetry between rows and columns). Start analogously from (25) of Lemma 2. Now apply the bra-ket $\langle 1|\cdot|v\rangle$ directly. One has $\langle 1|\,\check R_{n+j}(y_j/y_{j+1}) = (1-ty_j/y_{j+1})\,\langle 1|$ by the same argument as above; but one also has from (28)

$$\check R_j(y_j/y_{j+1})\,|v\rangle = \begin{cases}(1-t)\,|v\rangle + (1-y_j/y_{j+1})\,|vs_j\rangle & vs_j > v\\ (y_j/y_{j+1})(1-t)\,|v\rangle + t(1-y_j/y_{j+1})\,|vs_j\rangle & vs_j < v\end{cases}$$

which leads to the desired identity. □

These relations can be used to determine the $F_v$ inductively. More precisely, consider first the computation of $F_{w_0}$ where $w_0$ is the longest element of $\mathcal{S}_n^{(1)}$. There is a unique lattice path with such connectivity, of the form

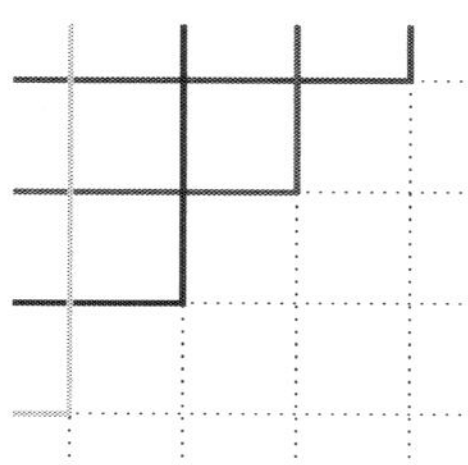

from which we conclude

$$F_{w_0} = (1-t)^n t^{\frac{n(n-1)}{2}} \prod_{\substack{1\le i,j\le n\\ i+j<n+1}} (1-x_i/y_j) \prod_{\substack{1\le i,j\le n\\ i+j>n+1}} (1-tx_i/y_j) \tag{32}$$

One can then apply *either* the second relation or the fourth relation of Proposition 6 to define the $F_w$ inductively.

## 3. The shuffle algebra

This section is devoted to the trigonometric Feigin–Odesskii shuffle algebra [**FO97**]. It was studied notably in [**FHH$^+$09, FT11, SV13, Neg14**]. We will review some known results and derive several formulas which are required for proving Theorems 1-3. We will follow the conventions of [**FHH$^+$09**] and define the shuffle algebra $\mathcal{A}$ which is slightly different from the algebra A in §1.1. The precise connection between $\mathcal{A}$ and A is given in Remark 1. For the treatment of the shuffle algebra $\mathcal{A}$ it is convenient to use three parameters $q_1, q_2, q_3$, expressed via $q$ and $t$ as

$$q_1 = q^{-1} \qquad q_2 = qt^{-1} \qquad q_3 = (q_1q_2)^{-1} = t \tag{33}$$

Let us outline the goals of this section:

- Discuss the shuffle algebra $\mathcal{A}^+$, its commutative subalgebra $\mathcal{A}$ and the isomorphism $\Upsilon$ of Theorem 1
- Give a characterization of the distinguished basis elements $\epsilon_\lambda$ of the shuffle algebra $\mathcal{A}$ via specializations which will be used in §4 to prove Theorem 1
- Introduce a special shuffle element $\kappa_n$ which will be used in §4 to prove Theorem 2 and in §5.1 to prove Theorem 3

**3.1. The shuffle algebra $\mathcal{A}^+$.** The shuffle algebra $\mathcal{A}^+$ is a vector space whose elements are symmetric rational functions of the form

$$P(x_1,\ldots,x_n) = \frac{p(x_1,\ldots,x_n)}{\prod_{1\le i<j\le n}(x_i-x_j)^2} \qquad p(x_1,\ldots,x_n) \in \mathbb{F}[x_1^{\pm 1},\ldots,x_n^{\pm 1}]^{\mathcal{S}_n} \tag{34}$$

where $p(x_1,\ldots,x_n)$ satisfies the *wheel condition*

(35)

$$p(x_1,\ldots,x_n)=0 \quad \text{if} \quad (x_i,x_j,x_k)=(x,q_1x,q_1q_2x) \quad \text{or} \quad (x_i,x_j,x_k)=(x,q_2x,q_1q_2x)$$

The number of arguments $(x_1,\ldots,x_n)$ gives a grading

$$\mathcal{A}^+ = \bigoplus_{n\ge 0} \mathcal{A}^+_n$$

We set $\mathcal{A}^+_0 = \mathbb{F}$ and $\mathcal{A}^+_1 = \mathbb{F}[x^{\pm 1}]$. The shuffle algebra $\mathcal{A}^+$ is closed under the *shuffle product* $*$, defined for two elements $F \in \mathcal{A}^+_k$ and $G \in \mathcal{A}^+_l$ by

(36)

$$(F*G)(x_1,\ldots,x_{k+l}) = \frac{1}{k!l!}\operatorname{Sym}\left[F(x_1,\ldots,x_k)G(x_{k+1},\ldots,x_{k+l})\prod_{\substack{1\le i\le k\\ 1\le j\le l}}\omega(x_{k+j},x_i)\right]$$

where Sym is the symmetrization over all arguments $x_i$

$$\operatorname{Sym}[f(x_1,\ldots,x_n)] = \sum_{w\in\mathcal{S}_n} f(x_{w(1)},\ldots,x_{w(n)}) \tag{37}$$

and $\omega(x,y)$ is defined as follows

$$\omega(x,y) = \frac{(x-q_1y)(x-q_2y)(x-q_3y)}{(x-y)^3} \tag{38}$$

It was shown in [**Neg14**] that the algebra $\mathcal{A}^+$ is generated by the elements $x^i \in \mathcal{A}^+_1$ for $i \in \mathbb{Z}$.

**3.2. The commutative subalgebra $\mathcal{A}$.** Consider a subalgebra $\mathcal{A} \subset \mathcal{A}^+$ of the elements $P \in \mathcal{A}_n^+$ for which the two limits

$$\lim_{\xi \to 0} P(\xi x_1, \ldots, \xi x_r, x_{r+1}, \ldots, x_n), \tag{39}$$

$$\lim_{\xi \to \infty} P(\xi x_1, \ldots, \xi x_r, x_{r+1}, \ldots, x_n) \tag{40}$$

exist and coincide for all fixed $r = 1, \ldots, n$. This subalgebra splits into components of fixed degree in the same way as $\mathcal{A}^+$

$$\mathcal{A} = \bigoplus_{n \in \mathbb{Z}_{\geq 0}} \mathcal{A}_n$$

REMARK 1. The shuffle algebras $\mathcal{A}_n$ and $\mathrm{A}_n$ are isomorphic. For $P(x_1, \ldots, x_n) \in \mathcal{A}_n$ and $Q(x_1, \ldots, x_n) \in \mathrm{A}_n$, s.t. $Q \mapsto P$, we have

$$P(x_1, \ldots, x_n) = \frac{1}{V_n(x_1, \ldots, x_n)} Q(x_1, \ldots, x_n) \tag{41}$$

where $V_n = V_n(x_1, \ldots, x_n)$ is defined by

$$V_n := \prod_{1 \leq i \neq j \leq n} (1 - x_i/x_j) \tag{42}$$

PROOF. The product $V_n^{-1} Q(x_1, \ldots, x_n)$ is of the form (34) and the wheel conditions of $\mathcal{A}_n^+$ and $\mathrm{A}_n$ coincide. The fact that the limiting conditions (39) and (40) of $\mathcal{A}_n$ are equivalent to the degree constraints (2) of $\mathrm{A}_n$ was shown in [**FHH$^+$09**]. □

Another important result of [**FHH$^+$09**] is the following proposition.

PROPOSITION 7. *The algebra $(\mathcal{A}, *)$ is commutative and the dimension of the graded subspace $\mathcal{A}_n$ is equal to the number of partitions of $n$.*

There are several distinguished bases in the space $\mathcal{A}$. These bases are built using the shuffle elements

$$\epsilon_n(x_1, \ldots, x_n; q_k) = \prod_{1 \leq i < j \leq n} \frac{(x_i - q_k x_j)(x_i - q_k^{-1} x_j)}{(x_i - x_j)^2} \qquad k = 1, 2, 3 \tag{43}$$

In the following we will abbreviate them as $\epsilon_n(q_k) = \epsilon_n(x_1, \ldots, x_n; q_k)$. Due to commutativity the products $\epsilon_i(q_k) * \epsilon_j(q_k)$ can be ordered such that $i > j$. Thus for a partition $\lambda$ the elements

$$\epsilon_\lambda(q_k) = \epsilon_{\lambda_1}(q_k) * \epsilon_{\lambda_2}(q_k) * \cdots * \epsilon_{\lambda_{\ell(\lambda)}}(q_k) \tag{44}$$

give three bases, one for each $k = 1, 2, 3$.

**3.3. Macdonald operators and the isomorphism $\Upsilon$.** Define the Heisenberg algebra $\mathfrak{h}$ with generators $a_{\pm n}$, $n > 0$, satisfying the relations

$$[a_m, a_n] = \delta_{m,-n} m \frac{1 - q^{|m|}}{1 - t^{|m|}} \tag{45}$$

We have two commutative subalgebras generated by $\{a_{\pm n}\}_{n>0}$ and containing the elements $a_{\pm\lambda} = a_{\pm\lambda_1} \ldots a_{\pm\lambda_{\ell(\lambda)}}$. Define the Fock space $\mathcal{F}$ by introducing the vacuum vector $|\emptyset\rangle$ which is annihilated by the positive modes $a_n |\emptyset\rangle = 0$, for $n > 0$. The action of the negative modes creates a basis

$$|a_\lambda\rangle = a_{-\lambda} |\emptyset\rangle \tag{46}$$

for all partitions $\lambda$.

Let $x = (x_1, x_2, \dots)$ be the alphabet for the ring of symmetric functions $\Lambda$ with the base field $\mathbb{F}$. Recall the power sum basis of $\Lambda$

$$p_\lambda = p_{\lambda_1} \dots p_{\lambda_{\ell(\lambda)}} \qquad p_n = x_1^n + x_2^n + \dots \tag{47}$$

The Fock space $\mathcal{F}$ is isomorphic to $\Lambda$ where the isomorphism is explicitly given by

$$|a_\lambda\rangle \mapsto p_\lambda$$

In $\Lambda$ a symmetric function is considered in the power sum basis and the above isomorphism replaces it with a vector in the Fock space. A distinguished basis in $\Lambda$ is given by the Macdonald symmetric functions $P_\lambda$ (see [**Mac98**]). We view it as a basis in the Fock space denoted by $|P_\lambda\rangle$.

The Macdonald operators are defined as the operators which act diagonally on $P_\lambda$. The *first* Macdonald operator $E$ [**Mac98**, Ch. VI] acts as follows

$$E\,|P_\lambda\rangle = \sum_{i\geq 1}(q^{\lambda_i} - 1)t^{-i}\,|P_\lambda\rangle \tag{48}$$

This operator is realized on the Fock space with the help of the vertex operator [**Shi06**]

$$\eta(z) := \exp\left(\sum_{n>0} \frac{1-t^{-n}}{n} a_{-n} z^n\right) \exp\left(-\sum_{n>0} \frac{1-t^n}{n} a_n z^{-n}\right) \tag{49}$$

The Fourier modes of this operator $\eta_n$ ($n \in \mathbb{Z}$) are given by

$$\eta(z) = \sum_{n\in\mathbb{Z}} \eta_n z^{-n}$$

and have a well defined action on $\mathcal{F}$. The operator $E$ is given by the zero mode [**Shi06**]

$$\eta_0 = (t-1)E + 1 \tag{50}$$

We can write the eigenvalue equation as follows

$$\frac{t^{-1}}{1-t^{-1}}\eta_0\,|P_\lambda\rangle = u_\lambda(e_1)\,|P_\lambda\rangle$$

where

$$u_\lambda(x_i) = q^{\lambda_i} t^{-i}$$

and $e_1$ is the first elementary symmetric function in the infinite alphabet $e_1 = x_1 + x_2 + \dots$. One can build higher order operators in a systematic way using the algebra $\mathcal{A}$ and the operators $\eta$.

The product of Heisenberg operators is *normally ordered* if all positive modes are on the right and negative modes on the left. The product $\eta(z)\eta(w)$ can be ordered at the expense of a rational function

$$\eta(z)\eta(w) = \frac{(1-w/z)(1-qt^{-1}w/z)}{(1-qw/z)(1-t^{-1}w/z)} : \eta(z)\eta(w) : \tag{51}$$

where the operators inside :: are normally ordered. For $f \in \mathcal{A}_n$ define the mapping $\mathcal{O} : \mathcal{A} \to \mathrm{End}_{\mathbb{F}}(\mathcal{F})$ by

(52)

$$\mathcal{O}(f) = \frac{1}{(t-1)^n n!} \oint \frac{\mathrm{d}z_1 \cdots \mathrm{d}z_n}{z_1 \cdots z_n} \frac{f(z_1, \dots, z_n)}{\epsilon_n(z_1, \dots, z_n; q_1)\epsilon_n(z_1, \dots, z_n; q_3)} : \eta(z_1) \dots \eta(z_n) :$$

where the integration contours are given by $|z_1| = \cdots = |z_n| = 1$ and $|t|, |q| < 1$. It was shown in [**FHH+09**] that $\mathcal{O}$ is compatible with the shuffle product, i.e.

$$\mathcal{O}(f * g) = \mathcal{O}(f)\mathcal{O}(g) \tag{53}$$

Since $\mathcal{A}$ is commutative this implies the commutativity of the operators $[\mathcal{O}(f), \mathcal{O}(g)] = 0$ for any two elements $f, g \in \mathcal{A}$. The map $\mathcal{O}$ was used in [**FHH+09**] to define an isomorphism $\Upsilon' : \mathcal{A} \to \Lambda$:

PROPOSITION 8. *Given $f \in \mathcal{A}_n$, there exists an $S \in \Lambda_n$ such that for any partition $\lambda$,*

$$\mathcal{O}(f)\,|P_\lambda\rangle = u_\lambda(S)\,|P_\lambda\rangle \tag{54}$$

*and $\Upsilon' : f \mapsto S$ thus defined is a ring isomorphism from $\mathcal{A}$ to $\Lambda$.*

By inserting a $V_n^{-1}$ in (54), cf Remark 1, we obtain the isomorphism $\Upsilon$ from A to $\Lambda$ that is stated in the introduction and used in Theorem 1. The symmetric functions in $\Lambda$ which correspond to the shuffle elements $\epsilon_n(q_k)$, $k = 1, 2, 3$ are all given by a plethystic substitution applied to the elementary symmetric functions $e_n$. Recalling (18) we set

$$e_n^{(1)} = e_n \qquad e_n^{(2)} = \sigma_{q^{-1}}\sigma_{qt^{-1}}^{-1}(e_n) \qquad e_n^{(3)} = \sigma_{q^{-1}}\sigma_t^{-1}(e_n) \tag{55}$$

Note that the plethystic substitution in $e_n^{(3)}$ appears in the statement of Theorem 1. This element will play a special role. We have [**FHH+09**]

$$\Upsilon'(\epsilon_n(q_k)) = e_n^{(k)} \qquad k = 1, 2, 3 \tag{56}$$

or equivalently

$$\mathcal{O}(\epsilon_n(q_k))\,|P_\lambda\rangle = u_\lambda(e_n^{(k)})\,|P_\lambda\rangle \qquad k = 1, 2, 3 \tag{57}$$

The eigenvalue equation (57) for $k = 1$ shows that the operator $\mathcal{O}(\epsilon_n(q_1))$ coincides with the $n$th Macdonald operator [**Mac98**]. Thus one shows (57) by matching the Macdonald difference operators $E_n$ with $\mathcal{O}(\epsilon_n(q_1))$. Based on this result one can show the other two eigenvalue equations for $k = 2, 3$ using the Wronski relations

$$\sum_{j=0}^{n} (q_k^j - q_l^{n-j}) \frac{(1 - q_l)^j}{(1 - q_k)^j} \epsilon_{n-j}(q_k) * \epsilon_j(q_l) = 0 \qquad k, l = 1, 2, 3 \tag{58}$$

In [**FHH+09**] the authors proved these Wronski relations. Applying $\mathcal{O}$ to (58) with $l = 1, k = 2, 3$ gives a relation between the eigenvalues of $\mathcal{O}(\epsilon_n(q_1))$ and $\mathcal{O}(\epsilon_n(q_k))$ which determines the eigenvalue in (57) for $k = 2, 3$.

We finish this section by giving a lemma which will be used at a later stage.

LEMMA 6. *The following identities hold*

(59)

$$(x_1 + \cdots + x_n)\,\epsilon_n(q_k) = \frac{1}{1 - q_k}\left(x_1 * \epsilon_{n-1}(q_k) - q_k \epsilon_{n-1}(q_k) * x_1\right) \qquad k = 1, 2, 3$$

PROOF. For simplicity we fix $k = 1$. Let us denote the right hand side of (59) by $C_n$ and rewrite it using the definition of the shuffle product (36)

$$C_n = \frac{1}{1-q_1}\sum_{i=1}^{n} x_i \epsilon_{n-1}(q_1)[\hat{x}_i]$$
$$\left(\prod_{\substack{j=1\\ j\neq i}}^{n}\frac{(x_j-q_1x_i)(x_j-q_2x_i)(x_j-q_3x_i)}{(x_j-x_i)^3} - q_1\prod_{\substack{j=1\\ j\neq i}}^{n}\frac{(x_i-q_1x_j)(x_i-q_2x_j)(x_i-q_3x_j)}{(x_i-x_j)^3}\right)$$

where $\epsilon_{n-1}(q_1)[\hat{x}_i]$ depends on the $n-1$ arguments $(x_1, \ldots, x_{i-1}, x_{i+1}, \ldots, x_n)$. Noting that

$$\epsilon_{n-1}(q_1)[\hat{x}_i]\prod_{\substack{j=1\\ j\neq i}}^{n}\frac{(x_j-q_1x_i)}{(x_j-x_i)} = \epsilon_n(q_1)\prod_{\substack{j=1\\ j\neq i}}^{n}\frac{(x_j-x_i)}{(x_j-q_1^{-1}x_i)}$$
$$\epsilon_{n-1}(q_1)[\hat{x}_i]\prod_{\substack{j=1\\ j\neq i}}^{n}\frac{(x_i-q_1x_j)}{(x_i-x_j)} = \epsilon_n(q_1)\prod_{\substack{j=1\\ j\neq i}}^{n}\frac{(x_i-x_j)}{(x_i-q_1^{-1}x_j)}$$

leads to

$$C_n = \epsilon_n(q_1)\frac{1}{1-q_1}\sum_{i=1}^{n} x_i\left(\prod_{\substack{j=1\\ j\neq i}}^{n}\frac{(x_j-q_2x_i)(x_j-q_3x_i)}{(x_j-q_1^{-1}x_i)(x_j-x_i)} - q_1\prod_{\substack{j=1\\ j\neq i}}^{n}\frac{(x_i-q_2x_j)(x_i-q_3x_j)}{(x_i-q_1^{-1}x_j)(x_i-x_j)}\right)$$

The factor next to $\epsilon_n(q_1)$ does not have poles at $x_i = x_j$ or $x_i = q_1^{-1}x_j$ because of the symmetry in $x_i$ and vanishing residues at $x_i = q_1^{-1}x_j$ which can be easily verified. It means that this factor must be a symmetric polynomial of degree 1. Thus it is proportional to the elementary symmetric polynomial $e_1$ in $n$ arguments

$$\frac{1}{1-q_1}\sum_{i=1}^{n} x_i\left(\prod_{\substack{j=1\\ j\neq i}}^{n}\frac{(x_i-q_2x_j)(x_i-q_3x_j)}{(x_i-q_1^{-1}x_j)(x_i-x_j)} - q_1\prod_{\substack{j=1\\ j\neq i}}^{n}\frac{(x_j-q_2x_i)(x_j-q_3x_i)}{(x_j-q_1^{-1}x_i)(x_j-x_i)}\right) = C\sum_{i=1}^{n} x_i$$

We need to show that the proportionality factor $C = 1$. This can be verified by setting $x_i = 0$ for $i > 1$ in the above equation. This completes the proof of (59). □

**3.4. Shuffle elements of $\mathcal{A}$ and specializations.** In this section we look at the problem of determining the shuffle algebra elements of $\mathcal{A}$ recursively via specializations of the arguments $x_i$. This problem was investigated in [**Neg14**, **Neg16**] using a coproduct map of $\mathcal{A}^+$. Our treatment of this problem is appropriate for the goal of matching shuffle elements with partition functions in §4.

Recall that the shuffle elements of $\mathcal{A}$ have the form

$$P(x_1, \ldots, x_n) = \frac{p(x_1, \ldots, x_n)}{\prod_{1\leq i<j\leq n}(x_i - x_j)^2} \tag{60}$$

where $p(x_1, \ldots, x_n)$ is a symmetric polynomial. We start with a lemma which gives us information about the degree of $p(x_1, \ldots, x_n)$.

LEMMA 7. *Let $P(x_1,\dots,x_n)$ be an element of $\mathcal{A}_n$ as in (60). Let $x$ be an indeterminate and $c_1,\dots,c_r\in\mathbb{F}$ then $p(c_1x,\dots,c_rx,x_{r+1},\dots,x_n)\in\mathbb{F}[x,x_{r+1},\dots,x_n]$ is of the form*

$$x^{r(r-1)}G(x,x_{r+1},\dots,x_n;c_1,\dots,c_r) \tag{61}$$

*and $G(x,x_{r+1},\dots,x_n;c_1,\dots,c_r)$ is a polynomial of degree at most $2r(n-r)$ in $x$.*

PROOF. Consider the specialization $x_k=c_kx$ $(k=1,\dots,r)$ of (60)

$$P(c_1x,\dots,c_rx,x_{r+1},\dots,x_n)$$
$$=\frac{p(c_1x,\dots,c_rx,x_{r+1},\dots,x_n)}{x^{r(r-1)}\prod_{i=1}^r\prod_{j=r+1}^n(c_ix-x_j)^2\prod_{1\le i<j\le r}(c_i-c_j)^2\prod_{r+1\le i<j\le n}(x_i-x_j)^2}$$

The existence of the limit (39) implies that, for $x\to 0$, the polynomial in the numerator $p(c_1x,\dots,c_rx,x_{r+1},\dots,x_n)$ contains an overall factor $x^{r(r-1)}$ to compensate the same factor in the denominator. Thus we established the form (61) and

$$P(c_1x,\dots,c_rx,x_{r+1},\dots,x_n)$$
$$=\frac{G(x,x_{r+1},\dots,x_n;c_1,\dots,c_r)}{\prod_{i=1}^r\prod_{j=r+1}^n(c_ix-x_j)^2\prod_{1\le i<j\le r}(c_i-c_j)^2\prod_{r+1\le i<j\le n}(x_i-x_j)^2}$$

The existence of the limit (40) implies that, for $x\to\infty$, the degree in $x$ of the polynomial $G(x,x_{r+1},\dots,x_n)$ is controlled by $\prod_{i=1}^r\prod_{j=r+1}^n(c_ix-x_j)^2$, i.e. bounded by $2r(n-r)$. □

In the following we will only be interested in the situations when $c_i=q_k^{i-1}$. The next statement gives a recipe for a recursive computation of shuffle elements of $\mathcal{A}$.

PROPOSITION 9. *Let $r\in\{1,\dots,n\}$, and $P(x_1,\dots,x_n)$ be an element of $\mathcal{A}_n$ as in (60). Fix $k\in\{1,2,3\}$, then $P(x_1,\dots,x_n)$ is determined by the following specializations:*

(1.$r$) $\lim_{x_i\to 0}\cdots\lim_{x_1\to 0}P(x_1,\dots,x_i,x_{i+1},\dots,x_n)$, $i=1,\dots,r-1$.
(2.$r$) $P(x,q_kx,\dots,q_k^{r-1}x,x_{r+1},\dots,x_n)$.

PROOF. We fix $k',k''\in\{1,2,3\}$ such that $k,k',k''$ are all distinct. The proof is by induction on $r$. The case $r=1$ is trivial. Now assume the property true at $r\ge 1$, i.e. knowing (1.$r$) and (2.$r$) allows us to compute $P(x_1,\dots,x_n)$. Let us show that the specializations (1.$r+1$) and (2.$r+1$) allow us to compute the specializations (1.$r$) and (2.$r$). Since the specializations (1.$r$) are a subset of the specializations (1.$r+1$) we only need to focus on computing (2.$r$)

$$P(x,q_kx,\dots,q_k^{r-1}x,x_{r+1},\dots,x_n)$$

According to Lemma 7 with $c_i=q_k^{i-1}$ for $1\le i\le r$ we have

$$P(x,q_kx,\dots,q_k^{r-1}x,x_{r+1},\dots,x_n)$$
$$=\frac{G(x,x_{r+1},\dots,x_n)}{\prod_{i=1}^r\prod_{j=r+1}^n(q_k^{i-1}x-x_j)^2\prod_{1\le i<j\le r}(q_k^{i-1}-q_k^{j-1})^2\prod_{r+1\le i<j\le n}(x_i-x_j)^2}$$

where $G$ is a polynomial of degree $2r(n-r)$ in $x$. According to the wheel conditions (35), it has the following factorization:

$$G(x, x_{r+1}, \ldots, x_n) = \prod_{i=0}^{r-2} \prod_{j=r+1}^{n} (q_{k'} x_j - q_k^i x)(q_{k''} x_j - q_k^i x) H(x, x_r, \ldots, x_n)$$

with the degree of $H$ in $x$ being $2r(n-r) - 2(r-1)(n-r) = 2(n-r)$. To determine $H$, we therefore need to know its values at $2(n-r)+1$ distinct specializations of $x$. The value of $H$ at $x = 0$ gives one specialization and can be computed from the knowledge of

$$\lim_{x_r \to 0} \cdots \lim_{x_1 \to 0} P(x_1, \ldots, x_r, x_{r+1}, \ldots, x_n)$$

The specialization (2), using the fact that $P$ is a symmetric polynomial, is equivalent to computing $H$ at $x = q_k^{-r} x_j$ and $x = q_k x_j$, $j = r+1, \ldots, n$. This is the required number of specializations, which means $G$ is determined and with it $P(x, q_k x, \ldots, q_k^{r-1} x, x_{r+1}, \ldots, x_n)$. □

Let us apply this result to the computation of the basis elements $\epsilon_\lambda(q_k)$ with $\ell(\lambda) = m$ and $\lambda_1 + \cdots + \lambda_m = n$. For convenience we only consider the case $k = 3$ and denote by

$$\epsilon_\lambda = \epsilon_\lambda(q_3) = \epsilon_\lambda(t)$$

We apply Proposition 9 with $r = n$ and $k = 3$ to the problem of computing $\epsilon_\lambda$. This means that in order to determine $\epsilon_\lambda$ we need to know the values of

$$\lim_{x_i \to 0} \cdots \lim_{x_1 \to 0} \epsilon_\lambda(x_1, \ldots, x_i, x_{i+1}, \ldots, x_n), \qquad i = 1, \ldots, n-1 \tag{62}$$

$$\epsilon_\lambda(x, tx, \ldots, t^{n-1}x) \tag{63}$$

One can compute the limits in (62) using the following lemma.

LEMMA 8. *The elements $\epsilon_{\lambda_1,\ldots,\lambda_m}$, with $\lambda_1 + \cdots + \lambda_m = n$, satisfy the recurrence relation*

$$\lim_{x_n \to 0} \epsilon_{\lambda_1,\ldots,\lambda_m}(x_1, \ldots, x_{n-1}, x_n) = \sum_{i=1}^{m} \epsilon_{\lambda_1,\ldots,\lambda_i - 1,\ldots,\lambda_m}(x_1, \ldots, x_{n-1}) \tag{64}$$

PROOF. We start by writing $\epsilon_{\lambda_1,\ldots,\lambda_m}$ explicitly as a shuffle product of $\epsilon_{\lambda_l}$

$$\begin{aligned}\epsilon_{\lambda_1,\ldots,\lambda_m} = \sum_{w \in \mathcal{S}_n / \mathcal{S}_{\lambda_1,\ldots,\lambda_m}} & \prod_{l=1}^{m} \epsilon_{\lambda_l}(x_{w(1+\sum_{i=1}^{l-1}\lambda_i)}, \ldots, x_{w(\sum_{i=1}^{l}\lambda_i)}) \\ & \prod_{1 \le a < b \le m} \prod_{i \in I_a, j \in I_b} \omega(x_{w(j)}, x_{w(i)})\end{aligned} \tag{65}$$

where we defined the subsets $I_a$ by

$$I_a := \left\{ \sum_{s=1}^{a-1} \lambda_s + 1, \ldots, \sum_{s=1}^{a} \lambda_s \right\} \tag{66}$$

From definitions (43) and (38) we have two obvious identities

$$\epsilon_k(x_1, \ldots, x_{k-1}, 0) = \epsilon_{k-1}(x_1, \ldots, x_{k-1}) \qquad \omega(x, 0) = \omega(0, x) = 1$$

Fix $r \in \{1, \ldots, m\}$ and consider the summands of (65) with $w$ such that $n \in w(I_r)$. The factor $\epsilon_{\lambda_r}$ will be replaced by $\epsilon_{\lambda_r - 1}$ upon setting $x_n = 0$ while the factors of $\omega$ containing $x_n$ will disappear. The summation over $w \in \mathcal{S}_n / \mathcal{S}_{\lambda_1,\ldots,\lambda_m}$ (with

$n \in w(I_r)$) reduces to the summation over $w' \in \mathcal{S}_{n-1}/\mathcal{S}_{\lambda_1,\dots,\lambda_r-1,\dots,\lambda_m}$ which is equal to the shuffle product of $\epsilon_{\lambda_1},\dots,\epsilon_{\lambda_r-1},\dots,\epsilon_{\lambda_m}$ similarly to (65). Treating in this way every summand with $n \in w(I_r)$ for $r \in \{1,\dots,m\}$ gives rise to the formula (64). $\square$

It is clear that computing (62) can be done inductively using (64). The specializations in (63) are not difficult to compute and in fact follow from more general specializations considered in [**FHH$^+$09**]. We have

$$\epsilon_\lambda(x,tx,\dots,t^{n-1}x) = 0, \qquad \lambda \neq (1,1,\dots,1) \tag{67}$$

$$\epsilon_{(1,1,\dots,1)}(x,tx,\dots,t^{n-1}x) = \prod_{1\le i<j\le n} \frac{(qt^i-t^j)(t^{i+1}-qt^j)(t^i-t^{j+1})}{qt(t^i-t^j)^3} \tag{68}$$

Let us explain (68). We use the explicit expression (65) with $\lambda_i = 1$ in which case all $\epsilon_{\lambda_i} = 1$ and all sets $I_a$ have a single element. Due to the vanishing of $\omega(tx,x) = 0$ only one term in the summation is non-vanishing. This term must have $x$'s with fully ordered indices as follows

$$\prod_{1\le i<j\le n} \omega(x_i,x_j)|_{x_k=t^{k-1}x}$$

which is equal to the product on the right hand side of (68).

**3.5. The shuffle element $\kappa_n$.** The results of this section are gathered in the following proposition:

PROPOSITION 10. *Let $\kappa_0 = 1 \in \mathcal{A}_0$. Each of the following formulas uniquely determines the same sequence of shuffle algebra elements $\kappa_n \in \mathcal{A}_n$:*

$$\Upsilon'(\kappa_n) = \sigma_q(e_n) \tag{69}$$

$$\kappa_n = \sum_{j=0}^{n} q_k^{-n}(q_k-1)^{n-j}\kappa_j * \epsilon_{n-j}(q_k) \qquad k=1,2,3 \tag{70}$$

$$(x_1+\cdots+x_n)\,\kappa_n = \kappa_{n-1} * x_1 \tag{71}$$

*where in* (70) *any one choice $k \in \{1,2,3\}$ suffices.*

PROOF. It is clear that each of these equations uniquely determines the element $\kappa_n$, we only need to show their consistency. We can take (71) together with $\kappa_0 = 1$ as the definition of $\kappa_n$ and prove that it implies (70) and (69).

The derivation of (70) is recursive. The case $n = 0$ of (70) obviously holds. Using (71) we will derive (70) for $\kappa_n$ assuming that (70) holds for $n-1$. We act with $*\,x_1$ on both sides of (70) where $n$ is replaced by $n-1$

$$\kappa_{n-1} * x_1 = \sum_{j=0}^{n-1} q_k^{-n+1}(q_k-1)^{n-1-j}\kappa_j * \epsilon_{n-j-1}(q_k) * x_1 \tag{72}$$

On the left hand side we apply (71) and on the right hand side (59)

$$\begin{aligned}\sum_{i=1}^{n} x_i\kappa_n = &\sum_{j=0}^{n-1} q_k^{-n}(q_k-1)^{n-j}\kappa_j * \left(\sum_{i=1}^{n-j} x_i\epsilon_{n-j}(q_k)\right)\\ &+\sum_{j=0}^{n-1} q_k^{-n}(q_k-1)^{n-j-1}\kappa_j * x_1 * \epsilon_{n-j-1}(q_k)\end{aligned}$$

On the right hand side we apply (71) to $\kappa_j * x_1$

$$\sum_{i=1}^{n} x_i\kappa_n = \sum_{j=0}^{n-1} q_k^{-n}(q_k-1)^{n-j}\kappa_j * \left(\sum_{i=1}^{n-j} x_i\epsilon_{n-j}(q_k)\right) + \sum_{j=0}^{n-1} q_k^{-n}(q_k-1)^{n-j-1}\left(\sum_{i=1}^{j+1} x_i\kappa_{j+1}\right) * \epsilon_{n-j-1}(q_k)$$

In the last summation we can shift the index $j$

$$\sum_{i=1}^{n} x_i\kappa_n = \sum_{j=0}^{n-1} q_k^{-n}(q_k-1)^{n-j}\kappa_j * \left(\sum_{i=1}^{n-j} x_i\epsilon_{n-j}(q_k)\right) + \sum_{j=1}^{n} q_k^{-n}(q_k-1)^{n-j}\left(\sum_{i=1}^{j} x_i\kappa_j\right) * \epsilon_{n-j}(q_k) \tag{73}$$

From the definition of the shuffle product (36) it is easy to check that

$$\kappa_j * \left(\sum_{i=1}^{n-j} x_i\epsilon_{n-j}(q_k)\right) + \left(\sum_{i=1}^{j} x_i\kappa_j\right) * \epsilon_{n-j}(q_k) = \left(\sum_{i=1}^{n} x_i\right)\kappa_j * \epsilon_{n-j}(q_k)$$

Using this identity in (73) for every pair of summands with $j = 1, \ldots, n-1$ leads to

$$\sum_{i=1}^{n} x_i\kappa_n = \sum_{i=1}^{n} x_i q_k^{-n}(q_k-1)^{n}\kappa_0 * \epsilon_n(q_k) + \sum_{i=1}^{n} x_i q_k^{-n}\kappa_n * \epsilon_0(q_k) + \sum_{i=1}^{n} x_i \sum_{j=1}^{n-1} q_k^{-n}(q_k-1)^{n-j}\kappa_j * \epsilon_{n-j}(q_k)$$

After removing the common factor $\sum_{i=1}^{n} x_i$ and putting all terms together on the right hand side we obtain (70) for $\kappa_n$.

Next we show that (69) holds. For that we will use (70). It is sufficient to consider $k = 1$ in (70). We apply the map $\Upsilon'$ to this equation

$$\Upsilon'(\kappa_n) = \sum_{j=0}^{n} q^n(q^{-1}-1)^{n-j}\Upsilon'(\kappa_j)\Upsilon'(\epsilon_{n-j}(q^{-1})) \tag{74}$$

From (56) we have $\Upsilon'(\epsilon_{n-j}(q^{-1})) = e_{n-j}$. Let us denote by $e'_n = \Upsilon'(\kappa_n)$, then (74) becomes

$$e'_n = \sum_{j=0}^{n} q^j(1-q)^{n-j}e'_j e_{n-j} \tag{75}$$

We need to show that $e'_n = \sigma_q(e_n)$. Note that, together with the initial condition, (75) uniquely determines $e'_n$. Let us solve (75) using generating functions, set

$$E'(z) := \sum_{n=0}^{\infty} e'_n z^n \qquad E(z) := \sum_{n=0}^{\infty} e_n z^n = \exp\left(-\sum_{j>0}\frac{(-1)^j}{j}p_j z^j\right) \tag{76}$$

Multiply both sides of (75) by $z^n$ and sum over $n \geq 0$

$$E'(z) = E'(qz)E((1-q)z)$$

By iterating this equation we find

$$
\begin{aligned}
E'(z) &= \prod_{k=0}^{\infty} E((1-q)q^k z) = \exp\left(-\sum_{k\geq 0}\sum_{j>0} \frac{(-1)^j(1-q)^j}{j} p_j q^{jk} z^j\right) \\
&= \exp\left(-\sum_{j>0} \frac{(-1)^j(1-q)^j}{j(1-q^j)} p_j z^j\right) = \exp\left(-\sum_{j>0}\frac{(-1)^j}{j}\sigma_q(p_j) z^j\right) \\
&= \sum_{n=0}^{\infty} \sigma_q(e_n) z^n
\end{aligned}
$$

Comparing the last result with (76) shows that $e'_n = \sigma_q(e_n)$. □

## 4. Matching the partition functions with shuffle elements

We now investigate the partition functions associated to the elements $A_{\lambda_1} * \cdots * A_{\lambda_m}$ (where $A_n$ was defined in (12)) that were studied in the previous section on the shuffle algebra side.

PROPOSITION 11. *Given $\lambda_1, \ldots, \lambda_m \in \mathbb{Z}_{>0}$ with $n = \sum_{i=1}^m \lambda_i$, one has the limit*

$$
\lim_{x_n \to 0} x_n^{n-1} f(A_{\lambda_1} * \cdots * A_{\lambda_m}) = (-1)^{n-1} \prod_{i=1}^{n-1} x_i \sum_{i=1}^{m} f(A_{\lambda_1} * \cdots * A_{\lambda_i - 1} * A_{\lambda_m}) \tag{77}
$$

Note that by commutativity, $\lambda$ can be chosen to be a partition.

PROOF. We start from the definition of $A_{\lambda_1} * \cdots * A_{\lambda_m}$ (cf (13)):

$$
A_{\lambda_1} * \cdots * A_{\lambda_m} = \sum_{v \in \mathcal{S}_{\lambda_1,\ldots,\lambda_m}} \sum_{w \in \mathcal{S}^{\lambda_1,\ldots,\lambda_m}} (-1)^{|v|} t^{-(|v|+|w|)} T_w T_v T_{w^{-1}}
$$

By definition of the groups involved, one has $T_w T_v = T_{wv}$ (and similarly $T_v T_{w^{-1}} = T_{vw^{-1}}$, but we choose to use the first identity).

Plugging this into (27), we have

$$
\begin{aligned}
&\alpha_n^{-1} f(A_{\lambda_1} * \cdots * A_{\lambda_m}) \\
&\quad = \sum_{v \in \mathcal{S}_{\lambda_1,\ldots,\lambda_m}} \sum_{w \in \mathcal{S}^{\lambda_1,\ldots,\lambda_m}} (-1)^{|v|} t^{-(|v|+|w|)} \langle Z|_{y_i = q x_i} (T_w T_v T_{w^{-1}} \otimes S_n) \rangle \\
&\quad = \sum_{v \in \mathcal{S}_{\lambda_1,\ldots,\lambda_m}} \sum_{w \in \mathcal{S}^{\lambda_1,\ldots,\lambda_m}} (-1)^{|v|} t^{-(|v|+|w|)} \langle T_{w^{-1}} Z|_{y_i = q x_i} T_{wv} (1 \otimes S_n) \rangle
\end{aligned}
$$

Using (30), we can rewrite this as

$$
\begin{aligned}
&\alpha_n^{-1} f(A_{\lambda_1} * \cdots * A_{\lambda_m}) \\
&\quad = \sum_{v \in \mathcal{S}_{\lambda_1,\ldots,\lambda_m}} \sum_{w \in \mathcal{S}^{\lambda_1,\ldots,\lambda_m}} (-1)^{|v|} t^{-(|v|+|w|)} \langle 1 | T_{w^{-1}} Z|_{y_i = q x_i} T_{wv} | 1 \rangle \\
&\quad = \sum_{v \in \mathcal{S}_{\lambda_1,\ldots,\lambda_m}} \sum_{w \in \mathcal{S}^{\lambda_1,\ldots,\lambda_m}} (-t)^{-|v|} \langle w^{-1} | Z|_{y_i = q x_i} | v^{-1} w^{-1} \rangle
\end{aligned} \tag{78}
$$

This has the following interpretation: $\alpha_n^{-1} f(A_{\lambda_1} * \cdots * A_{\lambda_m})$ is the sum over $v$ and $w$ of partition functions of lattice paths, where the paths are labelled $w^{-1}(1), \ldots, w^{-1}(n)$ on the left side and $v^{-1}(w^{-1}(1)), \ldots, v^{-1}(w^{-1}(n))$ on the top side (with an extra weight $(-t)^{-|v|}$).

We can now take the limit $x_n \to 0$. In this paragraph we denote by $a = w^{-1}$, $b = v^{-1}w^{-1}$, which are two arbitrary permutations, and consider $\langle a| Z|_{y_i = qx_i} |b\rangle$ as $x_n \to 0$. We are only interested for the purposes of deriving (77) in the terms of order $x_n^{-n+1}$; from Lemma 3 we know that this is the minimum order in $x_n$ and in fact, as in the proof of the lemma, we know that the only dependence on $x_n$ comes from the last row and column of the diagram of $Z$. Let us examine these in more detail with our boundary conditions $\langle a|$ and $|b\rangle$. We look at the destiny of the path $b(n)$ which ends at the top right: to maximize the degree in $x_n^{-1}$, we must pick up a $x_n^{-1}$ at each vertex $(i,n)$, $i = 1, \ldots, n-1$; this means, according to (3) that the first type of vertex (a path making a bend when the right edge is either empty or has a higher labelled path) must be excluded. Since all right edges are empty in the last column, one can prove inductively that the paths must go straight through every vertex $(i,n)$, $i = 1, \ldots, n-1$, incurring a weight of $-q^{-1}x_i x_n^{-1}$. At vertex $(n,n)$, because both bottom and right edges are empty, we must have the first type of vertex in (3), incurring a $1-t$. From there, the path can only go straight left to its destination with label $a(n)$, incurring a weight $t^{n-1}$ and imposing $a(n) = b(n)$. Combining all this, we find

$$\begin{aligned}(79)\quad \langle a| Z(x_1, \ldots, x_n, qx_1, \ldots, qx_n) |b\rangle &= \delta_{a(n),b(n)}(1-t)t^{n-1}x_n^{-n+1}\prod_{i=1}^{n-1}(-q^{-1}x_i)\\ \times\langle a(1), \ldots, a(n-1)| Z(x_1, \ldots, x_{n-1}, qx_1, \ldots, qx_{n-1}) &|b(1), \ldots, b(n-1)\rangle + O(x_n^{-n+2})\\ &\qquad a, b \in \mathcal{S}_n^{(1)}\end{aligned}$$

We apply this formula to $f(A_{\lambda_1} * \cdots * A_{\lambda_m})$. We fix $w \in \mathcal{S}_{\lambda_1,\ldots,\lambda_m}$ and denote by $k = w^{-1}(n)$. Because of the form of $w$, $k$ must be the last element of its block $I_a$, $1 \le a \le m$ ($I_a$ was defined in (66)). According to (79), the summation can be restricted to $v$ such that $v(k) = k$. Consider now the permutation $w'$ with $w'(i) = w(i)$, $i = 1, \ldots, k-1$, $w'(i) = w(i+1)$, $i = k, \ldots, n-1$, and the permutation $v'$ with $v'(i) = i$ for $i = 1, \ldots, k-1$ and $v'(i) = v(i+1) - 1$ for $i = k, \ldots, n-1$. Because $k$ is last of its block, $|v'| = |v|$ and we have

$$\begin{aligned}\alpha_n^{-1} f(A_{\lambda_1} * \cdots * A_{\lambda_m}) &\overset{x_n \to 0}{\sim} (1-t)t^{n-1}x_n^{-n+1}\prod_{i=1}^{n-1}(-q^{-1}x_i)\\ &\sum_{v' \in \mathcal{S}_{\lambda_1,\ldots,\lambda_a - 1,\ldots,\lambda_m}} (-t)^{-|v'|} \sum_{w' \in \mathcal{S}^{\lambda_1,\ldots,\lambda_a - 1,\ldots,\lambda_m}}\\ &\big\langle w'^{-1}(1), \ldots, w'^{-1}(n-1)\big| Z(x_1, \ldots, x_{n-1}, qx_1, \ldots, qx_{n-1})\\ &\qquad \big|v^{-1}(w'^{-1}(1)), \ldots, v'^{-1}(w'^{-1}(n-1))\big\rangle\end{aligned}$$

Comparing with (78) and using $\alpha_n^{-1}\alpha_{n-1} = (1-t)(t/q)^{n-1}$, we finally find the desired limit (77). □

In what follows, we shall use the diagrammatic language. We shall repeatedly use the following simple identities:

LEMMA 9. *One has for* $k \leq \ell, m \geq 2$*:*

(80)
(81)
(82)

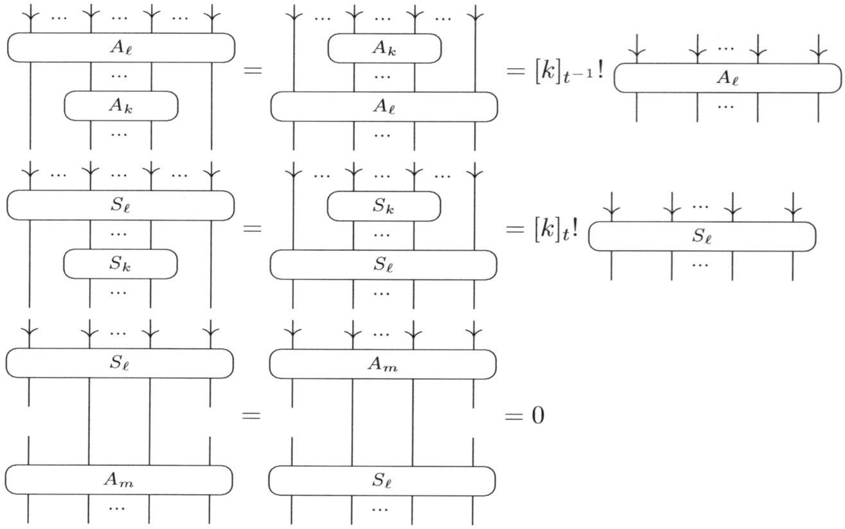


*where* $[k]_t! = \prod_{i=1}^{k} [i]_t$ *and* $[i]_t = 1 + \cdots + t^{i-1}$.

For general $c \in \mathrm{Z}(\mathcal{H}_n)$, $f(c)$ can be described as

(83)

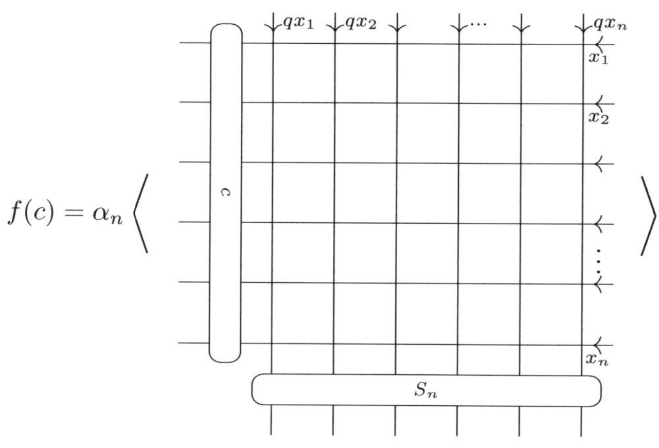


This immediately implies

LEMMA 10. *For any* $c \in \mathrm{Z}$, $f(c)$ *satisfies the wheel conditions*

$$f(c)(x, tx, qx, x_4, \dots, x_n) = f(c)(tqx, tx, qx, x_4, \dots, x_n) = 0$$

PROOF. $f(c)$ being symmetric in its arguments according to Proposition 4, we can pick any three variables and specialize them. Let us then consider the diagram above and set $x_{n-2} = x$, $x_{n-1} = tx$, $x_n = qx$. Zooming in on the bottom right corner, we have, using $\check{R}(t^{-1}) = (1-t)A_2$,

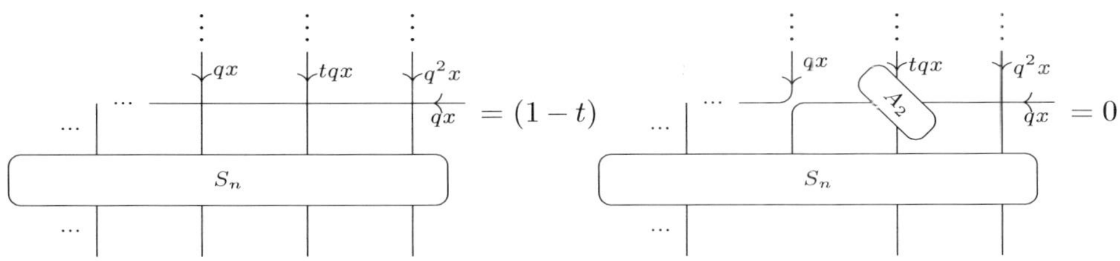


Switching rows and columns results similarly in the second wheel condition. $\square$

We now go back to $A_{\lambda_1} * \cdots * A_{\lambda_m}$. Starting from the expression

$$f(A_{\lambda_1} * \cdots * A_{\lambda_m}) \\ = \alpha_n \sum_{w \in \mathcal{S}^{\lambda_1,\ldots,\lambda_m}} t^{-|w|} \langle (T_w(A_{\lambda_1} \otimes \cdots \otimes A_{\lambda_m}) T_{w^{-1}} \otimes S_n) Z \rangle |_{y_i = qx_i,\ i=1,\ldots,n}$$

and using the cyclicity of $\langle \bullet \rangle$, we can describe the expression inside it as the diagram: (compare with (83))

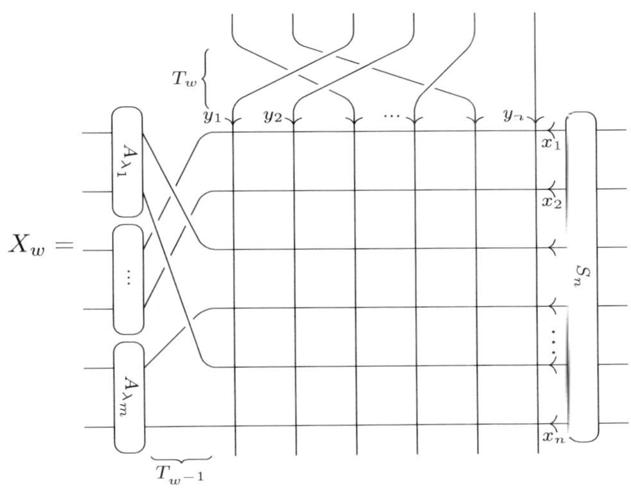

As a side remark, using the idempotency of $A_n$ and $S_n$, we can introduce the more symmetric diagram

(84) $\tilde{X}_w :=$ 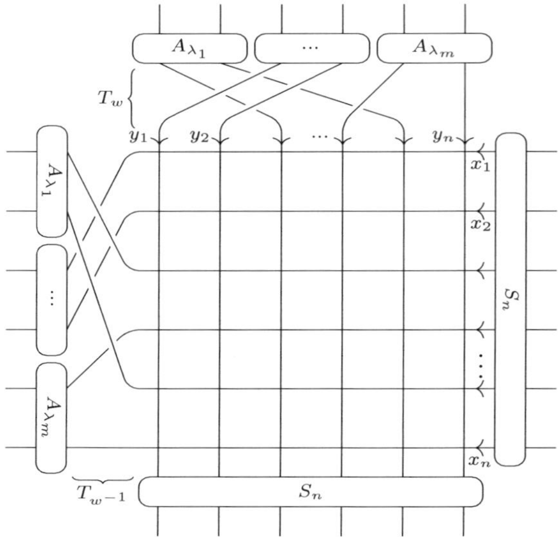

with $\tilde{X}_w = \langle X_w \rangle\, A_{\lambda_1} \otimes \cdots \otimes A_{\lambda_m} \otimes S_n$. We shall not use $\tilde{X}_w$ in what follows.

PROPOSITION 12. • *For $m < n$, one has the vanishing condition:*

$$f(A_{\lambda_1} * \cdots * A_{\lambda_m})(x, tx, \ldots, t^m x, x_{m+2}, \ldots, x_n) = 0 \tag{85}$$

• *The partition function $f(A_n)$ has the fully factorized form*

$$f(A_n) = t^{-\frac{n(n-1)}{2}} \prod_{\substack{i,j=1\\ i\neq j}}^{n} (1 - t x_i / x_j) \tag{86}$$

• *One has the specialization*

$$f(A_1 * \cdots * A_1)(1, t, \ldots, t^{n-1}) = [n]_{t^{-1}}! \prod_{1 \le i < j \le n} (q - t^{j-i})(q - t^{i-j+1}) \tag{87}$$

PROOF. We first consider $m = 1$, and investigate the effect of setting $x_1 = tx$, $x_2 = x$ in $X_w$, where here $w = 1$. Noting that $\check{R}_1(t) = (1-t)(1+T_1) = (1-t)S_2$, and using (81), we can insert a crossing at the right of the top two horizontal lines and then apply the Yang–Baxter equation repeatedly to move the crossing to the left, resulting in zero according to (82):

We conclude that $x_1 - t x_2$ divides $f(A_n)$, and by symmetry, $f(A_n)$ is a multiple of $\prod_{i\neq j}(1 - t x_i/x_j)$. The latter has degree range $[-(n-1), (n-1)]$ in each $x_i$, which is the maximum allowed according to (2), so only a constant in $\mathbb{F}$ remains to be determined. The latter is determined inductively by using Proposition 11.

We proceed identically for $1 < m < n$. We set $x_1 = t^m x, \ldots, x_m = tx, x_{m+1} = x$. We note that the "$R$-matrix associated to the longest element $w_0$ of $\mathcal{S}_{m+1}$", i.e., the product of $R$-matrices whose diagram reproduces (any) wiring diagram of $w_0$, with spectral parameters set to $t^m, \ldots, 1$, is nothing but $(1-t)^{\frac{m(m+1)}{2}} \prod_{i=1}^m [i]_t!\, S_{m+1}$ (see e.g. [**Las06**, Lem. 3] or [**IO09**, 3.1]), and move it across to the left using the Yang–Baxter equation:

$$X_1|_{x_1=tx,\ x_2=x} = \frac{1}{1+t}$$

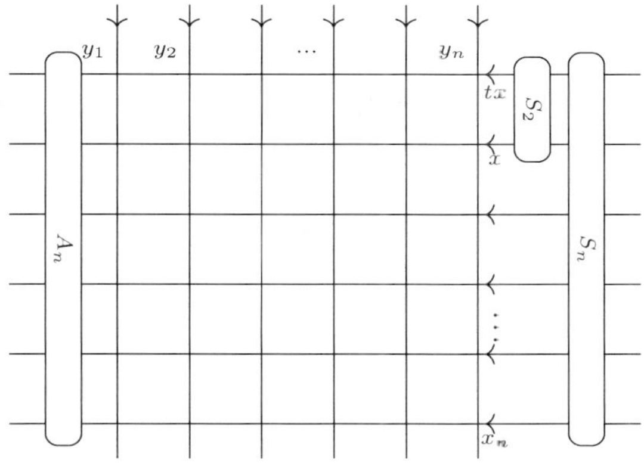


$$= \frac{1}{1+t}$$

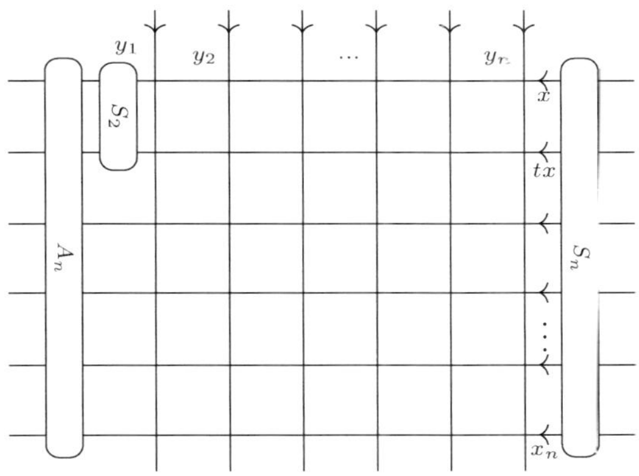


$$= 0$$

$$X_w|_{x_1=t^m x,\dots,x_{m+1}=x} = \frac{1}{[m+1]_t!}$$

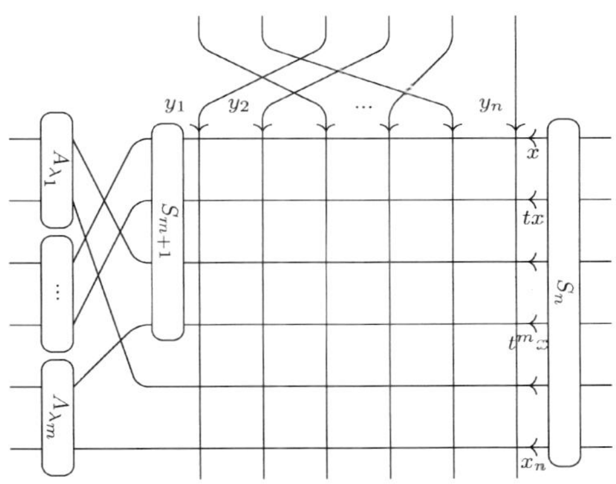


The pigeonhole principle tells us there will be two numbers between 1 and $m+1$ whose preimages under $w$ lie in the same block (i.e., are associated to the same part of $\lambda$). Since $w \in \mathcal{S}^{\lambda_1,\dots,\lambda_m}$, all numbers whose preimages are in the same block are consecutive, and so we can apply (82) to conclude that the result is zero. By summing over $w$ and using symmetry in the exchange of variables, we obtain the result (85) as stated in the Proposition.

The only case left is $m = n$, i.e., $A_1 * \cdots * A_1$. We set $x_1 = t^{n-1}x, \dots, x_{n-1} = tx, x_n = x$ but this time the result does not vanish. However, noting once again that $S_n$ is up to normalization the product of $R$-matrices associated to the longest

$$X_w|_{x_1=t^{n-1}x,\dots,x_n=x} =$$

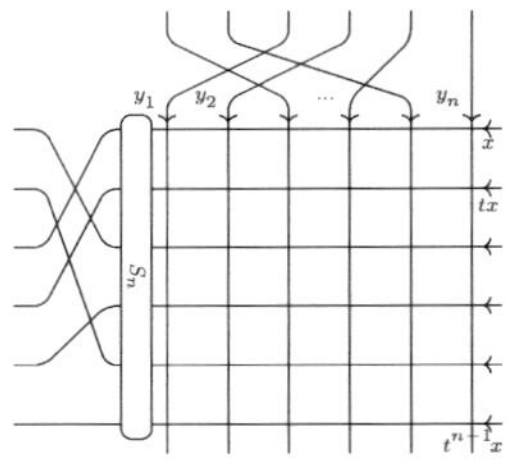


$$= t^{|w|}$$

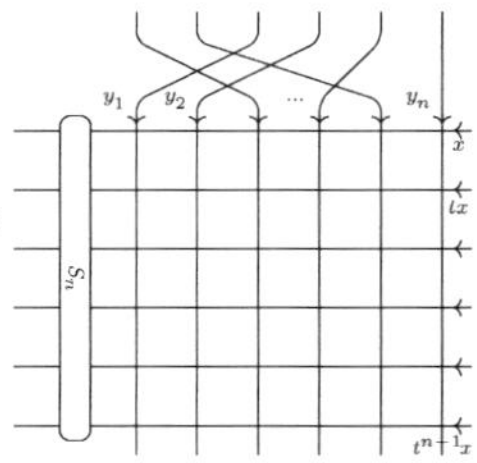


$$= t^{|w|-\frac{n(n-1)}{2}}$$

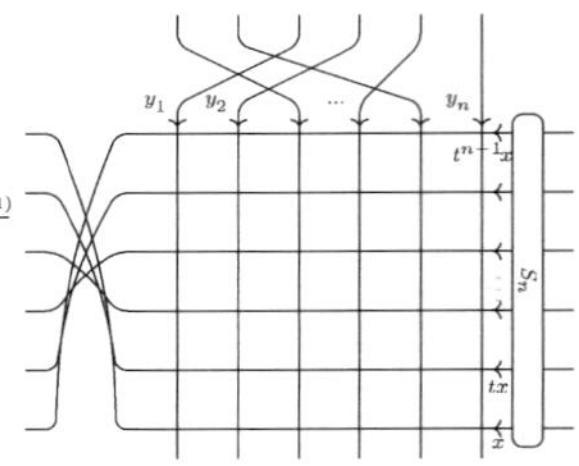


permutation $w_0$ of $\mathcal{S}_n$, we can move it to the left and absorb any crossing from $T_w$: where in the last line we find it convenient to reintroduce the longest element $w_0$, before moving back $S_n$ to the right. Performing the summation over $w$ (one could instead absorb the top crossings as well, but the result would be the same), we find

$$\sum_{w\in\mathcal{S}_n} t^{-|w|} X_w = t^{-\frac{n(n-1)}{2}}$$

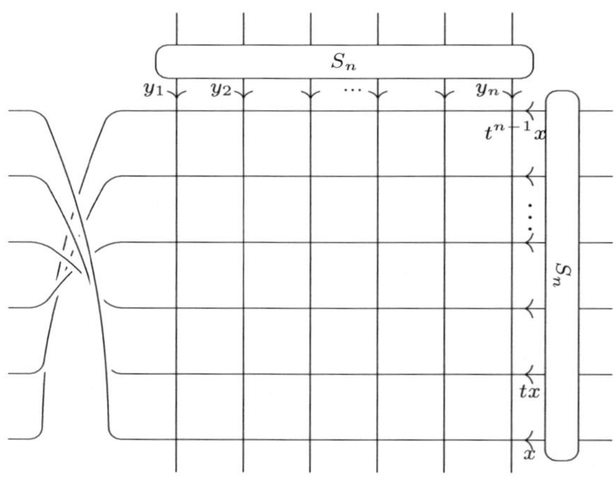


We then impose $y_i = qx_i$ and note that up to normalization, the $S_n$ at the top is nothing but the $R$-matrix associated to the longest element of $\mathcal{S}_n^{(2)}$; we can pull it to the bottom:

$$\sum_{w\in\mathcal{S}_n} t^{-|w|} X_w|_{y_i=qx_i} = t^{-\frac{n(n-1)}{2}}$$

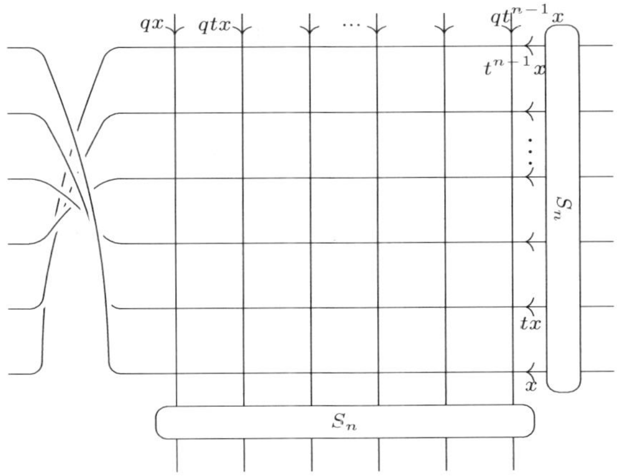


Finally, the bracket is

$$f(A_1 * \cdots * A_1) = \alpha_n t^{-\frac{n(n-1)}{2}} [n]_t!$$

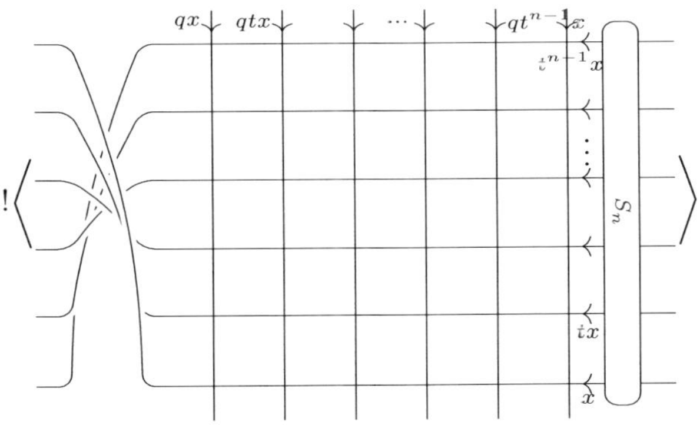


This is essentially the same situation that was considered at the end of §2. First we expand each $R$-matrix in the upper-left triangle as $\check{R}_i(u) = u(1-t) + t(1-u)T_i^{-1}$ and note that the only term that has a nonzero identity coefficient is

$$f(A_1 * \cdots * A_1) = \alpha_n [n]_t! (1-t)^n \prod_{\substack{1\leq i,j\leq n\\ i+j<n+1}} (1 - q^{-1} t^{n-i-j-1})$$

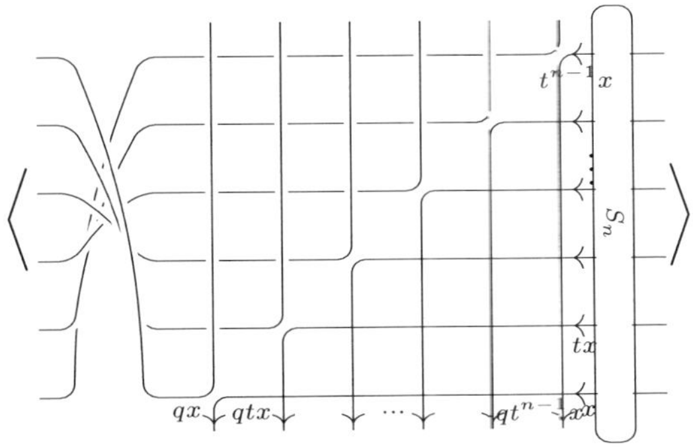

Secondly, we absorb all the crossings of the lower-right triangle into $S_n$, resulting in a weight of $\prod_{\substack{1\le i,j\le n\\ i+j>n+1}}(1-q^{-1}t^{n-i-j+2})$. Recombining and simplifying, we obtain (87). □

We are now in a position to prove Theorem 1. First, combining Lemma 3, Proposition 4 and Lemma 10, we find that $f(c)$, from its definition (27), is $\alpha_n$ times an element of $\mathrm{A}_n$. $\alpha_n$ contains a factor of $(1-t)^{-n}$, but it is compensated by the fact that any lattice path entering from the left and exiting from the top must have a bend, resulting in a factor of $(1-t)^n$ according to (3) in the expression (4) of $f(c)$. Therefore, $f$ is a map from $\mathrm{Z}(\mathcal{H}_n)$ to $\mathrm{A}_n$.

Next, we identify $f(A_{\lambda_1} * \cdots * A_{\lambda_m})$. We use the characterization of Proposition 9, so that $\epsilon_{\lambda_1,\dots,\lambda_m}$ is entirely determined by Lemma 8 and (68). On the other hand, it is easy to check that Proposition 12 implies the exact same relations for $V_n^{-1} f(A_{\lambda_1} * \cdots * A_{\lambda_m})$ (where $V_n$ is defined in (42), and implements the isomorphism $V_n \times$ from $\mathcal{A}_n$ to $\mathrm{A}_n$). Therefore,

$$f(A_{\lambda_1} * \cdots * A_{\lambda_m}) = V_n\, \epsilon_{\lambda_1,\dots,\lambda_m}$$

We finally compare $\Upsilon(f(A_{\lambda_1} * \cdots * A_{\lambda_m}))$ and $\Phi(A_{\lambda_1} * \cdots * A_{\lambda_m})$.

According to the ring map property of $\Upsilon$ and (56) applied to $\epsilon_r = \epsilon_r^{(3)}$,

$$\begin{aligned}\Upsilon(f(A_{\lambda_1} * \cdots * A_{\lambda_m})) &= \Upsilon(V_n\, \epsilon^{(3)}_{\lambda_1} * \cdots * \epsilon^{(3)}_{\lambda_m})\\ &= \prod_{i=1}^m \Upsilon(V_{\lambda_i}\, \epsilon^{(3)}_{\lambda_i})\\ &= \prod_{i=1}^m \sigma_{q^{-1}}\sigma_t^{-1}(e_{\lambda_i})\end{aligned}$$

On the other hand, from (17) and the ring map property of $\Phi$,

$$\Phi(A_{\lambda_1} * \cdots * A_{\lambda_m}) = \prod_{i=1}^m e_{\lambda_i}$$

We have therefore checked the statement of Theorem 1 on a basis of Z.

COROLLARY 1. *We have the following identity*

$$f(1_n) = V_n \kappa_n \tag{88}$$

PROOF. According to (69) and Proposition 8, $\Upsilon(V_n\kappa_n) = \Upsilon'(\kappa_n) = \sigma_q(e_n)$. On the other hand, according to Proposition 3, $\Phi(1_n) = \tilde{h}_n = \sigma_t(h_n) = \sigma_t\sigma_\infty(e_n)$. Applying Theorem 1, we conclude that (88) holds. □

Using the definition of the shuffle product (36) we can write

$$\begin{aligned}&(x_1+\cdots+x_n)\,\kappa_n(x_1,\dots,x_n)\\ &\quad= \sum_{i=1}^n x_i\kappa_{n-1}(x_1,\dots,\hat{x}_i,\dots,x_n)\prod_{\substack{j=1\\ j\ne i}}^n \frac{(x_i-q_1x_j)(x_i-q_2x_j)(x_i-q_3x_j)}{(x_i-x_j)^3}\end{aligned} \tag{89}$$

The recurrence relation for $f(1_n)$ stated in (5) follows from (88), (89) and (42) thus proving Theorem 2.

## 5. Application to the commuting scheme

**5.1. Proof of Theorem 3.** In this section, all schemes are over $\mathbb{C}$, as in the introduction.

PROPOSITION 13. *Assuming Conjecture 1, $K_n$ satisfies the following recurrence relation:*

(90)

$$(x_1 + \cdots + x_n)K_n = \sum_{i=1}^{n} x_i K_{n-1}[\hat{x_i}] \prod_{\substack{j=1\\ j\neq i}}^{n} \frac{(1-q_1x_j/x_i)(1-q_2x_j/x_i)(1-q_1q_2x_i/x_j)}{1-x_j/x_i}$$

*where $K_{n-1}[\hat{x_i}] = K_{n-1}(x_1, \ldots, x_{i-1}, x_{i+1}, \ldots, x_n)$.*

PROOF. For the purposes of this proof, we need more geometry than in the rest of the paper. We think of the multi-graded Hilbert series as the pushforward to a point in localized equivariant $K$-theory. Equivalently, the $K$-polynomial of $\mathfrak{C}_n$ is viewed as (the pushforward of) the $K_T$-class $[\mathcal{O}_{\mathfrak{C}_n}]$ in the ambient space $\mathfrak{gl}_n \times \mathfrak{gl}_n$. Note that $K_T(\mathfrak{gl}_n \times \mathfrak{gl}_n) \cong K_T(pt) \cong \mathbb{Z}[q_1^{\pm}, q_2^{\pm}, x_1^{\pm}, \ldots, x_n^{\pm}]$ where the variables $q_1, q_2, x_1, \ldots, x_n$ are the *inverses* of the equivariant parameters attached to the torus $T$.

Consider

$$\mathfrak{F}_n = \left\{(v, A, B) \in \mathbb{P}^{n-1} \times \mathfrak{gl}_n \times \mathfrak{gl}_n : [A, B] = 0 \text{ and } v \text{ eigenvector of } A \text{ and } B\right\}$$

There is a natural projective morphism

$$\mathfrak{F}_n \xrightarrow{p} \mathfrak{C}_n$$
$$(v, A, B) \mapsto (A, B)$$

which is generically $n$-to-1 (choice of one eigenvector among $n$). There is also the affine morphism $f : \mathfrak{F}_n \to \mathbb{P}^{n-1}$. Since all fibers are isomorphic by $GL_n$-action, the dimension of a fiber is $\dim \mathfrak{F}_n - \dim \mathbb{P}^{n-1} = n^2 + 1$.

The computation is in two parts:

- Computing $\pi_*[\mathcal{O}_{\mathfrak{F}_n}(1)]$ by equivariant localization (which relates it to $\pi_*[\mathcal{O}_{\mathfrak{C}_{n-1}}]$ using $(v, A, B) \mapsto (A|_{\mathbb{C}^n/v}, B|_{\mathbb{C}^n/v})$).
- Comparing $p_*[\mathcal{O}_{\mathfrak{F}_n}(1)]$ and $[\mathcal{O}_{\mathfrak{C}_n}]$.

5.1.1. *Equivariant localization.* Torus fixed points in $\mathbb{P}^{n-1}$ are basis vectors $e_i$, $i = 1, \ldots, n$. Consider such a fixed point, which up to $\mathcal{S}_n$ action, we can choose to be $v = e_n$. The fiber is

$$f^{-1}(v) = \{v\} \times X, \qquad X = \left\{\left(A = \begin{pmatrix} A' & 0 \\ \alpha & a \end{pmatrix}, B = \begin{pmatrix} B' & 0 \\ \beta & b \end{pmatrix}\right) : AB = BA\right\}$$

More explicitly, the equations are

$$A'B' = B'A' \tag{91}$$

$$a\beta + \alpha B' = b\alpha + \beta A' \tag{92}$$

In particular note that $(A', B') \in \mathfrak{C}_{n-1}$. There are $2n$ extra variables $\alpha, \beta, a, b$ and $n-1$ extra equations, and $\dim \mathfrak{C}_{n-1} + 2n - (n-1) = n^2 + 1 = \dim f^{-1}(v)$, so assuming Conjecture 1 (Cohen–Macaulay property for $\mathfrak{C}_{n-1}$, which implies it for $\mathfrak{C}_{n-1} \times \mathbb{A}^{2n+2}$), the unmixedness theorem asserts that the extra equations (92) form a regular sequence.

We then conclude by standard exact sequence arguments that

$$[\mathcal{O}_X] = [\mathcal{O}_{\mathfrak{C}_{n-1}}] \prod_{j=1}^{n-1} (1 - q_1 x_j/x_n)(1 - q_2 x_j/x_n)(1 - q_1 q_2 x_n/x_j) \tag{93}$$

where $[\mathcal{O}_X]$ is the $K_T$ class of $\mathcal{O}_X$ in $\mathfrak{gl}_n \times \mathfrak{gl}_n$, whereas $[\mathcal{O}_{\mathfrak{C}_{n-1}}]$ is the $K_T$ class of $\mathfrak{C}_{n-1}$ in $\mathfrak{gl}_{n-1} \times \mathfrak{gl}_{n-1}$ (where the $T$-action on $\mathfrak{gl}_{n-1} \times \mathfrak{gl}_{n-1}$ has a one-dimensional kernel leading to the obvious inclusion $K_{T/\mathbb{C}^\times} \subset K_T$, i.e., as a polynomial $[\mathcal{O}_{\mathfrak{C}_{n-1}}] = K_{n-1}$ is independent of $x_n$). The factors $1 - q_1 x_j/x_n$ (resp. $1 - q_2 x_j/x_n$) come from the zero column of $A$ (resp. $B$) whereas the factors $1 - q_1 q_2 x_n/x_j$ come from the equations (92).

We have thus managed to compute $[f^{-1}(v)] = [v][\mathcal{O}_X]$.

The weights at the tangent space of $e_i$ are $x_i/x_j$, $j \neq i$ (recalling that the $x_i$ are the inverses of the equivariant parameters). Pulling back the relation

$$1 = \sum_{i=1}^{n} \frac{[e_i]}{\prod_{\substack{j=1 \\ j \neq i}}^{n} (1 - x_j/x_i)}$$

in localized equivariant $K$-theory of $\mathbb{P}^{n-1}$ to $\mathfrak{F}_n$ and multiplying by $[\mathcal{O}_{\mathfrak{F}_n}(1)]$, we obtain

$$[\mathcal{O}_{\mathfrak{F}_n}(1)] = \sum_{i=1}^{n} x_i [e_i][\mathcal{O}_{\mathfrak{C}_{n-1}}](\hat{x}_i) \prod_{\substack{j=1 \\ j \neq i}}^{n} \frac{(1 - q_2 x_j/x_i)(1 - q_1 x_j/x_i)(1 - q_1 q_2 x_i/x_j)}{1 - x_j/x_i}$$

Finally, pushing forward to $\mathfrak{gl}_n \times \mathfrak{gl}_n$ leads to:

$$p_*[\mathcal{O}_{\mathfrak{F}_n}(1)] = \sum_{i=1}^{n} x_i \, [\mathcal{O}_{\mathfrak{C}_{n-1}}](\hat{x}_i) \prod_{\substack{j=1 \\ j \neq i}}^{n} \frac{(1 - q_2 x_j/x_i)(1 - q_1 x_j/x_i)(1 - q_1 q_2 x_i/x_j)}{1 - x_j/x_i} \tag{94}$$

5.1.2. *Pushforward.* Now let us try to compute directly the pushforward of $[\mathcal{O}_{\mathfrak{F}_n}(1)]$ under $p : \mathfrak{F}_n \to \mathfrak{C}_n$. Write $\mathfrak{C}_n = \operatorname{Spec} R$ and $\mathfrak{F}_n = \operatorname{Proj} S$ where $S_0 = R$. Since $\mathfrak{C}_n$ is affine, the pushforward is simply given in terms of the sheaf cohomology groups of $\mathcal{O}_{\mathfrak{F}_n}(1)$ (viewed as modules over $R$). We claim the following

$$H^i(\mathfrak{F}_n, \mathcal{O}_{\mathfrak{F}_n}(1)) \cong \begin{cases} S_1 & i = 0 \\ 0 & i > 0 \end{cases}$$

Proof. Let us first write the equations of $\mathfrak{F}_n$ explicitly:

$$AB = BA \qquad \text{minors of size 2 of } (v, Av) \text{ and } (v, Bv) = 0 \tag{95}$$

Now define the patches $U_i = \{v_i \neq 0\}$. Let us consider $U_1 \cap U_2$ inside $U_1$, $U_2$. Denote by $z = v_1/v_2$, $r_i = v_i/v_2$, $i > 2$ and collectively $r = (r_i)$. A section on $U_1$ is given by a polynomial $P(z, r, A, B)$ modulo the relations (95) rewritten in terms of $z, r, A, B$. Let us call these equations $q_i(z)$ where dependence on $r, A, B$ is suppressed. Note that the $q_i(z)$ are of degree less or equal to 2 in $z$. Similarly a section on $U_2$ is given by a polynomial $Q(z^{-1}, rz^{-1}, A, B)$ modulo the same equations expressed in terms of $z^{-1}, rz^{-1}, A, B$. The $\mathcal{O}(1)$ matching condition on $U_1 \cap U_2$ is

$$P(z) - zQ(z^{-1}) = \sum_i c_i(z) q_i(z) \tag{96}$$

where the $c_i(z)$ are Laurent polynomials in $z$. Now assume that $P$ is of degree greater than 1 in $z$. Then its top coefficient is equal to the top coefficient of $Q := \sum_i c_i^{pol}(z)q_i(z)$ where $c_i^{pol}(z)$ is the part of $c_i(z)$ with nonnegative powers of $z$. By subtracting $Q$ from $P$, we can reduce the degree of $P$ until it is at most 1.

This way, one can show that the restriction of a global section to $U_i$ is of degree in $v_i/v_j$ at most 1 for all $i \neq j$; this immediately implies that the $v_i$ generate the global sections of $\mathcal{O}(1)$. Linear independence is obvious (it can be tested at the generic point).

(96) also implies that the map $(P,Q) \mapsto P(z) - zQ(z^{-1})$ is surjective, so that $H^1(\mathcal{O}_{\mathfrak{F}_n}(1)) = 0$. Vanishing of higher cohomology follows similarly. □

Since $H^0(\mathcal{O}_{\mathfrak{F}_n}(1)) \cong S_1$ is a free $R$-module with basis $(v_1, \ldots, v_n)$ and higher sheaf cohomology vanishes, we conclude that

$$p_*[\mathcal{O}_{\mathfrak{F}_n}(1)] = [\mathcal{O}_{\mathfrak{C}_n}] \sum_{j=1}^{n} x_j \tag{97}$$

where $x_j$ is the weight of the basis element $v_j$.

Combining (94) and (97) leads to the recurrence relation (90). □

We are now in a position to prove Theorem 3 by relating the shuffle element $\kappa_n$ and the $K$-polynomial $K_n$. $K_n$ is uniquely determined by (90) and the initial condition $K_0 = 1$. Similarly $\kappa_n$ can be computed recursively with (89). By comparing (89) with (90) (and checking the trivial initial condition $\kappa_0 = K_0 = 1$) we conclude that

$$\kappa_n(x_1, \ldots, x_n) = \frac{t^{n(n-1)/2}}{V_n(x)} K_n(x_1, \ldots, x_n) \tag{98}$$

Combining (88) with (98) gives us the statement of Theorem 3.

**5.2. Proof of Theorem 4.** Theorem 4 involves computing the multidegree of $\mathfrak{C}_n$, which means one needs to work in equivariant cohomology. The situation is much simpler than in equivariant $K$-theory (which was needed for Theorem 3), and we only sketch the corresponding geometric argument. By abuse of notation, we use the same symbols $q_1, q_2, x_1, \ldots, x_n$ for the cohomology equivariant parameters. This is consistent with our passage from $K$-polynomial to multidegree, which was defined in the introduction as the substitution $q_i \mapsto 1 - q_i$ and $x_i \mapsto 1 - x_i$ and then expanding at leading order. We refer to this limit in what follows as the GRR (Grothendieck–Riemann–Roch) limit.

View the flag variety $\mathcal{B}_n$ as the variety of Borel subalgebras of $\mathfrak{gl}_n$ and consider $\tilde{\mathfrak{F}}_n = \{(\mathfrak{b}, A, B) \in \mathcal{B}_n \times \mathfrak{gl}_n \times \mathfrak{gl}_n : [A,B] = 0,\ A, B \in \mathfrak{b}\}$. There is a projection map $p$ from $\tilde{\mathfrak{F}}_n$ to $\mathfrak{C}_n$ which is generically $n!$-to-1 (generic matrices are diagonalizable); one then obtains by equivariant localization techniques (similar to §5.1.1) the formula

$$\begin{aligned} D_n &= \frac{1}{n!} \prod_{\substack{i,j=1\\ i\neq j}}^{n} (q_1 + x_i - x_j)(q_2 + x_i - x_j) \sum_{w \in S_n} w \left( \prod_{1 \le i < j \le n} \frac{q_1 + q_2 + x_i - x_j}{(x_j - x_i)(q_1 + x_i - x_j)(q_2 + x_i - x_j)} \right) \\ &= \operatorname{Sym} \left( \prod_{1 \le i < j \le n} \frac{(q_1 + q_2 + x_i - x_j)(q_1 - x_i + x_j)(q_2 - x_i + x_j)}{x_j - x_i} \right) \end{aligned} \tag{99}$$

where the symmetric group acts by permuting variables, and Sym denotes symmetrization w.r.t. to it.

One can think of this expression as the GRR limit of the shuffle element $\frac{1}{n!}f(A_1 * \cdots * A_1)$ (studied in §4). [2]

At $t = 1$, from (1), one has the identity in the center of the symmetric group algebras $A_1 * \cdots * A_1 = n!\, 1_n$, so that in the GRR limit, $\frac{1}{n!}f(A_1 * \cdots * A_1)$ and $f(1_n)$ coincide. As was already remarked in §1.4, expanding the weights (8), which correspond to $f(1_n)$, in the GRR limit, result in the weights (9). This leads to the formula of Theorem 4.

A few comments are in order. In [**DFZJ06, KZJ07**] another expression for the multidegree of the commuting variety was obtained, using apparently unrelated methods. There is actually a connection: consider Proposition 6 and take the GRR limit. One recovers one of the exchange relations of the work above (cf [**DFZJ06**, Eq. (3.10)] or [**KZJ07**, Prop. 6]). The expression for the multidegree of $\mathfrak{C}_n$ there is essentially a repeated application of these identities; it is not obvious if it is more or less useful than our (new) formula (99). (The extension of these ideas to $K$-theory will be investigated elsewhere [**KZJ**].)

The expression of Theorem 4, in contrast, has many desirable properties; for instance, it is division- and subtraction-free. Regarding positivity, let us only consider the bidegree with respect to scaling of the two matrices $A$ and $B$, i.e., set the $x_i$ to zero. On the one hand, (99) specializes to the bidegree being the constant term of the Laurent polynomial $\frac{1}{n!}\prod_{1\le i<j\le n}\frac{(u_i-u_j)(q_1-u_i+u_j)(q_2-u_i+u_j)(q_1+q_2+u_i-u_j)}{u_iu_j}$, which clearly involves subtractions. On the other hand, the expression in (10) is a polynomial in $q_1$ and $q_2$ with manifestly positive coefficients. If we further restrict to the ordinary degree, i.e., set $q_1 = q_2 = 1$, then we have the simple formula

$$\deg \mathfrak{C}_n = \sum_{\substack{\text{lattice paths } P\\ \text{on the } n\times n \text{ grid}\\ \text{with identity connectivity}}} 2^{\#\{(i,j):\text{ bends}\}-n} \tag{100}$$

(100) is also extremely effective computationally; here is the degree of $\mathfrak{C}_{12}$ (computed in a few seconds on a laptop):

$$\deg \mathfrak{C}_{12} = 1862632561783036151478238040096092649$$

(note that this is beyond the currently known entries of OEIS A029729).

**5.3. The $\mathcal{S}_3$ symmetry.** As an obvious corollary of (71) of Proposition 10, one has:

COROLLARY 2. *$\kappa_n$ possesses the $\mathcal{S}_3$ symmetry of permutations of $q_1, q_2, q_3$.*

In view of (88), the same holds for $f(1_n)$; this symmetry is however not obvious at the level of the partition function.

In a similar vein, assuming Conjecture 1, this also implies the symmetry property

$$K_n(q_1, q_2) = K_n(q_2, q_1) = (q_1q_2^2)^{n(n-1)/2} K_n(q_1, (q_1q_2)^{-1})$$

[2]It should be noted that the element $f(A_1 * \cdots * A_1)$ itself has a geometric interpretation [**Gin11**] as the Hilbert series of the coordinate ring of the normalization of the isospectral commuting variety $\mathfrak{X}_{\text{norm}}$. What we are finding here is that to go from $K$-theory to cohomology, instead of using the structure sheaf of $\mathfrak{C}_n$ one can use the combinatorially simpler structure sheaf of $\mathfrak{X}_{\text{norm}}$.

EXAMPLE 5. Here are the coefficients of $K_3|_{x_i=1}$:

| | | | | | | | | | | |
|---|---|---|---|---|---|---|---|---|---|---|
| | | | | | | | | | 1 | |
| | | | | 1 | | 1 | | $-8$ | | |
| | $q_1$ | | 1 | | $-16$ | | 29 | | 1 | |
| 1 | | $-8$ | | 29 | | $-27$ | | $-16$ | | 1 |
| | $q_2$ | | 1 | | $-16$ | | 29 | | 1 | |
| | | | | 1 | | 1 | | $-8$ | | |
| | | | | | | | | | 1 | |

It is not a priori obvious why the $K$-polynomial of the commuting scheme $\mathfrak{C}_n$ should possess the second symmetry property. Here we propose an "explanation": [3]

CONJECTURE 2. *Consider the triple commuting scheme*

$$\hat{\mathfrak{C}}_n := \{(A, B, C) \in \mathfrak{gl}_n^3 : \ [A, B] = [A, C] = [B, C] = 0\}$$

*and its grading w.r.t. scaling* $(A, B, C) \mapsto (q_1 A, q_2 B, q_3 C)$ *and conjugating by diagonal matrices. Then its $K$-polynomial $\hat{K}_n$ satisfies*

$$\hat{K}_n(q_1, q_2, (q_1 q_2)^{-1}) = K_n(q_1, q_2) K_n(q_1^{-1}, q_2^{-1})$$


## Acknowledgments

The authors thank Allen Knutson, Peter MacNamara, Andrei Negut, Gufang Zhao for discussions.


## References


[BW18] A. Borodin and M. Wheeler, *Coloured stochastic vertex models and their spectral theory*, 2018, arXiv:1808.01866.

[DFZJ06] P. Di Francesco and P. Zinn-Justin, *Inhomogeneous model of crossing loops and multidegrees of some algebraic varieties*, Comm. Math. Phys. **262** (2006), no. 2, 459–487, DOI 10.1007/s00220-005-1476-5. MR2200268

[FHH+09] B. Feigin, K. Hashizume, A. Hoshino, J. Shiraishi, and S. Yanagida, *A commutative algebra on degenerate* $\mathbb{CP}^1$ *and Macdonald polynomials*, J. Math. Phys. **50** (2009), no. 9, 095215, 42, DOI 10.1063/1.3192773. MR2566895

[FO97] Boris Feigin and Alexander Odesskii, *A family of elliptic algebras*, Internat. Math. Res. Notices **11** (1997), 531–539, DOI 10.1155/S1073792897000354. MR1448336

[FT11] B. L. Feigin and A. I. Tsymbaliuk, *Equivariant $K$-theory of Hilbert schemes via shuffle algebra*, Kyoto J. Math. **51** (2011), no. 4, 831–854, DOI 10.1215/21562261-1424875. MR2854154

[Gin11] Victor Ginzburg, *Isospectral commuting variety, the Harish-Chandra $\mathcal{D}$-module, and principal nilpotent pairs*, Duke Math. J. **161** (2012), no. 11, 2023–2111, DOI 10.1215/00127094-1699392. MR2957698


[3] This is to be compared with the following two toroidal algebra representations: the one coming from the Hilbert scheme of points in $\mathbb{C}^2$, which does not have manifest $S_3$ symmetry, and the one coming from the Hilbert scheme of points in $\mathbb{C}^3$, which does.

[IO09] A. P. Isaev and O. V. Ogievetsky, *Braids, shuffles and symmetrizers*, J. Phys. A **42** (2009), no. 30, 304017, 15, DOI 10.1088/1751-8113/42/30/304017. MR2521336

[Ize87] A. G. Izergin, *Partition function of a six-vertex model in a finite volume* (Russian), Dokl. Akad. Nauk SSSR **297** (1987), no. 2, 331–333. MR919260

[KBI97] V. E. Korepin, N. M. Bogoliubov, and A. G. Izergin, *Quantum inverse scattering method and correlation functions*, Cambridge Monographs on Mathematical Physics, Cambridge University Press, Cambridge, 1993, DOI 10.1017/CBO9780511628832. MR1245942

[Kor82] V. E. Korepin, *Calculation of norms of Bethe wave functions*, Comm. Math. Phys. **86** (1982), no. 3, 391–418. MR677006

[KZJ] A. Knutson and P. Zinn-Justin, work in progress.

[KZJ07] Allen Knutson and Paul Zinn-Justin, *A scheme related to the Brauer loop model*, Adv. Math. **214** (2007), no. 1, 40–77, DOI 10.1016/j.aim.2006.09.016. MR2348022

[Lam18] Jules Lamers, *The functional method for the domain-wall partition function*, SIGMA Symmetry Integrability Geom. Methods Appl. **14** (2018), Paper No. 064, 23, DOI 10.3842/SIGMA.2018.064. MR3817777

[Las06] A. Lascoux, *The Hecke algebra and structure constants of the ring of symmetric polynomials*, 2006, arXiv:math/0602379.

[Mac98] I. G. Macdonald, *Symmetric functions and Hall polynomials*, Oxford Mathematical Monographs, The Clarendon Press, Oxford University Press, New York, 1979. MR553598

[MT55] T. S. Motzkin and Olga Taussky, *Pairs of matrices with property L. II*, Trans. Amer. Math. Soc. **80** (1955), 387–401, DOI 10.2307/1992996. MR86781

[Neg14] Andrei Negut, *The shuffle algebra revisited*, Int. Math. Res. Not. IMRN **22** (2014), 6242–6275, DOI 10.1093/imrn/rnt156. MR3283004

[Neg16] Andrei Neguţ, *The $\frac{m}{n}$ Pieri rule*, Int. Math. Res. Not. IMRN **1** (2016), 219–257, DOI 10.1093/imrn/rnv110. MR3514062

[Pop08] Vladimir L. Popov, *Irregular and singular loci of commuting varieties*, Transform. Groups **13** (2008), no. 3-4, 819–837, DOI 10.1007/s00031-008-9018-9. MR2452617

[Ram97] Arun Ram, *Seminormal representations of Weyl groups and Iwahori-Hecke algebras*, Proc. London Math. Soc. (3) **75** (1997), no. 1, 99–133, DOI 10.1112/S0024611597000282. MR1444315

[Shi06] Jun'ichi Shiraishi, *A family of integral transformations and basic hypergeometric series*, Comm. Math. Phys. **263** (2006), no. 2, 439–460, DOI 10.1007/s00220-005-1504-5. MR2207651

[SV13] Olivier Schiffmann and Eric Vasserot, *The elliptic Hall algebra and the $K$-theory of the Hilbert scheme of $\mathbb{A}^2$*, Duke Math. J. **162** (2013), no. 2, 279–366, DOI 10.1215/00127094-1961849. MR3018956

[WW15] Jinkui Wan and Weiqiang Wang, *Frobenius map for the centers of Hecke algebras*, Trans. Amer. Math. Soc. **367** (2015), no. 8, 5507–5520, DOI 10.1090/S0002-9947-2014-06211-9. MR3347181

School of Mathematics and Statistics, University of Melbourne, Victoria 3010, Australia.

*Email address*: alexandr.garbali@unimelb.edu.au

School of Mathematics and Statistics, University of Melbourne, Victoria 3010, Australia.

*Email address*: pzinn@unimelb.edu.au

Contemporary Mathematics
Volume **780**, 2022
https://doi.org/10.1090/conm/780/15687

# The bar involution for quantum symmetric pairs – hidden in plain sight

Stefan Kolb

*To Jasper Stokman on his 50th birthday in gratitude and friendship*

Abstract. We show that all quantum symmetric pair coideal subalgebras $\mathcal{B}_{\mathbf{c}}$ of Kac-Moody type have a bar involution for a suitable choice of parameters $\mathbf{c}$. The proof relies on a generalized notion of quasi $K$-matrix. The proof does not involve an explicit presentation of $\mathcal{B}_{\mathbf{c}}$ in terms of generators and relations.

## 1. Introduction

Let $\mathfrak{g}$ be a symmetrizable Kac-Moody algebra and $\theta : \mathfrak{g} \to \mathfrak{g}$ an involutive Lie algebra automorphism of the second kind. The theory of quantum symmetric pairs, as developed by G. Letzter in the finite setting [**Let99**], studies certain coideal subalgebras $\mathcal{B}_{\mathbf{c}}$ of the quantized enveloping algebra $U = U_q(\mathfrak{g})$. The coideal subalgebras $\mathcal{B}_{\mathbf{c}}$ depend on a family of parameters $\mathbf{c}$. They are quantum group analogs of $U(\mathfrak{g}^\theta)$ where $\mathfrak{g}^\theta = \{x \in \mathfrak{g} \,|\, \theta(x) = x\}$.

In March 2012, C. Stroppel asked the author if there exists a bar involution for $\mathcal{B}_{\mathbf{c}}$. She conjectured that $\mathcal{B}_{\mathbf{c}}$ should allow an algebra automorphism $\overline{\phantom{x}}^B : \mathcal{B}_{\mathbf{c}} \to \mathcal{B}_{\mathbf{c}}$ which is the identity on the Letzter generators $B_i \in \mathcal{B}_{\mathbf{c}}$ and maps $q$ to $q^{-1}$. Unfortunately, back in 2012, this author did not sufficiently appreciate the question.

This changed drastically with the appearance of the preprint versions of [**BW18**] and [**ES18**] in October 2013. Both these papers contained a bar involution for an explicit example of $\mathcal{B}_{\mathbf{c}}$ of type AIII. For this example, H. Bao and W. Wang used the bar involution to construct an element $\mathfrak{X}$ (denoted by $\Upsilon$ in [**BW18**]), now called the quasi $K$-matrix for $\mathcal{B}_{\mathbf{c}}$, which intertwines the Lusztig bar involution $\overline{\phantom{x}}$ for $U$ and the new bar involution $\overline{\phantom{x}}^B$ for $\mathcal{B}_{\mathbf{c}}$ as follows

$$\overline{b}^B \mathfrak{X} = \mathfrak{X}\overline{b} \qquad \text{for all } b \in \mathcal{B}_{\mathbf{c}}. \tag{1.1}$$

Moreover, Bao and Wang used the quasi $K$-matrix to construct a natural family of $\mathcal{B}_{\mathbf{c}}$-module automorphisms $\mathcal{T}$ on the category of finite-dimensional $U$-modules. At this point it became clear that the family $\mathcal{T}$ was a hot candidate for a universal $K$-matrix which the author had been chasing previously in joint work with J. Stokman [**KS09**]. Subsequently, this was worked out jointly with M. Balagović [**BK19**].

2020 *Mathematics Subject Classification.* Primary 17B37; Secondary 53C35, 16T05, 17B67.
*Key words and phrases.* Quantum symmetric pairs, bar involution.

The line of argument, first proposed in [**BW18**] and then performed in [**BK15**], [**BK19**] in a general Kac-Moody setting, was as follows. First the existence of a bar involution on $\mathcal{B}_{\mathbf{c}}$ was established via an explicit presentation of $\mathcal{B}_{\mathbf{c}}$ in terms of generators and relations. Then it was proved that the defining relation (1.1) of the quasi $K$-matrix has an essentially unique solution.

Writing down $\mathcal{B}_{\mathbf{c}}$ in terms of generators and relations is a difficult problem in the general Kac-Moody setting. The quest for the bar involution for $\mathcal{B}_{\mathbf{c}}$ was often cited as an important motivation to find such a presentation of $\mathcal{B}_{\mathbf{c}}$, see for example the introductions of the papers [**CLW21**], [**dC21**]. However, G. Lusztig asked the author in January 2019 if there is a construction of the bar involution for $\mathcal{B}_{\mathbf{c}}$ which does not rely on a presentation of $\mathcal{B}_{\mathbf{c}}$ in terms of generators and relations. Lusztig pointed out that the construction of the bar involution for the positive part $U^+$ in [**Lus93**, 1.2.12] does not involve explicit knowledge of the quantum Serre relations for $U$.

The quasi $K$-matrix is the quantum symmetric pair analog of the quasi $R$-matrix for $U$. The quasi $R$-matrix for $U$ has a description which does not involve Lusztig's bar involution, see [**Lus93**, Theorem 4.1.2 (b)]. Hence it is natural to ask for a bar involution free description of the quasi $K$-matrix. Such a construction was given in the quasi-split case (and in a more general setting of Nichols algebras of diagonal type) in [**KY20**]. For suitable parameters $\mathbf{c}$ there exists an element $\mathfrak{X}$ satisfying a relation similar to (1.1), and the map

$$b \mapsto \mathfrak{X}\overline{b}\mathfrak{X}^{-1} \qquad \text{for all } b \in \mathcal{B}_{\mathbf{c}}$$

is a bar involution on $\mathcal{B}_{\mathbf{c}}$. The element $\mathfrak{X}$ in [**KY20**] is given by $\mathfrak{X} = (\varepsilon \otimes \mathrm{id})(\Theta^\theta)$ where $\varepsilon$ is the counit and the element $\Theta^\theta$ is defined by [**KY20**, (6.1)], see also [**KY20**, Remark 6.3].

Along a slightly different line, A. Appel and B. Vlaar observed in [**AV20**] that the construction of the quasi $K$-matrix in [**BK19**] can be reformulated without the bar involution $\overline{\phantom{x}}^B$. Recall that $\mathcal{B}_{\mathbf{c}}$ is defined for a (generalized) Satake diagram $(X, \tau)$, where $X$ is a subset of the nodes $I$ of the Dynkin diagram of $\mathfrak{g}$, and $\tau : I \to I$ is an involutive diagram automorphism. The pair $(X, \tau)$ needs to satisfy certain compatibility conditions, see Section 2.2. The Letzter generators of $\mathcal{B}_{\mathbf{c}}$ are given by $B_i = F_i$ for $i \in X$ and $B_i = F_i - c_i T_{w_X}(E_{\tau(i)})K_i^{-1}$ for $i \in I \setminus X$, see Section 3.1. With this notation, condition (1.1) is equivalent to the condition

$$B_i \mathfrak{X} = \mathfrak{X}\overline{B_i'} \qquad \text{for all } i \in I$$

where $B_i$ are the Letzter generators corresponding to the parameter family $\mathbf{c}$, and $B_i'$ are Letzter generators corresponding to a different parameter family $\mathbf{c}'$. One then obtains an algebra isomorphism $\Psi : \mathcal{B}_{\mathbf{c}'} \to \mathcal{B}_{\mathbf{c}}$ defined by $\Psi(b) = \mathfrak{X}\overline{b}\mathfrak{X}^{-1}$. This isomorphism satisfies $\Psi(B_i') = B_i$ and $\Psi(q) = q^{-1}$.

In [**AV20**] Appel and Vlaar are mostly concerned with the bar involution free formulation of the quasi $K$-matrix $\mathfrak{X}$. The main point of the present note, which we still believe has novelty to it, is the observation that the above argument provides a proof of the existence of the bar involution for all symmetrizable Kac-Moody algebras and all (generalized) Satake diagrams if the parameters are chosen such that $\mathbf{c}' = \mathbf{c}$, see Corollary 4.2. Although Appel and Vlaar get very close to this observation, they stop just short of the formulation of the existence of the bar involution in [**AV20**, Remark 7.2]. In particular, with the new perspective on the quasi $K$-matrix, the study of the quantum Serre relations for $\mathcal{B}_{\mathbf{c}}$ mentioned in [**AV20**, Remark 7.2] is not necessary anymore.

This note is organised as follows. In Section 2 we fix notation for quantized enveloping algebras and recall the notion of a generalized Satake diagram as formulated in [**RV20**] which generalizes the notion of an admissible pair in [**Kol14**, Definition 2.3]. We promote the view that future papers on quantum symmetric pairs should be written in the setting of generalized Satake diagrams. In Section 3 we recall the definition of $\mathcal{B}_{\mathbf{c}}$ and reprove the existence of the quasi $K$-matrix $\mathfrak{X}$. While this result is contained in [**AV20**, Theorem 7.4], we feel that it is worthwhile to reproduce it in the precise notations and setting of [**BK19**]. Section 4 contains the existence of the isomorphism $\Psi$. As a consequence we obtain the main message of this note, namely the general, relation-free existence proof for the bar involution on $\mathcal{B}_{\mathbf{c}}$, see Corollary 4.2

## 2. Preliminaries

**2.1. Quantum groups.** Let $\mathfrak{g}$ be a symmetrizable Kac-Moody algebra with generalized Cartan matrix $(a_{ij})_{i,j\in I}$ where $I$ is a finite set and Cartan subalgebra $\mathfrak{h}\subset\mathfrak{g}$. Let $\{d_i\,|\,i\in I\}$ be a set of relatively prime positive integers such that the matrix $(d_ia_{ij})$ is symmetric. Let $\Pi=\{\alpha_i\,|\,i\in I\}$ be the set of simple roots for $\mathfrak{g}$, let $\Phi$ be the root system, and let $Q=\mathbb{Z}\Pi$ be the root lattice. For $\beta=\sum_{i\in I}n_i\alpha_i\in Q$ we write $\mathrm{ht}(\beta)=\sum_{i\in I}n_i$. Consider the symmetric bilinear form $(\cdot,\cdot):Q\times Q\to\mathbb{Z}$ defined by $(\alpha_i,\alpha_j)=d_ia_{ij}$ for all $i,j\in I$. We denote $Q^+=\mathbb{N}_0\Pi$ where $\mathbb{N}_0=\{0,1,2,\dots\}$. Let $W$ be the Weyl group of $\mathfrak{g}$ which is generated by the simple reflections $s_i$ for $i\in I$.

Throughout this paper let $k$ be a field of characteristic zero and let $\mathbb{K}=k(q)$ be the field of rational functions in a variable $q$. We consider the quantized enveloping algebra $U=U_q(\mathfrak{g})$ as the $\mathbb{K}$-algebra with generators $E_i,F_i,K_i^{\pm1}$ for $i\in I$ subject to the relations given in [**Lus93**, 3.1.1]. For $\beta=\sum_{i\in I}n_i\alpha_i\in Q$ we write $K_\beta=\prod_{i\in I}K_i^{n_i}$. Let $U^+$ and $U^-$ denote the subalgebras of $U$ generated by $\{E_i\,|\,i\in I\}$ and $\{F_i\,|\,i\in I\}$, respectively. For any $\mu\in Q^+$ we write $U^+_\mu=\mathrm{span}_{\mathbb{K}}\{E_{i_1}\dots E_{i_\ell}\,|\,\sum_{j=1}^{\ell}\alpha_{i_j}=\mu\}$ where $\mathrm{span}_{\mathbb{K}}$ denotes the $\mathbb{K}$-linear span. Moreover, define $U^-_{-\mu}=\omega(U^+_\mu)$ where $\omega:U\to U$ denotes the algebra automorphism given in [**Lus93**, 3.1.3]. For any $i\in I$ let $T_i:U\to U$ denote the algebra automorphism denoted by $T''_{i,1}$ in [**Lus93**, 37.1]. By [**Lus93**, 39.4.3] the automorphisms $T_i$ satisfy braid relations. Hence, for $w\in W$ with reduced expression $w=s_{i_1}\dots s_{i_\ell}$ there is a well-defined automorphism $T_w=T_{i_1}\dots T_{i_\ell}$ of $U$. For any $i\in I$ let $r_i,{}_ir:U^+\to U^+$ denote the Lusztig-Kashiwara skew-derivations which are defined uniquely by the relation

$$xF_i-F_ix=\frac{1}{q_i-q_i^{-1}}\big(r_i(x)K_i-K_i^{-1}\,{}_ir(x)\big)\qquad\text{for all }x\in U^+\tag{2.1}$$

where $q_i=q^{d_i}$, see [**Lus93**, 3.1.6]. The maps $r_i$ and ${}_ir$ satisfy the relations $r_i(E_j)={}_ir(E_j)=\delta_{i,j}$ and the skew-derivation properties

$$r_i(xy)=q^{(\alpha_i,\nu)}r_i(x)y+xr_i(y),\qquad {}_ir(xy)={}_ir(x)y+q^{(\alpha_i,\mu)}x\,{}_ir(y)\tag{2.2}$$

for all $x\in U^+_\mu$, $y\in U^+_\nu$.

Recall that the bar involution for $U$ is the $k$-algebra automorphism $\overline{\phantom{x}} : U \to U$, $x \mapsto \overline{x}$ defined by

$$\overline{E_i} = E_i, \quad \overline{F_i} = F_i, \quad \overline{K_i} = K_i^{-1}, \quad \overline{q} = q^{-1}$$

for all $i \in I$.

To formulate the properties of the quasi $K$-matrix we need to work in an overalgebra $\mathscr{U}$ of $U$. Let $\mathcal{O}_{\mathrm{int}}$ denote the category of integrable $U$-modules in category $\mathcal{O}$ and let $\mathcal{F}or : \mathcal{O}_{\mathrm{int}} \to \mathcal{V}ect$ be the forgetful functor into the category of $\mathbb{K}$-vector spaces. We define $\mathscr{U} = \mathrm{End}(\mathcal{F}or)$ to be the set of natural transformations from $\mathcal{F}or$ to itself. By construction, $\mathscr{U}$ is an algebra which contains $U$ as a subalgebra. See [**BK19**, Section 3] for precise definitions and further details. Consider the elements of the product $\widehat{U^+} = \prod_{\mu \in Q^+} U^+_\mu$ as additive infinite sums of elements in $U^+$. Then $\widehat{U^+}$ is a subalgebra of $\mathscr{U}$, see [**BK19**, Example 3.2].

**2.2. Generalized Satake diagrams.** In [**Kol14**] quantum symmetric pairs were defined for involutive automorphisms of $\mathfrak{g}$ of the second kind. Such involutions are classified in terms of Satake diagrams $(X, \tau)$ where $X \subset I$ is a subset of finite type and $\tau : I \to I$ is an involutive diagram automorphism satisfying the compatibility conditions given in [**Kol14**, Definition 2.3]. Associated to the finite type subset $X$ is a finite type root system $\Phi_X \subset \Phi$ with parabolic Weyl group $W_X \subset W$. Let $w_X \in W_X$ be the longest element and let $\rho_X^\vee \in \mathfrak{h}$ be the half sum of positive coroots of $\Phi_X$. To be a Satake diagram, the pair $(X, \tau)$ has to satisfy the two conditions

(1) $\tau(\alpha_j) = -w_X(\alpha_j)$ for all $j \in X$;
(2) If $i \in I \setminus X$ and $\tau(i) = i$ then $\alpha_i(\rho_X^\vee) \in \mathbb{Z}$.

Condition (1) implies that the map $\Theta = -w_X \circ \tau : Q \to Q$ is an involutive automorphism of the root lattice. It was observed by V. Regelskis and B. Vlaar that condition (2) can be replaced by the weaker condition

(2') If $\tau(i) = i$ and $a_{ji} = -1$ then $\Theta(\alpha_i) \neq -\alpha_i - \alpha_j$.

Following [**RV20**], we call pairs $(X, \tau)$ as above, satisfying conditions (1), (2') generalized Satake diagrams. As explained in [**RV20**, Section 4] the construction of quantum symmetric pairs and the structure theory developed in [**Kol14**] remains valid for generalized Satake diagrams.

The diagram automorphism $\tau$ lifts to an Hopf-algebra automorphism of $U$. As in [**Lus93**, 3.1.3] let $\sigma : U \to U$ denote the $\mathbb{K}$-algebra antiautomorphism defined by

$$\sigma(E_i) = E_i, \qquad \sigma(F_i) = F_i, \qquad \sigma(K_i) = K_i^{-1} \qquad \text{for all } i \in I.$$

The following proposition was conjectured in [**BK15**, Conjecture 2.7] and was proved in [**BK15**, Proposition 2.5] up to a sign. The sign was confirmed by H. Bao and W. Wang in [**BW21**] by a subtle argument involving canonical bases for $U$. The arguments in [**BK15**], [**BW21**] do not involve condition (2) above, and hence also hold in the setting of generalized Satake diagrams, see also [**AV20**, Lemma 7.14].

PROPOSITION 2.1. *Let $(X, \tau)$ be a generalized Satake diagram for the symmetrizable Kac-Moody algebra $\mathfrak{g}$. Then the relation*

$$\sigma \circ \tau(r_i(T_{w_X}(E_i))) = r_i(T_{w_X}(E_i))$$

*holds in $U$ for all $i \in I \setminus X$.*

## 3. The quasi $K$-matrix, revisited

**3.1. Quantum symmetric pairs.** Let $(X,\tau)$ be an generalized Satake diagram. Set $\mathcal{M}_X = \mathbb{K}\langle E_j, F_j, K_j^{\pm 1} \,|\, j\in X\rangle$ and set $U^0_\Theta = \mathbb{K}\langle K_\beta \,|\, \beta\in Q, \Theta(\beta)=\beta\rangle$. Define a set of parameters

$$\mathcal{C} = \{\mathbf{c} = (c_i)_{i\in I\setminus X} \in (\mathbb{K}^\times)^{I\setminus X} \,|\, c_i = c_{\tau(i)} \text{ if } (\alpha_i, \Theta(\alpha_i)) = 0\}.$$

For any $\mathbf{c} = (c_i)_{i\in I\setminus X}\in \mathcal{C}$ we consider the subalgebra $\mathcal{B}_{\mathbf{c}}$ of $U_q(\mathfrak{g})$ generated by $\mathcal{M}_X$, $U^0_\Theta$ and the elements

$$B_i = F_i - c_i T_{w_X}(E_\tau(i))K_i^{-1} \qquad \text{for all } i\in I\setminus X. \tag{3.1}$$

To unify notation, we set $c_i = 0$ for $i\in X$ and extend (3.1) by writing $B_i = F_i$ if $i\in X$. Following [**Let99**], [**Kol14**] we call $\mathcal{B}_{\mathbf{c}}$ the quantum symmetric pair coideal subalgebra of $U_q(\mathfrak{g})$ corresponding to the generalized Satake diagram $(X,\tau)$

REMARK 3.1. The parameters $c_i$ were denoted by $c_i s(\tau(i))$ in [**BK19**]. The additional parameters $s(\tau(i))$ appeared in the construction of involutive automorphisms of $\mathfrak{g}$ of the second kind corresponding to the Satake diagram $(X,\tau)$. For the construction of quantum symmetric pairs, however, it is advantageous to suppress the parameters $s(\tau(i))$ in the notation.

REMARK 3.2. In [**Kol14**], quantum symmetric pairs depend on a second family of parameters $\mathbf{s}$ in a certain subset $\mathcal{S}\subset \mathbb{K}^{I\setminus X}$. The corresponding coideal subalgebras are then denoted by $\mathcal{B}_{\mathbf{c},\mathbf{s}}$. By [**Kol14**, Theorem 7.1] the algebra $\mathcal{B}_{\mathbf{c}}$ is isomorphic to $\mathcal{B}_{\mathbf{c},\mathbf{s}}$ as an algebra for all $\mathbf{s}$. As we only aim to establish the existence of a bar involution for $\mathcal{B}_{\mathbf{c},\mathbf{s}}$, it suffices to consider the case where $\mathbf{s}=(0,0,\dots,0)$. Moreover, it was noted in [**DK19**, 3.5], that for the construction of the quasi $K$-matrix there is no loss of generality in the restriction to the case $\mathbf{s}=(0,0,\dots,0)$.

**3.2. The quasi $K$-matrix $\mathfrak{X}$.** In [**AV20**, Section 7] A. Appel and B. Vlaar gave a reformulation of the results of [**BK19**, Section 6] which does not rely on the existence of a bar involution for $\mathcal{B}_{\mathbf{c}}$. As notations and conventions in [**AV20**] somewhat differ from those of the present note, we feel it is beneficial to recall this reformulation in the present setting. We begin with a reformulation of [**BK19**, Proposition 6.1]. Let $2\rho_X$ be the sum of positive roots in $\Phi_X$.

PROPOSITION 3.3. *Let $\mathfrak{X} = \sum_{\mu\in Q^+}\mathfrak{X}_\mu \in \widehat{U^+}$ with $\mathfrak{X}_\mu\in U^+_\mu$. The following are equivalent:*

(1) *For all $i\in I$ the relation*

$$B_i\mathfrak{X} = \mathfrak{X}\big(F_i - (-1)^{2\alpha_i(\rho_X^\vee)} q^{-(\alpha_i,\Theta(\alpha_i)-2\rho_X)} c_{\tau(i)} \overline{T_{w_X}(E_{\tau(i)})}K_i\big)$$

*holds in $\mathscr{U}$.*

(2) *The element $\mathfrak{X}$ satisfies the relations*

$$r_i(\mathfrak{X}_\mu) = (q_i - q_i^{-1})(-1)^{2\alpha_i(\rho_X^\vee)} q^{-(\alpha_i,\Theta(\alpha_i)-2\rho_X)} c_{\tau(i)} \mathfrak{X}_{\mu+\Theta(\alpha_i)-\alpha_i}\overline{T_{w_X}(E_{\tau(i)})}, \tag{3.2}$$

$${}_ir(\mathfrak{X}_\mu) = (q_i - q_i^{-1}) q^{-(\Theta(\alpha_i),\alpha_i)} c_i T_{w_X}(E_{\tau(i)})\mathfrak{X}_{\mu+\Theta(\alpha_i)-\alpha_i} \tag{3.3}$$

*for all $i\in I$.*

*If the equivalent conditions (1) and (2) hold, then the following also hold:*

(3) *For all $x\in U^0_\Theta\mathcal{M}_X$ one has $x\mathfrak{X} = \mathfrak{X}x$.*

(4) *For all $\mu\in Q^+$ such that $\mathfrak{X}_\mu\neq 0$ one has $\Theta(\mu) = -\mu$.*

PROOF. The equivalence of (1) and (2) follows from the property (2.1) of the skew derivations $r_i$ and ${}_ir$. Property (4) follows by induction over the height of $\mu$ and in turn implies property (3) just as in the proof of [**BK19**, Proposition 6.1]. □

Next we need [**BK19**, Proposition 6.3]. This proposition is a general statement about solving recursive relations of the type (3.2), (3.3) simultaneously, and hence holds in our setting. Let $\langle\, ,\rangle : U^- \otimes U^+ \to \mathbb{K}$ be the nondegenerate pairing considered in [**BK19**, Section 2.3].

PROPOSITION 3.4. ([**BK19**, Proposition 6.3]) *Let $\mu \in Q^+$ with $\mathrm{ht}(\mu) \geq 2$ and fix elements $A_i$, ${}_iA \in U^+_{\mu-\alpha_i}$ for all $i \in I$. Then the following are equivalent:*

(1) *There exists an element $\underline{X} \in U^+_\mu$ such that*

$$r_i(\underline{X}) = A_i \quad \text{and} \quad {}_ir(\underline{X}) = {}_iA \qquad \text{for all } i \in I. \tag{3.4}$$

(2) *The elements $A_i$, ${}_iA$ have the following two properties:*
   (a) *For all $i, j \in I$ one has*

$$r_i({}_jA) = {}_jr(A_i).$$

   (b) *For all $i \neq j \in I$ one has*

$$\frac{-1}{q_i - q_i^{-1}} \sum_{s=1}^{1-a_{ij}} \begin{bmatrix} 1-a_{ij} \\ s \end{bmatrix}_{q_i} (-1)^s \langle F_i^{1-a_{ij}-s} F_j F_i^{s-1}, A_j \rangle - \frac{1}{q_j - q_j^{-1}} \langle F_i^{1-a_{ij}}, A_j \rangle = 0. \tag{3.5}$$

*Moreover, if the system of equation (3.4) has a solution $\underline{X}$ then this solution is uniquely determined.*

We now translate [**BK19**, Section 6.4] into our setting. This is the crucial step where we also need Proposition 2.1. We hence give all the details. Fix $\mu \in Q^+$ and assume that a collection $(\mathfrak{X}_{\mu'})_{\mu'<\mu}$ with $\mathfrak{X}_{\mu'} \in U^+_{\mu'}$ and $\mathfrak{X}_0 = 1$ has already been constructed, and that this collection satisfies the relations (3.2),(3.3) for all $\mu' < \mu$ and all $i \in I$. Define

$$A_i = (q_i - q_i^{-1})(-1)^{2\alpha_i(\rho_X^\vee)} q^{-(\alpha_i, \Theta(\alpha_i) - 2\rho_X)} c_{\tau(i)} \mathfrak{X}_{\mu+\Theta(\alpha_i)-\alpha_i} \overline{T_{w_X}(E_{\tau(i)})} \tag{3.6}$$

$${}_iA = (q_i - q_i^{-1}) q^{-(\Theta(\alpha_i), \alpha_i)} c_i T_{w_X}(E_{\tau(i)}) \mathfrak{X}_{\mu+\Theta(\alpha_i)-\alpha_i}. \tag{3.7}$$

We have the following analog of [**BK19**, Lemma 6.7].

LEMMA 3.5. *The relation $r_i({}_jA) = {}_jr(A_i)$ holds for all $i, j \in I$.*

PROOF. Using the skew-derivation properties (2.2) and the assumptions on ${}_iA, A_i$ we calculate

$$\begin{aligned} r_i({}_jA) &= (q_j - q_j^{-1}) q^{-(\Theta(\alpha_j),\alpha_j)} c_j T_{w_X}(E_{\tau(j)}) r_i(\mathfrak{X}_{\mu+\Theta(\alpha_j)-\alpha_j}) \\ &\quad + (q_j - q_j^{-1}) q^{-(\Theta(\alpha_j),\alpha_j)} c_j q^{(\alpha_i, \mu+\Theta(\alpha_j)-\alpha_j)} r_i(T_{w_X}(E_{\tau(j)})) \mathfrak{X}_{\mu+\Theta(\alpha_j)-\alpha_j} \\ &= (q_j - q_j^{-1})(q_i - q_i^{-1})(-1)^{2\alpha_i(\rho_X^\vee)} q^{-(\Theta(\alpha_j),\alpha_j)} q^{-(\alpha_i,\Theta(\alpha_i)-2\rho_X)} c_j c_{\tau(i)} \cdot \\ &\qquad \cdot T_{w_X}(E_{\tau(j)}) \mathfrak{X}_{\mu+\Theta(\alpha_j)-\alpha_j+\Theta(\alpha_i)-\alpha_i} \overline{T_{w_X}(E_{\tau(i)})} \\ &\quad + (q_j - q_j^{-1}) q^{-(\Theta(\alpha_j),\alpha_j)} c_j q^{(\alpha_i, \mu+\Theta(\alpha_j)-\alpha_j)} r_i(T_{w_X}(E_{\tau(j)})) \mathfrak{X}_{\mu+\Theta(\alpha_j)-\alpha_j} \end{aligned}$$

and similarly

$$\begin{aligned}
{}_jr(A_i) &= (q_i - q_i^{-1})(-1)^{2\alpha_i(\rho_X^\vee)} q^{-(\alpha_i,\Theta(\alpha_i)-2\rho_X)} c_{\tau(i)}\, {}_jr(\mathfrak{X}_{\mu-\Theta(\alpha_i)-\alpha_i})\overline{T_{w_X}(E_{\tau(i)})}\\
&\quad + (q_i - q_i^{-1})(-1)^{2\alpha_i(\rho_X^\vee)} q^{(\alpha_j,\mu+\Theta(\alpha_i)-\alpha_i)-(\alpha_i,\Theta(\alpha_i)-2\rho_X)}\times\\
&\qquad\qquad \times c_{\tau(i)}\mathfrak{X}_{\mu+\Theta(\alpha_i)-\alpha_i}\, {}_jr(\overline{T_{w_X}(E_{\tau(i)})})\\
&= (q_i - q_i^{-1})(q_j - q_j^{-1})(-1)^{2\alpha_i(\rho_X^\vee)} q^{-(\alpha_i,\Theta(\alpha_i)-2\rho_X)} q^{-(\Theta(\alpha_j),\alpha_j)} c_{\tau(i)} c_j \times\\
&\qquad\qquad \times T_{w_X}(E_{\tau(j)})\mathfrak{X}_{\mu+\Theta(\alpha_i)-\alpha_i+\Theta(\alpha_j)-\alpha_j}\overline{T_{w_X}(E_{\tau(i)})}\\
&\quad + (q_i - q_i^{-1})(-1)^{2\alpha_i(\rho_X^\vee)} q^{(\alpha_j,\mu+\Theta(\alpha_i)-\alpha_i)-(\alpha_i,\Theta(\alpha_i)-2\rho_X)}\times\\
&\qquad\qquad \times c_{\tau(i)}\mathfrak{X}_{\mu+\Theta(\alpha_i)-\alpha_i}\, {}_jr(\overline{T_{w_X}(E_{\tau(i)})}).
\end{aligned}$$

Comparing the above two expressions we see that the relation $r_i({}_jA) = {}_jr(A_i)$ is equivalent to the relation

$$\begin{aligned}
(3.8)\quad & (q_j - q_j^{-1}) q^{-(\Theta(\alpha_j),\alpha_j)} c_j q^{(\alpha_i,\mu+\Theta(\alpha_j)-\alpha_j)} r_i(T_{w_X}(E_{\tau(j)}))\mathfrak{X}_{\mu+\Theta(\alpha_j)-\alpha_j}\\
&\quad = (q_i - q_i^{-1})(-1)^{2\alpha_i(\rho_X^\vee)} q^{(\alpha_j,\mu+\Theta(\alpha_i)-\alpha_i)-(\alpha_i,\Theta(\alpha_i)-2\rho_X)}\times\\
&\qquad\qquad \times c_{\tau(i)}\mathfrak{X}_{\mu+\Theta(\alpha_i)-\alpha_i}\, {}_jr(\overline{T_{w_X}(E_{\tau(i)})}).
\end{aligned}$$

By [**Lus93**, 1.2.14] we have ${}_jr(\overline{T_{w_X}(E_{\tau(i)})}) = q^{(\alpha_j,w_X(\alpha_{\tau(i)})-\alpha_j)}\overline{r_j(T_{w_X}(E_{\tau(i)}))}$. Hence both sides of (3.8) vanish unless $\tau(i) = j \in I\setminus X$, and in this case (3.8) can be rewritten as

$$\begin{aligned}
& q^{(\alpha_j,\Theta(\alpha_i)-\Theta(\alpha_j)-\alpha_i)+(\alpha_i,\mu)} r_i(T_{w_X}(E_i))\mathfrak{X}_{\mu+\Theta(\alpha_j)-\alpha_j}\\
&\quad = (-1)^{2\alpha_i(\rho_X^\vee)} q^{(\alpha_j,\mu-\alpha_i-\alpha_j)-(\alpha_i,\Theta(\alpha_i)-2\rho_X)}\mathfrak{X}_{\mu+\Theta(\alpha_j)-\alpha_j}\overline{r_j(T_{w_X}(E_j))}
\end{aligned}$$

From now on we assume $\tau(i) = j$ and hence $\Theta(\alpha_i) - \alpha_i = \Theta(\alpha_j) - \alpha_j$. Hence we may assume that $\mathfrak{X}_{\mu+\theta(\alpha_j)-\alpha_j} \neq 0$ and in this case the above equation is equivalent to the equation

$$q^{(\alpha_i,\mu)} r_i(T_{w_X}(E_i)) = (-1)^{2\alpha_i(\rho_X^\vee)} q^{(\alpha_{\tau(i)},\mu-\alpha_i)-(\alpha_i,\Theta(\alpha_i)-2\rho_X)}\overline{r_{\tau(i)}(T_{w_X}(E_{\tau(i)}))}.$$

By Proposition 3.3.(4) we have $(\alpha_i - \alpha_{\tau(i)}, \mu) = 0$ and hence the above equation can be rewritten as

$$\overline{r_i(T_{w_X}(E_i))} = (-1)^{2\alpha_i(\rho_X^\vee)} q^{(\alpha_i,\alpha_i-w_X(\alpha_i)-2\rho_X)} r_{\tau(i)}(T_{w_X}(E_{\tau(i)})).$$

By [**BK15**, Lemma 2.9], which also holds for generalized Satake diagrams, and by Proposition 2.1 the above equation does indeed hold. □

Next we obtain an analog of [**BK19**, Lemma 6.8]. We include the proof which is simplified along the lines of [**BK19**, Remark 6.9].

LEMMA 3.6. *For all $i \neq j \in I$ the elements $A_i, A_j$ given by* (3.6) *satisfy the relation* (3.5).

PROOF. As $A_i \in U^+_{\mu-\alpha_i}$ we may assume that $\mu = (1 - a_{ij})\alpha_i + \alpha_j$ and $\Theta(\mu) = -\mu$ as otherwise both terms of (3.5) vanish. By [**BK19**, Lemma 6.4] it suffices to consider the following two cases:

**Case 1: $\Theta(\alpha_i) = -\alpha_j$ and $a_{ij}$=0.** In this case $\mu = \alpha_i + \alpha_j$ and $(\alpha_i, \alpha_k) = (\alpha_j, \alpha_k) = 0$ for all $k \in X$, and hence

$$A_i = (q_i - q_i^{-1})c_j E_j, \qquad A_j = (q_i - q_i^{-1})c_i E_i.$$

Hence the left hand side of (3.5) vanishes.

**Case 2: $\Theta(\alpha_i) = -\alpha_i$ and $\Theta(\alpha_j) = -\alpha_j$.** In this case induction on $\mu'$ for $\mu' < \mu$ shows that $\mathfrak{X}_{\mu'} \neq 0$ implies that $\mu' \in \mathrm{span}_{\mathbb{N}_0}\{2\alpha_i, 2\alpha_j\}$, see [**BK19**, Lemma 6.5]. Hence for $\mu = (1 - a_{ij})\alpha_i + \alpha_j$ we have $\mathfrak{X}_{\mu+\Theta(\alpha_i)-\alpha_i} = \mathfrak{X}_{\mu-2\alpha_i} = 0$ and similarly $\mathfrak{X}_{\mu+\Theta(\alpha_j)-\alpha_j} = 0$. Hence $A_i = A_j = 0$ which implies (3.5) also in this case. □

With the above preparations the following theorem is proved just as in [**BK19**, Theorem 6.10] using Proposition 3.4.

THEOREM 3.7. *There exists a uniquely determined element* $\mathfrak{X} = \sum_{\mu \in Q^+} \mathfrak{X}_\mu \in \widehat{U^+}$ *with* $\mathfrak{X}_0 = 1$ *and* $\mathfrak{X}_\mu \in U^+_\mu$ *such that the equality*

$$B_i \mathfrak{X} = \mathfrak{X}\big(F_i - (-1)^{2\alpha_i(\rho_X^\vee)} q^{-(\alpha_i, \Theta(\alpha_i) - 2\rho_X)} c_{\tau(i)} \overline{T_{w_X}(E_{\tau(i)})} K_i\big) \tag{3.9}$$

*holds in* $\mathscr{U}$ *for all* $i \in I$. *Moreover, the element* $\mathfrak{X}$ *commutes with all elements of* $U^0_\Theta \mathcal{M}_X$.

REMARK 3.8. We call the element $\mathfrak{X} \in \mathscr{U}$ from the above theorem the quasi $K$-matrix corresponding to the quantum symmetric pair coideal subalgebra $\mathcal{B}_{\mathbf{c}}$. The quasi $K$-matrix is invertible in $\widehat{U^+} = \prod_{\mu \in Q^+} U^+_\mu$ because $\mathfrak{X}_0 = 1 \neq 0$.

## 4. The bar involution for quantum symmetric pairs, revisited

Recall that $\mathbb{K} = k(q)$. Following [**AV20**, Section 7.3] for any $\mathbf{c} = (c_i)_{i \in I \setminus X} \in \mathbb{K}^{I \setminus X}$ we define $\mathbf{c}' = (c'_i)_{i \in I \setminus X} \in \mathbb{K}^{I \setminus X}$ by

$$c'_i = (-1)^{2\alpha_i(\rho_X^\vee)} q^{(\alpha_i, \Theta(\alpha_i) - 2\rho_X)} \overline{c_{\tau(i)}}.$$

Observe that $\mathbf{c} \in \mathcal{C}$ if and only if $\mathbf{c}' \in \mathcal{C}$. For $\mathbf{c} \in \mathcal{C}$ let $\mathcal{B}_{\mathbf{c}'}$ denote the quantum symmetric pair coideal subalgebra corresponding to the parameters $\mathbf{c}'$. We denote the generators (3.1) for $\mathcal{B}_{\mathbf{c}'}$ by $B'_i$ to distinguish them from the corresponding generators $B_i$ of $\mathcal{B}_{\mathbf{c}}$. We immediately obtain the following consequence of Theorem 3.7.

COROLLARY 4.1. *For any* $\mathbf{c} \in \mathcal{C}$ *there exists a* $k$*-algebra isomorphism* $\Psi : \mathcal{B}_{\mathbf{c}'} \to \mathcal{B}_{\mathbf{c}}$ *such that*

$$\Psi|_{\mathcal{M}_X U^0_\Theta} = \overline{\phantom{x}}|_{\mathcal{M}_X U^0_\Theta} \qquad \text{and} \qquad \Psi(B'_i) = B_i \quad \text{for all } i \in I \setminus X.$$

*In particular, we have* $\Psi(q) = q^{-1}$ *and* $\Psi(K_\beta) = K_{-\beta}$ *for all* $\beta \in Q^\Theta$.

PROOF. With the above notation, Equation (3.9) can be rewritten as

$$B_i \mathfrak{X} = \mathfrak{X} \overline{B'_i} \qquad \text{for all } i \in I \setminus X \tag{4.1}$$

Hence there is a well-defined $k$-algebra homomorphism $\Psi : \mathcal{B}_{\mathbf{c}'} \to \mathcal{B}_{\mathbf{c}}$ such that $\Psi(b) = \mathfrak{X}\overline{b}\mathfrak{X}^{-1}$ for all $b \in \mathcal{B}_{\mathbf{c}'}$. Equation (4.1) implies that $\Psi(B'_i) = B_i$ for all $i \in I \setminus X$. Property (3) in Proposition 3.3 implies that $\Psi|_{\mathcal{M}_X U^0_\Theta} = \overline{\phantom{x}}|_{\mathcal{M}_X U^0_\Theta}$. □

In the special case that $\mathbf{c} = \mathbf{c}'$ the above corollary provides the desired bar involution for $\mathcal{B}_{\mathbf{c}}$.

COROLLARY 4.2. *Assume that the parameters* $\mathbf{c} = (c_i)_{i \in I \setminus X} \in \mathbb{K}^{I \setminus X}$ *satisfy the relation*

$$\overline{c_i} = (-1)^{2\alpha_i(\rho_X^\vee)} q^{-(\alpha_i, \Theta(\alpha_i) - 2\rho_X)} c_{\tau(i)} \tag{4.2}$$

*for all $i \in I \setminus X$. Then there exists a $k$-algebra automorphism*

$$\overline{\phantom{x}}^B : \mathcal{B}_{\mathbf{c}} \to \mathcal{B}_{\mathbf{c}}, \quad b \mapsto \overline{b}^B$$

*such that $\overline{\phantom{x}}^B|_{U^0_\Theta \mathcal{M}_X} = \overline{\phantom{x}}|_{U^0_\Theta \mathcal{M}_X}$ and $\overline{B_i}^B = B_i$ for all $i \in I \setminus X$. In particular, the automorphism $\overline{\phantom{x}}^B$ of $\mathcal{B}_{\mathbf{c}}$ satisfies the relation $\overline{q}^B = q^{-1}$.*

## References


[AV20] Andrea Appel and Bart Vlaar, *Universal k-matrix for quantum Kac-Moody algebras*, Preprint, arXiv:2007.09218v2 (2020), 67 pp.

[BK15] Martina Balagović and Stefan Kolb, *The bar involution for quantum symmetric pairs*, Represent. Theory **19** (2015), 186–210, DOI 10.1090/ert/469. MR3414769

[BK19] Martina Balagović and Stefan Kolb, *Universal K-matrix for quantum symmetric pairs*, J. Reine Angew. Math. **747** (2019), 299–353, DOI 10.1515/crelle-2016-0012. MR3905136

[BW18] Huanchen Bao and Weiqiang Wang, *A new approach to Kazhdan-Lusztig theory of type B via quantum symmetric pairs* (English, with English and French summaries), Astérisque **402** (2018), vii+134. MR3864017

[BW21] Huanchen Bao and Weiqiang Wang, *Canonical bases arising from quantum symmetric pairs of Kac-Moody type*, Compos. Math. **157** (2021), no 7, 1507–1537, DOI 10.1112/S0010437X2100734X. MR4277109

[CLW21] Xinhong Chen, Ming Lu, and Weiqiang Wang, *A Serre presentation for the ıquantum groups*, Transform. Groups **26** (2021), no. 3, 827–857, DOI 10.1007/s00031-020-09581-5. MR4309551

[dC21] Hadewijch De Clercq, *Defining relations for quantum symmetric pair coideals of Kac-Moody type*, J. Comb. Algebra **5** (2021), no. 4, 297–367, DOI 10.4171/jca/57. MR4344789

[DK19] Liam Dobson and Stefan Kolb, *Factorisation of quasi K-matrices for quantum symmetric pairs*, Selecta Math. (N.S.) **25** (2019), no. 4, Paper No. 63, 55, DOI 10.1007/s00029-019-0508-5. MR4021849

[ES18] Michael Ehrig and Catharina Stroppel, *Nazarov-Wenzl algebras, coideal subalgebras and categorified skew Howe duality*, Adv. Math. **331** (2018), 58–142, DOI 10.1016/j.aim.2018.01.013. MR3804673

[Kol14] Stefan Kolb, *Quantum symmetric Kac-Moody pairs*, Adv. Math. **267** (2014), 395–469, DOI 10.1016/j.aim.2014.08.010. MR3269184

[KS09] Stefan Kolb and Jasper V. Stokman, *Reflection equation algebras, coideal subalgebras, and their centres*, Selecta Math. (N.S.) **15** (2009), no. 4, 621–664, DOI 10.1007/s00029-009-0007-1. MR2565052

[KY20] Stefan Kolb and Milen Yakimov, *Symmetric pairs for Nichols algebras of diagonal type via star products*, Adv. Math. **365** (2020), 107042, 69, DOI 10.1016/j.aim.2020.107042. MR4065713

[Let99] Gail Letzter, *Symmetric pairs for quantized enveloping algebras*, J. Algebra **220** (1999), no. 2, 729–767, DOI 10.1006/jabr.1999.8015. MR1717368

[Lus93] George Lusztig, *Introduction to quantum groups*, Progress in Mathematics, vol. 110, Birkhäuser Boston, Inc., Boston, MA, 1993. MR1227098

[RV20] Vidas Regelskis and Bart Vlaar, *Quasitriangular coideal subalgebras of $U_q(\mathfrak{g})$ in terms of generalized Satake diagrams*, Bull. Lond. Math. Soc. **52** (2020), no. 4, 693–715, DOI 10.1112/blms.12360. MR4171396



School of Mathematics, Statistics and Physics, Newcastle University, Newcastle upon Tyne NE1 7RU, United Kingdom

*Email address*: stefan.kolb@newcastle.ac.uk

Contemporary Mathematics
Volume **780**, 2022
https://doi.org/10.1090/conm/780/15688

# Charting the $q$-Askey scheme

Tom H. Koornwinder

*Dedicated to Jasper Stokman on the occasion of his fiftieth birthday, in admiration and friendship*

ABSTRACT. Following Verde-Star [Linear Algebra Appl. **627** (2021), pp. 242–274] we label families of orthogonal polynomials in the $q$-Askey scheme together with their $q$-hypergeometric representations by three sequences $x_k, h_k, g_k$ of Laurent polynomials in $q^k$, two of degree 1 and one of degree 2, satisfying certain constraints. This gives rise to a precise classification and parametrization of these families together with their limit transitions. This is displayed in a graphical scheme. We also describe the four-manifold structure underlying the scheme.

## 1. Introduction

The Askey scheme [**2**, p.46], [**8**, p.184] and the $q$-Askey scheme [**8**, p.414] display in a graphical way the families of ($q$-)hypergeometric orthogonal polynomials as they occur as limit cases of the four-parameter top level families: Wilson and Racah polynomials for the Askey scheme and Askey–Wilson and $q$-Racah polynomials for the $q$-Askey scheme. By each arrow to the next lower level one parameter is lost. The bottom level families no longer depend on parameters. Since their introduction these schemes have been of great assistance to everybody who needs to do work with one or more of the families in the scheme.

These schemes are also expected and partially proven to exist in other contexts, parallel to the original schemes or generalizing them. These contexts are: (i) ($q$-)hypergeometric biorthogonal rational functions [**4**]; (ii) the nonsymmetric case [**12**], [**13**]; (iii) the $q = -1$ case starting with the Bannai–Ito polynomials [**3**, pp. 271–273], [**18**]; (iv) generalized (continuous) orthogonal systems [**9**]; (v) $q$-Askey scheme for root system $\mathrm{BC}_n$, see, among others, Stokman [**14**] and references given there.

Still some questions can be posed about the original schemes which, in the author's opinion, have not been answered in a satisfactory way until now:

(1) Are the schemes complete? For answering this question one first needs a precise criterium for inclusion of a family in the scheme. This criterium is usually that the orthogonal polynomials should satisfy a Bochner type

2020 *Mathematics Subject Classification.* Primary 33D45.

property, i.e., that they are eigenfunctions of a second order linear differential or ($q$-)difference operator of certain type. However, earlier classifications [**7**], [**20**] arrive, in the continuous case, at the Askey–Wilson polynomials being the most general family satisfying the requirements, but do not give an exhaustive classification of all such families. A related question is if all limits or specializations from one level to the level below are present in the scheme.

(2) Which families deserve an independent status in the scheme and which ones are just subfamilies of another family? Several families in the $q$-Askey scheme can be considered as a subfamily, obtained by restricting the parameters, of a family higher up in the scheme. Consider for instance the continuous dual $q$-Hahn polynomials [**8**, §14.3] and the continuous $q$-Jacobi polynomials [**8**, §14.10], both being subfamilies of the Askey–Wilson polynomials. Other subfamilies obtained by parameter restriction are not in the scheme, and do not even have a name. What makes the included subfamilies so particular?

(3) Is there a suitable reparametrization of the top level polynomials in the schemes such that all families lower in the scheme can be obtained by specialization of parameters? Many arrows in the schemes correspond to taking a limit to 0 or $\infty$ of rescaled polynomials, involving parameter dependent dilation or translation of the independent variable. It would be nice to simplify this and make it more uniform. The author [**11**] made an attempt in this direction for the Askey scheme. However, there the formulas for the reparametrization were quite tedious and not very conceptual.

This paper presents, for the $q$-Askey scheme, one possible way to answer these three questions in a systematic way. Following the ideas by Vinet & Zhedanov [**23**] and Verde-Star [**19**] one can try to classify monic orthogonal polynomials $u_n$ which not only satisfy the Bochner-type property that they are eigenfunctions of a second order linear $q$-difference operator $L$, so $Lu_n = h_n u_n$ with the $h_n$ distinct, but the $u_n$ should also have an expansion

$$u_n(x)=\sum_{k=0}^{n} c_{n,k}\, v_k(x), \quad v_k(x)=(x-x_0)(x-x_1)\dots(x-x_{k-1}) \ (k\geq 1), \quad v_0(x)=1.$$

So the $v_n$ are Newton type polynomials. Now replace the requirement on $L$ to be a second order $q$-difference operator by the assumption that it acts on the basis of polynomials $v_n$ as $Lv_n = h_n v_n + g_n v_{n-1}$. Then it follows that $c_{n,k} = \prod_{j=k}^{n-1} \frac{g_{j+1}}{h_n - h_j}$. Finally replace the orthogonality assumption by the property that $xu_n(x)$ is a linear combination of $u_{n+1}(x)$, $u_n(x)$ and (with nonvanishing coefficient) $u_{n-1}(x)$.

Verde-Star [**19**], whom we will follow in this paper, makes the Ansatz that $h_k$ and $x_k$ are Laurent polynomials in $q^k$ of degree 1 and that $g_k$ is a Laurent polynomial in $q^k$ of degree 2. The corresponding $3+3+5=11$ Laurent coefficients then satisfy one trivial relation because $g_0=0$ and two further relations implied by the three-term recurrence for the $u_n$. All families in the $q$-Askey scheme [**8**, p.414] are caught by giving the 11 Laurent coefficients, and hence the $x_k, h_k, g_k$, suitable values. The only exception is the continuous $q$-Hermite polynomial [**8**, §14.26]. It does not have an explicit expansion which fits into our framework. Apart from this case our method gives a positive answer to the question whether the $q$-Askey scheme is complete.

It turns out that almost always the distinction between two families in the scheme can be read off from their different patterns of vanishing Laurent coefficients (although this does not distinguish between a discrete family and its continuous analogue). These different patterns correspond with different types of $q$-hypergeometric representations. This answers, in a sense, the second question. Furthermore, if we draw an arrow from one family to another in the case that a suitable nonzero Laurent coefficient for the first family becomes zero for the second family, we recover all arrows in the existing $q$-Askey scheme and find a few more.

Finally, the question about the reparametrization can be answered by starting with the 11 Laurent coefficients, reduce them by a number of identifications to a four-manifold, and distinguish lower dimensional submanifolds by putting one or more suitable Laurent coefficients to zero.

The contents of this paper are as follows. In Section 2 we describe the general set-up, following Verde-Star [**19**], and we illustrate this for the case of the Askey–Wilson polynomials. In Section 3, the heart of this paper, we give the resulting scheme. Section 4 describes the manifold structure associated with the scheme. Finally Section 5 gives some further perspectives. There are two Appendices. The first one gives explicit data for families in the scheme. The second one gives some limit transitions, which are partially missing in [**8**, p.414].

Note. For definition and notation of $q$-shifted factorials and $q$-hypergeometric series we follow [**5**, §1.2]. We will only need terminating series:

$$
{}_r\phi_s\left(\begin{matrix} q^{-n}, a_2, \ldots, a_r \\ b_1, \ldots, b_s \end{matrix}; q, z\right) := \sum_{k=0}^{n} \frac{(q^{-n}; q)_k}{(q; q)_k} \frac{(a_2, \ldots, a_r; q)_k}{(b_1, \ldots, b_s; q)_k} \left((-1)^k q^{\frac{1}{2}k(k-1)}\right)^{s-r+1} z^k.
$$

Here $(b_1, \ldots, b_s; q)_k := (b_1; q)_k \ldots (b_s; q)_k$ with $(b; q)_k := (1-b)(1-qb)\ldots(1-q^{k-1}b)$ the $q$-shifted factorial.

For formulas on orthogonal polynomials in the $q$-Askey scheme we refer to [**8**, Chapter 14].

## 2. Askey–Wilson polynomials and Verde-Star's theorem

Let $u_n(x)$ be an Askey–Wilson polynomial, normalized such that it is monic in $x = z + z^{-1}$:

(2.1)

$$
u_n(x) = p_n\left(\tfrac{1}{2}x; a, b, c, d \,|\, q\right) = \frac{(ab, ac, ad; q)_n}{a^n\,(q^{n-1}abcd; q)_n}\, {}_4\phi_3\left(\begin{matrix} q^{-n}, q^{n-1}abcd, az, az^{-1} \\ ab, ac, ad \end{matrix}; q, q\right).
$$

We will write some properties of these polynomials in a conceptual form which we can next use more generally.

Formula (2.1) can be rewritten as

$$
u_n(x) = \sum_{k=0}^{n} c_{n,k}\, v_k(x), \tag{2.2}
$$

where

$$
v_k(x) = (x - x_0)(x - x_1)\ldots(x - x_{k-1}) \ \ (k \geq 1), \quad v_0(x) = 1, \tag{2.3}
$$

$$
x_k = aq^k + a^{-1}q^{-k}, \tag{2.4}
$$

and

$$c_{n,k}=\prod_{j=k}^{n-1}\frac{g_{j+1}}{h_n-h_j},\tag{2.5}$$

$$h_k=q^{-k}(1-q^k)(1-abcdq^{k-1}),\tag{2.6}$$

$$g_k=a^{-1}q^{-2k+1}(1-abq^{k-1})(1-acq^{k-1})(1-adq^{k-1})(1-q^k).\tag{2.7}$$

Note that (2.2) expands the Askey–Wilson polynomial in terms of Newton type polynomials (2.3) with nodes (2.4). The expansion coefficients (2.5) are expressed in terms of sequences $h_k$ and $g_k$ given by (2.6) and (2.7). Since we will not consider orthogonality, the only constraints to be imposed on $q,a,b,c,d\in\mathbb{C}$ are

$$q\neq 0,\quad 1\notin q^{\mathbb{Z}},\quad a\neq 0,\quad 1\notin abcdq^{\mathbb{Z}_{\geq 0}}.$$

These constraints make $x_k,h_k,g_k$ well-defined and they let $h_n\neq h_j$ for $n>0$, $0\leq j<n$.

According to [**8**, (14.1.7)] there is an explicit second order $q$-difference operator $L$ such that

$$Lu_n=h_nu_n,\quad n\geq 0.\tag{2.8}$$

By (2.2) we can also characterize $L$ by its action on the basis of polynomials $v_n$:

$$Lv_0=h_0v_0,\qquad Lv_n=h_nv_n+g_nv_{n-1},\quad n>0.\tag{2.9}$$

The $h_k$, $x_k$, $g_k$ have the form

$$\begin{aligned}h_k&=a_0+a_1q^k+a_2q^{-k},\qquad x_k=b_0+b_1q^k+b_2q^{-k},\\ g_k&=d_0+d_1q^k+d_2q^{-k}+d_3q^{2k}+d_4q^{-2k},\quad \sum_{i=0}^4 d_i=0.\end{aligned}\tag{2.10}$$

Furthermore, we see from (2.4), (2.6), (2.7) and (2.10) that

$$d_3=q^{-1}a_1b_1,\quad d_4=qa_2b_2.\tag{2.11}$$

Now consider arbitrary sequences $h_k,x_k,g_k$ $(k\geq 0)$. Assume $g_0=0$ and $h_n\neq h_j$ for $n>0$, $0\leq j<n$. Let monic polynomials $v_n$ be given by (2.4) and let monic polynomials $u_n$ of degree $n$ be expanded in terms of the $v_k$ by (2.2) for certain coefficients $c_{n,k}$. Let $L$ be a linear operator on the space of polynomials. Then any two of the three formulas (2.5), (2.8) and (2.9) implies the third formula.

The $q$-case of a recent more general result by Verde-Star [**19**, Theorem 6.1] can be formulated as follows.

THEOREM 2.1. *Let $q\neq 0$, $1\notin q^{\mathbb{Z}}$. Let $h_k,x_k,g_k$ have the form* (2.10). *Assume that $h_n\neq h_j$ for $n>0$, $0\leq j<n$, or equivalently $a_2\notin a_1q^{\mathbb{Z}_{>0}}$. Let the Newton type polynomials $v_k$ have the form* (2.3) *($v_k(x)=(x-b_0)^k$ allowed) and let the monic polynomials $u_n$ of degree $n$ be defined by* (2.2) *and* (2.5). *Then the polynomials $u_n$ satisfy a three-term recurrence relation*

$$xu_n(x)=u_{n+1}(x)+A_nu_n(x)+B_nu_{n-1}(x),\quad n\geq 1.\tag{2.12}$$

*iff* (2.11) *holds.*

By [**19**, (5.5), (5.6)] the coefficients $A_n$ and $B_n$ in (2.12) are given by

$$A_n = x_n + \frac{g_{n+1}}{h_n - h_{n+1}} - \frac{g_n}{h_{n-1} - h_n}, \tag{2.13}$$

$$B_n = \frac{g_n}{h_{n-1} - h_n} \times \left( \frac{g_{n-1}}{h_{n-2} - h_n} - \frac{g_n}{h_{n-1} - h_n} + \frac{g_{n+1}}{h_{n-1} - h_{n+1}} + x_n - x_{n-1} \right). \tag{2.14}$$

For $n = 0$ (2.12) and (2.13) degenerate to

$$u_1(x) = x - A_0, \qquad A_0 = x_0 - \frac{g_1}{h_1 - h_0}.$$

Verde-Star claims that all families in the $q$-Askey scheme [**8**, Chapter 14], except for the continuous $q$-Hermite polynomials, can be obtained in this way. We will make this concrete in the next section.

In Theorem 2.1 it is allowed that $B_n = 0$ for all $n$. This degenerate case will certainly happen if $g_n = 0$ for all $n$. Then $A_n = x_n$ and $u_n = v_n$, clearly not belonging to a family of orthogonal polynomials. We will not include this case in our classification.

It is also possible that the $B_n$ are zero because the second factor on the right-hand side of (2.14) is zero. This case will be included in our classification.

Finally we may have that $g_n$ vanishes only for some values of $n$. Let then $n = N + 1$ the lowest value of $n$ for which $g_n = 0$. Then $c_{n,k} = 0$ if $N < k < n$. If we only consider $u_n$ for $n \leq N$ we obtain one of the finite systems of orthogonal polynomials in the $q$-Askey scheme.

Note that a classification according to Theorem 2.1 does not use the usual Bochner type criterium [**20**] of finding all families of orthogonal polynomials which are eigenfunctions of a suitable second order $q$-difference operator. Instead it classifies families of polynomials satisfying a three-term recurrence relation which have an expansion of specific type in terms of Newton type polynomials of a specific type. Then there is also an eigenvalue equation (2.9), involving an operator $L$ defined by (2.8). For each family it can be shown in an ad hoc way that this operator $L$ can be written as the second order $q$-difference operator given in [**8**, Chapter 14]. But without having done this computation one already sees that the obtained numbers $h_n$ are the eigenvalues of $L$ given in [**8**].

REMARK 2.2. As sketched in [**23**, §3.3], if we assume that the $u_n$ satisfy a three-term recurrence relation (2.12) and if we assume (2.10) only for the $h_k$, then (2.10) for the $x_k$ and $g_k$ will follow.

REMARK 2.3. The polynomials $u_n$ can be renormalized (under assumptions on the $x_k$) as polynomials $U_n$ given by (3.6). In this form, and with $g_{N+1} = 0$ for some $N$, these polynomials also occur in some of Terwilliger's papers, in particular, [**17**, (10)], [**16**, (85)]. Our $x_i, h_i, g_i$ correspond to Terwilliger's $\theta_i, \theta_i^*, \varphi_i$, respectively. By [**16**, Defs. 7.1, 8.1, 14.1, Theor. 23.2] any Leonard system gives rise to a three-term recurrence relation, of which renormalized solutions have the mentioned form. See [**17**, §5] for explicit values of $\theta_i, \theta_i^*, \varphi_i$.

REMARK 2.4. Geronimus raised the problem to classify orthogonal polynomials $u_n$ and Newton polynomials $v_k$ which satisfy (2.2) with $c_{n,k} = a_{n-k}b_k$. For an exposition and follow-up of this problem see [**1**, §§3, 4]

## 3. The $q$-Verde-Star scheme

Let us again give the data leading to polynomials $u_n$ in the $q$-Askey scheme according to Theorem 2.1:

$$u_n(x) = \sum_{k=0}^{n} c_{n,k}\, v_k(x), \quad v_k(x) = \prod_{j=0}^{k-1}(x - x_j), \quad c_{n,k} = \prod_{j=k}^{n-1} \frac{g_{j+1}}{h_n - h_j}, \tag{3.1}$$

$$\begin{aligned} &x_k = b_2 q^{-k} + b_0 + b_1 q^k, \quad h_k = a_2 q^{-k} + a_0 + a_1 q^k, \\ &g_k = d_4 q^{-2k} + d_2 q^{-k} + d_0 + d_1 q^k + d_3 q^{2k}, \end{aligned} \tag{3.2}$$

$$\sum_{i=0}^{4} d_i = 0, \quad d_3 = q^{-1} a_1 b_1, \quad d_4 = q a_2 b_2, \tag{3.3}$$

$$a_2 \ne a_1 q^{\mathbb{Z}_{>0}}, \quad \text{in particular,} \quad a_1 \text{ or } a_2 \ne 0, \qquad d_i \ne 0 \text{ for some } i. \tag{3.4}$$

So everything is determined by the 11 parameters $a_0, a_1, a_2, b_0, b_1, b_2, d_0, d_1, d_2, d_3, d_4$. There are several invariances:

(1) If $a_0 \to a_0 + \tau$ then $h_k \to h_k + \tau$.
(2) If $a_0, a_1, a_2$ and $d_0, d_1, d_2, d_3, d_4$ are multiplied by $\mu \ne 0$ then $h_k, g_k$ are multiplied by $\mu$.
(3) If $b_0 \to b_0 + \sigma$ and $x \to x + \sigma$ then $x_k \to x_k + \sigma$ and $u_n(x) \to u_n(x + \sigma)$.
(4) If $b_0, b_1, b_2$ and $d_0, d_1, d_2, d_3, d_4$ are multiplied by $\rho \ne 0$ then $x_k, g_k$ are multiplied by $\rho$, $v_k(x) \to \rho^k v_k(\rho^{-1}x)$ and $u_n(x) \to \rho^n u_n(\rho^{-1}x)$.

In each case, what is not mentioned remains unchanged. In items 1 and 2 there is no effect on the $u_n(x)$. Also the translations and dilations of the independent variable of $u_n$ by items 3 and 4 are not considered as essential changes of a family of orthogonal polynomials. So the above four items give rise to four degrees of freedom in the 11 parameters. Together with the three constraints (3.3) on the parameters, there are four essential parameters left, in agreement with the number of four parameters of the Askey–Wilson polynomials.

There are two further remarkable operations which can be performed on the 11 parameters:

$q \leftrightarrow q^{-1}$ *exchange*: $a_1 \leftrightarrow a_2$, $b_1 \leftrightarrow b_2$, $d_1 \leftrightarrow d_2$, $d_3 \leftrightarrow d_4$.

$x \leftrightarrow h$ *duality*: $a_0 \leftrightarrow b_0$, $a_1 \leftrightarrow b_1$, $a_2 \leftrightarrow b_2$; assume also that $b_2 \ne b_1 q^{\mathbb{Z}_{>0}}$, in particular, $b_1$ or $b_2 \ne 0$. This relates $u_n$ given by (3.1) to its *dual* $\widetilde{u}_n$ given by

$$\widetilde{u}_n(x) = \sum_{k=0}^{n} \widetilde{c}_{n,k}\, \widetilde{v}_k(x), \quad \widetilde{v}_k(x) = \prod_{j=0}^{k-1}(x - h_j), \quad \widetilde{c}_{n,k} = \prod_{j=k}^{n-1} \frac{g_{j+1}}{x_n - x_j}. \tag{3.5}$$

If we put

$$U_n(x) = \prod_{j=0}^{n-1} \frac{h_n - h_j}{g_{j+1}} \times u_n(x) = \sum_{k=0}^{n} \frac{\prod_{j=0}^{k-1}(h_n - h_j) \times \prod_{j=0}^{k-1}(x - x_j)}{\prod_{j=1}^{k} g_j}, \tag{3.6}$$

$$\widetilde{U}_m(x) = \prod_{j=0}^{m-1} \frac{x_m - x_j}{g_{j+1}} \times \widetilde{u}_m(x) = \sum_{k=0}^{m} \frac{\prod_{j=0}^{k-1}(x_m - x_j) \times \prod_{j=0}^{k-1}(x - h_j)}{\prod_{j=1}^{k} g_j} \tag{3.7}$$

then (see also [**23**, (1.9)])

$$U_n(x_m) = \widetilde{U}_m(h_n). \tag{3.8}$$

For classification purposes we arrange the 11 parameters in an array

$$\begin{array}{ccccc} & b_2 & b_0 & b_1 & \\ d_4 & d_2 & d_0 & d_1 & d_3 \\ & a_2 & a_0 & a_1 & \end{array} \tag{3.9}$$

It will turn out that only the vanishing of some of these parameters determines the families in the scheme. Let • denote any parameter value (which may be zero) and o a zero parameter value. So we can represent Askey–Wilson by (3.9) with all entries given by •. The distribution of • and o in an array (3.9) has to satisfy the following rules:

(1) If $b_1$ or $a_1$ is o then $d_3$ is o; if $b_2$ or $a_2$ is o then $d_4$ is o
(because of the second and third formula in (3.3)).
(2) $b_0$ and $a_0$ are always •
(because $h_k \to h_k + \tau$ and $x_k \to x_k + \sigma$ are allowed).
(3) In the second row there are no o ones between two • ones
(because it will turn out that only the most left and the most right nonzero $d_i$ determines the family).
(4) In the third row there are at least two • ones
(because of rule 2 and the first part of (3.4)).
(5) In the second row there are at least two • ones
(because of the first part of (3.3) and the second part of (3.4)).
(6) Flipping a • into a o causes an arrow between the symbols.
(This determines a limit case where a parameter tends to zero. If $b_2$ or $a_2$ becomes white, then also $d_4$. If $b_1$ or $a_1$ becomes white, then also $d_3$.)
(7) Reflection with respect to the central column in the black-white array of form (3.9) means $q \leftrightarrow q^{-1}$ exchange.
(8) Reflection with respect to the middle row means $x \leftrightarrow h$ duality (only possible if there are at least two • ones in the first row).

In Figure 1 we give half of the scheme according to these rules. It has to be complemented with the scheme obtained from the present one by reflecting each diagram with respect to its middle column and preserving all arrows.

Let us number the rows in the scheme from top to bottom by 1 to 5. In each row list the successive diagrams from left to right by a, b, .... Adding a prime to this notation means a $q \to q^{-1}$ exchange for the corresponding diagram. For instance **3c'** is diagram **3c** reflected with respect to its central column. Note that **1a** = **1a'** and **3e** = **3e'**. For all other diagrams in Figure 1 the primed counterpart is different and not in Figure 1.

Note also the following $x \leftrightarrow h$ dualities:
**1a**, **3c**, **4b**, **5a** are self-dual;
**2a** $\leftrightarrow$ **2b**, **3a** $\leftrightarrow$ **3d**, **3b** $\leftrightarrow$ **3b'**, **4a** $\leftrightarrow$ **4f**, **4c** $\leftrightarrow$ **4d'** are dual pairs.

The diagrams in Figure 1 correspond with families in the $q$-Askey scheme as given in the list below (numbers given with these families apply to the corresponding section numbers in [**8**, Chapter 14]).

**1a.** Askey–Wilson (1), $q$-Racah (2)

**2a.** continuous dual $q$-Hahn (3), dual $q$-Hahn (7)

**2b.** big $q$-Jacobi (5), $q$-Hahn (6)

**3a.** Al-Salam–Chihara (8), dual $q$-Krawtchouk (17)

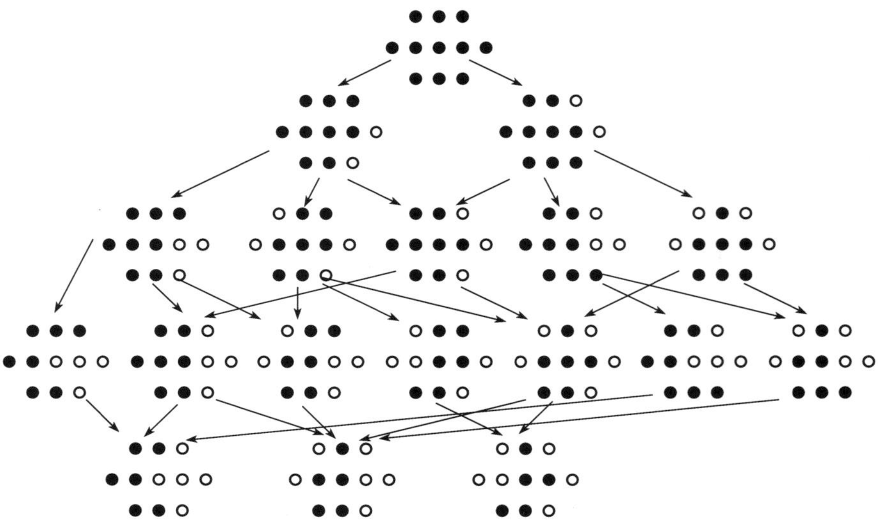

FIGURE 1. The $q$-Verde-Star scheme

**3b.** big $q$-Laguerre (11), $q^{-1}$-Meixner (13), affine $q$-Krawtchouk (16), quantum $q^{-1}$-Krawtchouk (14) with $v_k(x) = \prod_{j=0}^{k-1}(x - b_1 q^j)$.

**3c.** big $q$-Laguerre (11), $q^{-1}$-Meixner (13), affine $q$-Krawtchouk (16), quantum $q^{-1}$-Krawtchouk (14) with $v_k(x) = \prod_{j=0}^{k-1}(x - b_2 q^{-j})$.

**3d.** little $q$-Jacobi (12), $q$-Krawtchouk (15) with $v_k(x) = \prod_{j=0}^{k-1}(x - b_2 q^{-j})$.

**3e.** little $q$-Jacobi (12), $q$-Krawtchouk (15) with $v_k(x) = x^k$.

**4a.** continuous big $q$-Hermite (18)

**4b.** $u_n(x) = x^n(bx^{-1};q)_n, \quad v_k(x) = (-1)^k q^{\frac{1}{2}k(k-1)}(x;q)_k$.

**4c.** Al-Salam–Carlitz I (24), $q^{-1}$-Al-Salam–Carlitz II (25)

**4d.** little $q$-Laguerre (20), $q^{-1}$-Laguerre (21), $q^{-1}$-Charlier (23), $v_k(x) = x^k(x^{-1};q)_k$.

**4e.** little $q$-Laguerre (20), $q^{-1}$-Laguerre (21), $q^{-1}$-Charlier (23), $\quad v_k(x) = x^k$.

**4f.** $q^{-1}$-Bessel (22), $\quad v_k(x) = (-1)^k q^{\frac{1}{2}k(k-1)}(x;q)_k$.

**4g.** $q$-Bessel (22), $\quad v_k(x) = x^k$.

**5a.** $u_n(x) = x^n, \quad v_k(x) = (-1)^k q^{\frac{1}{2}k(k-1)}(x;q)_k$.

**5b.** $u_n(x) = x^n(x^{-1};q)_n, \quad v_k(x) = x^k$.

**5c.** $q^{-1}$-Stieltjes–Wigert (27)

Figure 1 should be complemented with a similar scheme, where each diagram is replaced by its primed counterpart and the arrows are preserved. There are a few arrows from a diagram in the one scheme to a diagram in the other scheme:

**2b → 3b'**: 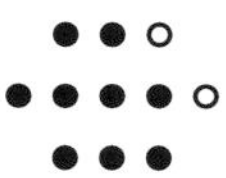 → 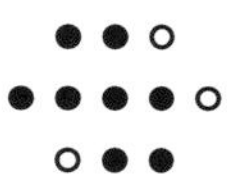 **2b' → 3b**: 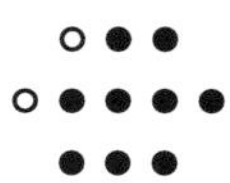 → 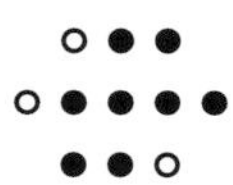

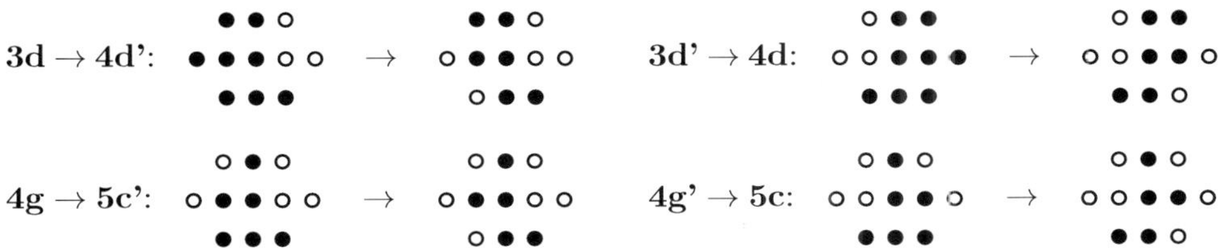


Remarks.

(1) For each diagram in Figure 1 the explicit expressions of $x_k, h_k, g_k$ for one (continuous) family belonging to this diagram or its primed counterpart are given in Appendix A.

(2) Since, the first row of a diagram determines the kind of the Newton type polynomials $v_k$ involved, it can happen that one family occurs twice in the scheme because it can be expanded in two different kinds of $v_k$. See **3b**, **3c** (big $q$-Laguerre, affine $q$-Krawtchouk), **3d**, **3e** (little $q$-Jacobi, $q$-Krawtchouk), **4d**, **4e** (little $q$-Laguerre, $q^{-1}$-Laguerre, $q^{-1}$-Charlier), **4f'**, **4g** ($q$-Bessel). A family may also have expansions in different $v_k$, which are still of the same kind. This will not be recognized by our scheme. For instance, with Askey–Wilson polynomials we may exchange the parameter $a$ with one of the three other parameters $b, c, d$.

(3) The cases **4b**, **5a**, **5b** are degenerate in the sense that $u_n$ turns out to be a Newton type polynomial itself which is expanded in terms of different Newton type polynomials $v_k$.

(4) Most diagrams in the scheme correspond to both a continuous and a discrete family of orthogonal polynomials.

(5) Figure 1, when compared with the $q$-Askey scheme on [**8**, p.414], misses some families. The reason is that they are special cases of larger families, obtained by restriction of parameter values, but such that our black-white diagrams do not recognize these restrictions. This concerns (numbers mean again section numbers in [**8**, Chapter 14]) continuous $q$-Jacobi (10) as subfamily of Askey–Wilson, continuous $q$-Laguerre (19) as subfamily of Al-Salam–Chihara, and discrete $q$-Hermite I, II (28, 29) as subfamilies of Al-Salam–Carlitz I, II. Similarly, continuous $q$-Hahn (4) (Askey–Wilson $p_n(x; a, b, c, d \,|\, q)$ with $a, c$ and $b, d$ pairs of complex conjugates such that $\arg a = \arg c$) and $q$-Meixner–Pollaczek (9) (Al-Salam–Chihara $Q_n(x; a, b \,|\, q)$ with $a, b$ as a pair of complex conjugates) are not in Figure 1. (The notation in [**8**, §§14.4, 14.9] for these two classes of polynomials is confusing.)

(6) The continuous $q$-Hermite polynomials are also missing in our scheme because their expansion falls outside the scope of Theorem 2.1.

(7) If, in the $q$-Askey scheme on [**8**, p.414], the families mentioned in the previous two items are omitted, together with the arrows to and from those families, then all further arrows in that scheme are also present in our scheme. However, we have some more arrows whcih are missing in [**8**, p.414]. These are (see Appendix B):

**2a** → **3b** and **2a** → **3c**: continuous dual $q$-Hahn → big $q$-Laguerre,

**3a** → **4c**: Al-Salam–Chihara → Al-Salam–Carlitz I

## 4. The $q$-Verde-Star scheme as a four-manifold

Fix $q \neq 0$ such that $1 \notin q^{\mathbb{Z}}$. We will sketch how the $q$-Verde-Star scheme can be made into a (complex) four-manifold having specific submanifolds of lower dimension 3, 2, 1 and 0. We will ignore the case of a finite system, where $g_{N+1} = 0$ for some $N$.

Let us start with a six-manifold with seven coordinates $a_1, a_2, b_1, b_2, d_0, d_1, d_2$ such that $d_0 + d_1 + d_2 + q^{-1}a_1b_1 + qa_2b_2 = 0$. Also assume that $a_2 \notin a_1q^{\mathbb{Z}_{>0}}$ and $(d_0, d_1, d_2, a_1b_1, a_2b_2) \neq (0, 0, 0, 0, 0)$. Now we make two one-parameter identifications. Let nothing change if $a_1$, $a_2$, $d_0$, $d_1$, $d_2$ are multiplied by the same nonzero constant or if $b_1$, $b_2$, $d_0$, $d_1$, $d_2$ are multiplied by the same nonzero constant. Then there are several possibilities to put two out of the five coordinates $a_1$, $a_2$, $b_1$, $b_2$, $d_0$ equal to 1. The three untouched coordinates among these five, together with $d_1$ or $d_2$ will then provide local coordinates for our four-manifold. In the generic case all choices are allowed. However, we can regard the six families in the bottom row of Figure 1 and its $q \leftrightarrow q^{-1}$ complement as points in our four-manifold. In a neighbourhood of each of these points we can make a special choice of four coordinates such that the point has all coordinates zero and such that any family in the scheme from which that point is reachable via arrows is a submanifold obtained by putting some of the coordinates equal to zero. Below we give the details for the three families in the bottom row of Figure 1.

$a_2 = b_2 = 1$

| | $a_1$ | $b_1$ | $d_0$ | $d_1$ |
|---|---|---|---|---|
| **1a** | ● | ● | ● | ● |
| **2a** | ○ | ● | ● | ● |
| **2b** | ● | ○ | ● | ● |
| **3a** | ○ | ● | ● | ○ |
| **3c** | ○ | ○ | ● | ● |
| **3d** | ○ | ● | ○ | ● |
| **4a** | ○ | ● | ○ | ○ |
| **4b** | ○ | ○ | ● | ○ |
| **4f** | ● | ○ | ○ | ○ |
| **5a** | ○ | ○ | ○ | ○ |

$a_2 = d_0 = 1$

| | $a_1$ | $b_1$ | $b_2$ | $d_1$ |
|---|---|---|---|---|
| **1a** | ● | ● | ● | ● |
| **2a** | ○ | ● | ● | ● |
| **2b** | ● | ○ | ● | ● |
| **2b'** | ● | ● | ○ | ● |
| **3a** | ○ | ● | ● | ○ |
| **3b** | ○ | ● | ○ | ● |
| **3c** | ○ | ○ | ● | ● |
| **3e** | ● | ○ | ○ | ● |
| **4b** | ○ | ○ | ● | ○ |
| **4c** | ○ | ● | ○ | ○ |
| **4e** | ○ | ○ | ○ | ● |
| **4g** | ● | ○ | ○ | ○ |
| **5b** | ○ | ○ | ○ | ○ |

$a_2 = d_0 = 1$

| | $a_1$ | $b_1$ | $b_2$ | $d_2$ |
|---|---|---|---|---|
| **1a** | ● | ● | ● | ● |
| **2a** | ○ | ● | ● | ● |
| **2b** | ● | ○ | ● | ● |
| **2b'** | ● | ● | ○ | ● |
| **3b** | ○ | ● | ○ | ● |
| **3c** | ○ | ○ | ● | ● |
| **3e** | ● | ○ | ○ | ● |
| **3d'** | ● | ○ | ● | ○ |
| **4d** | ○ | ● | ○ | ○ |
| **4e** | ○ | ○ | ○ | ● |
| **4g'** | ● | ○ | ○ | ○ |
| **5e** | ○ | ○ | ○ | ○ |

## 5. Further perspectives

In a next paper the author will express the coefficients in the relations defining the Zhedanov algebra associated with a family in the $q$-Askey scheme (see [**6**, (3.2)] with $R = 1 - \frac{1}{2}(q + q^{-1})$) in terms of the 11 parameters $a_0, a_1, a_2, b_0, b_1, b_2, d_0, d_1, d_2, d_3, d_4$. It will turn out that vanishing properties of these coefficients are also a way to distinguish between the families, although the resulting scheme is slightly different from the scheme in Figure 1.

Verde-Star [**19**] introduces polynomials $u_n$ and $v_k$ associated with sequences $x_k$, $h_k$, $g_k$ as in our §2, but only assuming that $x_k$ and $h_k$ are solutions of a certain four-term difference equation and $g_k$ is a solution of a certain six-term difference equation. As special cases he has the $q$-case, where $x_k$, $h_k$, $g_k$ have the form (2.10), the $q = 1$ case, and the $q = -1$ case. Earlier, in a somewhat different approach, these three cases were examined by Vinet & Zhedanov [**23**]. The author is also planning to write a paper where the $q = 1$ case will be treated systematically and in full detail, just as the $q$-case is treated in the present paper. We will also deal there with the corresponding Zhedanov algebra (see [**6**, (3.2)] with $R = 1$). There will also be need of a systematic and detailed treatment of the $q = -1$ case. Much material about this is already available in papers by Vinet & Zhedanov and coauthors, see for instance [**21**], [**22**].

Since the labeling of orthogonal polynomials in the ($q$-)Askey scheme is by the sequences $x_k$, $h_k$, $g_k$ or by the parameters occurring in their expansions is so clean, these data may be helpful for recognizing polynomials in these schemes from the coefficients in the three-term recurrence relation, assuming that it would be possible to obtain these data from these coefficients. See Tcheutia [**15**] for recent work on this recognition problem by different methods.

Finally, an approach as in the present paper may be tried in other situations where (part of) a $q$-Askey scheme occurs, see the examples mentioned in the second paragraph of the Introduction.

## Appendix A. Explicit data for the families in Figure 1

For each diagram in Figure 1 we give the data of one (continuous) family belonging to that diagram or its primed counterpart. Bold numbers like **1a** follow the convention explained in connection with Figure 1. Numbers in brackets apply to the corresponding section numbers in [**8**, Chapter 14].

**1a.** Askey–Wilson (1): $u_n(x) = k_n^{-1} p_n(\frac{1}{2}x; a, b, c, d \,|\, q)$,
$x_k = aq^k + a^{-1}q^{-k}$, $h_k = q^{-k}(1 - q^k)(1 - abcdq^{k-1})$,
$g_k = q^{1-2k}a^{-1}(1 - abq^{k-1})(1 - acq^{k-1})(1 - adq^{k-1})(1 - q^k)$,
$k_n = (q^{n-1}abcd; q)_n$.

**2a.** continuous dual $q$-Hahn (3): $u_n(x) = p_n(\frac{1}{2}x; a, b, c \,|\, q)$,
$x_k = aq^k + a^{-1}q^{-k}$, $h_k = q^{-k} - 1$,
$g_k = q^{-2k+1}a^{-1}(1 - abq^{k-1})(1 - acq^{k-1})(1 - q^k)$.

**2b.** big $q$-Jacobi (5): $u_n(x) = k_n^{-1} P_n(x; a, b, c; q)$,
$x_k = q^{-k}$, $h_k = (1 - q^{-k})(-1 + q^{k+1}ab)$,
$g_k = q^{1-2k}(1 - aq^k)(1 - cq^k)(1 - q^k)$, $k_n = \frac{(q^{n+1}ab;q)_n}{(qa;q)_n(qc;q)_n}$.

**3a.** Al-Salam–Chihara (8): $u_n(x) = Q_n(\frac{1}{2}x; a, b \,|\, q)$,
$x_k = aq^k + a^{-1}q^{-k}$, $h_k = q^{-k} - 1$, $g_k = q^{-2k+1}a^{-1}(1 - abq^{k-1})(1 - q^k)$.

**3b.** big $q$-Laguerre (11): $u_n(x) = k_n^{-1} P_n(x; a, b; q)$, $v_k(x) = x^k(qax^{-1}; q)_k$,
$x_k = aq^{k+1}$, $h_k = q^{-k} - 1$, $g_k = -q^{1-k}b(1 - aq^k)(1 - q^k)$, $k_n = \frac{1}{(qa;q)_n(qb;q)_n}$.

**3c.** idem, $v_k(x) = (-1)^k q^{-\frac{1}{2}k(k-1)}(x; q)_k$,
$x_k = q^{-k}$, $h_k = q^{-k} - 1$, $g_k = q^{1-2k}(1 - aq^k)(1 - bq^k)(1 - q^k)$.

**3d.** little $q$-Jacobi (12): $u_n(x) = k_n^{-1} p_n(x; a, b; q)$, $v_k(x) = (-b)^{-k} q^{-\frac{1}{2}k(k+1)}(qbx; q)_k$,
$x_k = q^{-k-1}b^{-1}$, $h_k = (1 - q^{-k})(-1 + q^{k+1}ab)$, $g_k = (1 - q^{-k})(1 - b^{-1}q^{-k})$,
$k_n = (-1)^n q^{-\frac{1}{2}n(n-1)} \frac{(abq^{n+1};q)_n}{(aq;q)_n}$.

**3e.** idem, $v_k(x) = x^k$,
$x_k = 0, \quad h_k = (1 - q^{-k})(-1 + q^{k+1}ab), \quad g_k = (1 - q^{-k})(1 - aq^k)$.

**4a.** continuous big $q$-Hermite (18): $u_n(x) = H_n(\frac{1}{2}x; a \,|\, q)$,
$x_k = aq^k + a^{-1}q^{-k}, \quad h_k = q^{-k} - 1, \quad g_k = q^{1-2k}a^{-1}(1 - q^k)$.

**4b.** $u_n(x) = x^n(bx^{-1}; q)_n\,, \quad x_k = q^{-k}, \quad h_k = q^{-k} - 1, \quad g_k = (1 - q^{-k})(b - q^{1-k})$.

**4c.** Al-Salam–Carlitz I (24): $u_n(x) = U_n^{(a)}(x; q)$,
$x_k = q^k, \quad h_k = q^{-k} - 1, \quad g_k = a(1 - q^{-k})$.

**4d.** little $q$-Laguerre (20): $u_n(x) = k_n^{-1}p_n(x; a; q), \quad v_k(x) = x^k(x^{-1}; q)_k$,
$x_k = q^k, \quad h_k = 1 - q^{-k}, \quad g_k = a(q^k - 1), \quad k_n = \frac{(-1)^n q^{-\frac{1}{2}n(n-1)}}{(aq;q)_n}$.
Note that $q^{-1}$-Laguerre and little $q$-Laguerre can be essentially identified with each other by [**8**, p.521].

**4e.** idem, $v_k(x) = x^k, \quad x_k = 0, \quad h_k = 1 - q^{-k}, \quad g_k = q^{-k}(1 - aq^k)(1 - q^k)$.

**4f'.** $q$-Bessel (22): $u_n(x) = k_n^{-1}y_n(x; a; q) \quad v_k(x) = x^k(x^{-1}; q)_k$,
$x_k = q^k, \quad h_k = (1 - q^{-k})(1 + aq^k), \quad g_k = aq^{k-1}(q^k - 1)$,
$k_n = (-1)^n q^{-\frac{1}{2}n(n-1)}(-aq^n; q)_n\,.$

**4g.** idem, $v_k(x) = x^k, \quad x_k = 0, \quad h_k = (1 - q^{-k})(1 + aq^k), \quad g_k = q^{-k} - 1$.

**5a.** $u_n(x) = x^n, \quad x_k = q^{-k}, \quad h_k = q^{-k} - 1, \quad g_k = q^{1-2k}(1 - q^k)$.

**5b.** $u_n(x) = k_n^{-1}\,{}_1\phi_0(q^{-n}; ; q, qx) = x^n(x^{-1}; q)_n\,,$
$x_k = 0, \quad h_k = q^{-k} - 1, \quad g_k = 1 - q^{-k}, \quad k_n = (-1)^n q^{-\frac{1}{2}n(n-1)}$.

**5c'.** Stieltjes–Wigert (27): $u_n(x) = k_n^{-1}S_n(x; q)$,
$x_k = 0, \quad h_k = q^k - 1, \quad g_k = q^{-k} - 1, \quad k_n = \frac{(-1)^n q^{n^2}}{(q;q)_n}$.

## Appendix B. Some explicit limit transitions

**2a** $\to$ **3b**: continuous dual $q$-Hahn $\to$ big $q$-Laguerre (missing in [**8**, §§14.3, 14.11]).

$$\lim_{a\to 0} a^n p_n\big(\tfrac{1}{2}a^{-1}x; a, a^{-1}bq, a^{-1}cq \,|\, q\big) = (bq, cq; q)_n\, P_n(x; b, c; q), \tag{B.1}$$

where continuous dual $q$-Hahn ([**8**, (14.3.1)] together with symmetry in $a, b, c$) and monic big $q$-Laguerre [**8**, (14.11.1)] are respectively represented as

$$p_n\big(\tfrac{1}{2}x; a, b, c \,|\, q\big) = \frac{(ab, bc; q)_n}{b^n}\, {}_3\phi_2\left(\begin{matrix} q^{-n}, bz, bz^{-1} \\ ab, bc \end{matrix}; q, q\right), \tag{B.2}$$
$$x = z + z^{-1}, \quad ab, ac, bc < 1,$$

$$(bq, cq; q)_n\, P_n(x; b, c; q) = (-c)^n q^{\frac{1}{2}n(n+1)}(bq; q)_n \tag{B.3}$$
$$\times\ {}_2\phi_1\left(\begin{matrix} q^{-n}, bqx^{-1} \\ bq \end{matrix}; q, c^{-1}x\right), \quad 0 < bq < 1,\ c < 0.$$

Here and elsewhere in this Appendix, when we mention conditions on the parameters, these are such that the coefficient $B_n$ in (2.12) is positive, also assuming $A_n$ real. This assures that the polynomials are orthogonal. The conditions above, where the parameters are assumed real, can be obtained from [**8**, (14.3.5), (14.11.4)]. By these conditions the passage to the limit in (B.1) can be made while keeping the polynomials orthogonal.

**2a** $\to$ **3c**: The same limit (B.1) also holds with other $q$-hypergeometric representations [**8**, (14.3.1), (14.11.1)]:

$$\text{(B.4)}\quad p_n\big(\tfrac12 x;a,b,c\,|\,q\big)=\frac{(ab,ac;q)_n}{a^n}\,{}_3\phi_2\left(\begin{matrix}q^{-n},az,az^{-1}\\ ab,ac\end{matrix};q,q\right),\qquad x=z+z^{-1},$$

$$\text{(B.5)}\quad (bq,cq;q)_n\,P_n(x;b,c;q)=(bq,cq;q)_n\,{}_3\phi_2\left(\begin{matrix}q^{-n},0,x\\ bq,cq\end{matrix};q,q\right).$$

**3a** $\to$ **4c**: Al-Salam–Chihara $\to$ Al-Salam–Carlitz I (missing in [**8**, §§14.8, 14.24]).

$$\text{(B.6)}\qquad \lim_{a\to\infty}(a)^{-n}Q_n\big(\tfrac12 ax;a,ab\,|\,q\big)=U_n^{(b)}(x;q),$$

where Al-Salam–Chihara [**8**, (14.8.1)] and Al-Salam–Carlitz I [**8**, (14.24.1)] are respectively represented as

$$\text{(B.7)}\quad Q_n\big(\tfrac12 x;a,b\,|\,q\big)=\frac{(ab;q)_n}{a^n}\,{}_3\phi_2\left(\begin{matrix}q^{-n},az,az^{-1}\\ ab,0\end{matrix};q,q\right),\ x=z+z^{-1},\ ab<1,$$

$$\text{(B.8)}\quad U_n^{(b)}(x;q)=(-b)^n q^{\frac12 n(n-1)}\,{}_2\phi_1\left(\begin{matrix}q^{-n},x^{-1}\\ 0\end{matrix};q,qb^{-1}x\right),\ b<0.$$

**3a** $\to$ **4b**: Al-Salam–Chihara $\to x^n(bx^{-1};q)_n$.

$$\text{(B.9)}\quad \lim_{a\to 0}a^nQ\big((2a)^{-1}x;a,a^{-1}b\,|\,q\big)=(b;q)_n\,{}_2\phi_1\left(\begin{matrix}q^{-n},x\\ b\end{matrix};q,q\right)=x^n(bx^{-1};q)_n,$$

where Al-Salm–Chihara is given by (B.7) and the second equality in (B.9) is [**5**, (II.6)].

**2b** $\to$ **3d**: big $q$-Jacobi $\to$ little $q$-Jacobi [**8**, p.442, Remarks].

$$\text{(B.10)}\quad \lim_{d\to 0}(qa)^{-n}\frac{(qa;q)_n(-qad;q)_n}{(q^{n+1}ab;q)_n}P_n(x;a,b,1,d;q)=(-1)^nq^{\frac12 n(n-1)}\frac{(qb;q)_n}{(q^{n+1}ab;q)_n}p_n(x;b,a;q),$$

where big and little $q$-Jacobi are respectively represented by [**8**, p.442, (14.5.1) and Remarks]

$$\text{(B.11)}\quad \begin{aligned}&P_n(x;a,b,c,d;q)\\ &=P_n(ac^{-1}qx;a,b,-ac^{-1}d;q)={}_3\phi_2\left(\begin{matrix}q^{-n},q^{n+1}ab,qac^{-1}x\\ qa,-qac^{-1}d\end{matrix};q,q\right),\\ &\qquad c,d>0,\ -q^{-1}cd^{-1}<a<q^{-1},\ -q^{-1}c^{-1}d<b<q^{-1},\end{aligned}$$

$$\text{(B.12)}\quad p_n(x;a,b;q)=(-qb)^{-n}q^{-\frac12 n(n-1)}\frac{(qb;q)_n}{(qa;q)_n}\,{}_3\phi_2\left(\begin{matrix}q^{-n},q^{n+1}ab,qbx\\ qb,0\end{matrix};q,q\right),\quad 0<a<q^{-1},\ b<q^{-1}.$$

**2b** → **3e**: The same limit (B.10) also holds with other $q$-hypergeometric representations [**10**, (2.39), (2.37)], [**8**, (14.12.1)]:

$$P_n(x;a,b,c,d;q) = \left(-\frac{ad}{bc}\right)^n \frac{(qb;q)_n(-qbcd^{-1};q)_n}{(qa;q)_n(-qac^{-1}d;q)_n}\, {}_3\phi_2\left(\begin{matrix} q^{-n}, q^{n+1}ab, -qbd^{-1}x \\ qb, -qbcd^{-1} \end{matrix}; q, q\right), \tag{B.13}$$

$$p_n(x;a,b;q) = {}_2\phi_1\left(\begin{matrix} q^{-n}, q^{n+1}ab \\ qa \end{matrix}; q, qx\right). \tag{B.14}$$

**3e** → **4g**: little $q$-Jacobi → $q$-Bessel [**8**, (14.12.14)].

$$\lim_{b\to-\infty} p_n(x;-q^{-1}ab^{-1},b;q) = y_n(x;a;q), \tag{B.15}$$

where the little $q$-Jacobi polynomial is given by (B.14) and the $q$-Bessel function by [**8**, (14.22.1)]

$$y_n(x;a;q) = {}_2\phi_1\left(\begin{matrix} q^{-n}, -aq^n \\ 0 \end{matrix}; q, qx\right), \qquad a > 0. \tag{B.16}$$

**3d'** → **4f'**: The same limit (B.15) also holds with other $q$-hypergeometric representations for little $q$-Jacobi and $q$-Bessel:

$$p_n(x;a,b;q) = (-1)^n q^{\frac12 n(n+1)} a^n \frac{(bq;q)_n}{(aq;q)_n}\, {}_3\phi_1\left(\begin{matrix} q^{-n}, abq^{n+1}, x^{-1} \\ qb \end{matrix}; q, a^{-1}x\right), \tag{B.17}$$

$$y_n(x;a;q) = (-1)^n q^{n^2} a^n\, {}_3\phi_0\left(\begin{matrix} q^{-n}, -aq^n, x^{-1} \\ - \end{matrix}; q, -a^{-1}x\right). \tag{B.18}$$

Formula (B.17) folllows from (B.14) by [**5**, (III.8)] and formula (B.18) follows from (B.16) by taking the limit $c \to 0$ in [**5**, (III.8)].

**4a** → **5a**: continuous big $q$-Hermite → $x^n$.

$$\lim_{a\to 0} a^n H_n\big((2a)^{-1}x;a\,|\,q\big) = {}_2\phi_1\left(\begin{matrix} q^{-n}, x \\ 0 \end{matrix}; q, q\right) = x^n, \tag{B.19}$$

where continuous big $q$-Hermite is given by [**8**, (14.18.1)]

$$H_n\big(\tfrac12 x;a\,|\,q\big) = a^{-n}\, {}_3\phi_2\left(\begin{matrix} q^{-n}, az, az^{-1} \\ 0, 0 \end{matrix}; q, q\right), \qquad x = z + z^{-1}, \tag{B.20}$$

and the second equality in (B.19) is [**5**, (II.6)].

**4e** → **5b**: little $q$-Laguerre → $x^n(x^{-1};q)_n$.

$$\lim_{a\to 0}(-1)^n q^{\frac12 n(n-1)}(aq;q)_n\, p_n(x;a;q) = (-1)^n q^{\frac12 n(n-1)}\, {}_1\phi_0\left(\begin{matrix} q^{-n} \\ - \end{matrix}; q, qx\right) = x^n(x^{-1};q)_n. \tag{B.21}$$

Here little $q$-Laguerre is given by [**8**, (14.20.1)]

$$p_n(x;a;q) = {}_2\phi_1\left(\begin{matrix} q^{-n}, 0 \\ qa \end{matrix}; q, qx\right), \tag{B.22}$$

and the second equality in (B.22) follows from [**5**, (II.4)].

## Acknowledgement

I thank Paul Terwilliger and the referees for helpful comments.

## References

[1] W. A. Al-Salam, *Characterization theorems for orthogonal polynomials*, Orthogonal polynomials (Columbus, OH, 1989), NATO Adv. Sci. Inst. Ser. C: Math. Phys. Sci., vol. 294, Kluwer Acad. Publ., Dordrecht, 1990, pp. 1–24, DOI 10.1007/978-94-009-0501-6_1. MR1100286

[2] Richard Askey and James Wilson, *Some basic hypergeometric orthogonal polynomials that generalize Jacobi polynomials*, Mem. Amer. Math. Soc. **54** (1985), no. 319, iv+55, DOI 10.1090/memo/0319. MR783216

[3] Eiichi Bannai and Tatsuro Ito, *Algebraic combinatorics. I: Association schemes*, The Benjamin/Cummings Publishing Co., Inc., Menlo Park, CA, 1984. MR882540

[4] Fokko J. van de Bult and Eric M. Rains, *Limits of elliptic hypergeometric biorthogonal functions*, J. Approx. Theory **193** (2015), 128–163, DOI 10.1016/j.jat.2014.06.009. MR3324567

[5] George Gasper and Mizan Rahman, *Basic hypergeometric series*, 2nd ed., Encyclopedia of Mathematics and its Applications, vol. 96, Cambridge University Press, Cambridge, 2004. With a foreword by Richard Askey, DOI 10.1017/CBO9780511526251. MR2128719

[6] Ya. I. Granovskiĭ, I. M. Lutzenko, and A. S. Zhedanov, *Mutual integrability, quadratic algebras, and dynamical symmetry*, Ann. Physics **217** (1992), no. 1, 1–20, DOI 10.1016/0003-4916(92)90336-K. MR1173277

[7] F. Alberto Grünbaum and Luc Haine, *The q-version of a theorem of Bochner*, J. Comput. Appl. Math. **68** (1996), no. 1-2, 103–114, DOI 10.1016/0377-0427(95)00262-6. MR1418753

[8] Roelof Koekoek, Peter A. Lesky, and René F. Swarttouw, *Hypergeometric orthogonal polynomials and their q-analogues*, Springer Monographs in Mathematics, Springer-Verlag, Berlin, 2010. With a foreword by Tom H. Koornwinder, DOI 10.1007/978-3-642-05014-5. MR2656096

[9] Erik Koelink and Jasper V. Stokman, *The Askey-Wilson function transform scheme*, Special functions 2000: current perspective and future directions (Tempe, AZ), NATO Sci. Ser. II Math. Phys. Chem., vol. 30, Kluwer Acad. Publ., Dordrecht, 2001, pp. 221–241, DOI 10.1007/978-94-010-0818-1_9. MR2006290

[10] Tom H. Koornwinder, *q-Special functions, a tutorial*, `arXiv:math/9403216v2` [math.CA], 1994, modifed 2013.

[11] Tom H. Koornwinder, *The Askey scheme as a four-manifold with corners*, Ramanujan J. **20** (2009), no. 3, 409–439, DOI 10.1007/s11139-009-9208-7. MR2574781

[12] Tom H. Koornwinder and Fethi Bouzeffour, *Nonsymmetric Askey-Wilson polynomials as vector-valued polynomials*, Appl. Anal. **90** (2011), no. 3-4, 731–746, DOI 10.1080/00036811.2010.502117. MR3413409

[13] Tom H. Koornwinder and Marta Mazzocco, *Dualities in the q-Askey scheme and degenerate DAHA*, Stud. Appl. Math. **141** (2018), no. 4, 424–473, DOI 10.1111/sapm.12229. MR3879966

[14] Jasper V. Stokman, *On BC type basic hypergeometric orthogonal polynomials*, Trans. Amer. Math. Soc. **352** (2000), no. 4, 1527–1579, DOI 10.1090/S0002-9947-99-02551-9. MR1694379

[15] Daniel Duviol Tcheutia, *Recurrence equations and their classical orthogonal polynomial solutions on a quadratic or a q-quadratic lattice*, J. Difference Equ. Appl. **25** (2019), no. 7, 969–993, DOI 10.1080/10236198.2019.1627346. MR3996961

[16] Paul Terwilliger, *Two linear transformations each tridiagonal with respect to an eigenbasis of the other; the TD-D canonical form and the LB-UB canonical form*, J. Algebra **291** (2005), no. 1, 1–45, DOI 10.1016/j.jalgebra.2005.05.033. MR2158508

[17] Paul Terwilliger, *Two linear transformations each tridiagonal with respect to an eigenbasis of the other; comments on the parameter array*, Des. Codes Cryptogr. **34** (2005), no. 2-3, 307–332, DOI 10.1007/s10623-004-4862-7. MR2128338

[18] Satoshi Tsujimoto, Luc Vinet, and Alexei Zhedanov, *Dunkl shift operators and Bannai-Ito polynomials*, Adv. Math. **229** (2012), no. 4, 2123–2158, DOI 10.1016/j.aim.2011.12.020. MR2880217

[19] Luis Verde-Star, *A unified construction of all the hypergeometric and basic hypergeometric families of orthogonal polynomial sequences*, Linear Algebra Appl. **627** (2021), 242–274, DOI 10.1016/j.laa.2021.06.012. MR4282678

[20] Luc Vinet and Alexei Zhedanov, *Generalized Bochner theorem: characterization of the Askey-Wilson polynomials*, J. Comput. Appl. Math. **211** (2008), no. 1, 45–56, DOI 10.1016/j.cam.2006.11.004. MR2386827

[21] Luc Vinet and Alexei Zhedanov, *A 'missing' family of classical orthogonal polynomials*, J. Phys. A **44** (2011), no. 8, 085201, 16 pp.

[22] Luc Vinet and Alexei Zhedanov, *A limit $q = -1$ for the big $q$-Jacobi polynomials*, Trans. Amer. Math. Soc. **364** (2012), no. 10, 5491–5507, DOI 10.1090/S0002-9947-2012-05539-5. MR2931336

[23] Luc Vinet and Alexei Zhedanov, *Hypergeometric orthogonal polynomials with respect to Newtonian bases*, SIGMA **12** (2016), Paper No. 048, 14 pp.

Korteweg-de Vries Institute, University of Amsterdam, P.O. Box 94248, 1090 GE Amsterdam, The Netherlands

*Email address*: thkmath@xs4all.nl

Contemporary Mathematics
Volume **780**, 2022
https://doi.org/10.1090/conm/780/15689

# Filtered deformations of elliptic algebras

Eric M. Rains

Abstract. One of the difficulties in doing noncommutative projective geometry via explicitly presented graded algebras is that it is usually quite difficult to show flatness, as the Hilbert series is uncomputable in general. If the algebra has a regular central element, one can reduce to understanding the (hopefully more tractable) quotient. If the quotient is particularly nice, one can proceed in reverse and find *all* algebras of which it is the quotient by a regular central element (the *filtered deformations* of the quotient). We consider in detail the case that the quotient is an *elliptic algebra* (the homogeneous endomorphism ring of a vector bundle on an elliptic curve, possibly twisted by translation). We explicitly compute the family of filtered deformations in many cases and give a (conjecturally exhaustive) construction of such deformations from noncommutative del Pezzo surfaces. In the process, we also give a number of results on the classification of exceptional collections on del Pezzo surfaces, which are new even in the commutative case.

Contents



## 1. Introduction

The origin of the present work was an attempt to produce elliptic analogues of the algebras of [**9**], with a particular focus on understanding the associated noncommutative analogues of del Pezzo surfaces. (Note that for our purposes, a "(degenerate) del Pezzo surface" is a smooth projective surface with an *almost* ample anticanonical bundle; that is, the anticanonical morphism is allowed to contract a finite configuration of $-2$-curves.) The algebras of [**9**] were given by very simple explicit presentations (they are generated by three elements that multiply to 1 and have specified minimal polynomials), and are difficult to generalize as such, but

2020 *Mathematics Subject Classification.* 16S38;14H52,14J26,14F08.

they also have (several) filtrations such that the associated gradeds have nice *geometric* descriptions. In particular, the associated graded algebras are constructed from the multiplication-by-$q$ map on a nodal curve of genus 1, and thus *their* elliptic analogues are relatively easy to describe. One is thus left with the problem of understanding their "filtered deformations", i.e., for which filtered algebras does one obtain the given associated graded. Although the problem of classifying filtered deformations of a given graded algebra is almost certainly not computable in general, it is still relatively straightforward to do non-rigorous calculations and thus formulate conjectures. In particular, we (P. Etingof and the author) found that the (putative) moduli spaces classifying such deformations were often quite nice: not only were they frequently rational, they were in fact quite frequently weighted projective spaces or products thereof.

Although the resulting conjectures were quite striking, it was not clear how to prove them rigorously; similarly, although the resulting algebras were clearly noncommutative analogues of del Pezzo surfaces, it was unclear how to prove anything much about them. As a result, the project was shelved for quite some time. In the interim, the author developed a completely different approach to noncommutative rational surfaces [**19**–**21**], extending beyond the del Pezzo case and giving a great deal of control over the corresponding representation theory (i.e., the categories and derived categories of sheaves), making much of the original motivation moot. This is not entirely the case, however: the work of [**7**] in particular suggests that there is independent interest in having representations of such surfaces via graded algebras. The object of the present work is therefore not so much to construct noncommutative del Pezzo surfaces as filtered deformations, but rather to construct filtered deformations via noncommutative del Pezzo surfaces.

As the title indicates, the algebras we are deforming are "elliptic algebras", which are constructed as follows. The simplest version is simply the graded homogeneous coordinate ring of an elliptic curve: if $E$ is an elliptic curve and $\mathcal{L}$ is a line bundle, then there is a corresponding graded algebra

$$B_{E,\mathcal{L}} = \bigoplus_{n\in\mathbb{Z}} \Gamma(E;\mathcal{L}^n),$$

with multiplication induced by the obvious map $\mathcal{L}^m \otimes \mathcal{L}^n \cong \mathcal{L}^{m+n}$. This has a noncommutative analogue (the twisted homogeneous coordinate rings of [**2**]), in which we twist by a translation $\tau \in \operatorname{Aut}(E)$:

$$B_{E,\mathcal{L},\tau} = \bigoplus_{n\in\mathbb{Z}} \Gamma(E;\mathcal{L}_n),$$

where

$$\mathcal{L}_1 = \mathcal{L}; \quad \mathcal{L}_n = \mathcal{L} \otimes (\tau^{-1})^*\mathcal{L}_{n-1},$$

with multiplication induced by

$$\mathcal{L}_m \otimes (\tau^{-m})^*\mathcal{L}_n \cong \mathcal{L}_{m+n}.$$

A somewhat cleaner description is based on the observation that the functor

$$\Psi : M \mapsto \mathcal{L} \otimes (\tau^{-1})^* M$$

is actually an autoequivalence of coh($E$). (In fact, any autoequivalence is of this form, except that $\tau$ need not be a translation.) We may thus instead define

$$B_\Psi = \bigoplus_{n\in\mathbb{Z}} \Gamma(E; \Psi^n \mathcal{O}_E),$$

with multiplication given by

$$x_m \cdot x_n = \Psi^m(x_n) \circ x_m$$

where we view $x_n$ as a homomorphism $\mathcal{O}_E \to \Psi^n \mathcal{O}_E$.

Although the filtered deformations of these algebras are enough to cover the examples of [**7**], analogues of the algebras of [**9**] require a further generalization.[1] For the commutative version, there is a generalization involving a vector bundle $V$, now with

$$B_{E,\mathcal{L},V} = \bigoplus_{n\in\mathbb{Z}} \operatorname{Hom}(V, V \otimes \mathcal{L}^n).$$

This also has a twisted version,

$$B_{V,\Psi} := \bigoplus_{n\in\mathbb{Z}} \operatorname{Hom}(V, \Psi^n V).$$

Note that unlike the instance $V = \mathcal{O}_E$, this may fail to be nonnegatively graded, but will always be bounded below.

The basic question we study in the present work is: What are the filtered deformations of $B_{V,\Psi}$? Although a full answer appears out of reach for the moment, we do give some partial results that narrow things down. In one direction, we show that when $\tau = 1$, any filtered deformation arises as the endomorphism ring of a vector bundle $V^+$ on a deformation of the appropriate cone over $E$, with the property that $V^+$ is both rigid ($\operatorname{Ext}^1(V^+, V^+) = 0$) and unobstructed ($\operatorname{Ext}^2(V^+, V^+) = 0$). Thus the problem essentially reduces to understanding such bundles on del Pezzo surfaces. (In principle, one could have a nontrivial filtered deformation of $B_{V,\mathcal{L}}$ living on an undeformed cone, but experiment suggests this is not actually possible.) More generally, we classify such bundles on *noncommutative* del Pezzo surfaces, and recover the phenomenology alluded to earlier by showing that in many cases the resulting family of filtered deformations is a weighted projective space (or a product thereof).

The main gaps in obtaining a full classification are that (a) we only classify rigid, unobstructed bundles on *smooth* del Pezzo surfaces, while in general for $\tau = 1$ the center is the coordinate ring of a *singular* del Pezzo surface, (b) when $\dim(V) \neq 1$, the algebra is not a deformation of a del Pezzo surface per se, but rather of an Azumaya algebra on the del Pezzo surface (in particular, there is no a priori reason to expect it to have a well-defined structure sheaf!), and (c) although the results of [**19–21**] certainly construct many examples of noncommutative del Pezzo surfaces (obtaining the same family in multiple ways), and some of the results below strongly suggest that these examples are exhaustive, actually proving this is beyond the reach of present techniques.

Note that our focus in the present work has been on the general theory, with little attempt to give explicit presentations for any of the filtered deformations

[1] We should note here that although the algebras of [**9**] are analogues of double affine Hecke algebras, the elliptic double affine Hecke algebras of [**21**] are *not* in general filtered deformations of elliptic algebras, even in rank 1.

constructed below (especially for $\tau \neq 1$). Those interested in explicit presentations should instead view this work as a guide for future investigations, by showing where one expects the full family to be particularly nice, and letting one predict the degrees of the generators and relations.

The plan of the paper is as follows. In Section 2, we give a general discussion of filtered deformations, in particular not requiring the graded algebra to be non-negatively graded. The main results for elliptic algebras are (a) the moduli stack of filtered deformations is a closed substack of a weighted projective space, and (b) if $\tau$ is torsion, then the center of any filtered deformation of $B_{V,\Psi}$ is a filtered deformation of the center of $B_{V,\Psi}$. (This section also gives explicit computations of the moduli stack in some cases with $\tau = 1$, and briefly discusses some interesting phenomena arising when viewing $B_{V,\Psi}$ itself as a filtered deformation.)

Any actual *algorithm* for classifying filtered deformations requires understanding not just the generators and relations but also the *syzygies* of the relations, so in Section 3, we give an algorithm for computing resolutions of the algebras $B_{V,\Psi}$. One particularly nice feature is that qualitative information about the resolution (i.e., the description of each term as a sum of projective modules) can be computed directly from qualitative information about the algebra (the degree of $\mathcal{L}$ and the slopes and multiplicities of the constant-slope summands of $V$). In particular, one finds that for nearly every case in which $V$ is semistable, the algebra $B_{V,\Psi}$ is Koszul; in particular, it is a quadratic algebra with syzygies only in degree 3, and thus it is straightforward to write down equations cutting out the family of filtered deformations. This understanding of resolutions also enables us to control the cohomological properties of the deformations, in particular showing that the tautological sheaf associated to any filtered deformation is rigid and unobstructed.

This suggests in general that we should study rigid, unobstructed torsion-free sheaves on noncommutative del Pezzo surfaces, which is done in Section 5 after first reviewing the definition of and main results on such surfaces in Section 4. It is fairly straightforward to show that any such sheaf gives rise to a filtered deformation of an elliptic algebra, and thus in the absence of an understanding of *all* filtered deformations, we would at least like to understand all the deformations arising from del Pezzo surfaces, or equivalently to understand all such sheaves. In the commutative case, these were studied in [**13, 14**], and it was in particular shown that any such bundle arises in a natural way from an exceptional collection on the surface. It is straightforward to extend their arguments (especially in the simplified form we give below) to the noncommutative case, and thus the problem of understanding rigid, unobstructed, torsion-free sheaves reduces to that of understanding exceptional collections. In fact, the problem simplifies further: for the cases of interest, any sequence of exceptional bundles with the desired numerical data is automatically an exceptional collection, so one needs only classify exceptional bundles. These are, moreover, uniquely determined by their numerical data, so the only real question is existence. This turns out to be an open and closed condition, so that one can immediately reduce to the commutative case (and further to nondegenerate del Pezzo surfaces of degree 1, as well as to a related question on rational elliptic surfaces). Although this existence question remains open in most cases, we are at least able to settle it for all cases of rank at most 9, which is enough to cover the most interesting examples for filtered deformations.

Finally, in Section 6, we describe how to use the above theory to compute the corresponding moduli spaces as well as their images in the moduli spaces of filtered deformations. Since exceptional collections are unique when they exist, their moduli spaces are finite covers of the moduli spaces of surfaces (the covering arising from the fact that the numerical data breaks symmetries), and the maps from such spaces to moduli spaces of elliptic algebras are straightforward to compute. One finds in general that any family of algebras arising in this way is explicitly parametrized by the quotient of a subgroup of $E^n$ by some Weyl group acting linearly on $E^n$. This partly explains why these spaces are often weighted projective spaces, as results of [**16**] give a number of cases in which such quotients are weighted projective spaces; the remainder of the explanation involves showing that the natural line bundle on the moduli stack of filtered deformations is the same as the invariant line bundle arising in [**16**]. Using this, we describe the results of the classification in a large number of cases, as well as explaining some of the consequences arising when the result is a weighted projective space with generators of degree $> 1$. (For instance, Sklyanin's original deformation [**23**] of $\mathbb{P}^3$ arises in our theory from the fact that when $\deg(\mathcal{L}) = 4$, the space of deformations of $B_{\mathcal{O}_E,\Psi}$ has two generators of degree 2, which become the two central degree 2 elements of Sklyanin's algebra.)

## 2. Filtered deformations

We first set a convention: for our purposes, a graded algebra $B$ is a $\mathbb{Z}$-graded algebra (over a commutative ring $k$) in the usual sense such that (a) $B_m$ is a finitely generated $k$-module for all $m$, and (b) $B_m = 0$ for $m \ll 0$. (The algebras will typically be algebras over a field $k$ of characteristic 0, but will on occasion need to be algebras over a field of finite characteristic or a discrete valuation ring. We will mainly give arguments in characteristic 0, indicating in remarks where changes need to be made to include finite characteristic.)

Similarly, our convention for filtrations is that they are *ascending* filtrations, i.e., an assignment of a finitely generated submodule $A_{\le m} \subset A$ to each integer $m$ such that (a) $A_{\le m}A_{\le n} \subset A_{\le m+n}$, (b) $A_{\le m} \subset A_{\le n}$ for $m \le n$, (c) $A_{\le m} = 0$ for $m \ll 0$, and (d) $\bigcup_n A_{\le n} = A$.

Given a filtered algebra, there are two natural ways to produce a graded algebra: the *associated graded* $\operatorname{gr} A := \bigoplus_i (A_{\le i}/A_{\le i-1})$, and the *Rees algebra* $A^+ := \bigoplus_i A_{\le i}t^i$, where $t$ is an auxiliary variable. Similarly, a graded algebra may be viewed as a filtered algebra with $B_{\le i} = \bigoplus_{j\le i} B_j$; in that case, the associated graded is $B$ and the Rees algebra is $B[t]$.

DEFINITION 2.1. Let $B$ be a graded algebra. A *filtered deformation* of $B$ is a filtered algebra $A$ equipped with an isomorphism $\operatorname{gr} A \cong B$.

REMARK. Note that the isomorphism is part of the data.

The following fact will later allow us to reduce to the case of nonnegatively graded algebras (for the algebras of interest, at least!).

PROPOSITION 2.1. *Let $B$ be a graded algebra. For any homogeneous idempotent $e \in B$ and any filtered deformation $A$ of $B$, there is an idempotent $\hat{e} \in A$ with leading term $e$.*

PROOF. An idempotent has degree 0, and thus $\hat{e}$ should be in $A_{\le 0}$. This is a *subalgebra*, and our conventions ensure that it is finite-dimensional, and thus

Artinian. The ideal $A_{\leq -1}$ is nil, so any idempotent of $A_{\leq 0}/A_{\leq -1} = B_0$ lifts to $A_{\leq 0}$, and thus to $A$. □

There is a natural action of $\mathbb{G}_m$ on any graded algebra $B$ with $t \in \mathbb{G}_m$ acting by $t^i$ on $B_i$, and this acts on the space of filtered deformations by composing the isomorphism $\operatorname{gr} A \cong B$ with the induced automorphism of $B$.

Any filtered isomorphism $A \cong A'$ of filtered algebras induces an isomorphism $\operatorname{gr} A \cong \operatorname{gr} A'$, and thus if $A$ and $A'$ are both filtered deformations of $B$, the isomorphism induces an automorphism of $B$. The natural notion of equivalence on filtered deformations is an isomorphism such that the induced automorphism is the identity; we also consider the notion of a *weak* equivalence, in which the induced automorphism comes from the action of $\mathbb{G}_m$.

One example of a filtered deformation is the filtered algebra associated to $B$ itself; in general, we say that a filtered deformation is *trivial* if it is equivalent to this deformation.

DEFINITION 2.2. The *moduli stack of filtered deformations* of $B$ is the stack classifying nontrivial filtered deformations of $B$ up to weak equivalence.

REMARK. There is an even weaker notion of equivalence one may consider, in which one allows *all* automorphisms of $B$ rather than just $\mathbb{G}_m$. This is not as well-behaved in families, however, and leads to more complicated moduli stacks (esp. in our examples below where the present definition gives a weighted projective space).

PROPOSITION 2.2. *If $B$ is a finitely presented graded algebra over a field $k$, then the moduli stack of filtered deformations of $B$ is algebraic.*

PROOF. Fix a finite (homogeneous) presentation of $B$, and choose a basis of $B$ consisting of monomials in the generators. Given a filtered deformation $A$ of $B$, since $\operatorname{gr} A \cong B$, each generator $x_i$ of $B$ has a preimage inside $A_{\leq \deg(x_i)}$, which is unique modulo $A_{\leq \deg(x_i)-1}$. Choose a lift $\hat{x}_i$ of each generator. This induces a basis of $A$ compatible with the filtration by replacing $x_i$ with $\hat{x}_i$ in the chosen monomial basis of $B$.

If we make a similar substitution in a relation $\rho_i$ of $B$, then the result will be an element of $A_{\leq \deg(\rho_i)-1}$; the degree $\deg(\rho_i)$ term cancels since the relation holds in $\operatorname{gr} A$. This element is then uniquely expressible in terms of our basis of $A$, and thus gives us a unique lift of each relation.

In the other direction, if we start with a collection of deformed relations (i.e., adding lower-order terms from the chosen basis to each relation from the presentation of $B$), the condition that the corresponding finitely presented algebra is a filtered deformation of $B$ is closed. To see this, note that it is equivalent to ask that the chosen monomials in the generators of the filtered algebra be linearly independent. It follows that a given collection of deformed relations gives a filtered deformation iff for any combination of deformed relations in which only the "good" monomials appear, the coefficients all vanish. But this is clearly a collection of polynomial equations, and thus the filtered deformation locus is an intersection of closed subschemes, so is closed.

This shows that the moduli stack of filtered deformations *with chosen lifts of the generators* is a scheme (an affine scheme, in fact). Two such deformations are equivalent iff one can be obtained from the other by changing the lifts of the generators. Each generator $\hat{x}_i$ can be changed by adding an element of $A_{\leq \deg(x_i)-1}$, which is in turn given by a polynomial in the generators of lower degree. We thus see that if we order the generators $x_1, \dots, x_n$ by degree, then after lifting $x_1, \dots, x_{n-1}$, the set of possible lifts of $x_n$ is a torsor over a unipotent group scheme, so that forgetting the lift takes the quotient by that group, leaving the stack algebraic. We can then continue inductively, and thus find that the moduli stack of filtered deformations up to equivalence is algebraic.

Finally, we can remove the trivial deformations (which correspond to a single closed point of the stack) and quotient out by $\mathbb{G}_m$ to obtain the desired moduli stack, which is therefore still algebraic. □

REMARK. Note that in the above argument, if the degrees of the generators are $\le$ the degrees of the relations, then the moduli stack is a quotient of a scheme by a group of the form $\mathbb{G}_m \ltimes U$ where $U$ is unipotent. The above induction is needed in general, however, since the coordinatization of the set of lifts of a generator depends on the relations of lower degree, which in turn depend on the lifts of generators of up to that degree.

One caution here is that this argument is *highly* nonconstructive; indeed, it seems likely that even the question of whether $B$ admits a nontrivial filtered deformation is undecidable in general, even if we are given a finite presentation of $B$. (The situation improves dramatically if we are also given generators of the module of syzygies.)

This moduli stack is nicest under the following condition.

PROPOSITION 2.3. *Suppose that the finitely presented graded algebra $B$ admits no homogeneous derivations of negative degree. Then the moduli stack of filtered deformations of $B$ is a closed substack in a weighted projective space.*

PROOF. For each positive integer $d$, consider the effect of adding terms of degree $\deg(x_i) - d$ to each lifted generator $\hat{x}_i$. The resulting effect on the relations is trivial until one reaches degree $\deg(\rho_i) - d$, where it induces a linear transformation from the space of deformations of generators to the space of deformations of relations. This linear transformation is independent of the choice of filtered deformation of $B$, and may thus be computed using the trivial filtered deformation. In particular, we find that if we view the linear transformation as its own derivative, then the kernel is precisely the space of degree $-d$ derivations of $B$, and thus by hypothesis the linear transformation is injective.

We now choose for each positive integer $d$ a complement to the image of the linear transformation in the space of degree $-d$ deformations of relations. We then claim that every filtered deformation is uniquely equivalent to one in which the deformations of every degree lie in the chosen complement. This may be seen by induction on $d$: if the deformations of relative degree $> -d$ all lie in their respective subspaces, then (by injectivity and complementarity) there is a unique way to shift the generators by terms of relative degree $-d$ to make the relations lie in the complementary subspace in degree $-d$. Since this does not affect the terms of higher degree, the result follows.

It follows that the moduli stack of filtered deformations up to equivalence is a scheme, and is in fact $\operatorname{Spec}(M)$ for some (commutative) graded algebra $M$ (with grading coming from the $\mathbb{G}_m$ action). The desired moduli stack is then obtained by removing the origin (the unique point corresponding to the trivial deformation) and quotienting by $\mathbb{G}_m$, and is thus closed in the weighted projective space corresponding to the homogeneous generators of $M$. □

REMARK. Here we view weighted projective spaces as stacks; if all coordinates of degree prime to $l$ vanish, then the result will be an orbifold point with stabilizer containing $\mu_l$. We can recover the usual scheme versions by taking coarse moduli spaces. Note that in either interpretation, weighted projective spaces are proper, and thus so are their closed substacks/subschemes.

An important special case in which it is straightforward to control derivations is when $B$ is a commutative homogeneous coordinate ring.

LEMMA 2.4. *Let $X$ be an irreducible smooth projective scheme of positive dimension with ample line bundle $\mathcal{L}$, with associated homogeneous coordinate ring $B := \bigoplus_{i\ge 0}\Gamma(X;\mathcal{L}^i)$. For any integer $k > 0$, there is a short exact sequence*

$$0 \to \Omega_X^k \to V_k \to \Omega_X^{k-1} \to 0$$

*such that there is an isomorphism*

$$\mathrm{Der}^k(B) \cong \bigoplus_{i\in\mathbb{Z}} \mathrm{Hom}(V_k, \mathcal{L}^i)$$

*of graded vector spaces.*

PROOF. If we remove the cone point from $\mathrm{Spec}(B)$, the resulting scheme is a $\mathbb{G}_m$-torsor $X^+$ over $X$, so that an element of $\mathrm{Der}^k(B)$ thus determines an element of $\mathrm{Der}^k(\mathcal{O}_{X^+})$. This is, in fact, a $\mathbb{G}_m$-equivariant isomorphism

$$\mathrm{Der}^k(B) \cong \mathrm{Der}^k(\mathcal{O}_{X^+}).$$

Indeed, injectivity follows from the fact that $X^+$ still contains the generic point of $\mathrm{Spec}(B)$, while surjectivity follows by observing that any holomorphic section of $\mathcal{O}_{X^+}$ decomposes via the $\mathbb{G}_m$ action as a sum of homogeneous elements, each of which induces a section of the corresponding power of $\mathcal{L}$ and thus gives a homogeneous element of $B$.

Since $\pi : X^+ \to X$ is smooth of dimension 1, there is a short exact sequence

$$0 \to \pi^*\Omega_X \to \Omega_{X^+} \to \Omega_{X^+/X} \to 0,$$

with $\Omega_{X^+/X}$ a line bundle. This line bundle is in fact canonically trivial, via the map

$$\Omega_{X^+/X} \otimes \mathrm{Lie}(\mathbb{G}_m) \to \mathcal{O}_{X^+}$$

coming from the $\mathbb{G}_m$-torsor structure. We further find that the induced map

$$\Gamma(X^+;\mathcal{O}_{X^+}) \to H^1(X^+;\pi^*\Omega_X) \cong H^1(X;\pi_*\pi^*\Omega_X) \cong H^1(X;\bigoplus_i \Omega_X \otimes \mathcal{L}^i)$$

is homogeneous of degree 0, and thus corresponds to an element of $H^1(X;\Omega_X)$. In other words, there is a short exact sequence

$$0 \to \Omega_X \to V_1 \to \mathcal{O}_X \to 0$$

such that $\Omega_{X^+} \cong \pi^*V_1$. Taking exterior powers gives a similar exact sequence

$$0 \to \Omega_X^k \to V_k \to \Omega_X^{k-1} \to 0$$

for $V_k := \wedge^k V_1$. We then find

$$\begin{aligned}\operatorname{Der}^k(\mathcal{O}_{X^+}) &\cong \Gamma(X^+; \mathcal{H}om(\pi^* V_k, \mathcal{O}_{X^+}))\\ &\cong \Gamma(X; \pi_* \mathcal{H}om(\pi^* V_k, \pi^* \mathcal{O}_X))\\ &\cong \Gamma(X; \mathcal{H}om(V_k, \pi_*\pi^*\mathcal{O}_X))\\ &\cong \bigoplus_{i\in\mathbb{Z}} \operatorname{Hom}(V_k, \mathcal{L}^i) \qquad (2.1)\end{aligned}$$

as required. □

REMARK. Note that in this generality, $B$ can fail to be finitely generated.

COROLLARY 2.5. *Let $C$ be a smooth projective curve of positive genus, $\mathcal{L}$ an ample line bundle on $C$, and $B$ the corresponding homogeneous coordinate ring. Then any homogeneous element of* $\operatorname{Der}^k(B)$ *of negative degree vanishes.*

PROOF. For $d > 0$, the space of homogeneous elements of $\operatorname{Der}^k(B)$ of degree $-d$ is

$$\operatorname{Hom}(V_k, \mathcal{L}^{-d})$$

where $V_k$ is the sheaf from the Lemma, so that for $k = 1$ there is an exact sequence

$$0 \to \operatorname{Hom}(\mathcal{O}_C, \mathcal{L}^{-d}) \to \operatorname{Hom}(V_1, \mathcal{L}^{-d}) \to \operatorname{Hom}(\Omega_C, \mathcal{L}^{-d}),$$

with both spaces on either side of the one of interest vanishing by degree considerations. Vanishing for $k = 2$ follows similarly from $V_2 \cong \Omega_C$, while for $k > 2$, $V_k = 0$. □

COROLLARY 2.6. *With $B$ as above, the moduli stack of filtered deformations is a closed substack of a weighted projective space, and every filtered deformation of $B$ is commutative.*

PROOF. The absence of derivations of negative degree gives the first claim. For the second claim, suppose $A$ were a noncommutative filtered deformation of $B$, and view the Rees algebra of $A$ as a (flat) $k[t]$-module. This gives a 1-parameter family of noncommutative algebras degenerating when $t = 0$ to the commutative algebra $B$, and thus induces a nonzero Poisson bracket on $B$. Since computing the Poisson bracket involves dividing commutators by an appropriate positive power of $t$, this Poisson bracket on $B$ is homogeneous of negative degree. Since a Poisson bracket is in particular a 2-derivation, it must therefore vanish, giving a contradiction. □

This makes it straightforward to compute the moduli space of filtered deformations of commutative elliptic algebras.

EXAMPLE 2.1. Consider the case of an elliptic curve $E$, with ample line bundle $\mathcal{L} = \mathcal{O}_E(\infty)$ of degree 1 coming from the point at infinity. The corresponding commutative graded algebra is generated by elements $z$, $x$, $y$ of degree 1, 2, 3 respectively, satisfying a single relation of degree 6:

$$y^2 + a_1 xyz + a_3 yz^3 = x^3 + a_2 x^2 z^2 + a_4 xz^4 + a_6 z^6.$$

By the corollary, any filtered deformation of this algebra is commutative, and thus simply adds lower degree terms to the single relation; and since there is a single commutative relation, there are no further conditions to give a flat deformation. To get equivalence, we must quotient by changes of variable of the form

$$(y, x, z) \mapsto (y + b_1 x + c_1 z^2 + b_2 z + b_3, x + d_1 z + c_2, z + e_1).$$

This group acts faithfully, and thus to obtain the structure of the moduli stack, we need simply remove four generators of degree 1, two of degree 2, and one of degree 3. We thus see that in this case, the moduli stack is the weighted projective space $\mathbb{P}^{[122334456]}$ (i.e., with generators of degrees $1, 2, 2, 3, 3, 4, 4, 5, 6$). (These are the standard labels of the affine Dynkin diagram of type $E_8$, as explained below).

A similar calculation works if we instead take $\mathcal{L}$ to have degree 2, 3, or 4; in the first two cases, the cone is a hypersurface, while for 4, it is a complete intersection, and thus in no case are there any syzygies. The result in each case is thus again a weighted projective space ($\mathbb{P}^{[11222334]}$ ($E_7$); $\mathbb{P}^{[1112223]}$ ($E_6$); and $\mathbb{P}^{[111122]}$ ($D_5$) respectively).

For higher degrees, there are syzygies, so some actual calculation is required, but one can still easily compute the corresponding moduli space since there is no need to consider noncommutative deformations. One finds that the space is $\mathbb{P}^4$ for degree 5 and $\mathbb{P}^1 \times \mathbb{P}^2$ for degree 6. For degrees 7,8,9, the structure is somewhat more complicated (but see below), while for degree $> 9$, there are no nontrivial deformations [**18**, §9]. For degree $\geq 3$, it was shown op. cit. that any nontrivial deformation is rational and anticanonical (albeit possibly singular), i.e., a del Pezzo surface of the given degree. For degrees 3 and 4, this can also be seen by comparing the moduli stack as computed above to the substack coming from del Pezzo surfaces, and the same argument establishes the corresponding claim for degrees 1 and 2. (This can also be seen more directly: that the surface is Fano follows by adjunction from the fact that the curve at infinity has genus 1; that it is rational reduces (via successive blowups of points at infinity) to the degree 1 case, where it follows from the theory of elliptic surfaces.)

Note that the hypotheses above are quite strong, and in particular exclude the cases in which $E$ becomes singular. These are, of course, still quite interesting, but harder to deal with geometrically. (By contrast, it is often quite straightforward to write down explicit presentations in those cases!) As an example of what can go wrong, one has the following family in which the commutative deformation fails to be a del Pezzo surface in the usual sense (and one obtains deformations for $\mathcal{L}$ of arbitrarily large degree).

EXAMPLE 2.2. For $d \geq 3$, let $Z$ be the weighted projective space with generators $x$, $y$, $z$, $w$ of degrees $d-2$, $d-2$, 1, 1 respectively, and let $X_{ab}$ be the family of hypersurfaces of degree $d$ with equation

$$zwx + (az^2 + bw^2)y = z^d + w^d.$$

The curve $y = 0$ in $X_{ab}$ is isomorphic to the standard nodal cubic, and its image under the embedding corresponding to the invertible sheaf $O_Z(d-2)$ is a degree $d$ model in $\mathbb{P}^{d-1}$; to be precise, it is the image of $\mathbb{P}^1$ under the map $(z^d + w^d, zw^{d-1}, \dots, z^{d-1}w)$ that identifies the points $(0, 1)$ and $(1, 0)$. In other words, the $(d-2)$nd Veronese of the homogeneous coordinate ring of $X_{ab}$ is the Rees algebra of a filtered deformation of that nodal cubic. It is clear that this filtered deformation is nontrivial unless $a = b = 0$, and thus gives us a $\mathbb{P}^1$ of such deformations for all $d$. When $d = 10$, one can verify by a direct Magma computation that the reduced subscheme of the full deformation scheme is $\mathbb{P}^1$, and thus all deformations arise in this way. The presence of such deformations arises from the fact that the cone $X_{00}$ is not normal, and thus the filtered deformations themselves need not be normal (and, indeed, are not: $X_{ab}$ is singular along the curve

$z = w = 0$). These are examples of the non-normal del Pezzo surfaces studied in [**22**].

Now, with $E$ still an elliptic curve and $\mathcal{L}$ an ample line bundle, we can also obtain noncommutative graded algebras by taking

$$B = \bigoplus_i \operatorname{Hom}(V, V \otimes \mathcal{L}^i)$$

for $V$ a vector bundle on $E$. For simplicity, we assume that $B$ is nonnegatively graded; we will see below that this is not truly a constraint. This algebra is finite over its center, the usual homogeneous coordinate ring, and is in fact the ring of global sections of a sheaf of (Azumaya) algebras $\mathcal{B} := \mathcal{E}nd(\pi^* V)$ on the $\mathbb{G}_m$-torsor $E^+$ associated to $\mathcal{L}$.

For any associative algebra $A$, there is a natural map $\operatorname{Der}(A) \to \operatorname{Der}(Z(A))$ coming from the fact that any automorphism (and thus any infinitesimal automorphism) preserves the center.

LEMMA 2.7. *A derivation of $B$ that is trivial on $Z(B)$ is inner.*

PROOF. A derivation of $B$ which is trivial over $Z(B)$ may be viewed as an $\mathcal{O}_{\operatorname{Spec}(Z(B))}$-linear derivation on the corresponding sheaf of algebras, and thus induces an $\mathcal{O}_{E^+}$-linear derivation on $\mathcal{B}$ after removing the cone point, or equivalently a holomorphic family of derivations on the fibers of $\mathcal{B}$. Since those fibers are matrix algebras over fields, their derivations are necessarily inner. In particular, we may locally represent the given derivation as the commutator with a *traceless* matrix, and since this representation is canonical, it extends to a global such representation. As in the proof of Lemma 2.4, a global section of $\mathcal{B}$ is an element of $B$, and thus the original derivation of $B$ is inner. □

REMARK. This argument fails in finite characteristic, since it uses the fact that the map $\operatorname{Mat}_n \to \operatorname{Inn}\operatorname{Mat}_n$ is split by the trace, which fails when $n$ is a multiple of the characteristic. Of course, any global splitting will do, so it in particular suffices to have *some* idempotent in $\operatorname{End}(V)$ with invertible trace, or equivalently for $V$ to have a summand with invertible rank. (Though even this appears to be stronger than needed for the conclusion to hold.)

REMARK. An equivalent statement is that the natural map $\operatorname{HH}^1(B) \to \operatorname{HH}^1(Z(B))$ is injective.

COROLLARY 2.8. *The algebra $B$ has no homogeneous derivations of negative degree.*

PROOF. We have already seen that $Z(B)$ has no such derivations, and thus such a derivation of $B$ is trivial on $Z(B)$, so inner, and thus itself vanishes since $B$ is nonnegatively graded. □

REMARK. Note that $B$ is Noetherian since $\mathcal{B}$ is Noetherian, and is in particular finitely presented. We may thus conclude that the moduli stack of filtered deformations of $B$ is a closed substack of a weighted projective space.

The fact that filtered deformations remain commutative when $V = \mathcal{O}_E$ also has an analogue: the center of any filtered deformation of $B$ is a filtered deformation of $Z(B)$. To see this, we use ideas of Hayashi [**11**] (see also [**12**, §1] for an explicit discussion), in somewhat more formal terms. We again view the Rees algebra

$A^+$ as a flat $k[t]$-algebra. Call an element $a \in A^+$ of this algebra *m-central* if $[a, A^+] \subset t^m A^+$, or equivalently if its image in $A^+/t^m A^+$ is central; in particular, an element is 1-central iff its image in $B$ is central. Now, given an $m$-central element $a$, we may define a derivation $D_{a,m}(x) := t^{-m}[a, x]$ on $A^+$. Since this is $k[t]$-linear, the image of $D_{a,m}(x)$ mod $t$ depends only on the image of $x$ mod $t$, so that $D_{a,m}$ induces a derivation of $B$. We also find

$$D_{a+t^m b,m}(x) = D_{a,m}(x) + [b, x],$$

and thus see that there is an $m+1$-central element of the form $a + t^m b$ iff $D_{a,m}$ is an inner derivation. In other words, the obstruction to lifting an element of $Z(A^+/t^m A^+)$ to an element of $Z(A^+/t^{m+1}A^+)$ is an element of $\operatorname{Der}(B)/\operatorname{Inn}(B) = \operatorname{HH}^1(B)$. Note that since the product of $m$-central elements is $m$-central and $D_{ab,m}(x) = aD_{b,m}(x) + D_{a,m}(x)b$, the obstruction map

$$Z(A^+/t^m A^+) \to \operatorname{HH}^1(B)$$

is itself a derivation.

Now, let $m > 0$ be an integer, and suppose that every central element of $A^+/t^{m-1}A^+$ lifts to an $m$-central element of $A$ (this is vacuously true for $m = 1$). If $a_1$, $a_2$ are $m$-central elements that reduce to the same element of $B$, then

$$D_{a_1,m}(x) - D_{a_2,m}(x) = t^{-m}[a_1 - a_2, x] = t^{1-m}[t^{-1}(a_1 - a_2), x] = D_{t^{-1}(a_1-a_2),m-1}(x).$$

Now, $t^{-1}(a_1 - a_2)$ is $(m-1)$-central, and thus agrees modulo $t^{m-1}$ with an $m$-central element, so that $D_{t^{-1}(a_1-a_2),m-1}(x)$ is inner, implying that

$$D_{a_1,m}(x) - D_{a_2,m}(x)$$

is inner. We thus find that the obstruction map factors through a derivation

$$Z(B) \to \operatorname{HH}^1(B),$$

which since it divides by $t^m$ is homogeneous of degree $-m$. (As pointed out in [**10**, Ex. 2.17(4)], this map appears in a spectral sequence for computing the Hochschild cohomology of the filtered deformation.)

For any $a \in Z(B)$, the induced outer derivation of $B$ induces a derivation of $Z(B)$, giving a 2-derivation of $Z(B)$ of negative degree, which vanishes. Since a derivation of $B$ vanishing on $Z(B)$ is inner, we conclude that the obstruction map for degree $m$ vanishes. We thus find the following.

PROPOSITION 2.9. *Let $E$ be an elliptic curve, $\mathcal{L}$ an ample line bundle on $E$, and $V$ a vector bundle such that* $\operatorname{Hom}(V, V \otimes \mathcal{L}^{-1}) = 0$. *The center of any filtered deformation $A$ of $\bigoplus_i \operatorname{Hom}(V, V \otimes \mathcal{L}^i)$ is a filtered deformation of $\bigoplus_i \Gamma(\mathcal{L}^i)$.*

PROOF. Let $\bar{x} \in Z(B)$ be a homogeneous element, and consider the sequence of $m$-central lifts $x_1, x_2, x_3, \dots, x_m, \dots$ guaranteed by the above discussion. Then $x_{i+1} - x_i \in t^i A^+$, so that the image of $x_{i+1} - x_i$ in $A$ (i.e., setting $t = 1$) has degree $\le \deg(\bar{x}) - i$. In particular, we find that any sequence of $m$-central lifts is eventually constant, and thus gives a lift of $\bar{x}$ which is $m$-central for all $m$, thus central. This shows that $\operatorname{gr} Z(A) \supset Z(B)$, while the other inclusion is trivial, so that the claim follows. □

As with the commutative case, one can use these results to enable explicit computations of the moduli stacks. One issue is that it is more difficult to write down an explicit presentation of $B$, especially if one wishes to deal with all $E$ at once. A useful trick for this purpose involves the observation that $B$ itself has regular central elements of degree 1, so may be viewed as a filtered deformation, and the resulting algebra turns out to be independent of $E$. To be precise, if $w \in Z(B)$ is a degree 1 element corresponding to $d = \deg(\mathcal{L})$ distinct points of $E$, i.e., corresponding to an isomorphism $\mathcal{L} \cong \mathcal{L}(D)$ for a reduced divisor $D$, then the quotient $C := B/\langle w\rangle$ has Hilbert series of the form $\dim(B_0) + (n - \dim(B_0))t + nt^2 + nt^3 + \cdots$ where $n = d \operatorname{rank} \mathcal{E}nd(V)$. If $\tilde{C}$ denotes the algebra $\Gamma(D; \mathcal{E}nd(V))$, then every homogeneous component of $C$ may be identified with a subspace of $\tilde{C}$, and thus in particular dimension considerations tell us that $C$ agrees with $\tilde{C}$ in all degrees $> 1$. (Note that the identification with $\tilde{C}$ depends on a choice of trivialization of $\mathcal{L}(D)|_D$, or in other words on a system of uniformizers at the points of $D$. By adjunction, such a system is determined by a choice of holomorphic differential $\omega$ on $E$, so there is an essentially canonical choice of identification.) In degree 0, we of course have $C_0 = B_0 = \operatorname{End}(V) \subset \Gamma(D; \mathcal{E}nd(V))$, so it remains only to determine $C_1$. There is a natural pairing $C_0 \times C_1 \to \Gamma(\mathcal{L}(D))$ given by $(x, y) \mapsto \operatorname{Tr}(xy)$, which agrees with the $\Gamma(\mathcal{O}_D)$-valued trace pairing on $\tilde{C}$ if we take residues (with the same choice of holomorphic differential). Since the value is a global function, the sum of residues vanishes, and thus any element of $C_1$ must be orthogonal to $C_0$ under the sum-of-traces pairing on $\tilde{C}$. This cuts out a space of the correct dimension, so completely determines $C_1$.

In other words, $C$ is determined by the algebra $\tilde{C}$ and its subalgebra $B_0 \subset \tilde{C}$:

$$C_n = \begin{cases} B_0 & n = 0 \\ \{x : x \in \tilde{C} \mid \operatorname{Tr}(xB_0) = 0\} & n = 1 \\ \tilde{C} & n > 1 \end{cases}.$$

Here $\tilde{C}$ is (geometrically) just a sum of $d$ copies of $\operatorname{Mat}_{\dim(V)}$, while $B_0 = \operatorname{End}(V)$ depends only on the gross structure of $V$ (the slopes and multiplicities of the indecomposable summands, along with which summands have isomorphic stable constituents). It is then straightforward to determine a presentation of $C$, either by using the results for $B$ below or by using the fact that $C$ has regular central elements (and in particular has a canonical regular central element of degree 2, namely $1 \in \tilde{C} = C_2$), and the quotient by such an element is finite-dimensional.

EXAMPLE 2.3. It is interesting to consider the commutative case from this perspective. Here the algebra $\tilde{C}$ is just $k^d$, while $B_0$ is the diagonal copy of $k$, so that for $d > 2$, $C$ is the embedded coordinate ring of $d$ points in general position in $\mathbb{P}^{d-2}$. (For $d = 1$, it is $k[y, x]/(y^2 - x^3)$ with $\deg(x) = 2$, $\deg(y) = 3$, and for $d = 2$, it is $k[y, x]/(y^2 - x^2 y)$ with $\deg(x) = 1$, $\deg(y) = 2$.) The commutative filtered deformations of such algebras were studied in [**15**], and in particular shown to always give a (possibly degenerate) genus 1 curve on which the marked degree $d$ divisor is ample. Moreover, the structure of the moduli stack was determined in low degree, giving the weighted projective spaces $\mathbb{P}^{[46]}, \mathbb{P}^{[234]}, \mathbb{P}^{[1223]}, \mathbb{P}^{[11122]}$ in degrees $1, 2, 3, 4$, $\mathbb{P}^4$ in degree 5, and the Grassmannian $\operatorname{Gr}(2, 5)$ in degree 6. Not only does this let us write down general versions of $B$ in low degree, but it in fact lets us compute the moduli stack of deformations of $B$: a filtered deformation of $B$ is essentially just a line through the corresponding point of the moduli stack of filtered deformations of $C$. (The description is somewhat more involved when the moduli space for $C$ has orbifold points: in general, it corresponds to a homomorphism from the graded algebra parametrizing deformations of $C$ to $k[x,y]$ such that the induced rational map from $\mathbb{P}^1$ is nonconstant and $\infty$ maps to the point associated to $B$.) This recovers the results for $d \le 4$ above, as well as giving a more conceptual argument for $d = 5, 6$. (There are issues for $d = 1, 2$ in small characteristic, but these can be resolved by waiting until the last step to kill the unipotent freedom.) Interestingly, for $d > 1$, if we quotient by a further regular element of degree 1, the deformation spaces remain nice: $\mathbb{P}^{[4]}, \mathbb{P}^{[23]}, \mathbb{P}^{[1122]}, \mathbb{P}^6$ respectively for $2 \le d \le 5$, while for $d = 6$ the stack of deformations is the isotropic Grassmannian $\mathrm{OGr}^+(5,10)$. (The arguments are, sadly, not conceptual, but simply involve explicitly computing the deformation space, and then, after guessing based on the Hilbert series, computing the explicit isomorphism for $d = 6$. It would be particularly interesting to have a conceptual construction of the resulting sheaf of filtered algebras over $\mathrm{OGr}^+(5,10)$!)

EXAMPLE 2.4. Consider the case that $V$ is a stable bundle of rank 2 and degree 1 and the ample bundle is $\mathcal{L}(\infty)$. Then $B$ has Hilbert series $1 + 4t/(1-t)^2$, while $Z(B)$ has Hilbert series $1 + t/(1-t)^2$. In this case, we get a particularly nice algebra $D$ if we quotient $C$ by the regular central element of degree 2: $D$ is simply the exterior algebra in three generators! The moduli stack of filtered deformations of that exterior algebra is then straightforward to compute, and one finds that the space of filtered deformations is the weighted projective space $\mathbb{P}^{[222222]}$: any filtered deformation of the exterior algebra in three generators is equivalent to a Clifford algebra. (There is a $\mathbb{P}^5$ worth of Clifford algebras, but Clifford algebras have an additional automorphism acting as $-1 \in \mathbb{G}_m$ on the associated graded.) Since a filtered deformation $A$ of $B$ has center a filtered deformation of $Z(B)$, the quotient of the Rees algebra $A^+$ by its two degree 1 elements and one degree 2 element is again the exterior algebra, and thus we may view the filtered deformation of $B$ as a family of filtered deformations of the exterior algebra. We thus see that it is given by a map from the weighted projective space $\mathbb{P}^{[112]}$ to the moduli stack of Clifford algebras, or more precisely by a collection of six homogeneous polynomials of degree 2 in those generators. Of course, the specializations of those polynomials to $t = 0$ are determined by $B$ (that is, together with the specific choices of degree 1 and 2 elements), and thus there are only two free parameters for each of the polynomials, one of degree 1 and one of degree 2. Of those, one each is eliminated by translations of the degree 2 central element of $B$, and one more of degree 1 is eliminated by the translations of the degree 1 central element. We thus conclude that for *any* elliptic curve and semistable bundle of rank 2, degree 1, the moduli stack of filtered deformations of the corresponding graded algebra is the weighted projective space $\mathbb{P}^{[1111122222]}$ ($D_8$). (One caveat: the above calculation assumes 2 is invertible, both because the bundle has even rank, breaking the proof of Lemma 2.7, and because the exterior algebra over $\bar{\mathbb{F}}_2$ has derivations of negative degree. One can, however, verify by a more complicated calculation that the claims remain true in characteristic 2.)

REMARK. For $V$ stable of slope $1/2$ and $\mathcal{L}$ of degree $d$, the algebra $C$ is equal to $\mathrm{Mat}_2^d$ in degree $> 1$, and the subspace of $d$-tuples of total trace 0 in degree 1, and in particular its center is the algebra considered in Example 2.3. There is an action of $\mathrm{GL}_2^d$ on the corresponding deformation space (of filtered deformations respecting the center), and we find (computationally) the following descriptions after taking that into account:

$$d = 1 : S^4(V_1)/(\mathbb{G}_m/\langle \pm 1\rangle) \tag{2.2}$$

$$d = 2 : (S^2(V_1) \otimes S^2(V_2))/\mathbb{G}_m \tag{2.3}$$

$$d = 3 : \mathrm{Gr}(2, V_1 \otimes V_2 \otimes V_3), \tag{2.4}$$

where $V_1$,... are the fundamental representations of the respective copies of $\mathrm{GL}_2$. In each case, this presumably has a geometric modular interpretation in terms of rank 2 vector bundles (or torsion-free sheaves) over the center, but it appears that there is an alternate interpretation in terms of hyperelliptic curves of genus 1 with additional structure. Indeed, for $d = 1$, $d = 2$, the given representation of $\mathrm{GL}_2^d$ is precisely the representation used in [**5**] in connection with 2-Selmer groups of elliptic curves with $d - 1$ marked points, while for $d = 3$, the space considered there is $V_1 \otimes V_2 \otimes V_3 \otimes V_4$, and thus our space is the GIT quotient of their space by the "extra" copy of $\mathrm{GL}_2$. Similarly, for the case $\mathrm{rank}(V) = 3$, $d = 1$, the deformation space is two disjoint copies of $S^3(V_1)/\mathbb{G}_m$ (with $V_1$ the fundamental representation of $\mathrm{GL}_3$), corresponding to cubic plane curves (and thus related to 3-Selmer groups). (One obtains two copies as there are two possible slopes modulo 1 with denominator 3.) The remaining representations considered in [**5**] thus suggest three more deformation spaces associated to stable bundles:

$$\mathrm{rank}(V) = 3,\ d = 2 : \mathrm{Gr}(3, V_1 \otimes V_2) \tag{2.5}$$

$$\mathrm{rank}(V) = 4,\ d = 1 : \mathrm{Gr}(2, S^2(V_1)) \tag{2.6}$$

$$\mathrm{rank}(V) = 5,\ d = 1 : \mathrm{Gr}(5, S^2(V_5)); \tag{2.7}$$

in the first two cases, one expects two components of that form, while in the third case, one expects a total of four components, two of which are of the given form. Note that in each case, the resulting prediction for the deformation spaces of the resulting elliptic algebras agrees with the computation in Section 5 below. It would be interesting to know if there are any other instances with nice deformation space.

EXAMPLE 2.5. Similarly, if $V$ is a sum of two nonisomorphic degree 0 line bundles, then $C$ is generated (over $B_0 \cong k^2$) by elements $F$ of degree 1 and $G$ of degree 2 interacting with the two idempotents of $B_0$ as $e_1F = Fe_2$, $e_1G = Ge_2$, and satisfying relations $FG = GF$, $G^2 = F^4$. (These both correspond to the matrix $\begin{pmatrix} 0 & 1 \\ 1 & 0 \end{pmatrix} \in \tilde{C}$.) Any filtered deformation is equivalent to one in which the relations involving the idempotents remain unchanged, and two such deformations are equivalent iff they are related by a shift $G \mapsto G + b_1F$. (We can only add lower terms that interact the same way with the idempotents.) An explicit calculation with syzygies of low degree tells us that the universal deformation of this algebra has the form

$$FG - GF = a_3(e_1 - e_2), \tag{2.8}$$

$$G^2 = F^4 + a_2F^2 + a_4. \tag{2.9}$$

In particular, $B$ corresponds to a point of the weighted projective space $\mathbb{P}^{[234]}$, and a filtered deformation of $B$ extends that point to a map from $\mathbb{P}^1$, from which it follows easily that the moduli stack of filtered deformations of $B$ is a weighted projective space with degrees $\mathbb{P}^{[11222334]}$ (again $E_7$). (This calculation again assumes 2 is invertible, but can be extended to work over $\mathbb{Z}$ by waiting until the last step to kill the unipotent symmetry.)

REMARK. Just as $A$ may be viewed as a line of filtered deformations of $C$, we may obtain filtered deformations of $A^+$ as either a line of filtered deformations of $B$ or a plane of filtered deformations of $C$. (This imposes the additional condition that $Z(A^+)$ lifts, which is not necessarily automatic.) This at the very least gives rise to a family of (algebras on) Fano 3-folds (embedded in such a way that $-K_X = \mathcal{O}(2)$), and we could hope to obtain noncommutative Fano 3-folds by replacing $B$ by a twisted elliptic algebra. (Indeed, the Sklyanin algebra **[23]** may be obtained by a variation on this idea, see below.) This can, of course, be iterated as long as enough degrees of freedom remain to allow the deformations to be nontrivial. When the moduli stack of deformations of $B$ is a weighted projective space, a particularly interesting variation is the *universal* deformation, in which one adjoins every generator of the weighted projective space to $B$. In the commutative case, the resulting scheme is independent of $E$, and is isomorphic to $\mathbb{P}^{[11222333445]}, \mathbb{P}^{[1111222233]}, \mathbb{P}^{[111111222]}, \mathbb{P}^7, \mathrm{Gr}(2,5)$ for $1 \le d \le 5$, so that we obtain noncommutative deformations of those spaces by twisting $B$. (It would be interesting to see if the other cases with nice moduli stacks give rise to recognizable algebras in the untwisted case.)

The claims about the center can be further generalized. Consider an elliptic algebra of the form $B_{V,\Psi} := \bigoplus_i \mathrm{Hom}(V, \Psi^i V)$, where $\Psi$ is an autoequivalence of $\mathrm{coh}(E)$ of the form $M \mapsto \mathcal{L} \otimes (\phi^{-1})^* M$ with $\deg(\mathcal{L}) > 0$. If $\phi$ has finite order, so that $\Psi^d$ is twisting by a line bundle, then $B$ has a large center, namely the homogeneous coordinate ring of the quotient curve $E/\langle\phi\rangle$ relative to the norm of $\mathcal{L}$, with degrees multiplied by the order of $\phi$. When $\phi$ is a translation by a $d$-torsion point $q$, $E' := E/\langle\phi\rangle$ is again an elliptic curve, and thus its homogeneous coordinate ring has no derivations of negative degree. (Note that this could fail when $\phi$ is not a translation, since then $E'$ has genus 0. It would be interesting to understand the filtered deformations in such cases, but they fall outside the scope of the present work.)

Let $E'^+$ be the $\mathbb{G}_m$-torsor associated to the norm line bundle $N_{E/E'}(\mathcal{L})$ on $E'$, modified by pulling back the action of $\mathbb{G}_m$ through the $d$-th power map. (This is to correct for the fact that the degrees in $Z(B)$ are all multiples of $d$, and in particular the degree $dm$ elements are sections of $N_{E/E'}(\mathcal{L})^m$.) Then $B$ is the ring of global sections of a corresponding sheaf of algebras $\mathcal{B}$ on $E'^+$. This sheaf is $\mathbb{G}_m$-equivariant in a somewhat complicated way, but the (geometric) fibers are again just matrix algebras. Indeed, the $\mu_d$-invariant subalgebra of $\mathcal{B}$ is the direct image of the endomorphism ring of a vector bundle on a $\mathbb{G}_m$-torsor over $E$, so is locally a matrix algebra over a cyclic degree $d$ cover of $E'^+$. Relative to this interpretation, the other congruence classes correspond to spaces of linear transformations which are semilinear w.r.to the appropriate power of $\phi$. We thus find that $\mathcal{B}$ is locally a matrix algebra over a *cyclic* algebra, and is thus an Azumaya algebra.

It is then straightforward to extend the above arguments to show that $B$ has no negative derivations. We can actually extend this to the case that $\tau$ has infinite order, but this will require us to control derivations in finite characteristic, where the argument of Lemma 2.7 can fail. Luckily, if we modify the conclusion somewhat, we obtain a result that holds in general.

LEMMA 2.10. *Let $B$ be an elliptic algebra (possibly in finite characteristic), twisted by a torsion point of exact order $d$. Then any derivation of $B$ of degree not a multiple of $d$ is inner.*

PROOF. Let $D$ be a derivation of $B$ of degree $\delta$, with $\delta$ not a multiple of $d$. Since every element of $Z(B)$ has degree a multiple of $d$, it follows immediately that $D$ is trivial on $Z(B)$. We may then argue as in the proof of Lemma 2.7 that $D$ is locally inner; in other words, there is a collection of representations $D = \mathrm{ad}(z_i^{-1}x_i)$ with $z_i \in Z(B)$, $x_i \in B$, such that the open subsets $U_i := [z_i \neq 0]$ cover $E'$.

Now, since $\mathrm{ad}(z_i^{-1}x_i) = \mathrm{ad}(z_j^{-1}x_j)$, the system of pairs must also satisfy $z_jx_i - z_ix_j \in Z(B)$ for all $i, j$. Since $\deg(z_i^{-1}x_i) = \delta$, we see that

$$\deg(z_jx_i - z_ix_j) \equiv \delta \pmod{d},$$

which (again using the fact that elements $Z(B)$ have degree a multiple of $d$) forces $z_jx_i = z_ix_j$. But this implies that the local representations glue to give a global representation, so that $D$ is inner as required. □

LEMMA 2.11. *Let $B$ an elliptic algebra twisted by a translation $\tau$. If $B$ has a nonzero derivation of degree $-d$ with $d > 0$, then $\tau^d = 1$.*

PROOF. Suppose otherwise, so that $B$ has such a derivation, but $\tau^d \neq 1$. If the field of definition $K$ is infinitely generated, then we can replace it by the field generated by (a) the coefficients of some Weierstrass equation for $E$, (b) the coordinates of the point by which $\tau$ is a translation, and (c) the coordinates of the points corresponding to the determinants of stable constituents of $V$, so that we may assume $K$ is finitely generated. (See [**4**] or the next section for a discussion of the structure of $V$.)

If $K$ is finite, then $\tau$ has finite order, with $d$ not a multiple of $\mathrm{ord}(\tau)$, and thus Lemma 2.10 gives a contradiction. Otherwise, let $v$ be a valuation of $K$, with associated discrete valuation ring $R$ having residue field $k$ such that (a) $E_k$ is smooth, (b) $\tau_k^d \neq 1$, and (c) the nonisomorphic stable constituents of $V$ have nonisomorphic reductions. (Such a place exists because we are imposing finitely many open conditions.) Then $B_{V_R,\Psi_R}$ is a flat extension of $B_{V,\Psi}$, so that the dimension of the space of derivations of degree $-d$ is upper semicontinuous, implying that $B_k$ itself has a nonzero derivation of degree $-d$. Since $\dim(k) = \dim(K) - 1$, the claim follows by induction on the dimension. □

REMARK. A similar reduction shows that any elliptic algebra is Noetherian, by a direct generalization of the proof of [**1**, Thm. 8.3].

This is enough to establish the following for an arbitrary elliptic algebra in characteristic 0.

THEOREM 2.12. *Let $B$ be an elliptic algebra over a field of characteristic 0, twisted by an arbitrary translation. Then the moduli space of filtered deformations of $B$ is a closed substack of a weighted projective space.*

PROOF. If $\tau$ has finite order, then the arguments of Lemma 2.7 and Corollary 2.8 carry over mutatis mutandis to show that $B$ has no negative derivations. Otherwise, $\mathrm{ord}(\tau) = \infty$, and we may simply replace Lemma 2.7 by Lemma 2.11. □

The argument of Proposition 2.9 also carries over.

THEOREM 2.13. *Let $B$ be an elliptic algebra of characteristic 0 twisted by translation by a torsion point. The center of any filtered deformation of $B$ is a filtered deformation of $Z(B)$.*

## 3. Resolutions of elliptic algebras

Let $E/k$ be an elliptic curve over an algebraically closed field $k$, let $V$ be a vector bundle on $E$, and let $\Psi : \mathrm{coh}(E) \to \mathrm{coh}(E)$ be an autoequivalence of the form $M \mapsto \mathcal{L}\otimes(\tau^{-1})^* M$ with $\tau \in E \subset \mathrm{Aut}(E)$ and $\mathcal{L}$ an ample line bundle of degree $d > 0$. (In fact, most of the arguments below work even if $\tau$ is not a translation, since they only depend on how $\Psi$ affects slopes; the only exception is in the explicit calculations of minimal resolutions, where minimality also depends on how it affects determinants.) As mentioned above, these ingredients determine a graded algebra $B_{V,\Psi}$ such that for $n \in \mathbb{Z}$,

$$(B_{V,\Psi})_n = \mathrm{Hom}(V, \Psi^n V),$$

with the obvious composition. Since $\mathcal{L}$ is ample, this algebra is bounded below, and we are interested in its filtered deformations.

Any summand of $V$ determines an idempotent of $B_{V,\Psi}$, which lifts (nonuniquely) to any filtered deformation by Proposition 2.1. It follows immediately that applying a power of $\Psi$ to a summand of $V$ does not change the space of filtered deformations—it simply shifts the degrees of the corresponding off-diagonal blocks. Since any vector bundle on an elliptic curve is a sum of semistable bundles (see [**4**], which also completely determines the structure of indecomposable bundles) and $\Psi$ adds $d$ to the slope of a bundle, this freedom allows us to assume that every summand of $V$ has slope in $[0, d)$ (or, more generally, in any desired half-open interval of length $d$). Since a nonzero morphism between semistable bundles forces the slope of the codomain to be at least the slope of the domain, this constraint on $V$ implies that $(B_{V,\Psi})_n = 0$ for $n < 0$.

This, of course, is not quite canonical; if $V$ has $c$ different slopes (modulo $d$), then there are $c$ inequivalent choices of nonnegative gradings on $B_{V,\Psi}$. The most natural way to resolve this involves replacing $B_{V,\Psi}$ by a category. For each slope $\mu \in \mathbb{Q}$, let $V_{[\mu]}$ denote the direct sum of all slope $\mu$ indecomposable summands of the modules $\Psi^n V$, so that the construction of the previous paragraph involves replacing $V$ by $\bigoplus_{\mu\in[\alpha,\alpha+d)} V_{[\mu]}$. The slopes for which $V_{[\mu]} \neq 0$ form a discrete subset of $\mathbb{Q}$, a union of finitely many cosets of $d\mathbb{Z}$, and thus we may choose an ordered bijection between $\mathbb{Z}$ and the set of such slopes. If $i \mapsto \mu_i$ is that bijection, then we may define $V_i := V_{[\mu_i]}$. We then let $\mathcal{B}_{V,\Psi}$ be the subcategory of $\mathrm{coh}\, E$ determined by $V_i$; i.e., it is the category with objects indexed by integers and Hom spaces given by

$$\mathcal{B}_{V,\Psi}(i, j) = \mathrm{Hom}(V_i, V_j).$$

(This is a nonnegatively graded $\mathbb{Z}$-algebra in the sense of [**6**].) It is then straightforward to see that the category of modules over this category (i.e., contravariant functors to the category of vector spaces) is isomorphic to the category of graded $B_{V,\Psi}$-modules. Note that since $\Psi$ induces an autoequivalence shifting the objects by $c$, this category induces a graded algebra which in degree $d$ is

$$\bigoplus_{1\le i,j\le c} \mathcal{B}_{V,\Psi}(i, j + d).$$

This graded algebra agrees with the naïve grading on $B_{V,\Psi}$, up to rescaling the grading by $c$ and shifting the grading on off-diagonal blocks relative to the natural idempotents, and thus gives the same space of filtered deformations. (The only caveat is that the additional terms added to the relations must remain in the same congruence class modulo $c$, but this is actually forced by consistency with the degree 0 involutions.) With this in mind, where $B_{V,\Psi}$ appears below, we will always take this canonical grading.

The theory of minimal resolutions, though most familiar in the case of connected graded algebras, applies more generally (following [**8**]) to nonnegatively graded algebras which are finite-dimensional in every degree. In particular, any $B_{V,\Psi}$-module has such a minimal resolution, and thus the same applies to $\mathcal{B}_{V,\Psi}$. The one caveat here is that when $B_0$ is not semisimple, the minimal resolution of a finitely graded module need not be convergent; that is, there may be degrees such that the restriction of the complex to those degrees gives an infinite resolution.

LEMMA 3.1. *Let $B$ be a nonnegatively graded $k$-algebra such that $B_0$ is finite-dimensional. Then the minimal resolution of $B_0$ as a graded $B$-module is convergent iff each $B_n$ has finite homological dimension as a $B_0$-module. Moreover, if this holds, then for any bounded below graded $B$-module $M$, $M$ has convergent minimal resolution iff each $M_n$ has finite homological dimension.*

PROOF. First suppose that $B_n$ and $M_n$ all have finite homological dimension, and WLOG assume that $M_n = 0$ for $n < 0$ and $M_0 \neq 0$. Then the restriction of the minimal resolution of $M$ to degree 0 gives a (minimal!) projective resolution of $M_0$ as a $B_0$-module, which thus has finite length. In particular, projective modules of degree 0 appear in only finitely many terms, and thus we may consider the partial resolution truncated at the last point where they occur. In each degree, the terms of the resulting complex have finite homological dimension (being homogeneous components of $B$ or $M$) and thus, since the complex is exact except at the very left, the kernel also has finite homological dimension. This gives us a new graded module $M'$ satisfying the same hypotheses but with a larger lower bound on its degrees, and thus the result follows by induction. This gives the "if" part of the second claim, and thus (taking $M = B_0$) of the first claim as well.

Now, supposing still that the components $B_n$ all have finite homological dimension, let $M$ be a bounded below module with at least one homogeneous component of infinite homological dimension. Then the same argument of the previous paragraph allows us to reduce to the case that $M_n = 0$ for $n < 0$ but $M_0$ has infinite homological dimension. (Here we use the fact that in a partial resolution by modules of finite homological dimension, the kernel has finite dimension iff the cokernel does.) But then the minimal resolution in degree 0 is a projective $B_0$-module resolution of $M_0$, which is infinite by assumption, preventing convergence.

Finally, if $B$ has a homogeneous component of infinite homological dimension, let $n$ be the minimal such degree. Then the degree $n$ part of the minimal resolution of $B_0$ is a resolution of $B_n$ by sums of summands of components $B_i$ for $i < n$, and thus of terms having finite homological dimension. Since $B_n$ has infinite homological dimension, this forces the resolution to have infinite length. □

This property holds for the canonical grading on $B_{V,\Psi}$, and in fact we have something stronger.

LEMMA 3.2. *Suppose that $V$, $W$ are indecomposable bundles with $\mu(V) < \mu(W)$. Then* $\mathrm{Hom}(V,W)$ *is free as a* $\mathrm{End}(V) \otimes_k \mathrm{End}(W)^{op}$*-module.*

PROOF. We have $\mathrm{End}(V) \cong k[x]/(x^n)$ and $\mathrm{End}(W) \cong k[y]/(y^m)$ for suitable positive integers $m$ and $n$, and thus we need to show that $\mathrm{Hom}(V,W)$ is a free $k[x,y]/(x^n,y^m)$-module, or equivalently that

$$\dim_k \mathrm{Hom}(V,W) = mn \dim_k(\mathrm{Hom}_{\mathrm{End}(V)}(k, \mathrm{Hom}(V,W) \otimes_{\mathrm{End}(W)} k)),$$

or in other words

$$\dim_k \mathrm{Hom}(V,W) = mn \dim_k(\mathrm{Hom}(V \otimes_{\mathrm{End}(V)} k, W \otimes_{\mathrm{End}(W)} k)).$$

Since $V \otimes_{\mathrm{End}(V)} k$ is the stable constituent of $V$, and similarly for $W \otimes_{\mathrm{End}(W)} k$, this certainly applies at the level of Euler characteristics, and the difference in slopes forces $\mathrm{Ext}^1$ to vanish (by duality). □

LEMMA 3.3. *Suppose that $V$, $W$ are semistable bundles with $\mu(V) < \mu(W)$. Then* $\mathrm{Hom}(V,W)$ *is projective as a* $\mathrm{End}(V) \otimes_k \mathrm{End}(W)^{op}$*-module.*

PROOF. Since each of $V$ and $W$ splits naturally as a direct sum over nonisomorphic stable constituents, we may assume that each is built from a single stable bundle. In particular, $V$ has a unique isomorphism class $V_{\max}$ of indecomposable summands of maximal rank, and one has

$$V \cong V_{\max} \otimes_{\mathrm{End}(V_{\max})} M_V$$

for some faithful $\mathrm{End}(V_{\max})$-module $M_V$, with a similar description applying to $W$. We then find that

$$\mathrm{Hom}(V,W) \cong \mathrm{Hom}_{\mathrm{End}(V_{\max})}(M_V, \mathrm{Hom}(V_{\max}, W_{\max}) \otimes_{\mathrm{End}(W_{\max})} M_W).$$

Since $\mathrm{Hom}(V_{\max}, W_{\max})$ is free, it remains only to show that

$$\mathrm{Hom}_{\mathrm{End}(V_{\max})}(M_V, \mathrm{End}(V_{\max})) \otimes_k M_W$$

is projective as an $\mathrm{End}(V) \otimes_k \mathrm{End}(W)^{op}$-module.

Here $M_W$ is a faithful module over $\mathrm{End}(W_{\max}) \cong k[x]/(x^n)$, and (since $x$ is central in $\mathrm{End}(W)$) we have $\mathrm{End}(W) \cong \mathrm{End}_{k[x]/(x^n)}(M_W)$. In particular, $M_W \cong \bigoplus_i k[x]/(x^{m_i})$ with $n = m_1 \geq m_2 \geq \cdots$, and thus there is an idempotent cutting out a copy of $k[x]/(x^n)$, so that the corresponding submodule of $\mathrm{End}_{k[x]/(x^n)}(M_W)$ is $M_W$, implying projectivity. The claim for $V$ is analogous (relative now to the left $\mathrm{End}(V)$-module structure on $\mathrm{End}(V)$). □

PROPOSITION 3.4. *The homogeneous components of $B_{V,\Psi}$ are projective modules over $(B_{V,\Psi})_0$.*

PROOF. This reduces to showing that the Hom spaces $\mathcal{B}_{V,\Psi}(i,j)$ are projective over the relevant endomorphism ring. If $j = i$, this is immediate, while if $j > i$, we have in fact shown that it is projective over the tensor product of endomorphism rings. □

We thus see that minimal resolutions of $B_{V,\Psi}$-modules (or, equivalently, $\mathcal{B}_{V,\Psi}$-modules) are convergent as long as they are bounded below and have homogeneous components of finite homological dimension. Let $P_i$ denote the projective $\mathcal{B}_{V,\Psi}$-module given by the contravariant functor

$$j \mapsto \mathcal{B}_{V,\Psi}(j,i),$$

and let $S_i$ be the module agreeing with $P_i$ in degree $i$ but 0 in all other degrees. Since $S_i(i) = \operatorname{End}(V_i)$, $S_i$ has a convergent projective resolution.

An important property of the $\mathbb{Z}$-algebras $\mathcal{B}_{V,\Psi}$ is that they are Gorenstein in a suitable sense.

PROPOSITION 3.5. *The space*

$$\operatorname{Ext}^p(S_i, P_j)$$

*vanishes unless $i = j$ and $p = 2$, in which case there is a canonical isomorphism*

$$\operatorname{Ext}^2(S_i, P_i) \cong \operatorname{Ext}^1(V_i, V_i).$$

PROOF. The module $S_i$ has a convergent projective resolution

$$\cdots \to Q_{i+2} \to Q_{i+1} \to P_i \to S_i.$$

For each $j > i$, projectivity of $Q_j$ implies that it is the module of global sections of a vector bundle $W_j$ on $E$ (this is true for any summand of any $P_j$) and thus the complex

$$\cdots \to W_{i+2} \to W_{i+1} \to V_i$$

of sheaves on $E$ is exact. Now, $R\operatorname{Hom}(S_i, P_j)$ is represented by the complex

$$\operatorname{Hom}(P_i, P_j) \to \operatorname{Hom}(Q_{i+1}, P_j) \to \operatorname{Hom}(Q_{i+2}, P_j) \to \cdots,$$

or equivalently

$$\operatorname{Hom}(V_i, V_j) \to \operatorname{Hom}(W_{i+1}, V_j) \to \operatorname{Hom}(W_{i+2}, V_j) \to \cdots.$$

Since the double complex

$$R\operatorname{Hom}(V_i, V_j) \to R\operatorname{Hom}(W_{i+1}, V_j) \to R\operatorname{Hom}(W_{i+2}, V_j) \to \cdots$$

represents $R\operatorname{Hom}(0, V_j)$, it is acyclic, which, since every column has magnitude contained in $[0, 1]$ (as a $R\operatorname{Hom}$ between sheaves on $E$), implies a quasi-isomorphism between the complex of Hom spaces and a shift of the complex

$$\operatorname{Ext}^1(V_i, V_j) \to \operatorname{Ext}^1(W_{i+1}, V_j) \to \operatorname{Ext}^1(W_{i+2}, V_j) \to \cdots,$$

or (after fixing an isomorphism $H^1(\mathcal{O}_E) \cong k$)

$$\operatorname{Hom}(V_j, V_i)^* \to \operatorname{Hom}(V_j, W_{i+1})^* \to \operatorname{Hom}(V_j, W_{i-2})^* \to \cdots.$$

This is dual to the complex

$$\cdots \to \operatorname{Hom}(V_j, W_{i+2}) \to \operatorname{Hom}(V_j, W_{i+1}) \to \operatorname{Hom}(V_j, V_i),$$

or equivalently

$$\cdots \to \operatorname{Hom}(P_j, Q_{i+2}) \to \operatorname{Hom}(P_j, Q_{i+1}) \to \operatorname{Hom}(P_j, P_i),$$

which (up to a shift) represents $R\operatorname{Hom}(P_j, S_i) \cong \operatorname{Hom}(P_j, S_i)$. This vanishes unless $i = j$, when it equals $\operatorname{End}(V_i)$.

We thus conclude that $\operatorname{Ext}^p(S_i, P_j) = 0$ unless $i = j$ and (taking into account the shift) $p = 2$, in which case one has a canonical isomorphism

$$\operatorname{Ext}^2(S_i, P_j) \cong \operatorname{Hom}_k(\operatorname{End}(V_i), H^1(\mathcal{O}_E)) \cong \operatorname{Ext}^1(V_i, V_i)$$

as required. $\square$

Now, let $\mathcal{B}^+$ be the $\mathbb{Z}$-algebra obtained from $\mathcal{B}_{V,\Psi}$ by adjoining a central element $t$ of degree $c$. In other words,

$$\mathcal{B}^+(i,j) = \bigoplus_{l\ge 0} \mathcal{B}_{V,\Psi}(i, j-cl),$$

with $t \in \mathcal{B}^+(i,i+c)$ the element corresponding to $1 \in \mathcal{B}_{V,\Psi}(i,i)$. Let $P_i^+$ be the natural family of projective $\mathcal{B}^+$-modules, with associated modules $S_i^+$, and note that we still have $\mathcal{B}^+(i,i) = \operatorname{Hom}(V_i, V_i)$.

COROLLARY 3.6. *The space*

$$\operatorname{Ext}^p(S_i^+, P_j^+)$$

*vanishes unless $j = i - c$ and $p = 3$, in which case there is a canonical isomorphism*

$$\operatorname{Ext}^3(S_i^+, P_{i-c}^+) \cong \operatorname{Ext}^1(V_i, V_i).$$

PROOF. Since $P_j^+$ restricts to $\bigoplus_{l\ge 0} P_{j-lc}$, we have

$$\operatorname{Ext}^p(S_i[t], P_j^+) \cong \bigoplus_{l\ge 0} \operatorname{Ext}^p(S_i, P_{j-lc}),$$

where $S_i[t]$ is the induced module. We may thus use the short exact sequence

$$0 \to S_i[t][-c] \to S_i[t] \to S_i^+ \to 0$$

to obtain a long exact sequence

$$\cdots \to \operatorname{Ext}^p(S_i^+, P_j^+) \to \bigoplus_{l\ge 0} \operatorname{Ext}^p(S_i, P_{j-lc}) \to \bigoplus_{l\ge -1} \operatorname{Ext}^p(S_i, P_{j-lc}) \to \cdots$$

This immediately gives an isomorphism

$$\operatorname{Ext}^p(S_i^+, P_j^+) \cong \operatorname{Ext}^{p-1}(S_i, P_{j+c}),$$

so the formula reduces to the analogous formula for $\mathcal{B}$. □

Given a graded algebra (or $\mathbb{Z}$-algebra) $A$, there is an associated category $\operatorname{Proj}(A)$, the quotient of the category of all graded modules by the subcategory of "torsion" modules (in which every element generates a finite-dimensional submodule). The Ext groups in that category can then be computed as limits of Ext groups in the category of graded modules, per [**3**], subject to a certain technical condition ("property $\chi$") which is satisfied for $B_{V,\Psi}$ by Thm 4.5 op. cit. and thus for the Rees algebra of any filtered deformation by Thm. 8.8 op. cit. Indeed, in our setting, the only nontrivial hypothesis between the two theorems is that $\Psi$ is "ample" relative to $V$: for any coherent $M$, there exists $n$ such that $\operatorname{Ext}^p(\Psi^{-n}V, M) = 0$ for $p > 0$ and $\operatorname{Hom}(\Psi^{-n}V, M) \otimes_k \Psi^{-n}V \to M$ is surjective. Both facts reduce immediately to the case that both $V$ and $M$ are indecomposable, with vanishing of $\operatorname{Ext}^1$ an immediate consequence via duality of the eventual inequality in slopes, and global generation proved in Corollary 3.15 below.

THEOREM 3.7. *Let $\bar{P}_i^+$ denote the object of* $\operatorname{Proj}(\mathcal{B}^+)$ *corresponding to $P_i^+$. Then $\operatorname{Ext}^p(\bar{P}_i^+, \bar{P}_j^+) = 0$ unless $p = 0$ and $j \ge i$ or $p = 2$ and $j \le i - c$, with*

$$\operatorname{Hom}(\bar{P}_i^+, \bar{P}_j^+) \cong \operatorname{Hom}(V_i, V_j)$$

*and*

$$\operatorname{Ext}^2(\bar{P}_i^+, \bar{P}_j^+) \cong \operatorname{Ext}^1(V_i, V_{j+c}).$$

PROOF. We have

$$\operatorname{Ext}^p(\bar{P}_i^+, \bar{P}_j^+) \cong \lim_{n\to\infty} \operatorname{Ext}^p((P_i^+)_{\geq n}, P_j^+).$$

Now consider the short exact sequence

$$0 \to (P_i^+)_{\geq n} \to P_i^+ \to P_i^+/(P_i^+)_{\geq n} \to 0.$$

Since $P_i^+$ is projective, $R\operatorname{Hom}(P_i^+, P_j^+) = \operatorname{Hom}(V_i, V_j)$ concentrated in degree 0. Moreover, since the degree $l$ component of $P_i^+$ is a free $\operatorname{End}(V_l)$-module, so $P_i^+/(P_i^+)_{\geq n}$ is a finite extension of such modules, we find that for $n > j - c$,

$$R\operatorname{Hom}(P_i^+/(P_i^+)_{\geq n}, P_j^+) \cong \operatorname{Ext}^1(V_i, V_{j+c})$$

concentrated in degree 3. The desired claim follows immediately. □

For the next claim, we revert to one of the naïve nonnegative gradings of $B_{V,\Psi}$.

COROLLARY 3.8. *Let $A^+$ be the Rees algebra of a filtered deformation of $B_{V,\Psi}$, and let $\mathcal{O}(l)$ be the objects of* $\operatorname{Proj}(A^+)$ *corresponding to shifts of $A^+$. Then for $l \leq m$, $\operatorname{Ext}^p(\mathcal{O}(l), \mathcal{O}(m)) = 0$ for $p \neq 0$ and for $l > m$, $\operatorname{Ext}^p(\mathcal{O}(l), \mathcal{O}(m)) = 0$ for $p \neq 2$. Moreover, there is a canonical duality*

$$\operatorname{Ext}^2(\mathcal{O}(l), \mathcal{O}(m)) \cong \operatorname{Hom}_k(\operatorname{Hom}(\mathcal{O}(m), \mathcal{O}(l-1)), H^1(\mathcal{O}_E)).$$

*In particular, $\mathcal{O}(0)$ has no higher endomorphisms, and if $V$ is stable, then $\mathcal{O}(0)$ is exceptional.*

PROOF. The $\mathbb{G}_m$ action on filtered deformations gives rise to a family of graded algebras with generic fiber isomorphic to $A^+$ and special fiber associated to the trivial deformation $B^+ \cong B[t]$. By semicontinuity, it suffices to prove vanishing when $A^+$ is trivial, where it reduces to computing

$$R\operatorname{Hom}(\bigoplus_{0\leq i<c} \bar{P}^+_{i-lc}, \bigoplus_{0\leq i<c} \bar{P}^+_{i-mc})$$

in $\operatorname{Proj}(\mathcal{B}^+)$. Note that since each case has precisely one nonvanishing cohomology space, the corresponding sheaves on $\mathbb{A}^1 = \mathbb{G}_m \cup \{0\}$ are flat.

For the duality between nonvanishing Ext spaces, we note that $\operatorname{coh}(E) \cong \operatorname{Proj}(B_{V,\Psi})$ embeds in $\operatorname{Proj}(A^+)$ as the subcategory on which the natural transformation $-(-1) \to \mathrm{id}$ associated to the central element of $A^+$ vanishes. This embedding factors into a pair $i_*i^*$ of adjoint functors with $i^* : \operatorname{Proj}(A^+) \to \operatorname{coh}(E)$ exact, so that we have distinguished triangles

$$M(-1) \to M \to Ri_*i^*M \to$$

for any $M$ and thus distinguished triangles

$$R\operatorname{Hom}(M, N(-1)) \to R\operatorname{Hom}(M,N) \to R\operatorname{Hom}_E(i^*M, i^*N) \to .$$

In particular, taking $M = N = \mathcal{O}(l)$ gives a canonical isomorphism

$$\operatorname{Ext}^2(\mathcal{O}(l), \mathcal{O}(l-1)) \cong \operatorname{Ext}^1(\Psi^l V, \Psi^l V)$$

and thus a canonical trace

$$\operatorname{Ext}^2(\mathcal{O}(l), \mathcal{O}(l-1)) \to H^1(\mathcal{O}_E)(\cong k).$$

This induces pairings in the usual way, and it remains only to show that the pairings are perfect. But this follows from another semicontinuity argument together with the fact they they are perfect in $\operatorname{Proj}(B^+)$. □

REMARK. Note that the trace form produces two pairings:

$$\mathrm{Hom}(\mathcal{O}(l),\mathcal{O}(m))\otimes \mathrm{Ext}^2(\mathcal{O}(m),\mathcal{O}(l-1))\to \mathrm{Ext}^2(\mathcal{O}(l),\mathcal{O}(l-1))\to k$$

and

(3.1)

$$\begin{aligned}&\mathrm{Ext}^2(\mathcal{O}(m),\mathcal{O}(l-1))\otimes \mathrm{Hom}(\mathcal{O}(l),\mathcal{O}(m))\\ \cong\ &\mathrm{Ext}^2(\mathcal{O}(m),\mathcal{O}(l-1))\otimes \mathrm{Hom}(\mathcal{O}(l-1),\mathcal{O}(m-1))\to \mathrm{Ext}^2(\mathcal{O}(m),\mathcal{O}(m-1))\to k.\end{aligned}$$

The corresponding pairings in $\mathrm{Proj}(B^+)$ agree (since the analogous pairings agree in $\mathrm{Proj}(B_{V,\Psi})\cong \mathrm{coh}(E)$), and thus they differ by a compatible family of automorphisms of $\mathrm{Hom}(\mathcal{O}(l),\mathcal{O}(m))$ agreeing with 1 on the associated graded, or in other words by an automorphism of $A^+$ acting as 1 on the associated graded. But this implies that the pairings agree: if this automorphism were nontrivial, it would induce a nonzero derivation of $B_{V,\Psi}$ of negative degree. It follows that (up to a choice of holomorphic differential on $E$) $-(-1)[2]$ is a Serre functor on $\mathrm{Proj}(A^+)$, since the relevant duality and compatibility relations hold on generators. (In particular, $S[-2]=-(-1)$ is an abelian autoequivalence, so that $\mathrm{Proj}(A^+)$ is Gorenstein.)

If the components of the slope decomposition of $V$ have commutative endomorphism rings, we can be more precise about the structure of the minimal resolution. To understand this, we first recall some facts about derived autoequivalences of elliptic curves. In addition to the abelian autoequivalences, which consist of automorphisms of $E$ along with translations and twists by line bundles, there is an additional autoequivalence $\Phi_{\mathcal{O}_E}$ which fits into a distinguished triangle

$$\mathcal{O}_E\otimes_k R\mathrm{Hom}(\mathcal{O}_E,M)\to M\to \Phi_{\mathcal{O}_E}M\to .$$

(We may either use Fourier-Mukai theory or a suitable dg-enhancement to see that this cone can be made functorial.) The inverse is also a cone:

$$\Phi_{\mathcal{O}_E}^{-1}M\to M\to \mathrm{Hom}_k(\mathrm{Hom}(M,\mathcal{O}_E),\mathcal{O}_E)\to,$$

which can be seen by observing that

$$R\mathrm{Hom}(\mathcal{O}_E,\Phi_{\mathcal{O}_E}M)\cong \mathrm{Ext}^1(\mathcal{O}_E,\mathcal{O}_E)\otimes R\mathrm{Hom}(\mathcal{O}_E,M)\cong R\mathrm{Hom}(M,\mathcal{O}_E)^*$$

Twisting by the line bundle $\mathcal{O}_E(z)$ (where $z$ is the identity) has a similar description:

$$\mathcal{O}_z\otimes_k R\mathrm{Hom}(\mathcal{O}_z,M)\to M\to M(z)\to$$

so that we may call it $\Phi_{\mathcal{O}_z}$. (These functors generate a central extension of $\mathrm{SL}_2(\mathbb{Z})$.) The following is then standard (essentially Atiyah's classification of vector bundles [**4**]):

PROPOSITION 3.9. *Any semistable sheaf on $E$ is the image of a torsion sheaf under a composition of the functors $\Phi_{\mathcal{O}_E}$ and $\Phi_{\mathcal{O}_z}$ and their inverses.*

REMARK. If the semistable sheaf $M$ is not already torsion, then we twist by a suitable power of $\mathcal{O}_E(z)$ to put its slope in $(0,1]$ and observe that applying $\Phi_{\mathcal{O}_E}$ gives a sheaf of smaller rank, so that iterating eventually produces a torsion sheaf.

REMARK. Since $\Phi_{\mathcal{O}_E}$ and $\Phi_{\mathcal{O}_z}$ both preserve the degree 0 divisor class $c_1(M)-\deg(M)[z]$ and a stable sheaf is uniquely determined by its determinant, the torsion sheaf that results is uniquely determined, even though the derived equivalence getting there is not.

REMARK. Note that since the action of $\Psi_{\mathcal{O}_E}$ and $\Phi_{\mathcal{O}_z}$ preserves $\gcd(\operatorname{rank}(M), \deg(M))$, we conclude that $M$ has $\gcd(\operatorname{rank}(M), \deg(M))$ stable constituents (counted with multiplicity), by reduction to the torsion case.

LEMMA 3.10. *Suppose that $M \in \operatorname{coh}(E)$ is such that $\operatorname{End}(M)$ is commutative. Then $M$ is semistable, and each indecomposable summand of $M$ is S-equivalent to a power of a different isomorphism class of stable sheaves.*

PROOF. Let $M = \bigoplus M_i$ be the decomposition of $M$ into indecomposable summands. We clearly have $\operatorname{Hom}(M_i, M_j) = 0$ for $i \neq j$, and thus $\mu(M_i) = \mu(M_j)$. Since an indecomposable sheaf is semistable, it follows immediately that $M$ is semistable. Moreover, each indecomposable sheaf is an iterated extension of a unique stable constituent, and the vanishing of $\operatorname{Hom}(M_i, M_j)$ forces the stable constituents to be nonisomorphic. □

COROLLARY 3.11. *A sheaf $M$ on $E$ has commutative endomorphism ring iff it is the image of the structure sheaf of an effective divisor $D_M$ under an element of the group of derived autoequivalences generated by $\Phi_{\mathcal{O}_E}$ and $\Phi_{\mathcal{O}_z}$.*

PROOF. Since the structure sheaf of an effective divisor is commutative, the image of such a sheaf under a derived autoequivalence also has commutative endomorphism ring. Conversely, if $M$ has commutative endomorphism ring, then the torsion sheaf it is derived equivalent to must also have commutative endomorphism ring, so is the structure sheaf of a 0-dimensional subscheme as required. □

REMARK. With this in mind, we call sheaves on $E$ with commutative endomorphism ring "divisorial".

Now, given a divisorial sheaf $M$, consider the functor $\Phi_M$ on $D^b \operatorname{coh}(E)$ fitting into the distinguished triangle

$$M \otimes^{\mathbf{L}}_{\operatorname{End}(M)} R\operatorname{Hom}(M, N) \to N \to \Phi_M(N) \to .$$

(Again, we either use a Fourier-Mukai kernel or a dg-enhancement to ensure that this functor is well-defined.)

PROPOSITION 3.12. *If $M$ is divisorial, then the functor $\Phi_M$ is a derived autoequivalence, and if $N$ is a semistable sheaf then either $\Phi_M(N)$ or $\Phi_M(N)[-1]$ is a semistable sheaf.*

PROOF. The conjugate of the functor $\Phi_M$ under a derived autoequivalence has the same form, and thus to see that it is a derived autoequivalence, it suffices to consider the case that $M$ is the structure sheaf of a divisor. But in that case, $\Phi_{\mathcal{O}_D} N \cong N(D)$, which is clearly an autoequivalence. It also follows from this that the functor $M \otimes_{\operatorname{End}(M)} -$ is exact, and thus we may take the underived tensor product above.

For any sheaf $N$, we then have an exact sequence

$$0 \to h^{-1}\Phi_M(N) \to M \otimes_{\operatorname{End}(M)} \operatorname{Hom}(M, N) \to N \to h^0\Phi_M(N) \\ \to M \otimes_{\operatorname{End}(M)} \operatorname{Ext}^1(M, N) \to 0$$

with $h^p\Phi_M(N) = 0$ for $p \notin \{-1, 0\}$. If $N$ is stable, then it is simple ($\operatorname{End}(N) \cong k$), and the same applies to $\Phi_M(N)$, forcing $\Phi_M(N)$ to be a shift of a simple (and thus stable!) sheaf. (Since $\operatorname{Ext}^2$ vanishes in $\operatorname{coh}(E)$, any complex in $D^b_{\operatorname{coh}}(E)$

is formal, and thus its endomorphism ring contains at least one idempotent for each nonvanishing cohomology sheaf.) Since there are only two possible shifts, we can distinguish whether it is a sheaf or a shift from its class in the *numerical* Grothendieck group, so that the claim immediately extends to semistable sheaves. □

COROLLARY 3.13. *If the divisorial sheaf $M$ has stable constituents $M_1,\dots,M_m$ (with multiplicity, and in any order), then $\Phi_M \cong \Phi_{M_1} \circ \cdots \circ \Phi_{M_m}$. Moreover, $\Phi_M$ depends only on* $\operatorname{rank}(M)$ *and* $\det(M)$.

PROOF. When $M$ is torsion, the claims are immediate; here the second claim simply says that the autoequivalence $N \mapsto N(D)$ depends only on the isomorphism class of the line bundle $\mathcal{O}_E(D)$. The claim in general follows by conjugating by the appropriate element of the group generated by $\Phi_{\mathcal{O}_E}$ and $\Phi_{\mathcal{O}_z}$. □

REMARK. Note that the claim as stated descends to smooth genus 1 curves over non-closed fields, where the functor $\Phi_{\mathcal{O}_z}$ is not defined.

The second claim can be restated as saying that $\Phi_M$ only depends on the class of $M$ in $K_0(E)$. If $M$ is stable, then we clearly have

$$[\Phi_M(N)] = [N] - \chi(M,N)[M],$$

where $\chi(M,N) := \sum_{i\in\mathbb{Z}}(-1)^i \dim \operatorname{Ext}^i(M,N)$ is the Mukai pairing on $K_0$. More generally, if $M$ has $m = \gcd(\operatorname{rank}(M),\deg(M))$ stable constituents, then we can use the factorization along with the fact that $\chi(M_i,M_j)=0$ (since they are stable sheaves of the same slope) to find that

$$[\Phi_M(N)] = [N] - m^{-1}\chi(M,N)[M].$$

Since

$$\chi(M,N) = \operatorname{rank}(M)\deg(N) - \deg(M)\operatorname{rank}(N),$$

this lets us explicitly compute the action of $\Phi_M$ on the Grothendieck group.

COROLLARY 3.14. *Suppose that the divisorial sheaf $M$ has rank $r$, degree $d$, and $m = \gcd(r,d)$ stable constituents. Then the autoequivalence $\Phi_M$ acts on the Grothendieck group of $E$ by*

$$\begin{aligned}\operatorname{rank}(\Phi_M(N)) &= \frac{m+dr}{m}\operatorname{rank}(N) - \frac{r^2}{m}\deg(N)\\ \deg(\Phi_M(N)) &= \frac{d^2}{m}\operatorname{rank}(N) + \frac{m-dr}{m}\deg(N)\end{aligned} \tag{3.2}$$

*and*

$$\det(\Phi_M(N)) \cong \det(N)\otimes\det(M)^{-(r\deg(N)-d\operatorname{rank}(N))/m}.$$

REMARK. As a function of $m$ with $r/m$, $d/m$ fixed, this gives a homomorphism from $\mathbb{Z}$ to $\mathrm{SL}_2(\mathbb{Z})$; in particular, the inverse is just the image under negating $m$, $r$, $d$.

In the following statement, and below, the convention is that the slope of a (nonzero) torsion sheaf is *positive* infinity. (This is compatible with the definition of semistability: a semistable sheaf of positive rank has no torsion subsheaf but plenty of torsion quotients.)

COROLLARY 3.15. *Let $V$ be a divisorial vector bundle with* $\operatorname{rank}(V) = r$, $\deg(V) = d$. *Then for any semistable sheaf $M$, the morphism*

$$V \otimes_{\operatorname{End}(V)} \operatorname{Hom}(V, M) \to M$$

*is surjective if*

$$\mu(M) > \mu(V) + \frac{\gcd(r,d)}{r^2},$$

*and is otherwise injective. In either case, the (co)kernel is semistable, of slope $\le \mu(V) - \frac{\gcd(r,d)}{r^2}$ iff the morphism is surjective.*

PROOF. The strict lower bound on the slope of $M$ ensures that $\operatorname{rank}(\Phi_V(M)) < 0$, and thus $\Phi_V(M)[-1]$ is a semistable sheaf fitting into a short exact sequence

$$0 \to \Phi_V(M)[-1] \to V \otimes_{\operatorname{End}(V)} \operatorname{Hom}(V, M) \to M \to 0.$$

If the opposite inequality holds, then $\operatorname{rank}(\Phi_V(M)) > 0$, so that $\Phi_V(M)$ is a semistable sheaf and the morphism is injective. Finally, if equality holds, then $\operatorname{rank}(M) > 0$ and eliminating $\deg(M)$ gives (assuming $V$ has $m$ stable constituents)

$$\deg(\Phi_V(M)) = \frac{m}{r^2} \operatorname{rank}(M),$$

which since $\operatorname{rank}(\Phi_V(M)) = 0$ again forces $\Phi_V(M)$ to be a semistable sheaf.

If $\mu(M) > \mu(V)$, the map vanishes, so the cokernel is certainly semistable. When $\mu(M) < \mu(V)$, the (co)kernel is a shift of $\Phi_V(M)$, so is semistable by Proposition 3.12; when equality holds (forcing injectivity), this fails, but the morphism is between semistable bundles of the same slope, and thus the cokernel must also have that slope (which also agrees with the slope of $\Phi_V(M)$).

Since $\mu(\Phi_V(M))$ is a linear fractional function of $\mu(M)$, the surjective region in terms of $\mu(\Phi_V(M))$ is the image of the interval $(\mu(V) + \frac{\gcd(r,d)}{r^2}, \infty]$ under that LFT, giving the stated result. □

PROPOSITION 3.16. *Let $V$ be a divisorial vector bundle on $E$, and suppose $M$ is a coherent sheaf such that $(\Phi_V\Psi)^n M[-n]$ is a sheaf for all $n$. Then $M$ has a natural resolution of the form*

$$\cdots \to \Psi^{-n} V \otimes_{\operatorname{End}(V)} M_n \to \cdots \to \Psi^{-1} V \otimes_{\operatorname{End}(V)} M_1 \to M \to 0$$

*for suitable* $\operatorname{End}(V)$*-modules $M_i$, and the corresponding complex of $B_{V,\Psi}$-modules is exact in all positive $B$-degrees.*

PROOF. For the first claim, we first refine to say that there is an exact sequence

$$0 \to \Psi^{-n}(\Phi_V\Psi)^n M[-n] \to \Psi^{-n} V \otimes_{\operatorname{End}(V)} M_n \to \cdots \to \Psi^{-1} V \otimes_{\operatorname{End}(V)} M_1 \to M \to 0.$$

for all $n$. This is straightforward: it is trivial for $n = 0$, and we may increase $n$ using the short exact sequence

$$\begin{aligned} 0 \to \Psi^{-n-1}(\Phi_V\Psi)^{n+1} M[-n-1] \to \Psi^{-n-1} V \otimes_{\operatorname{End}(V)} \operatorname{Hom}(V, \Psi(\Phi_V\Psi)^n M[-n]) \\ \to \Psi^{-n}(\Phi_V\Psi)^n M[-n] \to 0 \end{aligned} \tag{3.3}$$

obtained by applying $\Psi^{-n-1}$ to the short exact sequence associated to $\Phi_V(\Psi(\Phi_V\Psi)^n M[-n])$.

To see that this also gives a resolution of the (truncated!) $B$-module associated to $M$, observe that every term (except possibly $M$ itself) of the sequence for $n$ is acyclic for $R\operatorname{Hom}(V, \Psi^{n+1}-)$. For the intermediate terms, this is immediate from the fact that $\mu(\Psi^i V) > \mu(V)$, while for the first term it follows from the fact that

$(\Phi_V\Psi)^{n+1}M[-n-1]$ is a sheaf. In particular, the spectral sequence computing the (vanishing!) hypercohomology collapses immediately, and thus all homology groups vanish as required. □

This is particularly powerful for the following reason: the condition that $(\Phi_V\Psi)^n M[-n]$ should be a sheaf is essentially *numerical* in nature. Indeed, it follows easily (by considering indecomposable summands of $M$) that this holds iff it holds for every stable constituent of $M$. But for a stable sheaf, we know that $(\Phi_V\Psi)^n M[-n]$ is a sheaf iff $\operatorname{rank}((\Phi_V\Psi)^j M[-j]) > 0$ for $1 \le j \le n$, iff $(-1)^j \operatorname{rank}((\Phi_V\Psi)^j M) > 0$ for $1 \le j \le n$.

COROLLARY 3.17. *With $V$, $M$ as above, suppose that $M$ is a semistable bundle. Then the above conclusion holds iff the power series*

$$\frac{1 + (r^2\mu(M) + (m - dr))t/m}{1 - (\deg(\mathcal{L})r^2/m - 2)t + t^2}$$

*has positive coefficients, where $r = \operatorname{rank}(V)$, $d = \deg(V)$, and $m = \gcd(r, d)$ is its number of stable constituents.*

PROOF. We need $(-1)^j \operatorname{rank}((\Phi_V\Psi)^j M) > 0$ for $j \ge 1$, or equivalently that the power series

$$\sum_{j\ge 1} (-t)^j \operatorname{rank}((\Phi_V\Psi)^j M)$$

has positive coefficients. Since $\Phi_V\Psi$ acts linearly on the Grothendieck group, we may rephrase this as saying that all nonconstant coefficients of

$$\operatorname{rank}(\sum_{j\ge 0} (-t)^j (\Phi_V\Psi)^j [M]) = \operatorname{rank}((1 + t\Phi_V\Psi)^{-1}[M])$$

are positive. The claim then follows by inverting the relevant $2\times 2$ matrix and dividing by $\operatorname{rank}(M)$ to rewrite it in terms of the slope. □

REMARK. In the case that $M$ is torsion, we obtain a resolution iff

$$\frac{1}{1 - (\deg(\mathcal{L})r^2/m - 2)t + t^2}$$

has positive coefficients, or equivalently when

$$\deg(\mathcal{L})r^2 \ge 4m.$$

LEMMA 3.18. *If $\tau \ge 2$, then the power series $(1 + \alpha t)/(1 - \tau t + t^2)$ has positive coefficients iff $\alpha \ge -(\tau + \sqrt{\tau^2 - 4})/2$.*

PROOF. We may write $\tau = \lambda + 1/\lambda$ with $\lambda = (\tau + \sqrt{\tau^2 - 4})/2 \ge 1$. We then find that

$$\frac{1 - \lambda t}{1 - (\lambda + 1/\lambda)t + t^2} = \frac{1}{1 - t/\lambda}$$

has positive coefficients, as (apart from the constant term) does $\frac{t}{1-(\lambda+1/\lambda)t+t^2}$. Moreover, the coefficients of the latter are unbounded, while the coefficients of the former go to 0, and thus the coefficients of

$$\frac{1 - (\lambda + \epsilon)t}{1 - (\lambda + 1/\lambda)t + t^2}$$

are either always positive (if $\epsilon \le 0$) or eventually negative (if $\epsilon > 0$), implying the desired result. □

REMARK. If $\tau \in \{-1, 0, 1\}$, positivity fails for any $\alpha$.

THEOREM 3.19. *Let $V$ be a divisorial vector bundle on $E$ with rank $r$, degree $d$ and $m = \gcd(r, d)$ stable constituents. Then $B_{V,\Psi}$ is Koszul unless $m = r$ (so $r|d$) and* $\deg(\mathcal{L})r \leq 3$.

PROOF. If we apply the corollary to the case $M = V$, we find that the relevant power series is
$$\frac{1+t}{1-(\deg(\mathcal{L})r^2/m-2)t+t^2},$$
which has positive coefficients as long as the denominator has positive real roots. This gives the condition
$$\deg(\mathcal{L})(r/m)^2 m \geq 4$$
which holds unless $r|d$ and $\deg(\mathcal{L})r \leq 3$.

The resulting complex of $B_{V,\Psi}$-modules is exact in all positive degrees, and thus gives a *free* resolution of $(B_{V,\Psi})_0$. Since this resolution is manifestly linear, the claim follows. □

Something similar applies to minimal resolutions in the category $\mathcal{B}_{V,\Psi}$ more generally, again with the assumptions that all objects are divisorial. We may in fact generalize somewhat: we choose for each $i \in \mathbb{Z}$ a divisorial sheaf $M_i$. If $\mu(M_i)$ is an increasing function of $i$, this gives a category as before, but even if not, we can still define a *dg*-category:
$$\mathcal{B}_{\vec{M}}(i,j) = \begin{cases} 0, & i > j \\ \operatorname{End}(M_i), & i = j \\ R\operatorname{Hom}(M_i, M_j) & i < j. \end{cases}$$
(Here, of course, $R\operatorname{Hom}(M_i, M_j)$ should be replaced by the appropriate complex.) Let $P_i$ be the associated free modules (which are no longer projective, per se), and $S_i$ be defined analogously to the $B_{V,\Psi}$ case.

Suppose now that $N$ is a semistable sheaf, and consider the truncated module:
$$M_{N,\geq 0} : i \mapsto \begin{cases} 0 & i > 0 \\ R\operatorname{Hom}(M_i, N) & i \leq 0. \end{cases}$$
over this dg-category. The computation of the minimal resolution of $M_{N,\geq 0}$ begins with the morphism
$$P_0 \otimes_{\operatorname{End}(M_0)} R\operatorname{Hom}(M_0, N) \to M_{N,\geq 0}.$$
apart from replacing $R\operatorname{Hom}(M_0, N)$ itself by a minimal resolution (as an $\operatorname{End}(M_0)$-module). Let $M'$ be the corresponding cone, so that we have for each $d \in \mathbb{Z}$ a distinguished triangle
$$R\operatorname{Hom}(P_{-d}, P_0) \otimes_{\operatorname{End}(M_0)} R\operatorname{Hom}(M_0, N) \to R\operatorname{Hom}(P_{-d}, M_{N,\geq 0}) \to M'(-d) \to .$$
For $d < 0$, the first two terms vanish, while for $d = 0$ the map becomes
$$\operatorname{End}(M_0) \otimes_{\operatorname{End}(M_0)} R\operatorname{Hom}(M_0, N) \cong R\operatorname{Hom}(M_0, M_{N,\geq 0}),$$
so that $M'(-d) = 0$ for $d \leq 0$, while for $d > 0$ we may rewrite the distinguished triangle as
$$R\operatorname{Hom}(M_{-d}, M_0) \otimes_{\operatorname{End}(M_0)} R\operatorname{Hom}(M_0, N) \to R\operatorname{Hom}(M_{-d}, N) \to M'(-d) \to .$$

If we pull the tensor product inside the first $R\operatorname{Hom}$, we see that the first morphism is the image under $R\operatorname{Hom}(M_{-d},-)$ of the natural morphism

$$M_0 \otimes_{\operatorname{End}(M_0)} R\operatorname{Hom}(M_0,N) \to N,$$

and thus we find that

$$M'(-d) \cong R\operatorname{Hom}(M_{-d},\Phi_{M_0}N).$$

In other words, we have a distinguished triangle

$$P_0 \otimes_{\operatorname{End}(M_0)} R\operatorname{Hom}(M_0,N) \to M_{N,\geq 0} \to M_{\Phi_{M_0}N,\geq 1} \to,$$

from which it is straightforward to inductively construct the minimal resolution of $M_{N,\geq 0}$. In particular, we find that the associated graded of the resulting filtered complex is the direct sum

$$\begin{aligned} &P_0 \otimes_{\operatorname{End}(M_0)} R\operatorname{Hom}(M_0,N)\\ &\oplus P_1 \otimes_{\operatorname{End}(M_1)} R\operatorname{Hom}(M_1,\Phi_{M_0}N)\\ &\oplus P_2 \otimes_{\operatorname{End}(M_2)} R\operatorname{Hom}(M_2,\Phi_{M_1}\Phi_{M_0}N)\\ &\oplus\cdots, \end{aligned} \tag{3.4}$$

again with each $R\operatorname{Hom}$ complex replaced by a suitable minimal resolution.

The one caveat above is that it is not a priori obvious that $R\operatorname{Hom}(M_0,N)$ admits a finite projective resolution. If $\mu(N) > \mu(M_0)$, then this is just $\operatorname{Hom}(M_0,N)$, which we have shown is already free, and similarly if $\mu(N) < \mu(M_0)$, we know by duality that $\operatorname{Ext}^1(M_0,N)$ is free. Thus only the case that equality holds is an issue. We may reduce to the case that $M_0$, $N$ are both structure sheaves of divisors, and from that immediately reduce to the case that they are structure sheaves of jets supported at the same point, and may then pass to the complete local ring. In other words, we need to know that

$$R\operatorname{Hom}_{k[[x]]}(k[[x]]/(x^n),k[[x]]/(x^m))$$

is represented by a finite complex of free $k[[x]]/(x^n)$-modules. Using the two-term injective resolution

$$0 \to k[[x]]/(x^m) \to k((x))/x^m k[[x]] \to k((x))/k[[x]] \to 0,$$

we obtain a two-term free resolution, given by the morphism

$$x^m : k[[x]]/(x^n) \to k[[x]]/(x^n).$$

(This morphism vanishes on $k[[x]]/(x)$, so this is the minimal resolution.) It follows more generally that if $N$ with $\mu(N) = \mu(M_0)$ is also divisorial, then $R\operatorname{Hom}(M_0,N)$ is represented by a two-term complex in which both terms are the same direct summand of $\operatorname{End}(M_0)$. Note that we can also obtain a (non-minimal) free resolution in this case, by taking the direct sum of this complex with a complex having morphism 1 on a complementary direct summand.

We thus obtain the following.

PROPOSITION 3.20. *The minimal resolution of the module $S_i$ is given by a filtered complex such that the associated graded is*

$$P_i \oplus P_{i-1} \otimes R\operatorname{Hom}(M_{i-1},M_i)[1] \oplus P_{i-2} \otimes R\operatorname{Hom}(M_{i-2},\Phi_{M_{i-1}}M_i)[1] \oplus \cdots.$$

*In particular, each $P_i$ either appears in a single cohomological degree, tensored with a free module, or in two consecutive degrees, tensored with the same direct summand of* $\operatorname{End}(M_i)$.

In particular, we can compute the structure of the minimal resolutions from purely numerical data. Indeed, the object

$$\Phi_{M_{j+1}} \cdots \Phi_{M_{i-1}} M_i$$

is always a shift of a sheaf (by induction using Proposition 3.12), with the amount of shift monotone in $j$ and increasing by at most 1 when we decrease $j$. Moreover, the jumps occur precisely when the pair (rank, deg) switches to the other side of $(0,0)$ in lexicographic ordering, so depend only on the image of $(\operatorname{rank}(M_i), \deg(M_i))$ under the relevant product of elements of $\mathrm{SL}_2(\mathbb{Z})$. The support of

$$R\operatorname{Hom}(M_j, \Phi_{M_{j+1}} \cdots \Phi_{M_{i-1}} M_i)$$

is then determined by adding 0, 1, or $[0,1]$ to that shift, depending on how the two slopes compare.

The above structure of the minimal resolution has the following consequence, using the fact that $R\operatorname{Hom}(P_i, S_j) = 0$ unless $i = j$, when $R\operatorname{Hom}(P_i, S_j) = \operatorname{End}(M_j)$.

PROPOSITION 3.21. *One has*

$$R\operatorname{Hom}(S_i, S_j) \cong \begin{cases} 0, & i < j \\ \operatorname{End}(M_i), & i = j \\ R\operatorname{Hom}_{\operatorname{End}(M_j)}(R\operatorname{Hom}(M_j, \Phi_{M_{j+1}} \cdots \Phi_{M_{i-1}} M_i), \operatorname{End}(M_j))[-1], & i > j. \end{cases}$$

REMARK. If we define a new sequence of (shifts of) divisorial sheaves by

$$M_i' = \Phi_{M_0} \cdots \Phi_{M_{i-1}} M_i$$

for $i \geq 0$ and

$$M_i' = \Phi_{M_{-1}}^{-1} \cdots \Phi_{M_i}^{-1} M_i$$

for $i \leq 0$, then one has a non-canonical isomorphism

$$R\operatorname{Hom}(M_i', M_j') \cong R\operatorname{Hom}_{\operatorname{End}(M_j')}(R\operatorname{Hom}(M_j', M_i'), \operatorname{End}(M_j'))[-1],$$

which together with the canonical isomorphism

$$R\operatorname{Hom}(M_j', M_i') \cong R\operatorname{Hom}(M_j, \Phi_{M_{j+1}} \cdots \Phi_{M_{i-1}} M_i)$$

for $i \geq j$ produces an isomorphism

$$R\operatorname{Hom}(S_i, S_j) \cong \begin{cases} 0, & i < j \\ \operatorname{End}(M_i'), & i = j \\ R\operatorname{Hom}(M_i', M_j'), & i > j. \end{cases}$$

Presumably this quasi-isomorphism of complexes can be made compatible with composition. This is certainly true in any of the cases to which Theorem 3.19 above applies, when the dg-category with Hom complexes $R\operatorname{Hom}(S_i, S_j)$ is essentially just a $\mathbb{Z}$-algebra version of the Koszul dual. The general conjecture is thus that if we relax $B_{\vec{M}}$ to allow (cohomological) *shifts* of divisorial sheaves, then the

Koszul dual of $B_{\vec{M}}$ is a dg-category of the same form, and if $B_{\vec{M}}$ is periodic up to derived autoequivalence (letting us turn the $\mathbb{Z}$-dg-algebra structure into a dg-algebra structure with an additional grading), then so is its Koszul dual.

EXAMPLE 3.1. Consider the case $V = \mathcal{O}_E$, $\deg(\mathcal{L}) = 1$ of our $B_{V,\Psi}$ construction. This is already not generated in degree 1, since $\Phi_{V_{-1}}(V_0)$ is a sheaf, the structure sheaf of a point. We do, however, get a short exact sequence

$$0 \to \Phi_{V_{-2}}\Phi_{V_{-1}}(V_0)[\pm 1] \to V_{-2} \oplus V_{-1} \to V_0 \to 0$$

and find that

$$\Phi_{V_{-2}}\Phi_{V_{-1}}(V_0)[\pm 1] \cong V_{-2} \otimes V_{-1}.$$

In particular, it has slope $-3$, so the standard resolution requires us to add the generic morphism from $V_{-3}$ to $V_0$. The resulting complex

$$V_{-3} \to V_{-3} \oplus V_{-2} \oplus V_{-1} \to V_0$$

represents a shift of $\Phi_{V_{-3}}\Phi_{V_{-2}}\Phi_{V_{-1}}V_0$, and the corresponding complex of $\mathcal{B}$-modules represents the truncation of this module to degree $\geq 4$ (i.e., to the functor taking $V_i$ to 0 if $i > -4$). Apart from a shift by $-3$, this is numerically the same as the case we started with, and we thus obtain a free resolution of $k$ in the following periodic form:

$$\cdots \to [-12, -11, -10, -9] \to [-9, -8, -7, -6] \to [-6, -5, -4, -3] \to [-3, -2, -1] \to [0],$$

where $[i, j, \dots, k]$ denotes $P_i \oplus P_j \oplus \cdots \oplus P_k$. If $\tau^2 = \mathrm{id}$, then this is minimal, and otherwise the minimal resolution has shape

$$\cdots \to [-14, -13, -12] \to [-12, -11, -10] \to [-8, -7, -6] \to [-6, -5, -4] \to [-2, -1] \to [0].$$

Similarly, for the case $V = \mathcal{O}_E$, $\deg(\mathcal{L}) = 2$, the resolution has shape

$$\cdots \to [-8, -7, -7, -6] \to [-6, -5, -5, -4] \to [-4, -3, -3, -2] \to [-2, -1, -1] \to [0] \to 0,$$

reducing to

$$\cdots \to [-9, -9, -8] \to [-8, -7, -7] \to [-5, -5, -4] \to [-4, -3, -3] \to [-1, -1] \to [0] \to 0$$

unless $\tau^2 = \mathrm{id}$. For $\deg(\mathcal{L}) = 3$, the resolution has shape

$$\begin{aligned} &\cdots \to [-7, -7, -7, -6] \to [-6, -5, -5, -5] \\ &\to [-4, -4, -4, -3] \to [-3, -2, -2, -2] \to [-1-1, -1] \to [0] \to 0 \end{aligned} \tag{3.5}$$

and is always minimal.

An interesting common property of these three cases is that, in each case, if we omit the highest-degree relation, the resulting algebra still has a nice resolution, and in particular has the same Hilbert series as a free polynomial ring in the generators, with the highest-degree relation corresponding to a *central* element of this ring. (This will follow once we understand the filtered deformations. In each case, the graded algebra of deformation parameters is a free polynomial ring with one generator of the same degree as the highest-degree relation. So if weWS set the other deformation parameters to 0, we get a deformation in which the lower-degree relations are undeformed while the top-degree relation has a (nonzero) scalar added. The commutator of the deformed relation with any generator is thus expressible in terms of the undeformed relations, giving the requisite centrality statement, and letting us recover the Hilbert series.)

The other two cases in which the algebra fails to be Koszul have $\deg(\mathcal{L}) = 1$ and $V$ a slope 0 divisorial bundle with rank 2 or 3. (This is generically a sum of line bundles of degree 0, except that when the line bundles become isomorphic, they should be replaced by the appropriate self-extension.) Here the free resolution has the same shape as in the previous two cases:

$$\cdots\to[-8,-7,-7,-6]\to[-6,-5,-5,-4]\to[-4,-3,-3,-2]\to[-2,-1,-1]\to[0]\to 0$$

or

$$\begin{aligned}&\cdots \to [-7,-7,-7,-6] \to [-6,-5,-5,-5]\\ &\to [-4,-4,-4,-3] \to [-3,-2,-2,-2] \to [-1-1,-1] \to [0] \to 0;\end{aligned} \tag{3.6}$$

the main difference is that minimality can fail in more (and more complicated) ways. Again, in both of these cases, there is a deformation with a single parameter of degree 4 or 3 as appropriate, thus expressing the algebra as a central quotient of another algebra with well-behaved Hilbert series.

One question suggested by the above arguments is: given that we can compute resolutions of $B_{V,\Psi}$ purely in terms of sheaves on $E$, can we do something similar for infinitesimal deformations? One could settle a number of cases of Conjecture 1 below (and in many ways the most interesting cases) if one could compute the negative degree part of $\mathrm{HH}^2(B_{V,\Psi})$. Certainly, having an explicit resolution of $B_{V,\Psi}$ makes it feasible to compute this group in special cases, but one would like to do so in general (or at least by hand).

## 4. Elliptic noncommutative del Pezzo surfaces

We have seen in Corollary 3.8 that any filtered deformation of an elliptic algebra $B_{V,\Psi}$ is associated to a sheaf on the corresponding compactification (suitably defined) with no higher endomorphisms. This suggests a way to construct such deformations by starting with a suitable sheaf on a (noncommutative) surface. Since (by the same Corollary) the compactification is Gorenstein with dualizing sheaf $\mathcal{O}(-1)$, this surface should be del Pezzo in a suitable sense. In general, of course, this would be a quite broad sense, since the cone over $B_{V,\Psi}$ satisfies the same conditions, but this at least suggests that we should obtain a large class of deformations by taking more familiar-looking del Pezzo surfaces. Of course, if the del Pezzo surfaces are *too* familiar (i.e., commutative), then this will constrain $\Psi$ to be the tensor product with a line bundle, and thus fail to explain why various properties of the moduli space of filtered deformations extend to more general $\Psi$.

Luckily, there is a theory of noncommutative surfaces available [**19**], including a large subclass coming from elliptic curves [**20**]. This gives two classes of direct constructions, both of which are rather involved. Note that in contrast to the commutative theory, there is no direct geometric interpretation of a noncommutative surface; what one actually studies in general is a category of (quasi-)coherent sheaves on such a surface (which in our case is a flat deformation of the category of (quasi-)coherent sheaves on a commutative surface). It turns out that the corresponding *derived* category is quite simple to describe. For simplicity, we will restrict our attention to coherent sheaves (and bounded complexes with coherent cohomology); in particular, here and below, all sheaves on noncommutative surfaces will be assumed coherent. Note that since the noncommutative surfaces we consider are Noetherian [**19**, Prop. 6.1], coherent sheaves are always Noetherian.

In general, if $X$ is a smooth projective rational surface, then the structure sheaf $\mathcal{O}_X$ is exceptional, and thus determines a semiorthogonal decomposition $(\mathcal{O}_X^\perp, \mathcal{O}_X)$ of $D^b_{\mathrm{coh}}(X)$. If $X$ has an anticanonical curve $Q$ (e.g., if $X$ is del Pezzo), with corresponding morphism $i: Q \to X$, then one has a functorial distinguished triangle

$$- \otimes \omega_X \to \mathrm{id} \to i_* i^* \to$$

and thus in particular a distinguished triangle

$$R\mathrm{Hom}(M, N \otimes \omega_X) \to R\mathrm{Hom}(M,N) \to R\mathrm{Hom}_Q(M|^L_Q, N|^L_Q) \to$$

for any objects $M$, $N \in D^b_{\mathrm{coh}}(X)$. (One caveat is that the natural transformation $- \otimes \omega_X \to \mathrm{id}$ is only determined up to a scalar by the choice of $Q$, with that scalar corresponding to a choice of isomorphism $H^1(\mathcal{O}_E) \cong H^2(\omega_X) \cong k$.) In particular, if $R\mathrm{Hom}(N,M) = 0$, then Serre duality gives $R\mathrm{Hom}(M, N \otimes \omega_X) = 0$ and thus $R\mathrm{Hom}(M,N) \cong R\mathrm{Hom}_Q(M|^L_Q, N^L|_Q)$. As a special case, if $M \in \mathcal{O}_X^\perp$, then $R\mathrm{Hom}(M, \mathcal{O}_X) \cong R\mathrm{Hom}_Q(M|^L_Q, \mathcal{O}_Q)$.

The key observation is that this lets one reconstruct $D^b_{\mathrm{coh}}X$ from the (dg-)subcategory $\mathcal{O}_X^\perp$ and the restriction-to-$Q$ functor on that subcategory. Moreover, the dg-functor

$$M \mapsto R\mathrm{Hom}_Q(M|^L_Q, \mathcal{O}_Q)$$

admits a natural flat deformation: given an invertible sheaf $q$ in the identity component of $\mathrm{Pic}(X)$, we may define $D^b_{\mathrm{coh}}(X_q)$ to be the dg-category obtained by "gluing" [**17**] $\mathcal{O}_X^\perp$ to $\mathcal{O}_X$ via the functor

$$M \mapsto R\mathrm{Hom}_Q(M|^L_Q, q) \cong R\mathrm{Hom}_Q(M|^L_Q \otimes q^{-1}, \mathcal{O}_Q).$$

(That this is the derived category of a noncommutative surface is shown in [**19**, Thm. 4.30].) Note that since the forward maps in the semiorthogonal decomposition come from $Q$, the functor $Li^*$ immediately extends to $D^b_{\mathrm{coh}}(X_q)$, with $Li^*\mathcal{O}_{X_q} \cong \mathcal{O}_Q$ and the restriction on $\mathcal{O}_X^\perp$ twisted by $q^{-1}$.

In particular, we see that the Grothendieck group of $D^b_{\mathrm{coh}}(X_q)$ is independent of $q$ (and largely independent of $X$ itself); this lets one define the rank, first Chern class, and Euler characteristic of a general object (by taking the corresponding homomorphisms on $K_0(X)$). Moreover, since this is a flat deformation, the Mukai pairing $\chi([M],[N]) := \chi R\mathrm{Hom}(M,N)$ is also independent of $q$, so that one immediately has an analogue of Hirzebruch-Riemann-Roch [**19**, Cor. 7.13] for objects in $D^b_{\mathrm{coh}}(X_q)$:

$$\begin{aligned}\chi(M,N) = &-\mathrm{rank}(M)\,\mathrm{rank}(N) + \mathrm{rank}(M)\chi(N)\\ &+ \mathrm{rank}(N)\chi(M) - c_1(M)\cdot(c_1(N) + \mathrm{rank}(N)Q).\end{aligned}$$

We also obtain Serre duality, i.e., the existence of a (canonical) dg-functor $S$ with a (suitably coherent) isomorphism

$$R\mathrm{Hom}(M,N) \cong R\mathrm{Hom}(R\mathrm{Hom}(N, SM), k).$$

(This follows from standard facts about the interactions of Serre functors with semiorthogonal decompositions.) Moreover, $Li^*$ has a right adjoint $Ri_*$ fitting into a functorial distinguished triangle

$$\theta \to \mathrm{id} \to Ri_* Li^* \to$$

with $\theta := S[-2]$ (the analogue of twisting by the anticanonical bundle). (This distinguished triangle is again not quite canonical, as the natural transformation $\theta \to$

id is only determined up to scalar multiple.) There is also [19, Prop. 5.1] an analogue of Cohen-Macaulay duality, an equivalence $\mathrm{ad}_q : (D^b_{\mathrm{coh}} X_q)^{\mathrm{op}} \cong D^b_{\mathrm{coh}}(X_{q^{-1}})$ satisfying $\mathrm{ad}_{q^{-1}} \mathrm{ad}_Q \cong \mathrm{id}$, acting on $K_0(X)$ as $R\operatorname{Hom}(-, \omega_X)$.

When $Q$ is smooth, the description of the derived category can be simplified further by observing that $\mathcal{O}_X^\perp$ admits a full exceptional collection. Indeed, in that case we can nearly always write $X$ as an iterated blowup of $\mathbb{P}^2$:

$$X \cong X_m \to X_{m-1} \to \cdots \to X_{-1} \cong \mathbb{P}^2$$

where each map $X_i \to X_{i-1}$ blows down a single $-1$-curve. (We take the sequence to end at $-1$ rather than 0 to be compatible with the general case discussed in [19], in which one must typically consider iterated blowups of ruled surfaces.) Then we have a full exceptional collection on $X$ of the form

$$(\mathcal{O}_{e_m}(-1), \mathcal{O}_{e_{m-1}}(-1), \ldots, \mathcal{O}_{e_0}(-1), \mathcal{O}_{\mathbb{P}^2}(-2), \mathcal{O}_{\mathbb{P}^2}(-1), \mathcal{O}_{\mathbb{P}^2}),$$

where we omit the pullback functors from $X_i$ to $X_m$. (The only case not covered by iterated blowups of $\mathbb{P}^2$ is when $X$ is an even Hirzebruch surface $F_0$ or $F_2$, in which case we have a full exceptional collection of the form

$$(\mathcal{O}_X(-s-f), \mathcal{O}_X(-s), \mathcal{O}_X(-f), \mathcal{O}_X),$$

where $f$ is the divisor class of a fiber of the ruling, while $s$ is the unique divisor class satisfying $s \cdot f = 1$, $s^2 = 0$. The analogue of the following discussion for that case is straightforward.)

We then see that $D^b_{\mathrm{coh}}(X_q)$ may be obtained from the sequence of restrictions of the above sheaves to $Q$: it is the category of perfect modules over a triangular dg-algebra obtained from

$$\begin{aligned} &R\operatorname{End}(\mathcal{O}_{e_m}(-1)|_Q \oplus \mathcal{O}_{e_{m-1}}(-1)|_Q \oplus \cdots \oplus \mathcal{O}_{e_0}(-1)|_Q \oplus \\ &\mathcal{O}_{\mathbb{P}^2}(-2)|_Q \otimes q^{-1} \oplus \mathcal{O}_{\mathbb{P}^2}(-1)|_Q \otimes q^{-1} \oplus \mathcal{O}_Q) \end{aligned} \tag{4.1}$$

by omitting all "backwards" morphisms and all nontrivial endomorphisms of summands. Since each summand is a simple coherent sheaf on $Q$, it is determined up to isomorphism by its class in $K_0(Q)$, and thus we see that $D^b_{\mathrm{coh}}(X_q)$ is determined by the restriction morphism $K_0(X) \to K_0(Q)$. In fact, for any morphism $\phi : K_0(X) \to K_0(Q)$ satisfying

$$\operatorname{rank}(\phi(M)) = \operatorname{rank}(M), \qquad \deg(\phi(M)) = c_1(M) \cdot Q, \qquad \phi([\mathcal{O}_X]) = [\mathcal{O}_Q],$$

there is a corresponding sequence of simple modules on $Q$ and thus a corresponding dg-category. Moreover, if we set

$$q = \det(\phi([\mathcal{O}_{\mathbb{P}^2}(-2)])) \otimes \det(\phi([\mathcal{O}_{\mathbb{P}^2}(-1)]))^{-2},$$

then the dg-category has the form $D^b_{\mathrm{coh}}(X_q)$. (Indeed, the surface $X$ is obtained from this data by first embedding $Q$ in $\mathbb{P}^2$ using the line bundle $(\det(\phi([\mathcal{O}_{\mathbb{P}^2}(-1)]))\otimes q)^{-1} \in \operatorname{Pic}^3(Q)$, then sequentially blowing up the points corresponding to $\det(\phi([\mathcal{O}_{e_i}(-1)])) \in \operatorname{Pic}^1(Q)$.)

If we apply $\theta$ to the sum of exceptional objects before restricting to $Q$, we of course obtain the same endomorphism dg-algebra, and thus (since the summands generate $D^b_{\mathrm{coh}}(Q)$) there is necessarily a derived autoequivalence $\xi$ of $Q$ such that $(\theta M)|^L_Q \cong \xi(M|^L_Q)$. Moreover, since the action of $\theta$ on $K_0(X_q)$ follows from the commutative case by flatness, we may deduce the action of $\xi$ on $K_0(Q)$. We thus find that $\xi$ has the form $\Psi^{-1}$ where $\Psi$ is an *abelian* autoequivalence of the kind considered above.

A choice of blowdown structure induces a basis of the Néron-Severi lattice of $X$ of the form $h, e_0, \dots, e_m$, with intersection form given by

$$h^2 = 1, \quad h \cdot e_i = 0, \quad e_i \cdot e_j = -\delta_{ij},$$

with $Q = 3h - e_0 - \cdots - e_m = -K_X$. (In the even Hirzebruch case, this becomes $s^2 = f^2 = 0$ and $s \cdot f = 1$ with $Q = 2s + 2f$.) Changing the blowdown structure by swapping two consecutive blowups acts on the basis as the reflection in $e_i - e_{i+1}$, while a quadratic transformation in the first three points acts as the reflection in $h - e_0 - e_1 - e_2$; in either case, the result corresponds to a blowdown structure iff the corresponding $-2$-class is ineffective.

One can show that any two blowdown structures are related by a sequence of such operations: see, e.g., [**19**, Prop. 8.10] for the analogous result for arbitrary blowups of noncommutative ruled surfaces. (Analogously, for even Hirzebruch surfaces, changing the marked ruling acts as the reflection in $s - f$.)

These reflections are the simple reflections of a natural Coxeter group action. The point is that since $Q^2 > 0$, the Hodge index theorem tells us that the lattice $Q^\perp := \{v : v \in \Lambda | v \cdot Q = 0\}$ is negative definite relative to the intersection form, so negating the pairing gives a positive definite integral lattice, which is in particular acted on by reflections in any vectors of square norm 2. For $Q^2 = 9 - m \le 6$, the lattice is in fact *generated* by such vectors, and one recovers familiar root lattices:

$$E_8, E_7, E_6, D_5, A_4, A_2A_1$$

(of rank $m = 9 - Q^2$). (For $m = 2$, one obtains the unique rank 2 even lattice of determinant 7, while for $m = 1$, the rank 1 lattice depends on the parity of the Hirzebruch surface $X$.) In the resulting finite root system, we can take the positive roots to be those which are lexicographically positive, in which case the simple roots are precisely $h - e_0 - e_1 - e_2$, $e_0 - e_1$,... One then finds that a simple root is effective iff its image in $\operatorname{Pic}^0(Q)$ is trivial.

Given this description of the derived category, we can then recover the actual category of coherent sheaves $\operatorname{coh}(X_q)$ from a suitable description of the $t$-structure. The simplest description involves first defining "line bundles", inductively defined [**19**, Defn. 5.1] by starting with $\mathcal{O}_{X_{-1,q}}(d)$ for $d \in \{-2, -1, 0\}$ (i.e., the three terms in the semiorthogonal decomposition) and taking the closure under $\theta^{\pm 1}$ and pullback from $X_{l-1,w}$ to $X_{l,q}$. (For the even Hirzebruch surface case, we note that the subcategories generated by the first two and last two exceptional sheaves in the above exceptional collection generate copies of $D^b_{\mathrm{coh}}(\mathbb{P}^1)$, and we take as our line bundles the images under powers of $\theta$ of line bundles on $\mathbb{P}^1$ in those two subcategories.) We caution that the resulting collection of sheaves depends on the choice of blowdown structure (and the choice of ruling when $X \cong \mathbb{P}^1 \times \mathbb{P}^1$) in general. (The precise dependence is that we may only apply a reflection in a simple root when that simple root is ineffective on the *noncommutative* surface, which happens unless the corresponding class in $\operatorname{Pic}^0(Q)$ is a power of $q$. In particular, each simple root is *generically* ineffective, but is effective over a dense collection of hypersurfaces in the moduli space of surfaces.)

Then [**19**, Lem. 5.7], an object $M \in D^b_{\mathrm{coh}}(X)$ is in $D^b_{\mathrm{coh}}(X)^{\ge 0}$ iff $R\operatorname{Hom}(L, M) \in D^b_{\mathrm{coh}}(k)^{\ge 0}$ for every line bundle $L$; this fully determines the $t$-structure, since a $t$-structure is determined by its nonnegative part. Moreover, line bundles are themselves in the heart of this $t$-structure, and in fact generate the heart. Furthermore, nonzero morphisms between line bundles are injective [**19**, Prop. 7.21]. (The construction of [**20**] is via a representation of the category of line bundles in which the morphisms are expressed as elliptic difference operators, and such operators cannot be zero divisors.) Furthermore, by the discussion following [**19**, Defn. 5.1], there is for any $D$ an abelian equivalence of the form $\operatorname{coh}(X_q) \cong \operatorname{coh}(X'_q)$ acting on $K_0$ as $-(D)$ and taking $\mathcal{O}_{X_q}(D')$ to $\mathcal{O}_{X'_q}(D'+D)$ for all $D'$. Note that since twisting by a line bundle changes the surface, there are far fewer graded algebras associated to a noncommutative surface than in the commutative case, and even fewer maps to noncommutative projective spaces.

In particular, we define an *elliptic noncommutative (degenerate) del Pezzo surface* of degree $d > 0$ to be an abelian category constructed from a map $K_0(X) \to K_0(Q)$ as above, where $Q$ is a smooth genus 1 curve and $K_0(X)$ is the Grothendieck group of a del Pezzo surface. (It will be convenient also to fix the scale of the natural transformation $\alpha : \theta \to \mathrm{id}$.) When $q = \mathcal{O}_Q$, such a surface corresponds to a pair $(X, Q)$, where $X$ is a usual (smooth!) degenerate del Pezzo surface and $Q$ is a smooth anticanonical curve. Note that since these are analogues of *smooth* surfaces, they are generally not Fano; in the commutative case, the anticanonical embedding contracts $-2$-curves, and the situation is worse in the noncommutative case. (In general, if there is *some* commutative del Pezzo surface with a $-2$-curve of class $D$, then there is a pure 1-dimensional sheaf of class $[\mathcal{O}_D(l)]$ on $X_q$ iff $\phi[\mathcal{O}_D(l)] = 0$, in which case $Q$ fails to be ample. And we have already seen that this happens for a dense set of $q$!)

We may then use our understanding of the Grothendieck group to extend various concepts from the commutative case: a sheaf is *torsion* if it has rank 0 (note that every sheaf has nonnegative rank), a divisor class is *effective* if it is the first Chern class of a torsion sheaf (and [**19**, Prop. 7.19] if both $D$ and $-D$ are effective, then $D = 0$), a divisor class is *nef* if it is in the dual cone to the monoid of effective divisors, and ample if it is in the interior of the nef cone, with [**19**, Thm. 9.10] ample divisors satisfying

$$\operatorname{Ext}^p(\mathcal{O}_X(-nD), M) = 0$$

for $p > 0$ and $\mathcal{O}_X(-nD) \otimes \operatorname{Hom}(\mathcal{O}_X(-nD), M) \to M$ surjective for $n \gg 0$. Note that in our case, $Q$ is always nef: the effective monoid is generated by components of $Q$ along with those $-1$- and $-2$-classes which are effective [**19**, Thm. 9.2], and thus since $Q$ is irreducible with $Q^2 \geq 0$, it is nef.

Given an ample divisor, there is a corresponding notion of Hilbert polynomial

$$p_D(M) := \chi(\mathcal{O}_X(-nD), M)$$

(which we note is 0 iff $M = 0$ and is otherwise positive for $n \gg 0$), which one may use to define (semi)stable sheaves. One then finds [**20**, Prop. 11.34] that the corresponding semistable moduli spaces are proper. (One caveat, however: although it is likely that semistable moduli spaces are projective in general, this has only been proved for rank $\leq 1$.)

In general, there are relatively few concepts for commutative surfaces that do not immediately carry over to these noncommutative surfaces. The most basic issue is that the *support* of a torsion sheaf is not well-defined in general. However, since the Chern class *is* well-defined, we can often work around this (e.g., replacing the condition that a sheaf is supported on an exceptional curve $e$ by the condition that its Chern class is a multiple of the divisor class $e$). A related issue is that noncommutative surfaces have relatively few points; indeed, on a typical noncommutative surface, the only 0-dimensional sheaves (i.e., with vanishing rank and first Chern class) are in the Serre subcategory generated by 0-dimensional sheaves on $Q$. We have already mentioned the dearth of embeddings in projective space, which breaks arguments involving taking hyperplane sections. In addition, the category of sheaves on a noncommutative surface is no longer a closed monoidal category (i.e., there is no internal Hom or tensor product).

Finally, and most significant for our purposes, although *line bundles* make sense on a noncommutative surface, there is no fully satisfying definition of *vector bundles*. For many purposes (and indeed below) it suffices to consider torsion-free sheaves. One could also consider *reflexive* sheaves, i.e., sheaves $M$ on $X_q$ such that $\operatorname{ad}_q M$ is a sheaf on $X_{q^{-1}}$. (As in the commutative case, such sheaves are necessarily torsion-free.) This has a number of pathologies in general (e.g., there can be reflexive sheaves of rank 1, trivial Chern class, and arbitrarily small Euler characteristic), but for present purposes those can be ignored, as they do not arise for *rigid* reflexive sheaves (i.e., with $\operatorname{Ext}^1(M, M) = 0$). In any event, although one can show that the sheaves we construct below are reflexive, we will not need this fact.

Note that when $M$ is torsion-free, $\alpha : \theta M \to M$ is injective, as the kernel is supported on $Q$, so torsion. In particular, we see that $M|_Q$ is a sheaf. (When $M$ is reflexive, we have $(\operatorname{ad} M)|_Q \cong \mathcal{H}om(M|_Q, \mathcal{O}_Q)$ and thus $M|_Q$ is a vector bundle.)

## 5. Filtered deformations from del Pezzo surfaces

We return to our considerations of filtered deformations. Let $X$ be a noncommutative elliptic del Pezzo surface, with chosen anticanonical curve $Q$, adjoint functors $i_* : \operatorname{coh}(Q) \to \operatorname{coh}(X)$, $i^* : \operatorname{coh}(X) \to \operatorname{coh}(Q)$, and a choice of natural transformation $\alpha : \theta \to \operatorname{id}$ giving the usual functorial distinguished triangle

$$\theta \to \operatorname{id} \to i_* Li^* \to .$$

Note that the choice of $\alpha$ can be encapsulated in the observation that there is a canonical natural transformation $\theta \otimes H^1(\mathcal{O}_Q) \to \operatorname{id}$. (In principle, one could develop much of the theory below even when $Q$ is singular, but we assume $Q$ smooth for simplicity. This, of course, is the only case that can produce filtered deformations of *elliptic* algebras, and is also (see Example 2.2 above) the only case in which we can hope to obtain *all* filtered deformations using such surfaces. In addition, the argument that the moduli stack is proper breaks down when $Q$ is singular, though the claim appears to still be valid; if so, one can obtain algebras in degenerate cases by taking limits.) Recall that $\theta^{-1}$ induces an autoequivalence of $\operatorname{coh}(Q)$ of the form $\Psi$ considered above, with associated line bundle of degree $Q^2$.

By Corollary 3.8, we expect there to be a relation between filtered deformations on elliptic algebras and homogeneous endomorphism rings of sheaves on del Pezzo surfaces with no higher endomorphisms, and this is indeed the case.

PROPOSITION 5.1. *Let $(X, Q)$ be a noncommutative elliptic del Pezzo surface, let $M \in \operatorname{coh}(X)$ be a torsion-free sheaf with $\operatorname{Ext}^1(M, M) = \operatorname{Ext}^2(M, M) = 0$ such that the slopes of $M|_Q$ are constrained to a half-open interval of length $Q^2$, and consider the graded algebra*

$$A_M^+ := \bigoplus_i \operatorname{Hom}(M, \theta^{-i} M).$$

*This algebra is nonnegatively graded, the element $\alpha_{\theta^{-1}M} : M \to \theta^{-1}M$ of degree 1 is regular and central, and the quotient by $\alpha_{\theta^{-1}M}$ is $B_{M|_Q,\theta^{-1}}$.*

PROOF. That $\alpha_{\theta^{-1}M}$ is regular and central in $A^+_M$ follows immediately from the facts that it is injective and comes from a natural transformation. For each $i \in \mathbb{Z}$, $\alpha$ induces a distinguished triangle

$$R\operatorname{Hom}(M, \theta^{1-i}M) \to R\operatorname{Hom}(M, \theta^{-i}M) \to R\operatorname{Hom}(M|_Q, \theta^{-i}M|_Q) \to .$$

For $i = 0$, we may use duality to compute $R\operatorname{Hom}(M, \theta M)$, and thus (using the hypotheses on $M$) we obtain isomorphisms

$$\operatorname{End}(M) \cong \operatorname{End}(M|_Q), \quad \operatorname{Ext}^2(M, \theta M) \cong \operatorname{Ext}^1(M|_Q, M|_Q).$$

Moreover, since $\alpha_{\theta^i M}$ is injective for all $i$, the assumption that $\operatorname{Ext}^2(M, M) = 0$ implies that $\operatorname{Ext}^2(M, \theta^i M) = 0$ for all $i \le 0$, and thus that $\operatorname{Hom}(M, \theta^i M) = 0$ for $i \ge 1$. In addition, the condition on $M|_Q$ ensures that $\operatorname{Ext}^1(M|_Q, \theta^i M|_Q) = 0$ for $i > 0$. We thus inductively find that $\operatorname{Ext}^1(M, \theta^i M) = 0$ for $i \le 0$, giving us the desired short exact sequence

$$0 \to \operatorname{Hom}(M, \theta^{1-i}M) \to \operatorname{Hom}(M, \theta^{-i}M) \to \operatorname{Hom}(M|_Q, \theta^{-i}M|_Q) \to 0$$

for all $i$. □

REMARK. Of course, the quotient by $\alpha_{\theta^{-1}M}$ is independent of the choice of scaling on $\alpha_{\theta^{-1}M}$. In addition, although $\theta^{-1}$ appears in the definition of $A^+_M$, we can safely omit it from the notation because $\theta$ is canonical (in both senses of the word).

Note that if we instead quotient by $\alpha_{\theta^{-1}M} - 1$, then the result will be a filtered algebra, and the Proposition tells us that the associated graded algebra is $B_{M|_Q,\theta^{-1}}$. In other words, the quotient is precisely a filtered deformation of the type we are looking for. The filtered deformation depends only mildly on the choice of scale on $\alpha$: the action of $\mathbb{G}_m$ by rescaling $\alpha$ agrees with the standard action on filtered deformations.

We are thus left with the question of classifying such sheaves. In the commutative case, sheaves with $\operatorname{Ext}^1(M, M) = 0$ on degenerate del Pezzo surfaces were studied in [**13**]; though the classification was only partial and some of the arguments do not apply in the noncommutative setting, we can use the ideas therein with some supplements to get a fairly complete understanding in general.

A useful simplification comes from the following fact.

PROPOSITION 5.2. *Suppose $M$, $N$ are sheaves on a noncommutative del Pezzo surface $(X, Q)$ such that $\operatorname{Ext}^1(N, M) = 0$ and $N$ is torsion-free. Then $\operatorname{Hom}(M, N) \to \operatorname{Hom}(M|_Q, N|_Q)$ is surjective.*

PROOF. We have the exact sequence

$$0 \to \operatorname{Hom}(M, \theta N) \to \operatorname{Hom}(M, N) \to \operatorname{Hom}(M|_Q, N|_Q) \to \operatorname{Ext}^1(M, \theta N),$$

but duality gives

$$\operatorname{Ext}^1(M, \theta N) \cong \operatorname{Ext}^1(N, M)^* = 0.$$

□

COROLLARY 5.3. *If $M$ is a torsion-free sheaf with $\operatorname{Ext}^1(M, M) = 0$, then $\operatorname{End}(M) \to \operatorname{End}(M|_Q)$ is surjective, and any idempotent in $\operatorname{End}(M|_Q)$ lifts to $\operatorname{End}(M)$.*

PROOF. Surjectivity is just the case $N = M$ above, and thus $\operatorname{End}(M|_Q)$ is the quotient of $\operatorname{End}(M)$ by the ideal $\alpha_M \circ \operatorname{Hom}(M, \theta M)$. Since idempotents lift through nil ideals, it remains to show that any element $\beta$ of the ideal is nilpotent.

Since $X$ is Noetherian and $M$ is coherent, the ascending chain $\ker(\beta^i)$ of subsheaves is eventually constant, so that the descending chain $\operatorname{im}(\beta^i) \subset M$ is also eventually constant, stabilizing at $N$. Since $\beta|_N$ is an automorphism and $\beta$ factors through $\alpha_M$, we see that $\alpha_N$ must be surjective, and thus $\operatorname{rank}(N) = 0$. But $M$ is torsion-free, and thus $N = 0$, so that $\beta$ is nilpotent as required. □

In general, we can write the sheaf $M|_Q$ as a direct sum $M|_Q = \bigoplus_{\mu \in \mathbb{Q} \cup \{\infty\}} V_\mu$ where each $V_\mu$ is a semistable sheaf of slope $\mu$. By the Corollary, when $M$ is rigid and torsion-free, this direct sum decomposition lifts (nonuniquely!) to $M$ and thus $M$ itself is a sum of torsion-free sheaves $M_\mu$ such that $(M_\mu)|_Q = V_\mu$. (Note that $\operatorname{rank}(M_\infty) = \operatorname{rank}(V_\infty) = 0$, and thus $M_\infty = 0$.) And, of course, each $M_\mu$ itself splits as a direct sum of sheaves with *indecomposable* restriction to $Q$.

The key observation of [**13**] is that such a summand $M_\mu$ is (slope) semistable in a suitable sense.

PROPOSITION 5.4. *Let $M$ be a torsion-free sheaf on $X$ such that $M|_Q$ is semistable. Then any nontrivial subsheaf $N \subset M$ satisfies*

$$\frac{c_1(N) \cdot Q}{\operatorname{rank}(N)} \le \frac{c_1(M) \cdot Q}{\operatorname{rank}(M)}.$$

*If $M|_Q$ is stable, equality implies $N = M$.*

PROOF. We may assume that $M/N$ is torsion-free, as otherwise replacing $N$ by the saturated subsheaf can only increase $c_1(N) \cdot Q$ without increasing $\operatorname{rank}(N)$. (Here we use the fact that $Q$ is nef!) But then (since $(M/N)|_Q$ is a sheaf) we have a short exact sequence

$$0 \to N|_Q \to M|_Q \to (M/N)|_Q \to 0$$

and thus by assumption have $\mu(N|_Q) \le \mu(M|_Q)$. The first claim then follows by observing that $\operatorname{rank}(N|_Q) = \operatorname{rank}(N)$ and $\deg(N|_Q) = c_1(N) \cdot Q$. Moreover, stability of $M|_Q$ implies that $\mu(M|_Q)$ has denominator $\operatorname{rank}(M|_Q)$, and thus $\mu(N|_Q) = \mu(M|_Q)$ only if $\operatorname{rank}(N) = \operatorname{rank}(M)$, only if $N = M$. □

Define $\mu_Q(M) := (c_1(M) \cdot Q)/\operatorname{rank}(M) = \mu(M|_Q)$, and say that the torsion-free sheaf $M$ is $\mu_Q$-semistable if $\mu_Q(N) \le \mu_Q(M)$ for nonzero $N \subset M$ (and similarly for $\mu_Q$-stable). In particular, we have just shown that (semi)stability of $M|_Q$ implies $\mu_Q$-(semi)stability of $M$.

COROLLARY 5.5. *If $M$, $N$ are $\mu_Q$-semistable torsion-free sheaves, then* $\operatorname{Hom}(M, N) = 0$ *if $\mu_Q(M) > \mu_Q(N)$ and* $\operatorname{Ext}^2(M, N) = 0$ *if $\mu_Q(M) < \mu_Q(N) + Q^2$ (and in particular if $M = N$).*

PROOF. If $\operatorname{Hom}(M, N) \ne 0$, then the image $I$ of such a homomorphism satisfies

$$\mu_Q(M) \le \mu_Q(I) \le \mu_Q(N),$$

since it is a quotient of $M$ and a subsheaf of $N$. By duality, if $\operatorname{Ext}^2(M, N) \ne 0$, then $\operatorname{Hom}(N, \theta M) \ne 0$, and thus $\mu_Q(N) \le \mu_Q(\theta M) = \mu_Q(M) - Q^2$. □

Recall that an exceptional collection in a triangulated category is a sequence $(E_1,\dots,E_n)$ of objects such that $R\operatorname{Hom}(E_i,E_i)\cong k$ and $R\operatorname{Hom}(E_i,E_j)=0$ for $i>j$. In the commutative case, the following is a special case of [**13**, Lem. 2.4.4]. (The proof below is somewhat more elementary, as we do not need to use the spectral sequence for Ext of filtered sheaves.)

PROPOSITION 5.6. *Let $M$ be a rigid torsion-free sheaf with $M|_Q$ indecomposable. Then there is an induced filtration $0=F_n\subset F_{n-1}\subset\cdots\subset F_0=M$ with subquotients $E_i=F_{i-1}/F_i$ such that $(E_1,\dots,E_n)$ is an exceptional collection.*

PROOF. By assumption, $\operatorname{Ext}^1(M,M)=0$, and slope considerations imply $\operatorname{Ext}^2(M,M)=0$, so that $\operatorname{End}(M)\cong\operatorname{End}_Q(M|_Q)\cong k[t]/(t^n)$ for some $n$. We may thus induce a filtration on $M$ by taking $F_i=\ker(t^{n-i})$, and see that the restriction of this filtration to $Q$ has the same description. In particular, $\operatorname{rank}(F_i)=(1-i/n)\operatorname{rank}(M)$, and thus each $E_i|_Q$ is isomorphic to the unique stable constituent $V$ of $M|_Q$. The action of $t$ certainly respects this filtration, and thus has an induced action on the associated graded. Since $t^{n-1-i}$ induces an injection $E_i\to E_n$, we see that $t:E_i\to E_{i+1}$ is injective for all $i$, giving a chain $E_1\subset\cdots\subset E_n$ of sheaves restricting to the same stable bundle $V$ on $Q$.

Slope considerations again tell us that $\operatorname{Ext}^2(E_i,E_j)=0$ for all $i,j$, and thus $\operatorname{Hom}(E_i,E_j)\to\operatorname{End}(V)\cong k$ is injective for all $i,j$. Moreover, the natural map $\operatorname{Ext}^1(E_i,E_{i+1})\to\operatorname{Ext}^1(V,V)\cong k$ is nonzero (since the corresponding subquotient of $M|_Q$ is indecomposable), and thus surjective, implying that $\operatorname{Ext}^2(E_i,\theta E_{i+1})\to\operatorname{Ext}^2(E_i,E_{i+1})=0$ is injective. We thus conclude that $\operatorname{Hom}(E_{i+1},E_i)=0$ so that $\operatorname{Hom}(E_i,E_j)=0$ for all $i>j$, and thus (reversing the above argument) $\operatorname{Ext}^1(E_i,E_j)\to\operatorname{Ext}^1(V,V)$ is surjective for $i<j$.

We thus conclude that the Ext spaces satisfy the inequalities

$$\begin{aligned}\dim\operatorname{Hom}(E_i,E_j)&=\delta_{i\le j}\\ \dim\operatorname{Ext}^1(E_i,E_j)&\ge\delta_{i<j}\\ \dim\operatorname{Ext}^2(E_i,E_j)&=0\end{aligned}$$

and thus

$$\chi(E_i,E_j)\le\delta_{i\le j}-\delta_{i<j}=\delta_{ij}.$$

Since

$$\chi(M,M)=\sum_{i,j}\chi(E_i,E_j)\le n=\chi(M,M),$$

the inequalities must be tight, and thus in particular $(E_1,\dots,E_n)$ is an exceptional collection as required. □

PROPOSITION 5.7. *If $M$, $N$ are sheaves with $\mu_Q(M)=\mu_Q(N)\in\mathbb{Q}\cup\infty$, then $\chi(M,N)=\chi(N,M)$. In particular, if $E$ and $F$ are exceptional sheaves of the same slope, then $\chi(E,F)=\chi(F,E)=1+\rho^2/2\le 1$, where $\rho=c_1(E)-c_1(F)$.*

PROOF. From the distinguished triangle

$$R\operatorname{Hom}(M,\theta N)\to R\operatorname{Hom}(M,N)\to R\operatorname{Hom}_Q(M|_Q,N|_Q)\to,$$

we obtain the identity

$$\chi_Q(M|_Q,N|_Q)=\chi(M,N)-\chi(M,\theta N).$$

Serre duality gives $\chi(M,\theta N) = \chi(N,M)$, and the fact that $\chi_Q$ is a symplectic pairing on $K_0(Q)$ implies $\chi_Q(M|_Q, N|_Q) = 0$ and thus $\chi(M,N) = \chi(N,M)$.

In the exceptional case, we have

$$-\rho^2 = \chi([M]-[N],[M]-[N]) = \chi(M,M)+\chi(N,N)-2\chi(M,N) = 2-2\chi(M,N)$$

and solving for $\chi(M,N)$ gives the desired result. The inequality then follows from the fact that $\rho$ is in the negative definite even lattice of divisors orthogonal to $Q$. □

LEMMA 5.8. *Let $M_1 \oplus M_2$ be a rigid torsion-free sheaf such that $M_1|_Q$, $M_2|_Q$ are indecomposable bundles with isomorphic stable constituents. If* $\operatorname{rank}(M_1) \geq \operatorname{rank}(M_2)$, *then (up to isomorphism) the exceptional collection associated to $M_1$ contains that associated to $M_2$.*

PROOF. As usual, slope considerations imply that $\operatorname{Ext}^2(M_i, M_j) = 0$. Let the two exceptional collections be $(E_1,\dots,E_{n_1})$ and $(E'_1,\dots,E'_{n_2})$, with $n_1 \geq n_2$. Then $\chi(E_i,E'_j) \leq \delta_{E_i\cong E'_j}$, and thus

$$n_2 = \dim \operatorname{Hom}(M_1,M_2) = \chi(M_1,M_2) = \sum_{i,j}\chi(E_i,E'_j) \leq \sum_{i,j}\delta_{E_i\cong E'_j} \leq n_2,$$

since each $E'_j$ is isomorphic to at most one $E_i$. It follows that both inequalities are tight, and in particular that every $E'_j$ is isomorphic to some $E_i$ as required. □

LEMMA 5.9. *Let $M_1\oplus M_2$ be a rigid torsion-free sheaf such that $M_1|_Q$ and $M_2|_Q$ are indecomposable bundles with $\mu(M_1|_Q) \leq \mu(M_2|_Q) < \mu(M_1|_Q) + Q^2$. Then the exceptional collections associated to $M_1$ and $M_2$ are contained in a common exceptional collection.*

PROOF. If $M_1|_Q$, $M_2|_Q$ have isomorphic stable constituents, then this follows from the previous Lemma. Otherwise, let the two exceptional collections again be $(E_1,\dots,E_{n_1})$ and $(E'_1,\dots,E'_{n_2})$. Then we claim that $(E_1,\dots,E_{n_1},E'_1,\dots,E'_{n_2})$ is an exceptional collection.

The slope constraint implies that $\operatorname{Hom}(M_2,M_1) = \operatorname{Ext}^2(M_2,M_1) = 0$ and (since they have the same slopes) $\operatorname{Hom}(E'_i,E_j) = \operatorname{Ext}^2(E'_i,E_j) = 0$ for all $i,j$. By hypothesis, $\operatorname{Ext}^1(M_2,M_1) = 0$, so that $\chi(M_2,M_1) = 0$. We then find that

$$0 = \chi(M_2,M_1) = \sum_{i,j}\chi(E'_i,E_j) \leq 0$$

and thus $\chi(E'_i,E_j) = 0$ for all $i,j$, implying $\operatorname{Ext}^1(E'_i,E_j) = 0$ as required. □

COROLLARY 5.10. *Let $M$ be a rigid torsion-free sheaf with slopes contained in a half-open interval of length $Q^2$. Then there is an exceptional collection $(E_1,\dots,E_n)$ and a filtration of $M$ such that the $i$th subquotient has the form $n_iE_i$.*

PROOF. The two Lemmas tell us that any two indecomposable summands of $M$ have compatible exceptional collections, and thus we immediately conclude that there is a single exceptional collection containing all of them; moreover, that exceptional collection can be split into blocks associated to the different stable constituents of $M|_Q$, and the existence of a filtration as described for each such block is straightforward. □

REMARK. Since the hypothesis ensures that $\operatorname{Ext}^2(M,M)=0$, this is (in the commutative case) a special case of [**13**, Thm. 2.5.1(2)]. The argument there can be extended to the noncommutative case, but we will not need this generalization, so omit the details.

Of course, the above is in a sense the converse of the result we want: we want to know when a given exceptional collection corresponds to a rigid, unobstructed torsion-free sheaf. In the indecomposable case, this is straightforward.

PROPOSITION 5.11. *Suppose $(E_1,\dots,E_n)$ is an exceptional collection on a noncommutative del Pezzo surface, such that each $E_i$ is a torsion-free sheaf and the restrictions to $Q$ are all isomorphic. Then $\dim\operatorname{Hom}(E_i,E_j)=\dim\operatorname{Ext}^1(E_i,E_j)=1$ for $i<j$, and there is (up to isomorphism) a unique filtered sheaf $0=F_n\subset\cdots\subset F_0=M$ with $F_{i-1}/F_i\cong E_i$, $\operatorname{Ext}^1(M,M)=0$, and $M|_Q$ indecomposable.*

PROOF. For $i<j$, since $R\operatorname{Hom}(E_j,E_i)=0$, we have $R\operatorname{Hom}(E_i,E_j)\cong R\operatorname{Hom}(E_i|_Q,E_j|_Q)$, and the sheaves on $Q$ are isomorphic stable sheaves. We then define a sequence $F_i$ by letting $F_{i-1}$ be the generic extension of $E_i$ by $F_i$, giving a sheaf $M$. That this satisfies $\operatorname{Ext}^1(F_i,F_i)=0$ for all $i$ follows as in [**13**], giving existence. Moreover, the only way to obtain a nonisomorphic sheaf with the same subquotients is to take a nongeneric extension in some step, but since $\operatorname{Ext}^1(E_i,F_i)\cong\operatorname{Ext}^1(E_i|_Q,F_i|_Q)\cong k$ at the first such step, this would make $F_0|_Q$ decomposable. □

THEOREM 5.12. *Let $(E_1,\dots,E_n)$ be an exceptional collection of torsion-free sheaves, with $\mu_Q(E_i)$ constant. Then for any sequence $(m_1,\dots,m_n)$ of nonnegative integers, there is a unique rigid sheaf $M$ having a filtration with subquotients $E_i^{m_i}$.*

PROOF. For existence, note that for any subset $S\subset\{1,\dots,n\}$, there is a corresponding sheaf $M_S$ obtained as in Proposition 5.11: at each step, take the generic extension. We then find as above that

$$\chi(M_S,M_T)=|S\cap T|.$$

Since $\operatorname{Ext}^2(M_T,M_S)=0$, we find that $\dim\operatorname{Hom}(M_S,M_T)\le\dim\operatorname{Hom}(M_S|_Q,M_T|_Q)$. If $S\subset T$ or $T\subset S$, then $\dim(M_S|_Q,M_T|_Q)=|S\cap T|$, and

$$|S\cap T|=\chi(M_S,M_T)\le\dim\operatorname{Hom}(M_S,M_T)\le|S\cap T|,$$

forcing $\dim\operatorname{Ext}^1(M_S,M_T)=0$ and $\dim\operatorname{Hom}(M_S,M_T)=|S\cap T|$. Existence then follows by taking the sum $\sum_i M_{S_i}^{c_i}$ for a suitable chain of subsets and multiplicities $c_i$. (The chain of subsets is determined by a permutation that sorts the sequence $\vec{m}$, and the $c_i$ are then differences of consecutive entries in the sorted sequence. This permutation need not be unique, but the sequence $c_i$ is independent of the choice of permutation, and when $c_i\neq 0$, $S_i$ is also uniquely determined.)

For uniqueness, we may use the fact that $M$ decomposes as a direct sum over different isomorphism classes of stable constituents to reduce to the case that $E_1|_Q\cong\cdots\cong E_n|_Q$, at which point we conclude from Lemma 5.8 that if both $M_S$ and $M_T$ appear, then $S$ and $T$ are comparable. □

REMARK. There is a slight additional technical issue if we want to perform this construction in families, namely the fact that over a field, we are taking *generic* extensions at various points in the process, and thus run into potential problems in constructing a global extension class that is nonzero whenever possible. Note that since the relevant $\mathrm{Ext}^1$ spaces are preserved by restriction to $Q$, it suffices to show that one can define $M_S|_Q$ globally for each $S$. Since this sheaf is semistable, applying a suitable derived autoequivalence of $D^b_{\mathrm{coh}}(E)$ turns it into a torsion sheaf ([**4**]; see Proposition 3.9 above), the fibers of which are isomorphic to structures sheaves of divisors. But such a torsion sheaf can certainly be defined globally: simply view the family of divisors as a divisor on the total space of the family and take its structure sheaf. Inverting the derived autoequivalence gives a global version of $M_S|_Q$ as required, and thus the required global extension classes needed to construct $M_S$ itself.

More generally, if we are given an exceptional collection consisting of torsion-free sheaves with nondecreasing $Q$-slopes contained in a half-open interval of length $Q^2$, and for each $\mu$ we construct a rigid sheaf $M_\mu$ from the corresponding subcollection, then $M = \bigoplus_\mu M_\mu$ is rigid and unobstructed. Indeed, we have $R\,\mathrm{Hom}(M_\nu, M_\mu) = 0$ for $\mu<\nu$, and thus

$$R\,\mathrm{Hom}(M_\mu, M_\nu) \cong R\,\mathrm{Hom}_Q(M_\mu|_Q, M_\nu|_Q) \cong \mathrm{Hom}(M_\mu|_Q, M_\nu|_Q),$$

so that in neither direction are there higher morphisms.

Thus the question of classifying filtered deformations arising from Proposition 5.1 reduces to one of classifying exceptional collections. The basic question here is: given a sequence of classes in $K_0(X)$ satisfying the obvious numerical conditions, is there a corresponding exceptional collection, and is it unique? Unfortunately, the main result of [**13**] on exceptional collections is not particularly useful in this regard: it says that any exceptional collection can be obtained from a certain standard collection via the braid group action, but it is unclear how to turn that into an existence test, and (worse yet) the arguments depend in a crucial way on the commutativity of the surface. (They involve restriction functors to $D^b_{\mathrm{coh}}(\mathbb{P}^1)$ coming from $-1$-curves.)

Although we do not have a complete answer, we can reduce the problem considerably.

PROPOSITION 5.13. *Suppose* $(E_1, \dots, E_n)$ *is a sequence of exceptional torsion-free sheaves such that* $\mu_Q(E_1) \le \mu_Q(E_2) \le \cdots \le \mu_Q(E_n) < \mu_Q(E_1) + Q^2$ *and* $\chi(E_j, E_i) = 0$ *for* $i < j$*; suppose moreover that if* $\mu_Q(E_i) = \mu_Q(E_j)$ *for* $i < j$*, then there is no sheaf of class* $[E_i] - [E_j]$*. Then* $(E_1, \dots, E_n)$ *is an exceptional collection.*

PROOF. Let $i < j$. Since $\mu_Q(E_j) < \mu_Q(E_i) + Q^2$, we find $\mathrm{Ext}^2(E_j, E_i) = 0$. Since $\chi(E_j, E_i) = 0$, it remains only to show that $\mathrm{Hom}(E_j, E_i) = 0$. If $\mu_Q(E_i) < \mu_Q(E_j)$, this follows from stability, while if the slopes agree, any morphism must be injective, and thus the cokernel is a sheaf of the given class in $K_0(X)$. □

REMARK. Note that as observed above, when the slopes agree, $c_1(E_i) - c_1(E_j)$ is a root of the lattice $Q^\perp \sim E_{9-Q^2}$. It follows that there is a sheaf of class $[E_i]-[E_j]$ iff $c_1(E_i) - c_1(E_j)$ is a *positive* root and $[E_i]|_Q \cong [E_j]|_Q$. In particular, if we arrange for $c_1(E_j) - c_1(E_i)$ to be a positive root for $i < j$, then $[E_i] - [E_j]$ can never be effective.

Uniqueness is even easier.

PROPOSITION 5.14. *If* $E$, $F$ *are exceptional torsion-free sheaves with* $[E] = [F] \in K_0(X)$*, then* $E \cong F$*.*

PROOF. Since $\chi(E,E)=\chi(F,F)=1$, we have $\chi(E,F)=\chi(F,E)=1$. Since $E$ is rigid and unobstructed, $k\cong \operatorname{End}(E)\cong \operatorname{End}_Q(E|_Q)$, and thus $E|_Q$ is stable, so that $E$ is $\mu_Q$-stable. Slope considerations imply that $\operatorname{Ext}^2(E,F)=\operatorname{Ext}^2(F,E)=0$ and thus there are nonzero morphisms in either direction. Stability forces the morphisms to be injective, and thus so is the composition in either direction, which must therefore be a nonzero multiple of the identity, implying $E\cong F$ as required. □

Thus to understand the exceptional collections corresponding to filtered deformations, it suffices to understand (a) which sequences of classes in $K_0(X)$ satisfy the numerical conditions (and positivity of roots, when relevant) and (b) when a given class in $K_0(X)$ is represented by an exceptional torsion-free sheaf.

LEMMA 5.15. *Let $M$ be a $\mu_Q$-stable torsion-free sheaf, and let $D\in \operatorname{NS}(X)$ be an ample divisor class. Then for all sufficiently large $n$, $M$ is $nQ+D$-stable.*

PROOF. Suppose otherwise, and for each $n\geq 0$, let $B_n$ be the maximally destabilizing subsheaf of $M$. Since $M$ is $\mu_Q$-stable, we must have $\mu_Q(B_n)<\mu_Q(M)$, and thus there is some $n'>n$ such that

$$n'\mu_Q(B_n)+\mu_D(B_n)<n'\mu_Q(M)+\mu_D(M),$$

and thus in particular

$$n'\mu_Q(B_{n'})+\mu_D(B_{n'})>n'\mu_Q(B_n)+\mu_D(B_n)$$

Since $B_n$ is maximally destabilizing for $nQ+D$, we also have

$$n\mu_Q(B_{n'})+\mu_D(B_{n'})\leq n\mu_Q(B_n)+\mu_D(B_n),$$

and thus $\mu_Q(B_n)<\mu_Q(B_{n'})$. We thus obtain an infinite sequence of strictly increasing slopes bounded by $\mu_Q(M)$, but since the denominators are bounded by $\operatorname{rank}(M)$, this is impossible. □

REMARK. If $M$ is only assumed to be $\mu_Q$-semistable, an analogous argument shows that either $M$ is still $nQ+D$-stable for $n\gg 0$ or $B_n$ eventually stabilizes, and thus by induction the corresponding Harder-Narasimhan filtration of $M$ eventually stabilizes.

PROPOSITION 5.16. *Let $R$ be a discrete valuation ring with residue field $k$ and field of fractions $K$, and let $X_R$ be a noncommutative del Pezzo surface over $R$. An exceptional torsion-free sheaf on either fiber of $X_R$ extends to a torsion-free sheaf which is exceptional on the other fiber.*

PROOF. An exceptional sheaf $E_k$ on $X_k$ is simple, so induces a point of the moduli space of simple sheaves. (It is shown in [**19**, Cor. 11.3] that this moduli problem is represented by an algebraic space, with Prop. 11.1 op. cit. implying that the obstruction theory is as expected.) Since $\operatorname{Ext}^2(E_k,E_k)=0$, that algebraic space is formally smooth (and thus smooth) at $E_k$, so that the point corresponding to $E_k$ extends to $R$. Since $\gcd(c_1(E)\cdot Q,\operatorname{rank}(E))=\gcd(\deg(E|_Q),\operatorname{rank}(E|_Q))=1$, there is a universal family. (The moduli stack of simple sheaves is a $\mathbb{G}_m$-gerbe over the moduli space of simple sheaves, so that the obstruction to the existence of a universal family is a class in the Brauer group. For any sufficiently ample $D$, this class is represented by the Azumaya algebra $\operatorname{End}(\operatorname{Hom}(\mathcal{O}_X(-D),E))$, and the gcd condition ensures that the degrees of those Azumaya algebras generate $\mathbb{Z}$, so the obstruction is trivial.) Thus this point corresponds to a simple sheaf $E_R$, which has exceptional generic fiber by semicontinuity of $\operatorname{Ext}^i$.

For the other direction, fix an ample divisor $D$ on $X_k$, so that an exceptional torsion-free sheaf $E_K$ on $X_K$ is $nQ+rD$-stable for some $n$, with $r = \operatorname{rank}(E_K)$. Since the semistable moduli functor is universally closed, there is thus an extension to a sheaf $E_R$ with $nQ+rD$-semistable special fiber $E_k$. If $\gcd(n,r)=1$, then the corresponding slope of $E_K$ has denominator $r$, and thus cannot be the slope of a sheaf of smaller rank, so that $E_k$ is actually *stable*. It follows that $\operatorname{End}(E_k)\cong k$, and thus that $\operatorname{Hom}(E_k,\theta E_k)=0$, giving $\operatorname{Ext}^2(E_k,E_k)=0$. Since $\chi(E_k,E_k)=\chi(E_K,E_K)=1$, it follows that $E_k$ is exceptional as required. □

It follows immediately that, given a class $[E]\in K_0(X)$ such that $\operatorname{rank}([E])>0$ and $\chi([E],[E])=1$, if *any* noncommutative del Pezzo surface (with the correct value of $Q^2$ and parity if $Q^2=8$; i.e., in a given component of the moduli stack of noncommutative surfaces) has an exceptional torsion-free sheaf of class $[E]$, then *every* noncommutative del Pezzo surface has such a sheaf. (Indeed, every $X$ can be obtained from a finite base change of the generic surface by a descending down a sequence of valuations of the corresponding field, and the existence of a universal family lets us descend through finite field extensions.) We thus in particular see that the existence question reduces to the case of commutative del Pezzo surfaces, which may even be assumed nondegenerate.

In fact, it reduces to the case of (commutative, nondegenerate) del Pezzo surfaces of degree 1. Let $\pi : Y \to X$ be a birational morphism with $Y$ a nondegenerate commutative del Pezzo surface of degree 1. If $[E]\in K_0(X)$ is the class of an exceptional bundle $E$, then $\pi^*[E]$ is the class of the exceptional bundle $\pi^*E$. Conversely, if $\pi^*[E]$ is the class of an exceptional bundle $E'$, then it descends to an exceptional bundle on $X$ by [**14**, Cor 3.2].

A fairly strong necessary condition for $[E]$ to be representable comes from the following.

LEMMA 5.17. *If $E$ is an exceptional bundle on $X$, then for any exceptional bundle $E'$ with $\mu(E')-Q^2<\mu(E)<\mu(E')$, $\chi(E',E)\le 0$.*

PROOF. Indeed, the conditions on the slopes ensure that $\operatorname{Hom}(E',E)=\operatorname{Ext}^2(E',E)=0$ and thus $\chi(E',E)=-\dim\operatorname{Ext}^1(E',E)\le 0$. □

Since the twist of an exceptional bundle by a line bundle remains exceptional, any given $E'$ induces a whole family of necessary conditions: $\chi(E'(-D),[E])\le 0$ for all $D\in Q^\perp=\{v\in \operatorname{NS}(X)|v\cdot Q=0\}$. Since (letting $r=\operatorname{rank}([E])$, $r'=\operatorname{rank}(E')$)

$$\chi(E'(-D),[E])-\frac{rr'}{2}(D-(D_{E'}/r'-D_E/r))^2$$

is independent of $D$, with $D_E := c_1(E)-\frac{c_1(E)\cdot Q}{Q^2}Q$, we see that $\chi(E'(-D),[E])$ is maximized when $D$ is the closest vector (relative to the negative of the intersection form) in $Q^\perp$ to $D_{E'}/r'-D_E/r$; if the maximum is positive, then $[E]$ is not representable. The resulting condition is particularly straightforward to compute when $E'$ ranges over line bundles, and is already quite strong in that case.

It turns out it is useful to reformulate the question in terms of one on the elliptic surface $\tilde{X}$ associated to the nondegenerate degree 1 del Pezzo surface $X$. (Note that nondegeneracy of $X$ corresponds to smoothness of the Weierstrass model of $\tilde{X}$.) Here, the representability condition turns out to be essentially trivial.

PROPOSITION 5.18. *Let $\tilde{X}$ be a smooth rational Weierstrass elliptic surface, and let $[E] \in K_0(\tilde{X})$ be a class with* $\operatorname{rank}([E]) > 0$, $\chi([E],[E]) = 1$, *and* $\gcd(\operatorname{rank}([E]), c_1([E]) \cdot Q) = 1$ *(where $Q$ denotes the divisor class of a fiber of $\tilde{X}$). Then $[E]$ is the class of an exceptional vector bundle on $\tilde{X}$, and any two bundles of class $[E]$ are isomorphic.*

PROOF. The derived autoequivalences of elliptic curves extend in a straightforward way to any Weierstrass curve, and thus in particular we can apply them to $\tilde{X}$, viewed as a family of such curves. Those derived autoequivalences act as $\mathrm{SL}_2(\mathbb{Z})$ on $(\operatorname{rank}(M), c_1(M) \cdot Q)$, and thus in particular there is a derived autoequivalence taking $[E]$ to a class with $\operatorname{rank}([E']) = 0$, $c_1([E']) \cdot Q = 1$, $\chi([E'],[E']) = 1$. Any such class is represented by a sheaf $\mathcal{O}_e(\chi([E'])-1)$ where $e = c_1([E'])$ is a section of the elliptic fibration, and thus we can apply the inverse autoequivalence to obtain an object $E$ representing $[E]$. Since a section $e$ cannot meet the singular point of any fiber, we find that the restriction of $E$ to any fiber is a shift of a vector bundle, and thus $E$ itself is a shift of a vector bundle. This establishes existence, with uniqueness similarly following from the rank 0 case. □

PROPOSITION 5.19. *Let $\pi : \tilde{X} \to X$ be the birational morphism contracting the identity section of the smooth rational Weierstrass elliptic surface $\tilde{X}$ with marked smooth fiber $Q$, and let $[E] \in K_0(X)$ be a class with* $\operatorname{rank}([E]) > 0$, $\chi([E],[E]) = 1$, *and* $\gcd(\operatorname{rank}([E]), c_1([E]) \cdot Q) = 1$. *Let $E'$ be the exceptional bundle on $\tilde{X}$ representing $\pi^*[E]$. Then $[E]$ is represented by an exceptional bundle iff $R\operatorname{Hom}(E', \mathcal{O}_e(-1)) = 0$, where $e$ is the identity section.*

PROOF. If $[E]$ is representable, then we must have $E' \cong \pi^* E$, and thus $R\operatorname{Hom}(E', \mathcal{O}_e(-1)) = 0$ as required. Conversely, if $R\operatorname{Hom}(E', \mathcal{O}_e(-1)) = 0$, then $E' \cong \pi^* E$ for some sheaf $E$ of class $[E]$, necessarily exceptional. □

REMARK. This reformulation appears particularly amenable to a computational approach: rather than asking about whether an object exists, it asks about the properties of an explicit object $E'$. Moreover $R\operatorname{Hom}(E', \mathcal{O}_e(-1)) = 0$ iff $E'|_e \cong \mathcal{O}_e(-1)^r$, so one need not even compute Ext spaces. In addition, it suffices to consider *any* rational elliptic surface, and the reduction of Proposition 5.16 works equally well when the discrete valuation ring $R$ has mixed characteristic, so that one may feel free to compute in finite characteristic. It is particularly tempting to work with the surface $y^2 = x^3 + z^6 + w^6$ over a suitable finite field. Not only is it easy to find prime fields over which this surface has rational Mordell-Weil group (any prime that splits completely in $\mathbb{Q}(2^{1/3})$ will do), but the Mordell-Weil group is the group of $C_6$-equivariant maps from the Fermat curve $t^6 + u^6 + v^6 = 0$ to the curve $E_0 : y^2 = x^3 - 1$ of $j$-invariant 0, suggesting that it should be possible to reduce to a question about equivariant bundles on the Fermat curve.

More generally, if $E$, $E'$ are two exceptional bundles on $\tilde{X}$ satisfying $R\operatorname{Hom}(E, E') = 0$, then we can apply the standard derived autoequivalences to put $E'$ in the form $\mathcal{O}_{e'}(\chi - 1)$, at which point translating by the corresponding section and twisting by a line bundle makes it $\mathcal{O}_e(-1)$, and thus the image of $E$ under the same operations gives rise to an exceptional bundle on $X$. In fact, this works equally well if $E$, $E'$ are bundles on some blowdown of $\tilde{X}$, since their pullbacks to $\tilde{X}$ still form an exceptional pair. This gives a powerful technique for constructing exceptional bundles on del Pezzo surfaces.

As an example, consider the case of two line bundles. We may as well take one of the bundles to be $\mathcal{O}_X$, and thus ask when $R\operatorname{Hom}(\mathcal{O}_X, \mathcal{O}_X(-D)) = 0$, with $D \cdot Q > 0$. In particular, we must have $0 = \chi(\mathcal{O}_X(-D)) = (D \cdot (D - Q))/2 + 1$, and since $\chi(\mathcal{O}_X(-D), \mathcal{O}_X) > 0$, we must have $\operatorname{Hom}(\mathcal{O}_X(-D), \mathcal{O}_X) \neq 0$, so that $D$ is effective. In other words, if $(\mathcal{O}_X(-D), \mathcal{O}_X)$ forms an exceptional pair, then $D$ must be the class of a rational curve on $X$. Conversely, if $D$ is the class of a rational curve, then $H^p(\mathcal{O}_D) \cong H^p(\mathcal{O}_X)$, and thus $R\Gamma(\mathcal{O}_X(-D)) = 0$ as required.

There is a *contravariant* derived autoequivalence $\Phi^*$ of $\tilde{X}$ that swaps $\mathcal{O}_{\tilde{X}}$ and $\mathcal{O}_e(-1)$, and thus preserves the triangulated subcategory of objects $M$ on $X$ with $R\Gamma(M) = 0$. This is the composition of the autoequivalence $\Psi$ of [**20**, Lem. 12.5] with Cohen-Macaulay duality, and thus we may explicitly compute

(5.1)
$$\operatorname{rank}(\Phi^* M) = c_1(M) \cdot Q$$
(5.2)
$$c_1(\Phi^* M) = -c_1(M) + (c_1(M) \cdot Q + \operatorname{rank}(M))(Q + e) + (c_1(M) \cdot e - \chi(M))Q$$
(5.3)
$$\chi(\Phi^* M) = -c_1(M) \cdot e.$$

On the subcategory $\mathcal{O}_X^{\perp}$ of $D^b_{\text{coh}}(X)$, the action simplifies to

$$\operatorname{rank}(\Phi^* M) = c_1(M) \cdot Q \tag{5.4}$$
$$c_1(\Phi^* M) = -c_1(M) + (c_1(M) \cdot Q + \operatorname{rank}(M))Q, \tag{5.5}$$
$$\chi(\Phi^* M) = 0, \tag{5.6}$$

where we note that the $Q$ here is the one on $X$, corresponding to $\tilde{Q} + e$ on $\tilde{X}$.

In particular, since $\mathcal{O}_X(-D)$ is an exceptional object on $X$ with $R\Gamma(\mathcal{O}_X(-D)) = 0$, applying $\Phi^*$ and shifting gives another such object, with invariants

$$\operatorname{rank}(E) = D \cdot Q, \quad c_1(E) = -D + ((D \cdot Q) - 1)Q, \quad \chi(E) = 0.$$

In this way, we obtain an exceptional bundle of slope $-1/r$ for all $r \geq 2$, and in two different ways for $r = 6$ (corresponding to the two types of rational curve of degree 6: bidegree $(1, 2)$ curves in $\mathbb{P}^1 \times \mathbb{P}^1$ and conics in $\mathbb{P}^2$).

Conversely, suppose we are given an exceptional bundle $E$ of slope $-1/r$, with Chern class of the form $-Q + v$ with $v \in \Lambda_{E_8}$. This has Euler characteristic $(r^2 - r + 2 - v^2)/2r$ (taking the positive inner product on $\Lambda_{E_8}$ rather than the intersection form), which we insist must be negative for any representative of the coset $v + r\Lambda_{E_8}$. We may thus just as well assume that $v/r$ is in the Voronoi cell around 0 (the closure of the set of points closer to 0 than to any other lattice vector), so that $v$ is a minimal coset representative. The maximum of $w^2$ over the Voronoi cell is 1, and thus $v^2 \leq r^2$, forcing $v^2 = r^2 - r + 2$ and $\chi = 0$. We thus obtain an exceptional pair $(E, \mathcal{O}_X)$ on $X$, and thus an exceptional triple $(\mathcal{O}_e(-1), \pi^* E, \mathcal{O}_{\tilde{X}})$ on $\tilde{X}$. Applying $\Phi^*$ turns $\pi^* E$ into (a shift of) a line bundle, and thus $E$ must have come from the above construction.

The above example is somewhat ad hoc, but there is a somewhat more systematic approach available. Consider the case of slope $-2/r$ ($r$ odd). Here $v^2 = r^2 - 2r + 5$, so we still must have $\chi = 0$, but it is difficult to control when $R\operatorname{Hom}(\mathcal{O}_X, V) = 0$ for $V$ an exceptional bundle of rank 2 and very small negative slope. However, since $v^2 > (r-1)^2$, so that $v/(r-1)$ must be outside the

Voronoi cell, we find that our putative bundle $E$ fits into an exceptional triple on $E$ of the form

$$(E, \mathcal{O}_X(-\alpha), \mathcal{O}_X)$$

where $\alpha$ is the highest root. Applying the Cohen-Macaulay dual gives a triple

$$(\omega_X, \omega_X(\alpha), E^D),$$

inducing a triple

$$(E^D, \mathcal{O}_X, \mathcal{O}_X(\alpha)),$$

in which we can swap $\mathcal{O}_X$ and $\mathcal{O}_X(\alpha)$. Lifting to $\tilde{X}$ and applying $\Phi^*$ then gives a four-term exceptional collection of the form

$$(\mathcal{O}_e(-1), \mathcal{O}_{e'}(-1), M, \mathcal{O}_X)$$

where $M$ has rank $r-2$ and slope $-r/(r-2)$ and (up to the action of $W(E_8)$) is the pullback of a sheaf on a degree 2 del Pezzo surface.

We could, of course, do this calculation entirely in the Grothendieck group to get a class $[M]$. Clearly, if $[E]$ is representable, then $[M]$ is representable, but it turns out that the converse applies as well. Indeed, if $[M]$ is representable, then since $\mu(M) \in (-2,0)$, we find that $R\operatorname{Hom}(\mathcal{O}_X, M) = 0$ and thus we indeed obtain a four-term exceptional collection, and can then invert the derived equivalences to deduce that $[E]$ is representable. Moreover, the representability of $[M]$ is equivalent to that of the class $[M](Q_X)$ of slope $-2/(r-2)$, and may thus be shown by induction. (To make this explicit, we must reduce $v/(r-2)$ to the fundamental alcove, but this requires only a small number of affine reflections for $r$ large since it is already quite close to the fundamental alcove!) We use the standard coordinatization of the fundamental alcove via inner products with the simple roots. (In terms of the basis of $\operatorname{Pic}(X)$ coming from $\mathbb{P}^2$, the simple roots are $(h-e_1-e_2-e_3, e_1-e_2, e_2-e_3, \dots, e_7-e_8)$.)

PROPOSITION 5.20. *An exceptional sheaf of slope $-2/r$ on a degree 1 del Pezzo surface has Chern class $-2Q+v$ where the image of $v/r$ in the fundamental alcove is either*

$$r^{-1}(0,(r-5)/2,0,0,1,0,0,0,0)$$

*or one of the following sporadic cases:*

$$3^{-1}(1,0,0,0,0,0,0,0,0),$$
$$9^{-1}(1,0,0,1,0,0,0,0,0),$$
$$11^{-1}(1,0,2,0,0,0,0,0,0),$$
$$13^{-1}(3,2,0,0,0,0,0,0,0),$$
$$15^{-1}(5,0,0,0,0,0,0,0,0)$$

PROOF. Since $v^2$ is close to $r^2$, $v/r$ must be close to the fundamental weight $(0,1/2,0,0,0,0,0,0,0)$, from which it is straightforward to verify that for $r \gg 0$, the only possibility is as stated. Moreover, one finds that for $r \geq 5$, the image of $(r/(r-2))v$ in the fundamental alcove has the same form, while for $r=5$, the image is $3^{-1}(1,0,0,0,0,0,0,0,0)$. The only other possibilities for $r \leq 17$ are as stated, with the sporadic instances for $r \in \{11,13,15\}$ each reducing to the sporadic instance of rank $r-2$, and the sporadic instance for $r=9$ reducing to the regular instance of rank 7. On the other hand, for the regular instance of rank $r \geq 9$, there is a unique orbit of $-1$-curves on $X$ (consisting of four orthogonal $-1$-curves) such that the slope $-(r-2)/r$ bundle has $\chi(E,\mathcal{O}_e(-1))=0$, and thus (by duality) at most one $W(E_7)$-orbit of numerical classes of slope $-(r+2)/r$ corresponding to exceptional bundles on the degree 2 del Pezzo surface. We thus see that the regular instance can account for at most one type of bundle of rank $r+2$, from which the claim follows by induction. □

REMARK. In both the slope $-1/r$ and $-2/r$ cases, one can verify that there are no other classes with $v^2=r^2-r+2$ or $v^2=r^2-2r+5$, and thus that the above necessary condition is sufficient for bundles of degree $\pm 1$, $\pm 2$ modulo $r$.

The interpretation via $\tilde{X}$ also gives rise to one further construction. If $E$ is an exceptional bundle on $X$ of slope $-d/r\in(0,1)$, then in addition to the pair $(\pi^*E,\mathcal{O}_e(-1))$ on $\tilde{X}$, we also obtain a pair $(\mathcal{O}_e,\pi^*E)$, which itself gives rise to an exceptional bundle $E'$ on $X$. This new bundle has slope $-d'/r\in(0,1)$ where $d'd=-1(r)$ and satisfies $v'=d'v$. As before, this construction not only gives a new bundle but a new necessary condition. In this way, the slope $-2/r$ construction gives rise to a construction for slope $(r-1)/2r$ ($r$ odd). Together with Cohen-Macaulay duality, this gives us a complete classification for slopes $-1/r$, $-2/r$, $(r-1)/2r$, $(r+1)/2r$, $(r-2)/r$, $(r-1)/r$. This accounts for every possibility for $r\le 9$ except $\{-3/8,-5/8\}$. For slope $-3/8$, there are two cases with $\chi\le 0$, but for one of those cases the putative bundle $E'$ of slope $-d'/r=-5/8$ violates the necessary condition. We thus see that the only possibility for slope $-3/8$ corresponds to $v$ with coordinates

$$8^{-1}(0,0,1,0,0,1,0,0),$$

and can verify that this reduces to the known bundle of slope $-3/5$.

For rank $\ge 10$, the above techniques no longer suffice to obtain a classification: there is a candidate of slope $-3/10$ with maximal Euler characteristic $-1$. (A direct computation on a suitable Weierstrass surface shows that no such bundle exists.)

## 6. Classifications

We now return to our original question: given a bundle $V$ and autoequivalence $\Psi$, understand the filtered deformations of $B_{V,\Psi}$. If as usual we constrain the slopes of $V$ to an interval of length $d$, then we can produce such a deformation from the following data: (1) a noncommutative del Pezzo surface $X$ with anticanonical curve $Q$ such that $\theta\otimes H^1(\mathcal{O}_Q)\cong\Psi^{-1}$ on $\mathcal{O}_Q$, so in particular $Q^2=d$, (2) an abelian equivalence $\psi:\operatorname{coh}(Q)\cong\operatorname{coh}(E)$, and (3) a rigid, unobstructed, torsion-free sheaf $M$ satisfying $\psi M|_Q\cong V$. The discussion above further allows us to replace $M$ by an exceptional collection $(E_1,\dots,E_n)$ such that the bundles $(\psi E_1|_Q,\dots,\psi E_n|_Q)$ are the stable constituents of $V$ (with appropriate multiplicities).

Now, a pair $(X,\psi)$ with $X$ a noncommutative del Pezzo surface and $\psi$ : $\operatorname{coh}(Q)\cong\operatorname{coh}(E)$ induces a homomorphism $K_0(X)\to K_0(E)$ by $[M]\mapsto[\psi M|_Q]$, and one can in fact recover the surface from this homomorphism. Indeed, $\psi$ as usual has the form $\psi(M)=f_*(M)\otimes\psi(\mathcal{O}_Q)$ where $f:Q\cong E$ is an automorphism of curves. Since $\psi(\mathcal{O}_Q)\cong\psi(\mathcal{O}_X|_Q)$, we can recover $[\psi(\mathcal{O}_Q)]$ (and thus, up to isomorphism, $\psi(\mathcal{O}_Q)$) from the homomorphism $K_0(X)\to K_0(E)$. Since

$$[f_*M|_Q]\cong[\psi M|_Q]-\operatorname{rank}(M|_Q)[\mathcal{L}],$$

we can thus compute the homomorphism $[M] \mapsto [f_* M|_Q]$ from the given homomorphism. But this homomorphism is precisely of the kind we considered in Section 4, so lets us recover $X$.

This works for any homomorphism $\phi : K_0(X) \to K_0(E)$, as long as for all $[M] \in K_0(X)$ and some $\delta \in \mathbb{Z}$ one has $\operatorname{rank}(\phi([M])) = \operatorname{rank}([M])$ and $\deg(\phi([M])) = c_1([M]) \cdot Q + \delta \operatorname{rank}([M])$. (It then follows that the new map $\phi'([M]) := \phi([M]) - \operatorname{rank}([M])(\phi([\mathcal{O}_X]) - [\mathcal{O}_Q])$ gives rise to a noncommutative surface $(X, Q)$ with a marked isomorphism $Q \cong E$, and we recover $\psi$ after twisting by a line bundle of class $\phi([\mathcal{O}_X])$.) We thus find that the moduli stack of pairs $(X, \psi)$ (with blow-down structure) consists of countably many copies of $E^{12-d}$, one for each value of $\deg(\phi([\mathcal{O}_X])) \in \mathbb{Z}$. (More precisely, if one allows $E$ to vary, one obtains the corresponding power of the universal curve over the moduli stack of elliptic curves, since one can act on $\phi$ by $\operatorname{Aut}(E) \ltimes E$.)

Note that the moduli space of possible $\psi$s for a given choice of $(X, Q)$ is 2-dimensional: there is a degree of freedom corresponding to a choice of marked point (to identify $Q$ with an elliptic curve) and a degree of freedom corresponding to the choice of $\psi(\mathcal{O}_Q)$. Thus, for instance, the moduli stack of pairs $(X, Q)$ with $X$ a noncommutative $\mathbb{P}^2$ may be identified with the moduli space of cubic plane curves with a choice of translation, or equivalently by the moduli stack of smooth genus 1 curves with marked line bundles of degrees 3 and 0. Making $Q$ an elliptic curve requires marking a point, at which point we may use $\operatorname{Pic}^2(E) \cong \operatorname{Pic}^0(E) \cong E$ to write the family as $E^2$, where $E$ is the universal elliptic curve. This determines how $\psi$ acts on structure sheaves of points, but it remains to specify its image on $\mathcal{O}_Q$, which is a point of $\operatorname{Pic}(E) \cong \mathbb{Z} \times E$, giving countably many copies of $E^3$.

There is a natural action of $W(E_{9-d})$ on $\operatorname{NS}(X)$, which extends to an action of $W(E_{9-d}) \ltimes \operatorname{NS}(X)$ on $K_0(X)$. This arises from the fact that line bundles span $K_0(X)$ (and, in fact, there is a basis of line bundles), and thus we can define an action by

$$[\mathcal{O}_X(D)] \mapsto [\mathcal{O}_X(wD + D_0)].$$

There is then an induced action of this group on our moduli stack, and we obtain a derived equivalence between the categories of coherent sheaves on any two surfaces in the same orbit. (Indeed, $W(E_{9-d})$ acts on commutative del Pezzo surfaces preserving the structure sheaf, while $\operatorname{NS}(X)$ corresponds to twists by line bundles. One caveat is that in the noncommutative setting, twisting by a line bundle changes the surface!) Similarly, there is an action of $\operatorname{Aut}(E) \ltimes E \times \operatorname{Pic}(E)$ on the other side, and this preserves the surface (since it only changes $\psi$).

Note that the autoequivalence $\Psi_X$ of $\operatorname{coh}(E)$ corresponding to the action of $\theta^{-1}$ depends only on $\phi([\mathrm{pt}])$, $\phi([\mathcal{O}_Q])$ and the integer $\delta = \deg(\phi([\mathcal{O}_X]))$. Indeed, the automorphism $\tau$ is translation by $q := \det \phi([\mathrm{pt}])$, so that $\Psi_X(\mathcal{O}_x) \cong \mathcal{O}_{x+q}$, while the line bundle is determined by the equation

$$[\mathcal{L}] + \delta q + \phi([\mathcal{O}_X]) = \Psi_X(\phi([\mathcal{O}_X])) = \phi([\mathcal{O}_X(Q)]) = \phi([\mathcal{O}_X]) + \phi([\mathcal{O}_Q]) + dq,$$

giving

$$[\mathcal{L}] = (d - \delta)q + \phi([\mathcal{O}_Q]).$$

(One caveat, which we will discuss further below, is that the *isomorphism class* of $\mathcal{L}$ is fixed, but $\mathcal{L}$ itself is a more subtle issue.) Thus any given isomorphism class of $\Psi$ cuts out a codimension 2 subscheme of the moduli stack on which $\phi([\mathrm{pt}])$ and

$\phi([\mathcal{O}_Q])$ are fixed. For $d < 8$, the two classes [pt] and $[\mathcal{O}_Q]$ span a saturated sublattice of $K_0(X)$, so that this subscheme is itself a power of $E$, namely $E^{10-d}$; this also holds on the component for $d = 8$ corresponding to odd Hirzebruch surfaces. (For $\mathbb{P}^2$, the subscheme is a coset of $E[3] \times E$, while for $F_0/F_2$ it is a coset of $E[2] \times E^2$.)

It will be useful to give names for the coordinates in $E^{12-d}$. Recall that the expression of $X$ as an iterated blowup of $\mathbb{P}^2$ (i.e., such an expression of the underlying commutative surface) induces a basis of $\mathrm{NS}(X)$ of the form

$$h, e_0, \dots, e_{8-d}$$

which in turn induces a basis

$$[\mathcal{O}_X], [\mathcal{O}_h(-1)], [\mathcal{O}_{e_0}(-1)], \dots, [\mathcal{O}_{e_{8-d}}(-1)], [\mathrm{pt}]$$

of $K_0(X)$, so that $(X, \psi)$ is determined by

$$u := \det \phi[\mathcal{O}_X], \quad h := \det \phi[\mathcal{O}_h(-1)], \quad q := [\mathrm{pt}]$$

and

$$x_i := \det \phi[\mathcal{O}_{e_i}(-1)].$$

(An expression of $X$ as an iterated blowup of an even Hirzebruch surface gives a similar parametrization, but with $h$ and $e_0$ replaced by $s$ and $f$.)

Now, given a representable numerical exceptional collection $([E_1], \dots, [E_n])$ in $K_0(X)$ (satisfying the positive root condition where the slopes agree), there is an associated family $(E_1, \dots, E_n)$ of exceptional collections on the moduli stack, and the bundles $(E_1|_Q, \dots, E_n|_Q)$ can be read off (up to isomorphism) from the classes $\phi([E_i])$, since any stable sheaf on $E$ is determined by its class in $K_0(E)$. Thus fixing the bundles is again tantamount to fixing the values under the corresponding homomorphisms from $E^{12-d}$ to $E$. An exceptional collection always forms a saturated sublattice (the matrix of Mukai pairings has determinant 1), and thus fixing the bundles gives a family isomorphic to $E^{12-n-d}$.

We thus see in general that the family of surfaces with blowdown structure and choice of $\psi$ such that the given collection gives rise to a given $B_{V,\Psi}$ is the appropriate fiber of a homomorphism $E^{12-d} \to E^{n+2}$. (Of course, the fiber will be empty unless the values $\mathrm{rank}(\phi([E_i]))$, $\deg(\phi([E_i]))$ are all correct, a constraint on the numerical collection along with the integer $\delta$.) We must then take the union over all numerical collections and quotient by the action of $W(E_{9-d}) \ltimes \mathrm{NS}(X)$ to obtain the "true" moduli space. (Equivalently, we choose one representative from each orbit of valid numerical collections, and for each one, take the quotient of the relevant subscheme of $E^{12-d}$ by the action of the stabilizer, a subgroup of the affine Weyl group $W(E_{9-d}) \ltimes Q^\perp$.)

One caveat here is that reflections do not give *abelian* autoequivalences; they do not in general preserve the effective cone. In particular, it is not a priori obvious that $W(E_{9-d})$ preserves the graded algebra. Certainly, where none of the roots of $E_{9-d}$ are effective, $W(E_{9-d})$ acts as abelian equivalences, and thus there is no difficulty. In particular, for any coordinate on the appropriate weighted projective space, the two algebras we want to compare induce sections of the same line bundle on the relevant subscheme of $E^{12-d}$ that agree where no roots are effective (or, more precisely, where the given $w$ does not make any effective root negative). So the only way those coordinates can fail to agree is if the condition on roots *always* fails. But in that case (which since it holds for all $q$, must come from a case in which the underlying commutative surface has a $-2$-curve in which we wish to reflect), the reflection corresponding to such a root actually acts trivially on the entire component. It follows in either case that the map to the deformation *scheme* factors through the quotient by $W(E_{9-d}) \ltimes \mathrm{NS}(X)$. (Caveat: the image in the deformation *stack* is not the quotient stack: weighted projective spaces have cyclic stabilizers, unlike the action of $W(E_{9-d}) \ltimes \mathrm{NS}(X)$ on $E^{12-d}$. This is presumably related to the fact that the quotient by the reflection subgroup of the stabilizer of a point is regular at that point; quotienting by that normal subgroup gives a cyclic group in the cases of interest.)

It remains to understand the morphism from this family to deformation space. The missing piece here is that there is a natural line bundle on deformation space (it is a quotient by $\mathbb{G}_m$, after all), and thus we need to understand the pullback of this line bundle to the relevant family of surfaces. (The image will then be the corresponding embedding in weighted projective space.) It turns out that this is in fact the restriction of a line bundle on the ambient moduli space of pairs $(X, \psi)$. The main issue is the fact that when defining the moduli space, we only determined the various exceptional sheaves and $\mathcal{L}$ up to isomorphism. For the exceptional sheaves, this is not a problem: rescaling the exceptional sheaves by scalars simply acts on $B_{V,\Psi}$ by an inner automorphism. The line bundle is thus entirely determined by the failure of the induced autoequivalence $\Psi_X$ of $Q$ to be a pullback from the corresponding family over $E^2$.

We thus need to understand this autoequivalence better. For convenience, we consider its inverse instead. A pair $(X, \psi)$ determines an adjoint pair of functors $j_* : \operatorname{coh} E \to \operatorname{coh} X$, $j^* : \operatorname{coh} X \to \operatorname{coh} E$, and there is a functorial distinguished triangle

$$\Psi_X^{-1}[1] \to Lj^*j_* \to \mathrm{id} \to$$

(actually a direct sum decomposition of $Lj^*j_*$). Since $\Psi_X^{-1}M \cong (\Psi_X^{-1}\mathcal{O}_E) \otimes \tau_q^*M$, it remains only to compute $\Psi_X^{-1}\mathcal{O}_E$. This is a line bundle on $E \times E^{12-d}$, and is functorial in $E$, so is really a line bundle on the corresponding moduli stack. Such line bundles were classified in [**21**, Prop. 2.2], and have the form $\mathcal{L}_{Q,w}$ with "weight" $w \in \mathbb{Z}/12\mathbb{Z}$ (the isomorphism class of the pullback of the identity section) and "polarization" $Q$ (the class in the Néron-Severi group of the generic fiber). The polarization is a symmetric element of $\mathrm{End}(E \times E^{12-d}) \cong \mathrm{Mat}_{13-d}(\mathbb{Z})$, and thus may be expressed as a quadratic form $vQv^t/2$ in the coordinates, in our case $z$ on $E$ and $u, h, q, x_0, \dots, x_{8-d}$ on $E^{12-d}$. (Note that homomorphisms between powers of $E$ are generically given by integer matrices, and the pullback acts on the polarization in the obvious way. Also, the line bundle is relatively ample iff the polarization is positive definite.)

Composing $\psi$ with an equivalence $\mathrm{coh}(X) \cong \mathrm{coh}(X')$ does not change $Lj^*j_*$, so leaves $\Psi_X^{-1}$ alone. Not only does this imply that $\Psi_X^{-1}$ is well-defined as a functor on the category of sheaves on the universal curve over the parameter space, but that it is invariant under the action of $\mathrm{NS}(X)$. Similarly, if we compose with an autoequivalence of $\mathrm{coh}(E)$, this simply conjugates $\Psi_X^{-1}$ by that autoequivalence, and thus we know how $\Psi_X^{-1}\mathcal{O}_E$ must transform under that action on parameter space.

The action of $E \times \mathrm{Pic}(E)$ suffices to determine the dependence of $\Psi_X^{-1}\mathcal{O}_E$ on $u$ and $\delta$ (i.e., it tells us that the polarization has the form $uq - \delta(qz + q^2/2)$ plus terms independent of $u$ and $\delta$). The twist-by-$h$ symmetry implies that the polarization is invariant under $(h, u, \delta) \mapsto (h + q, u + 2q + h, \delta + 3)$, which together with the known dependence on $u$ and $\delta$ tells us that the $h$-dependent terms are precisely $-h^2/2+3hz$. Similarly, invariance under $(x_i-q,u+x_i,\delta+1)$ (i.e., twisting by $e_i$) implies that the $x_i$-dependent terms are precisely $x_i^2/2-x_iz$. Since the $z$-dependence is determined by the *isomorphism classes* of the fibers over $E^{12-d}$ (i.e., that they correspond to the point $-3h+x_0+\cdots+x_{8-d}+\delta q$ of $\operatorname{Pic}^{-d}(E)$), we thus conclude that

$$\Psi_X^{-1}\mathcal{O}_E\cong\mathcal{L}_{-dz^2/2+(3h-x_0-\cdots-x_{8-d}-\delta q)z-h^2/2+x_0^2/2+\cdots+x_{8-d}^2/2+uq-(\delta+a)q^2/2,w},$$

for some unknown $a\in\mathbb{Z}$ and $w\in\mathbb{Z}/12\mathbb{Z}$. (One can also easily verify that this is invariant under $W(E_{9-d})$.)

Since changing $a$ and $w$ merely twists by a line bundle pulled back from the curve parametrizing $q$, this is enough to tell us the following. (Note that the dependence on $h,\vec{x}$ is essentially just the intersection form, making the analogue for $F_0/F_2$ straightforward to write down.)

PROPOSITION 6.1. *If we fix $E$, $V$, $\Psi$, then for any family of filtered deformations of $B_{V,\Psi}$ constructed via Proposition* 5.1, *the pullback of the natural line bundle on deformation space is isomorphic to the restriction of* $\mathcal{L}_{-h^2/2+x_0^2/2+\cdots+x_{8-d}^2/2+uq,0}$.

PROOF. Twisting by the inverse of this bundle turns $\Psi_X^{-1}\mathcal{O}_E$ into

$$\mathcal{L}_{-dz^2/2+(3h-x_0-\cdots-x_{8-d}-\delta q)z-(\delta+a)q^2/2,w},$$

which is the pullback through the map $(z,3h-x_0-\cdots-x_{8-d},q)$ of a family of line bundles depending only on the isomorphism class of the autoequivalence $\Psi$. In particular, for fixed $\Psi$, twisting by the given line bundle makes the associated graded of the universal filtered deformation globally isomorphic to $B_{V,\Psi}$ as required. □

Since the polarization is not positive definite, this line bundle is not ample on all of $E^{12-d}$. On the fibers of $(q,3h-x_0-\cdots-x_{8-d})$, the polarization becomes positive *semi*definite, and becomes positive definite once we fix the image under $\psi$ of some exceptional sheaf of positive rank (as this fixes a linear combination of $u$ and the other parameters). We thus see that we indeed obtain an ample bundle on our family of deformations in this way, and thus a map to a weighted projective space, in such a way that the deformation parameters are homogeneous in the coordinates on weighted projective space.

As an example, let us consider the case that $V$ is semistable of rank $r$ and slope $m\in\mathbb{Z}$ (divisorial, for simplicity) while $\Psi$ increases degrees by $d$. Since changing $\deg\phi([\mathcal{O}_X])$ adds the same integer to all slopes in the exceptional collection, the corresponding exceptional collection must consist of line bundles of the same slope. It must therefore have the form

$$(\mathcal{O}_X(D-\alpha_1-\cdots-\alpha_{r-1}),\mathcal{O}_X(D-\alpha_1-\cdots-\alpha_{r-2}),\dots\mathcal{O}_X(D))$$

where $D$ is some divisor class and $\alpha_1,\dots,\alpha_{r-1}$ are the simple roots of a subsystem of type $A_{r-1}$ inside $E_{9-d}$. The orbits of such collections under $W(E_{9-d})\ltimes\operatorname{NS}(X)$ are in natural correspondence with the conjugacy classes of $A_{r-1}$ subsystems, and for each orbit, we may take the representative with $D=0$. (This forces us to take $\delta=m$ in the map $K_0(X)\to K_0(E)$, though of course in this case we can tensor $V$ by a line bundle to obtain an equivalent problem with $m=0$.)

For $r=1$, there is (trivially) a unique orbit of $A_0$ subsystems, and we find that, for $d\le 6$, the base of the corresponding family is the quotient of (a torsor over) $E^{9-d}$ by $W(E_{9-d})$, with line bundle corresponding to the minimal invariant polarization. (Minimality can be read off from the relation of the above polarization to the intersection form, so that any root evaluates to 1.) Per Looijenga [**16**], such a quotient for an irreducible Weyl group is isomorphic to a weighted projective space, in which the degrees of the generators are the labels on the affine Dynkin diagram. We thus obtain the following descriptions of the quotient subschemes (corresponding to $E_8$, $E_7$, $E_6$, $D_5$, $A_4$, $A_2A_1$:

$$\mathbb{P}^{[122334456]}, \mathbb{P}^{[11222334]}, \mathbb{P}^{[1112223]}, \mathbb{P}^{[111122]}, \mathbb{P}^4, \mathbb{P}^2 \times \mathbb{P}^1.$$

For $d = 7$, the Weyl group is too small, so we obtain a quotient of $E^2$ by $\mathrm{Sym}_2$, while for $d = 8$, there are two components (corresponding to the two parities of Hirzebruch surface), one isomorphic to $E$ (the odd case), and one isomorphic to $E[2] \times E/A_1$ (the even case). Finally, for $d = 9$, we obtain $E[3]$.

For $r = 2$, $d \leq 5$, there is again a unique orbit of $A_1$ subsystems, and thus we must replace $E_{9-d}$ by the stabilizer of a root in $E_{9-d}$. This gives ($E_7$, $D_6$, $A_5$, $A_3A_1$)

$$\mathbb{P}^{[11222334]}, \mathbb{P}^{[1111222]}, \mathbb{P}^5, \mathbb{P}^3 \times \mathbb{P}^1$$

for $d \leq 4$. For $d = 5$, the rank is too small, so we have an elliptic factor, and for $d = 6$, there are two orbits of roots, one of which has an elliptic factor. For $d = 7$, we have a unique orbit of roots, with an elliptic factor, while for $d = 8$ only the even case has a root, making the moduli space in that case a torsor over $E[2]$, and for $d = 9$ no configuration exists.

For $r = 3$, we similarly obtain

$$\mathbb{P}^{[1112223]}, \mathbb{P}^5, \mathbb{P}^2 \times \mathbb{P}^2$$

for $d \leq 3$; for $d = 4$, $d = 5$, there are unique orbits but with a rank deficit. For $d = 6$, there is a unique orbit, and the space has the form $E[3] \times \mathbb{P}^1$. For $d > 6$, no configuration exists.

For $r = 4$, we have

$$\mathbb{P}^{[111122]}, \mathbb{P}^3 \times \mathbb{P}^1,$$

for $d = 1, 2$, a unique orbit with a rank deficit for $d = 3$, two orbits for $d = 4$ (one with a rank deficit, and one with a torsion factor), and two orbits for $d = 5$ (the maximum), again with a rank deficit.

For $r = 5$, we have $\mathbb{P}^4$ for $d = 1$, a unique orbit with a rank deficit for $d = 2, 3$, two orbits with a rank deficit for $d = 4$, and for $d = 5$ obtain $E[5]$.

For $r = 6$, we have $\mathbb{P}^2 \times \mathbb{P}^1$ for $d = 1$, two orbits (one with a rank deficit, the other with torsion) for $d = 2$, two orbits (with torsion) for $d = 3$, and nothing for $d > 3$. For $r = 7$, we have a rank deficit for $d = 1, 2$ and nothing for $d > 2$, while for $r = 8$ we have two components (one with a rank deficit and one with torsion) for $d = 1$ and $E[2]$ for $d = 2$, and for $r = 9$ we have $E[3]$ for $d = 1$. (In general, if $r + d > 10$, then no configuration exists, since the rank of the subsystem is too large.)

The astute reader will have noticed a symmetry above: the description of the given portion of deformation space is invariant under swapping $r$ and $d$. Indeed, if we blow up $d$ further points, then we obtain an exceptional collection on a (deformed!) elliptic surface, and a derived equivalence swaps the roles of the $r$ line bundles and the $d$ exceptional curves. Such a symmetry holds in general for the semistable case, with the one caveat that it will in general change the slope from $a/r$ to $b/r$ with $ab = 1(r)$.

In many of the above cases, the image of the family in weighted projective space contains an orbifold point, which leads to additional interesting families. In general, if the space for $B_{V,\Psi}$ passes through an orbifold point of degree $g$, then the corresponding deformation extends to one of $B_{V',\Psi'}$ with

$$V' = V \oplus \Psi(V) \oplus \cdots \oplus \Psi^{g-1}(V), \quad \Psi' = \Psi^g.$$

We can then further deform $V'$, $\Psi'$.

For instance, since the family for $d = r = 1$ is $\mathbb{P}^{[122334456]}$, we have orbifold points of order $2 \le g \le 6$. The result corresponds to taking $V$ to be a sum of line bundles of slopes $0, 1, \dots, g-1$, with $\Psi$ increasing degrees by $g$. The differences between the Chern classes of the corresponding line bundles on $X$ (now a del Pezzo surface of degree $g$) must then be rational curves. It is straightforward to classify such configurations, and for $g = 2, 3, 4$, we find that the family is the Looijenga-style quotient associated to $E_6$, $D_4$, $A_2$ respectively. For $g = 5$, there is a unique filtered deformation, while for $g = 6$, the *generic* such $V$ has no filtered deformations, but there is a codimension 2 subfamily of $V$s with unique filtered deformations. Indeed, in this case, we find that the exceptional collection is actually a *full* exceptional collection, and thus $\Psi$ is determined by $V$.

Of course, the cases corresponding to orbifold points are special cases of this scenario, and have an additional symmetry corresponding to shifting the slopes by 1 (and applying $\Psi$ or $\Psi^{-1}$ as needed). This relates to the fact that for $d \le 6$, the group $W(E_{9-d}) \ltimes \mathrm{NS}(X)$ has the affine Weyl group (or product of two affine Weyl groups, for $d = 6$) as a normal subgroup, and thus we have an isomorphism

$$W(E_{9-d}) \ltimes \mathrm{NS}(X) \cong \mathbb{Z} \ltimes \tilde{W}(E_{9-d}),$$

where the splitting is induced by the requirement that $\mathbb{Z}$ preserves the set of simple roots. In other words, the group $W(E_{9-d}) \ltimes \mathrm{NS}(X)$ contains an element $\pi$ corresponding to a diagram automorphism. Since $\tilde{W}(E_{9-d})$ does not change slopes, $\pi$ must change the slope (increasing it by 1, say). Moreover, $\pi^d$ commutes with the affine Weyl group, and thus (since it increases slopes by $d$) is translation by $Q$. We can then verify that the divisor classes $\pi(0), \dots, \pi^{d-1}(0)$ all correspond to rational curves, and thus induce a configuration as in the previous paragraph. The orbifold case then corresponds to the case that the point in the moduli space is invariant under $\pi$.

For $d = 2$, $r = 1$, we have decimations for $g \in \{2, 3, 4\}$. In each case, there is a single configuration (up to the Weyl group action and choice of parity for $2g = 8$). For $g = 2$, $g = 3$, the result is a Looijenga quotient, with Weyl groups $D_4$ and $A_1$ respectively, while for $g = 4$, the result is again a full exceptional collection (in an even Hirzebruch surface), and thus there is a unique deformation on the appropriate codimension 2 subscheme.

For $d = 1$, $r = 2$, we again have $D_4$ and $A_1$ for $g \in \{2, 3\}$ and a unique deformation over a codimension 2 subscheme of the base for $g = 4$, with the only difference from the $(d, r) = (2, 1)$ case being that the surfaces have degree $g$ rather than $2g$. (In particular, for $g = 4$, we obtain a full exceptional collection of line bundles on a quartic del Pezzo surface.)

For $d = 3$, $r = 1$, the $g = 2$ case is the Looijenga quotient of type $A_2$, while the $g = 3$ case is a full exceptional collection on $\mathbb{P}^2$. The $g = 3$ case corresponds to the fact that the standard noncommutative deformation (the three generator Sklyanin algebra [**1**]) of $k[x,y,z]$ is obtained by removing the degree 3 equation from the $d=3$, $r=1$ elliptic algebra.

For $d=1$, $r=3$, the $g=2$ case is again the Looijenga quotient of type $A_2$, while the $g=3$ case is a full exceptional collection on a noncommutative cubic surface.

For $d=4$, $r=1$, $g=2$, the exceptional collection lives on an even Hirzebruch surface, and $V$ imposes a single (saturated!) condition on $\Psi$. So deformations only exist on a codimension 1 subscheme of parameter space, but when they exist are classified by $\mathbb{P}^1 \cong E/A_1$. Each such deformation corresponds to removing a quadratic relation from the algebra with $d=4$, $r=1$; since the scheme is $\mathbb{P}^1$ embedded by $O(1)$, the possibilities span a 2-dimensional subspace of the dual. Removing both relations gives Sklyanin's deformation [**23**] of $k[x,y,z,w]$ (with the two removed relations corresponding to the 2-dimensional space of central elements of degree 2.)

For $d=1$, $r=4$, $g=2$, we again have a $\mathbb{P}^1$ of deformations on a codimension 1 subscheme. Again, we could also remove both relations, which gives something best described as an Azumaya algebra on a noncommutative Fano 3-fold; it is unclear how to characterize that 3-fold, however.

Finally, for $d=r=2$, $g=2$, deformations only exist on a codimension 1 subscheme, while the general nonempty fiber is the Looijenga quotient of type $A_3$, a.k.a. $\mathbb{P}^3$.

We now turn to semisimple examples with non-integer slope. Note that in the case of a single bundle, the process of testing representability of the candidate exceptional classes involves first putting them in the fundamental alcove. It is of course quite straightforward to find all candidates in the fundamental alcove, and for each candidate surviving the Euler characteristic test, we can read off the stabilizer from its standard coordinates. This works particularly well for $r=d=1$; for larger values of $r$, we need only classify configurations of $r-1$ roots that have intersection 1 (modulo the denominator) with the fundamental representative and $-1$ with each other, while for larger values of $d$, we need only find $d-1$ exceptional curves that are orthogonal (modulo the denominator) and to each other. (And when both are larger, the roots must be orthogonal to the $-1$-curves.)

For $\mu=-1/2$, we obtain the following nice cases (omitting cases where there is either a rank deficit or torsion) as $r$, $d$ vary:

$$\begin{pmatrix} D_8 & A_7 & A_5A_1 \\ A_7 & A_3A_3 & \\ A_5A_1 & & \end{pmatrix}$$

In the case $r=9$, $d=1$ (or vice versa), there is a unique combinatorial configuration, with the corresponding subscheme of parameter space a torsor over $E[10]$. This case is primarily of note because there actually are *no* bundles of slope $-1/2$ on a noncommutative $\mathbb{P}^2$, and thus this case can only arise by taking $\delta \neq 0$.

In the $r=d=1$ case, there are orbifold points, so we can again take a decimation. It turns out that for bundles of slope $-3/2$, $-1/2$ (on a degree 2 del Pezzo surface) to form an exceptional collection, it is necessary for their Chern classes to differ by $Q$. This means that a deformation will only exist on a codimension 1 subscheme of parameter space, and when it does will be classified by the quotient corresponding to $A_7$ (i.e., the same as for $r=1$, $d=2$).

For slope $-1/3$ (or equivalently $-2/3$), we obtain the nice cases

$$\begin{pmatrix} A_8 & A_5A_2 \\ A_5A_2 & \end{pmatrix}.$$

For slope $-1/4$, the $r = d = 1$ case is nice, with corresponding Weyl group $A_7A_1$. For slope $-1/5$, it appears that no cases are nice, but for $r = d = 1$, slope $-2/5$, the $r = d = 1$ case is nice, with group $A_4A_4$.

One caveat here is that we have not shown that every invariant section of a power of the ample bundle can be expressed as a polynomial in the coefficients of the associated filtered deformation. For sections of the ample bundle itself, this would reduce to a calculation of the infinitesimal deformation of $\mathrm{Proj}(B[t])$ associated to the $\mathbb{G}_m$ action on the general filtered deformation in the family; this, together with analogous calculations for orbifold points, should be enough to establish the result in general. (Note that this would also establish that the map from the quotient scheme to the deformation scheme is an embedding.)

The fact that precisely these weighted projective spaces arise when computing moduli spaces of filtered deformations of random elliptic algebras suggests the following.

CONJECTURE 1. *Any nontrivial filtered deformation of any elliptic algebra $B_{V,\Psi}$ has Rees algebra of the form $A^+_M$ for some rigid, unobstructed, torsion-free sheaf $M$ on a noncommutative del Pezzo surface.*

REMARK. Since the moduli stack is closed in a weighted projective space, the Hilbert series of the moduli stack is upper semicontinuous. As a result, the argument of Lemma 2.11 would let us reduce to the case that $\tau$ has finite order (so the construction involves Azumaya algebras on del Pezzo surfaces), at the cost of having to work in finite characteristic. We also immediately deduce that if the claim holds for a given algebra $B$, then it holds for the generic algebra of the same shape.

In the cases where this would make the deformation space a weighted projective space, there is a possible strategy for proving this: show that for each degree, the space of (nontrivial) infinitesimal deformations agrees with the putative number of generators. This implies that the deformation space is *contained* in the conjectured weighted projective space, and therefore equal.

Although one could hope to compute the desired Hochschild cohomology spaces by hand, this does at the very least simplify the calculation when done by computer. In particular, for the line bundle cases with $r = 1$ and $d \in \{1, 2\}$, we could carry out the calculation not just for generic parameters, but for *general* parameters: that is, for any $E$ and any $\Psi$ that increases degrees by 1 or 2, every filtered deformation of the algebra $B_{\mathcal{O}_E,\Psi}$ arises from the surface construction. We have also checked the conjecture for *random* instances (over a moderately large finite field) in every slope 0 case (including those where the family is not nice), as well as for a few of the cases with noninteger slope; as mentioned, this implies the conjecture holds generically in those cases as well.

Although the general twisted case appears out of reach at the moment, there is some hope for the untwisted case, since any deformation has a large center. The first step would be to show that the filtered deformation arises from a sheaf on a *singular* del Pezzo surface, which would reduce to the following (which would follow from the main conjecture, as would the analogous statement when twisted by a torsion point, since the center of a noncommutative del Pezzo surface with $q$ torsion is a commutative del Pezzo surface).

CONJECTURE 2. *Any filtered deformation of an untwisted elliptic algebra* $B_{V,\Psi}$ *which induces a trivial deformation of the center is trivial.*

REMARK. This holds for $d = 1$ if $V$ is stable of rank 2 or a sum of two degree 0 line bundles, as in both cases we showed above (Examples 2.4 and 2.5 respectively) that the moduli space is the correct weighted projective space.

The other missing ingredients are showing that the resulting algebra on a singular del Pezzo surface lifts to a flat sheaf of algebras on the minimal desingularization, and showing that the result is still the endomorphism algebra of a rigid, unobstructed object. (In the twisted case, one would also need to show that one obtains the correct class in the Brauer group of the affine del Pezzo surface; this should not be an issue in the untwisted case where the algebra remains Azumaya on the compactification, since the compactification has trivial Brauer group.)

## Acknowledgments

A great many thanks are due to P. Etingof, not only for suggesting several versions of the above problem (both the original version in which $V$ was a sum of line bundles of degree 0 and more recent versions in which $V$ is semistable) but also for a large number of discussions of filtered deformations over the years, including some recent pointers regarding Poisson structures on centers. The author would also like to thank the organizers of the workshop "Hypergeometry, Integrability and Lie Theory" at the Lorentz Center in Leiden for providing the necessary impetus to finish and write up the present work. In addition, the author would like to thank the referee for extensive comments which led to many improvements in the current version of this paper.

## References

[1] M. Artin, J. Tate, and M. Van den Bergh, *Some algebras associated to automorphisms of elliptic curves*, The Grothendieck Festschrift, Vol. I, Progr. Math., vol. 86, Birkhäuser Boston, Boston, MA, 1990, pp. 33–85. MR1086882

[2] M. Artin and M. Van den Bergh, *Twisted homogeneous coordinate rings*, J. Algebra **133** (1990), no. 2, 249–271, DOI 10.1016/0021-8693(90)90269-T. MR1067406

[3] M. Artin and J. J. Zhang, *Noncommutative projective schemes*, Adv. Math. **109** (1994), no. 2, 228–287, DOI 10.1006/aima.1994.1087. MR1304753

[4] M. F. Atiyah, *Vector bundles over an elliptic curve*, Proc. London Math. Soc. (3) **7** (1957), 414–452, DOI 10.1112/plms/s3-7.1.414. MR131423

[5] Manjul Bhargava and Wei Ho, *Coregular spaces and genus one curves*, Camb. J. Math. **4** (2016), no. 1, 1–119, DOI 10.4310/CJM.2016.v4.n1.a1. MR3472915

[6] A. I. Bondal and A. E. Polishchuk, *Homological properties of associative algebras: the method of helices* (Russian, with Russian summary), Izv. Ross. Akad. Nauk Ser. Mat. **57** (1993), no. 2, 3–50, DOI 10.1070/IM1994v042n02ABEH001536; English transl., Russian Acad. Sci. Izv. Math. **42** (1994), no. 2, 219–260. MR1230966

[7] Leonid Chekhov, Marta Mazzocco, and Vladimir Rubtsov, *Quantised Painlevé monodromy manifolds, Sklyanin and Calabi-Yau algebras*, Adv. Math. **376** (2021), Paper No. 107442, 52, DOI 10.1016/j.aim.2020.107442. MR4178914

[8] Samuel Eilenberg, *Homological dimension and syzygies*, Ann. of Math. (2) **64** (1956), 328–336, DOI 10.2307/1969977. MR82489

[9] Pavel Etingof, Alexei Oblomkov, and Eric Rains, *Generalized double affine Hecke algebras of rank 1 and quantized del Pezzo surfaces*, Adv. Math. **212** (2007), no. 2, 749–796, DOI 10.1016/j.aim.2006.11.008. MR2329319

[10] Pavel Etingof and Travis Schedler, *Poisson traces and D-modules on Poisson varieties*, Geom. Funct. Anal. **20** (2010), no. 4, 958–987, DOI 10.1007/s00039-010-0085-4. With an appendix by Ivan Losev. MR2729282

[11] Takahiro Hayashi, *Sugawara operators and Kac-Kazhdan conjecture*, Invent. Math. **94** (1988), no. 1, 13–52, DOI 10.1007/BF01394343. MR958588

[12] V. Kac and A. Radul, *Poisson structure for restricted Lie algebras*, The Gelfand Mathematical Seminars, 1996–1999, Gelfand Math. Sem., Birkhäuser Boston, Boston, MA, 2000, pp. 77–84. MR1731634

[13] S. A. Kuleshov, *Exceptional and rigid sheaves on surfaces with anticanonical class without base components*, J. Math. Sci. (New York) **86** (1997), no. 5, 2951–3003, DOI 10.1007/BF02355112. Algebraic geometry, 2. MR1604186

[14] S. A. Kuleshov and D. O. Orlov, *Exceptional sheaves on Del Pezzo surfaces* (Russian, with Russian summary), Izv. Ross. Akad. Nauk Ser. Mat. **58** (1994), no. 3, 53–87, DOI 10.1070/IM1995v044n03ABEH001609; English transl., Russian Acad. Sci. Izv. Math. **44** (1995), no. 3, 479–513. MR1286839

[15] Yankı Lekili and Alexander Polishchuk, *A modular compactification of $\mathcal{M}_{1,n}$ from $A_\infty$-structures*, J. Reine Angew. Math. **755** (2019), 151–189, DOI 10.1515/crelle-2017-0015. MR4015231

[16] Eduard Looijenga, *Root systems and elliptic curves*, Invent. Math. **38** (1976/77), no. 1, 17–32, DOI 10.1007/BF01390167. MR466134

[17] Dmitri Orlov, *Smooth and proper noncommutative schemes and gluing of DG categories*, Adv. Math. **302** (2016), 59–105, DOI 10.1016/j.aim.2016.07.014. MR3545926

[18] Henry C. Pinkham, *Deformations of algebraic varieties with $G_m$ action*, Astérisque, No. 20, Société Mathématique de France, Paris, 1974. MR0376672

[19] E. M. Rains. *The birational geometry of noncommutative surfaces.* `arXiv:1907.11301`, 2019.

[20] E. M. Rains. *The noncommutative geometry of elliptic difference equations.* `arXiv:1607.08876`, 2016.

[21] Eric M. Rains, *Elliptic double affine Hecke algebras*, SIGMA Symmetry Integrability Geom. Methods Appl. **16** (2020), Paper No. 111, 133, DOI 10.3842/SIGMA.2020.111. MR4170709

[22] Miles Reid, *Nonnormal del Pezzo surfaces*, Publ. Res. Inst. Math. Sci. **30** (1994), no. 5, 695–727, DOI 10.2977/prims/1195165581. MR1311389

[23] E. K. Sklyanin, *Some algebraic structures connected with the Yang-Baxter equation* (Russian), Funktsional. Anal. i Prilozhen. **16** (1982), no. 4, 27–34, 96. MR684124

Department of Mathematics, California Institute of Technology, Pasadena, CA 91125

Contemporary Mathematics
Volume **780**, 2022
https://doi.org/10.1090/conm/780/15690

# Pseudo-symmetric pairs for Kac-Moody algebras

Vidas Regelskis and Bart Vlaar

*To Jasper Stokman on the occasion of his 50th birthday*

Abstract. Lie algebra involutions and their fixed-point subalgebras give rise to symmetric spaces and real forms of complex Lie algebras, and are well-studied in the context of symmetrizable Kac-Moody algebras. In this paper we study a generalization. Namely, we introduce the concept of a pseudo-involution, an automorphism which is only required to act involutively on a stable Cartan subalgebra, and the concept of a pseudo-fixed-point subalgebra, a natural substitute for the fixed-point subalgebra. In the symmetrizable Kac-Moody setting, we give a comprehensive discussion of pseudo-involutions of the second kind, the associated pseudo-fixed-point subalgebras, restricted root systems and Weyl groups, in terms of generalizations of Satake diagrams.

Contents



## 1. Introduction

Given a set, possibly with additional structure, its involutive automorphisms and corresponding fixed-point subsets are natural objects of interest. In the case of a Lie algebra[1] $\mathfrak{g}$, an involutive automorphism $\theta : \mathfrak{g} \to \mathfrak{g}$ and its fixed-point subalgebra $\mathfrak{g}^\theta$ give rise to the notion of a symmetric pair $(\mathfrak{g}, \mathfrak{g}^\theta)$ and a linear decomposition of $\mathfrak{g}$ in terms of $\pm 1$-eigenspaces: $\mathfrak{g} = \mathfrak{g}^\theta \oplus \mathfrak{g}^{-\theta}$. These have been extensively studied

2020 *Mathematics Subject Classification.* Primary 17B22, 17B40, 17B67; Secondary 16B30, 17B37, 20F55.

*Key words and phrases.* Kac-Moody algebras, automorphism group, symmetric pairs, restricted Weyl group.

The second-named author was supported by the UK Engineering and Physical Sciences Research Council, grant number EP/R009465/1.

[1] We always work over an algebraically closed field of characteristic 0. Lie algebra automorphisms are always understood to fix this field pointwise.

for semisimple finite-dimensional Lie algebras, see e.g. [**Sa60**, **Ar62**, **He78**], for instance as part of the study of the associated symmetric space $G/G^\theta$ or the real form of $\mathfrak{g}$ defined in terms of the semi-involution associated to $\theta$; key tools for this study are the restricted root system and the restricted Weyl group associated to $\theta$. In these works the involutions are described combinatorially in terms of Satake diagrams, which are particular decorations of the Dynkin diagram associated to $\mathfrak{g}$ (there are also other descriptions in terms of so-called Kac or Vogan diagrams, see e.g. [**Ka69**, **OV94**, **CH04**, **CZ17**]).

Kac-Moody algebras, which we always assume to be symmetrizable and indecomposable, generalize simple finite-dimensional Lie algebras as well as their (suitably extended) loop algebras [**Ka68**, **Mo68**, **Ka74**, **PK83**, **Ka90**]. Involutive automorphisms for Kac-Moody algebras were studied for the affine case in [**Ba86**, **Le88**] and more generally in [**KW92**, Ch. 5]. In the infinite-dimensional case, a crucial distinction exists between those automorphisms $\theta$ which preserve the two conjugacy classes of Borel subalgebras and those which swap them; they are said to be *of the first* and *of the second kind* and the corresponding real forms are called *almost compact* and *almost split*, respectively. Involutions of the second kind, which are naturally described by Kac-Moody versions of Satake diagrams, were studied in [**Be89**] and, with a particular focus on the associated real forms, in [**BBBR95**, **BR06**]. They were revisited in the context of q-deformations of their fixed-point subalgebras in [**Ko14**, Sec. 2]. For involutions of the first kind, see e.g. [**Na92**, **BR03**, **BR07**].

Such classifications of involutions rely on detailed knowledge of the structure of the automorphism group in general, see e.g. [**PK83**, **Ba86**, **KW92**, **Gü10**, **CZ17**]. A natural generalization of the study of involutive automorphism and their fixed-point subalgebras is obtained when we replace "involutive" by "finite-order", see e.g. [**Ka69**, **JZh01**, **HG09**].

**1.1. Pseudo-involutions and pseudo-fixed-point subalgebras.** In this paper we pursue a rather different direction. Namely, we consider the following two questions for an arbitrary Lie algebra $\mathfrak{g}$, defined over an algebraically closed field $\mathbb{F}$ of characteristic 0, and a distinguished subalgebra $\mathfrak{t} \subseteq \mathfrak{g}$:

(1) Are there (non-involutive) automorphisms $\theta : \mathfrak{g} \to \mathfrak{g}$ which stabilize $\mathfrak{t}$ such that the restriction $\theta|_\mathfrak{t}$ is an involution? Can we classify them?

(2) What is a reasonable substitute $\mathfrak{k} \subseteq \mathfrak{g}$ for the fixed-point subalgebra $\mathfrak{g}^\theta$? Since $\theta|_\mathfrak{t}$ is still an involution, it is natural to require that $\mathfrak{k} \cap \mathfrak{t} = \mathfrak{t}^\theta$.

Furthermore, it is natural to replace $\mathfrak{t}$, $\theta$ and $\mathfrak{k}$ in this setup by their $\mathsf{Aut}(\mathfrak{g})$-conjugacy classes.

To our best knowledge this problem has not been studied in great detail. Of course, whether this is interesting depends on the choice of $\mathfrak{t}$; in particular, $\mathfrak{t}$ should not be too small. If $\mathfrak{t}$ is a Cartan subalgebra[2], we believe that the problem above is of interest for many different $\mathfrak{g}$, given the role that Cartan subalgebras play in representation theory.

Slightly modifying the above setup, we propose the following generalization of an involutive Lie algebra automorphism.

[2]In this paper by Cartan subalgebra we mean a maximal $\mathsf{ad}$-diagonalizable subalgebra of $\mathfrak{g}$. Since $\mathbb{F}$ is algebraically closed, Cartan subalgebras are abelian. If in addition $\mathfrak{g}$ is finite-dimensional its Cartan subalgebras are precisely its self-normalizing nilpotent subalgebras.

DEFINITION 1.1. We call $\theta \in \mathsf{Aut}(\mathfrak{g})$ a *pseudo-involution* if there exists a $\theta$-stable Cartan subalgebra $\mathfrak{t} \subseteq \mathfrak{g}$ such that $\theta|_{\mathfrak{t}}$ is an involution, i.e. $\theta^2|_{\mathfrak{t}} = \mathsf{id}_{\mathfrak{t}}$.

Immediately we see that involutive automorphism of $\mathfrak{g}$ are pseudo-involutions if they stabilize a Cartan subalgebra. We expect that the resulting linear decomposition

$$\mathfrak{t} = \mathfrak{t}^{\theta} \oplus \mathfrak{t}^{-\theta}, \tag{1.1}$$

although weaker than the corresponding decomposition for $\mathfrak{g}$, is strong enough to serve as a natural minimal condition for extensions of results in harmonic analysis on symmetric spaces to more general homogeneous spaces.

To address question (ii) we propose the following generalization of the notion of a symmetric pair (associated to $\theta$), which relies on the $\mathfrak{t}$-root space decomposition[3] of $\mathfrak{g}$. Denote by $\Phi^{(\mathfrak{t})} \subset \mathfrak{t}^*\backslash\{0\}$ the $\mathfrak{t}$-root system and by $\mathfrak{g}_{\alpha}^{(\mathfrak{t})} \subset \mathfrak{g}$ the $\mathfrak{t}$-root space associated to $\alpha \in \Phi^{(\mathfrak{t})}$.

DEFINITION 1.2. Let $\theta \in \mathsf{Aut}(\mathfrak{g})$ be a pseudo-involution and let $\mathfrak{t} \subseteq \mathfrak{g}$ be a $\theta$-stable Cartan subalgebra such that $\theta|_{\mathfrak{t}}$ is an involution. We call a subalgebra $\mathfrak{k} \subseteq \mathfrak{g}$ a *pseudo-fixed-point subalgebra* if

$$\mathfrak{k} \cap \mathfrak{t} = \mathfrak{t}^{\theta}, \tag{1.2}$$

$$\dim\Big(\mathfrak{k} \cap \big(\mathfrak{g}_{\alpha}^{(\mathfrak{t})} + \theta(\mathfrak{g}_{\alpha}^{(\mathfrak{t})})\big)\Big) = \dim(\mathfrak{g}_{\alpha}^{(\mathfrak{t})}) \qquad \text{for all } \alpha \in \Phi^{(\mathfrak{t})}. \tag{1.3}$$

We call the pair $(\mathfrak{g}, \mathfrak{k})$ a *pseudo-symmetric pair* (associated to $\theta$).

The two conditions in Definition 1.2 can be motivated by showing that they hold for fixed-point subalgebras of (suitable) involutions. Indeed, let $\theta$ be an involutive automorphism of $\mathfrak{g}$ with $\theta$-stable Cartan subalgebra $\mathfrak{t}$, so that $\theta(\mathfrak{g}_{\alpha}^{(\mathfrak{t})}) = \mathfrak{g}_{\theta^*(\alpha)}^{(\mathfrak{t})}$ for all $\alpha \in \Phi^{(\mathfrak{t})}$, and set $\mathfrak{k} = \mathfrak{g}^{\theta}$. We observe that (1.2) is trivially satisfied. If $\theta^*(\alpha) \neq \alpha$ then the sum $\mathfrak{g}_{\alpha}^{(\mathfrak{t})} + \theta(\mathfrak{g}_{\alpha}^{(\mathfrak{t})})$ is direct and it is straightforward to show that a basis of $\mathfrak{g}_{\alpha}^{(\mathfrak{t})}$ can be $\theta$-symmetrized to obtain a basis of $\mathfrak{g}_{\alpha}^{(\mathfrak{t})} + \theta(\mathfrak{g}_{\alpha}^{(\mathfrak{t})})$. If $\theta^*(\alpha) = \alpha$ then (1.3) is equivalent to $\mathfrak{g}_{\alpha}^{(\mathfrak{t})} \subseteq \mathfrak{g}^{\theta}$ and hence one needs the additional assumption that $\theta$ fixes pointwise $\mathfrak{g}_{\alpha}^{(\mathfrak{t})}$, cf. [**KW92**, 5.15]. We obtain that, given an involutive automorphism $\theta$ with stable Cartan subalgebra $\mathfrak{t}$, the fixed-point subalgebra $\mathfrak{g}^{\theta}$ is pseudo-fixed-point if $\theta$ fixes pointwise all stable $\mathfrak{t}$-root spaces, i.e. all $\mathfrak{g}_{\alpha}^{(\mathfrak{t})}$ such that $\alpha \in (\Phi^{(\mathfrak{t})})^{\theta^*}$.

In the rest of the paper we study pseudo-involutions of the second kind and their pseudo-fixed-point subalgebras of a Kac-Moody algebra $\mathfrak{g}$. To make the representation-theoretic role for Cartan subalgebras more concrete we remark here that important categories of $\mathfrak{g}$-modules are defined in terms of the standard Cartan subalgebra $\mathfrak{h}$, notably the category $\mathcal{O}$ introduced by Kac [**Ka74**] as a Kac-Moody generalization of the category introduced in [**Ge70, BGG71**].

Since every involution $\theta$ of the second kind of a Kac-Moody algebra has a stable Cartan subalgebra (see e.g. [**KW92**, Lem. 5.7]), involutions of the second kind are examples of pseudo-involutions. Moreover, by [**KW92**, Cor. 5.19], $\mathfrak{g}^{\theta}$ is a pseudo-fixed-point subalgebra.

[3] By $\mathfrak{t}$-root space we shall always mean a root space which is not associated to $0 \in \mathfrak{t}^*$, i.e. not $\mathfrak{t}$ itself.

Recall that $\mathfrak{g}$ is defined in terms of combinatorial datum (generalized Cartan matrix and associated Dynkin diagram). In terms of particular decorations of Dynkin diagrams, which we call *compatible decorations*, see Definition 2.1, in Theorem 2.5 we classify pseudo-involutions of $\mathfrak{g}$ of the second kind. It is intermediate between (1) a weaker statement about semisimple automorphisms of the second kind of $\mathfrak{g}$ given in [**KW92**, 4.38-4.39] in terms of more general combinatorial datum and (2) a classification of involutions of the second kind of $\mathfrak{g}$, see [**KW92**, Rmk. 5.33], [**BBBR95**], [**Ko14**, App. A] in terms of Satake diagrams (more special combinatorial datum). The key ingredients of the proof of Theorem 2.5 are basic decomposition results for $\mathsf{Aut}(\mathfrak{g})$, see e.g. [**KW92**, Sec. 4] and a handy result about twisted involutions in Coxeter groups due to Springer, see [**Sp85**]. We give a refinement in Corollary 2.7 and point to a possible strengthening of this result in Conjecture 2.8. Although these classifications follow straightforwardly from existing results on Kac-Moody algebras and Coxeter groups, we believe that due to the novelty of the concept of pseudo-involution, they qualify as main results of this paper.

If a compatible decoration satisfies an additional constraint then it is called a *generalized Satake diagram*, see Definition 3.4. Precisely in this case it gives rise to a pseudo-fixed-point subalgebra $\mathfrak{k}$, see Theorem 3.8, which is a direct Kac-Moody generalization of [**RV20**, Thm. 3.3], written in the language of pseudo-fixed-point subalgebras. From this an Iwasawa decomposition associated to the pair $(\mathfrak{g}, \mathfrak{k})$ readily follows, see Corollary 3.10; such a statement was not included in [**RV20**]. Note that Iwasawa decompositions are well-known for symmetric spaces in the Kac-Moody setting, see e.g. [**PK83**, Cor. 4(b)] (written in terms of the Kac-Moody group $G$).

The restricted Weyl group and restricted root system can be considered for any involutive automorphism of the root system of $\mathfrak{g}$; in fact the restricted Weyl group and root system can be studied without direct reference to $\mathfrak{g}$, see e.g. [**Sch69, Lu76, He84, Hé91, Lu95**] and also [**Lu03**, Ch. 25], [**DK19**, Sec. 2] and [**RV20**, Sec. 2]. Generalized Satake diagrams already arose in the work [**He84**] on restricted root systems associated to involutions of finite root systems. In this paper we consider the Kac-Moody generalization of this and show that generalized Satake diagrams unify various approaches to the restricted root system and Weyl group. The key results are Theorem 4.24, Corollary 4.25, Theorem 4.29 and Theorem 4.30.

**1.2. Applications in the quantum deformed setting.** If $\mathfrak{g}$ is a symmetrizable Kac-Moody algebra, the universal enveloping algebra $U\mathfrak{g}$ possesses a q-deformation $U_q\mathfrak{g}$ (Drinfeld-Jimbo quantum group), see e.g. [**Dr87, Lu94**]. We briefly discuss the role of pseudo-involutions and pseudo-fixed-point subalgebras in the context of reflection equations (quartic braid relations) in representations of $U_q\mathfrak{g}$. Note that the canonical Cartan subalgebra $\mathfrak{h} \subset \mathfrak{g}$ plays an important role: since $U_q\mathfrak{h}$ is generated by all invertible elements of $U_q\mathfrak{g}$, only automorphisms of $\mathfrak{g}$ which stabilize $\mathfrak{h}$ can be lifted to algebra automorphisms of $U_q\mathfrak{g}$.

Given an involutive automorphism $\theta$ of $\mathfrak{g}$, consider the fixed-point subalgebra $\mathfrak{k} = \mathfrak{g}^\theta$. For finite-dimensional $\mathfrak{g}$ Letzter identified in [**Le99, Le02, Le03**] a one-sided coideal subalgebra $U_q\mathfrak{k} \subseteq U_q\mathfrak{g}$ which q-deforms $U\mathfrak{k}$ (cf. [**NS95, NDS97**] for alternative approaches to $U_q\mathfrak{k}$). A crucial step in Letzter's work was to "arrange" $\theta$ (by conjugacy) so that it stabilizes $\mathfrak{h}$, fixes pointwise all $\theta$-stable positive root spaces and sends other positive root spaces to negative root spaces, see [**Le99**, Eqns. (3.1)-(3.3)].

In addition, there is a $\mathsf{B}_2$-analogue of quasitriangularity for these subalgebras: there exists an element $\mathcal{K}$ in the centralizer of $U_q\mathfrak{k}$ (in a completion of $U_q\mathfrak{g}$) which satisfies the type $\mathsf{B}_2$-braid relation. The construction of this element was described by Balagović and Kolb in [**BK19**], building upon an earlier construction in [**BW18**] due to Bao and Wang for a particular $\mathfrak{k}$ in the case $\mathfrak{g} = \mathfrak{sl}_N$. These works place quantum symmetric pairs in the context of *reflection equation algebras*, see [**KS92, KS93, KS09**]. In [**RV20**] we showed that the above construction for $\mathfrak{g}$, with minor adjustments, is applicable in a larger setting. The Lie-theoretic underpinning of this is precisely given by pseudo-involutions and their pseudo-fixed-point subalgebras.

In [**Ko14**], Kolb provided the Kac-Moody generalization of Letzter's construction of $U_q\mathfrak{k}$ of $\mathfrak{k} = \mathfrak{g}^\theta$ in the case that $\theta$ is an involution of the second kind, using a combinatorial approach to such $\theta$ indicated in [**KW92**, Sec. 5] and discussed in [**BBBR95**]. Recall that Kac-Moody algebras of affine type are (extensions of) loop algebras of finite-dimensional simple Lie algebras. Indeed, in the affine case suitable interwiners for $U_q\mathfrak{k}$ are known, see e.g. [**DM03, RV16**], to satisfy the *reflection equation* (boundary Yang-Baxter equation) with spectral parameter [**Ch84, Sk88, Ch92**]. Therefore pseudo-fixed-point subalgebras are relevant to the study of (quantum) integrability in the presence of a boundary.

In [**AV20**] universal versions of these interwiners were constructed by extending the approach from [**BK19**] to the Kac-Moody setting. Moreover, Kolb in [**Ko21**] showed that the existence of the bar involution for $U_q\mathfrak{k}$ follows from the construction in [**AV20**] which does not rely on explicit presentations of $U_q\mathfrak{k}$. The works [**AV20, Ko21**] deal with quantizations of pseudo-fixed-point subalgebras $\mathfrak{k}$ of Kac-Moody algebras, not just fixed-point subalgebras of involutions. Therefore the present paper provides a Lie-theoretic context for the q-deformed constructions in [**AV20, Ko21**].

It is also intended to serve as a platform supporting generalizations, in terms of a homogeneous framework for the entire range of pseudo-symmetric pairs, of advanced representation-theoretic applications, both undeformed and q-deformed. Thus far these have mainly focused on quasi-split symmetric pairs; we mention [**SV15, RSV15a, RSV15b, RSV18**], [**SR20, RS20, St21**] and [**ES18, BW18, LW20**] as illustrative examples of various flavours.

**1.3. Outline.** This paper is organized as follows. Section 2 discusses pseudo-involutions in the Kac-Moody setting in terms of compatible decorations. Subsections 2.1-2.7 are used to introduce background material, mainly drawn from [**KW92**] and [**BBBR95**], and fix notation.

In Section 3 we look at pseudo-fixed-point subalgebras of Kac-Moody algebras in terms of generalized Satake diagrams, thereby providing Kac-Moody generalizations of the finite-dimensional work in [**RV20**, Sec. 2 and 3].

In Section 4 we survey, combine and extend various approaches to the corresponding restricted root system and restricted Weyl group, creating a comprehensive catalogue of results in this area.

Finally, in Appendix A we provide a classification of generalized Satake diagrams of finite and affine types, combining and adding to existing lists, featuring a new easy-to-use notation. We hope that these tables prove to be a useful reference for the community.

In this paper we will follow (almost precisely) Carter's "Dynkin notation" for the affine root systems, see [**Ca05**, App.]. We have the following correspondence with the customary notation due to Kac, see e.g. [**Ka90**, Tables Aff 1-3]:

| (Dynkin) | $\widehat{\mathsf{A}}_{n\geq 1}$ | $\widehat{\mathsf{B}}_{n\geq 3}$ | $\widehat{\mathsf{B}}^{\vee}_{n\geq 3}$ | $\widehat{\mathsf{C}}_{n\geq 2}$ | $\widehat{\mathsf{C}}^{\vee}_{n\geq 2}$ | $\widehat{\mathsf{C}}'_{n\geq 1}$ | $\widehat{\mathsf{D}}_{n\geq 4}$ |
|---|---|---|---|---|---|---|---|
| (Kac) | $\mathsf{A}^{(1)}_n$ | $\mathsf{B}^{(1)}_n$ | $\mathsf{A}^{(2)}_{2n-1}$ | $\mathsf{C}^{(1)}_n$ | $\mathsf{D}^{(2)}_{n+1}$ | $\mathsf{A}^{(2)}_{2n}$ | $\mathsf{D}^{(1)}_n$ |
| (Dynkin) | $\widehat{\mathsf{E}}_{6,7,8}$ | $\widehat{\mathsf{F}}_4$ | $\widehat{\mathsf{F}}^{\vee}_4$ | $\widehat{\mathsf{G}}_2$ | $\widehat{\mathsf{G}}^{\vee}_2$ | | |
| (Kac) | $\mathsf{E}^{(1)}_{6,7,8}$ | $\mathsf{F}^{(1)}_4$ | $\mathsf{E}^{(2)}_6$ | $\mathsf{G}^{(1)}_2$ | $\mathsf{D}^{(3)}_4$ | | |

where for the classical Lie types we have indicated the usual constraints on $n$ (the rank of the generalized Cartan matrix) to avoid low-rank coincidences.

## 2. Pseudo-involutions in terms of compatible decorations

This section gives a basic and, we hope, pedagogical account of the classification of pseudo-involutions of the second kind of Kac-Moody algebras in terms of decorations of Dynkin diagrams. We begin by reviewing some theory of Kac-Moody algebras and their automorphisms, for which our main references are [**KW92**] and [**BBBR95**].

**2.1. Generalized Cartan matrices and Dynkin diagrams.** Given a finite index set $I$, let $A = (a_{ij})_{i,j\in I}$ be a generalized Cartan matrix, i.e. $a_{ii} = 2$ for all $i \in I$, $a_{ij} \in \mathbb{Z}_{\leq 0}$ for all distinct $i, j \in I$ and $a_{ij} = 0$ if and only if $a_{ji} = 0$. We will always assume that $A$ is symmetrizable, i.e. there exist setwise-coprime positive integers $\epsilon_i$ $(i \in I)$ such that $\epsilon_i a_{ij} = \epsilon_j a_{ji}$ for all $i, j \in I$. The group of *diagram automorphisms* of $A$ is the finite group

$$\mathsf{Aut}(A) := \{\tau : I \to I \text{ invertible} : a_{\tau(i)\tau(j)} = a_{ij} \text{ for all } i, j \in I\}. \tag{2.1}$$

Let $J \subseteq I$. The principal submatrix $A_J := (a_{ij})_{i,j\in J}$ is also a symmetrizable generalized Cartan matrix. We set

$$J^{\perp} := \{i \in I : a_{ij} = 0 \text{ for all } j \in J\}. \tag{2.2}$$

We call $K \subseteq J$ a *component* of $J$ if $J \subseteq K \cup K^{\perp}$. We call $J$ *connected* if it is nonempty and has no nonempty proper components. From now on we assume $I$ is connected; hence $(\epsilon_i)_{i\in I}$ is uniquely determined by $A$.

We say that $A$ is *of finite type* if $\det(A_J) > 0$ for all $J \subseteq I$ and *of affine type* if $\det(A) = 0$ and $\det(A_J) > 0$ for all $J \subset I$. Recall that if $a_{ij}a_{ji} \leq 4$ for all $i, j \in I$, which is satisfied if $A$ is of finite or affine type, the Dynkin diagram associated to $(I, A)$ is an oriented multigraph with vertex set $I$ with the edges determined as follows. Distinct vertices $i, j \in I$ are connected by $\max(|a_{ij}|, |a_{ji}|)$ edges; if $\epsilon_i = \epsilon_j$ we do not assign an orientation to the edges between $i$ and $j$ and otherwise these edges point to the node with the smaller value of $\epsilon$.

**2.2. Braid group and Weyl group.** Associated to $A$ is the Artin-Tits braid group

$$\mathsf{Br} = \Big\langle \{T_i\}_{i\in I} : \underbrace{T_iT_j\cdots}_{m_{ij}} = \underbrace{T_jT_i\cdots}_{m_{ij}} \quad \text{if } i \neq j \Big\rangle \tag{2.3}$$

where $m_{ij} = 2, 3, 4, 6$ if $a_{ij}a_{ji} = 0, 1, 2, 3$, respectively, and $m_{ij} = \infty$ if $a_{ij}a_{ji} \geq 4$. The assignment $T_i \mapsto s_i$ extends to a surjective group map from $\mathsf{Br}$ to the *Weyl group*

$$W = \left\langle \{s_i\}_{i \in I} \,:\, s_i^2 = 1, \, (s_i s_j)^{m_{ij}} = 1 \text{ if } i \neq j \right\rangle. \tag{2.4}$$

The pair $(W, \{s_i\}_{i \in I})$ is a Coxeter system and hence associated to it is the length function $\ell$ and the notion of reduced expression, see e.g. [**Bo68**, Ch. IV, §1.1]. We define a map $T : W \to \mathsf{Br}$ by $T(s_{i_1} \cdots s_{i_\ell}) = T_{i_1} \cdots T_{i_\ell}$ if $s_{i_1} \cdots s_{i_\ell}$ is reduced; hence $T(ww') = T(w)T(w')$ for all $w, w' \in W$ such that $\ell(ww') = \ell(w) + \ell(w')$.

The action of $\mathsf{Aut}(A)$ on $\mathsf{Br}$ defined by $\tau(T_i) = T_{\tau(i)}$ descends to an action of $\mathsf{Aut}(A)$ on $W$ which respects $\ell$ so that $\tau(T(w)) = T(\tau(w))$ for all $\tau \in \mathsf{Aut}(A)$, $w \in W$.

**2.3. Minimal realization and bilinear forms.** Recall that $\mathbb{F}$ is an algebraically closed field of characteristic 0. We fix a *minimal realization* $(\mathfrak{h}, \Pi, \Pi^\vee)$ of $A$ over $\mathbb{F}$; that is, $\mathfrak{h}$ is a $(|I| + \mathsf{corank}(A))$-dimensional $\mathbb{F}$-linear space and $\Pi^\vee = \{h_i\}_{i \in I} \subset \mathfrak{h}$ and $\Pi = \{\alpha_i\}_{i \in I} \subset \mathfrak{h}^*$ are linearly independent subsets such that $\alpha_j(h_i) = a_{ij}$ for all $i, j \in I$. There is a unique symmetric bilinear $\mathbb{F}$-valued form $(\ ,\ )$ on $\mathfrak{h}' := \mathsf{Sp}_{\mathbb{F}}\Pi^\vee$ with the property $(h_i, h_j) = \epsilon_j^{-1} a_{ij}$ for all $i, j \in I$. By [**Ka90**, 2.1] one may choose any complementary subspace $\mathfrak{h}''$ to $\mathfrak{h}'$ in $\mathfrak{h}$ and extend $(\ ,\ )$ to a nondegenerate symmetric bilinear $\mathbb{F}$-valued form on $\mathfrak{h}$ with the properties

$$(h, h_j) = \epsilon_j^{-1}\alpha_j(h), \qquad (h', h'') = 0 \qquad \text{for all } h \in \mathfrak{h}, \, j \in I \text{ and } h', h'' \in \mathfrak{h}''. \tag{2.5}$$

Consider the linear map $\nu : \mathfrak{h} \to \mathfrak{h}^*$ defined by $\nu(h)(h') = (h, h')$ for all $h, h' \in \mathfrak{h}$; it satisfies $\nu(h_i) = \epsilon_i^{-1}\alpha_i$ for all $i \in I$. Since $(\ ,\ )$ is nondegenerate, $\nu$ is an isomorphism. We now define a symmetric bilinear form on $\mathfrak{h}^*$ by $(\lambda, \mu) = (\nu^{-1}(\lambda), \nu^{-1}(\mu))$, which satisfies $(\alpha_i, \alpha_j) = \epsilon_i a_{ij}$ for all $i, j \in I$. In the remainder of this paper, whenever we discuss orthogonality, self-adjointness or isometry, it will always be with respect to $(\ ,\ )$.

There are faithful linear isometric actions of $W$ on $\mathfrak{h}$ and $\mathfrak{h}^*$ determined by

$$s_i(h) = h - \alpha_i(h)h_i, \qquad s_i(\alpha) = \alpha - \alpha(h)\alpha_i \qquad \text{for } i \in I, \, h \in \mathfrak{h}, \, \alpha \in \mathfrak{h}^*. \tag{2.6}$$

The group $\mathsf{Aut}(A)$ acts via relabelling on $\mathfrak{h}'$; this can be extended to an action on $\mathfrak{h}$ by choosing $\mathfrak{h}''$ according to [**KW92**, 4.19]. The isomorphism $\nu$ allows us to define an action of $\mathsf{Aut}(A)$ on $\mathfrak{h}^*$ and the bilinear forms $(\ ,\ )$ are $\mathsf{Aut}(A)$-invariant.

**2.4. Kac-Moody algebra and roots.** Let $\mathfrak{g} = \mathfrak{g}(A)$ be the (indecomposable symmetrizable) Kac-Moody Lie algebra defined in terms of $A$ with Chevalley generators $e_i, f_i$ $(i \in I)$ and $h_i := [e_i, f_i]$, see [**KW92**, 1.1 and 1.2]. The derived subalgebra of $\mathfrak{g}$ is the $\mathsf{corank}(A)$-codimensional subalgebra $\mathfrak{g}'$ generated by all $e_i$ and $f_i$ $(i \in I)$. The centre $\mathfrak{c}$ is contained in $\mathfrak{h}' = \mathfrak{h} \cap \mathfrak{g}'$. The action of $\mathsf{Aut}(A)$ on $\mathfrak{h}$ extends to an action by Lie algebra automorphisms of $\mathfrak{g}$ if we set $\tau(e_i) = e_{\tau(i)}$ and $\tau(f_i) = f_{\tau(i)}$ for all $\tau \in \mathsf{Aut}(A)$, $i \in I$.

We denote the subalgebra of $\mathfrak{g}$ generated by all $e_i$ by $\mathfrak{n}^+$ and the subalgebra generated by all $f_i$ by $\mathfrak{n}^-$. As identities of $\mathfrak{h}$-modules, we have the triangular decomposition $\mathfrak{g} = \mathfrak{n}^+ \oplus \mathfrak{h} \oplus \mathfrak{n}^-$ holds and the root space decomposition

$$\mathfrak{g} = \bigoplus_{\alpha \in \mathfrak{h}^*} \mathfrak{g}_\alpha, \qquad \mathfrak{g}_\alpha = \{x \in \mathfrak{g} \,:\, \forall h \in \mathfrak{h}, \, [h, x] = \alpha(h)x\}. \tag{2.7}$$

We call $\alpha \in \mathfrak{h}^*$ a *root* if $\mathfrak{g}_\alpha \neq \{0\}$ and define the root system $\Phi$ as the set of nonzero roots (note that $\mathfrak{g}_0 = \mathfrak{h}$). The *root lattice* is $Q = \mathbb{Z}\Phi = \mathbb{Z}\Pi$. We denote the positive cone of $Q$ by $Q^+ = \mathbb{Z}_{\geq 0}\Pi$. We set $\Phi^+ = \Phi \cap Q^+$ so that $\mathfrak{n}^+ = \bigoplus_{\alpha \in \Phi^+} \mathfrak{g}_\alpha$, $\Pi \subseteq \Phi^+$ and $\Phi = \Phi^+ \cup (-\Phi^+)$.

The action of $W$ on $\mathfrak{h}^*$ stabilizes $\Phi$. We call $\alpha \in \Phi$ *real* if $\alpha \in W(\Pi)$ and otherwise *imaginary*. If $\alpha = w(\alpha_i)$ for some $w \in W$, $i \in I$ then $\alpha^\vee := w(h_i) \in \mathfrak{h}$ is well-defined, $\Phi \cap \mathbb{Z}\alpha = \{\alpha, -\alpha\}$, $(\alpha, \alpha) > 0$, the root space $\mathfrak{g}_\alpha$ is 1-dimensional and for all $x \in \mathfrak{g}_\alpha$ the adjoint map $\mathsf{ad}(x) : \mathfrak{g} \to \mathfrak{g}$ is a locally nilpotent derivation.

**2.5. Kac-Moody group and triple exponentials.** Recall the Kac-Moody group $G$ and, for $\alpha \in W(\Pi)$, the map $\mathsf{exp} : \mathfrak{g}_\alpha \to G$, see [**KW92**, 1.3]. There is a group morphism $\mathsf{Ad} : G \to \mathsf{Aut}(\mathfrak{g})$ uniquely determined by $\mathsf{Ad}(\mathsf{exp}(x)) = \mathsf{exp}(\mathsf{ad}(x))$ for all $x \in \mathfrak{g}_\alpha$ with $\alpha \in \Phi$ real. There is also a group morphism from $\mathsf{Br}$ to $G$ given by *triple exponentials*

$$T_i \mapsto n_i := \mathsf{exp}(e_i)\,\mathsf{exp}(-f_i)\,\mathsf{exp}(e_i). \tag{2.8}$$

Denote by $N$ the subgroup of $G$ which is the image of this morphism; note that $\mathsf{Aut}(A)$ acts on $N$ by relabelling. We compose the map $T : W \to \mathsf{Br}$ defined in Section 2.2 with this morphism to obtain a map $n : W \to G$ with image $N$ which satisfies $n(s_i) = n_i$ for all $i \in I$ and $n(ww') = n(w)n(w')$ if $w, w' \in W$ are such that $\ell(ww') = \ell(w) + \ell(w')$. For all $i \in I$ and all $\tau \in \mathsf{Aut}(A)$ it follows that

$$\mathsf{Ad}(n_i)|_{\mathfrak{h}} = s_i, \qquad \tau \circ \mathsf{Ad}(n_i) = \mathsf{Ad}(n_{\tau(i)}) \circ \tau. \tag{2.9}$$

Let $\widetilde{H} := \mathsf{Hom}(Q, \mathbb{F}^\times)$ denote the group of characters on the root lattice $Q$. There is a group morphism $\mathsf{Ad} : \widetilde{H} \to \mathsf{Aut}(\mathfrak{g})$ given by

$$\mathsf{Ad}(\chi)(x) = \chi(\alpha)x, \qquad \chi \in \widetilde{H},\, x \in \mathfrak{g}_\alpha,\, \alpha \in \Phi, \tag{2.10}$$

which induces an action of $\widetilde{H}$ on $G$, so we may consider $\widetilde{H} \ltimes G$ and $\mathsf{Ad}(\widetilde{H} \ltimes G) < \mathsf{Aut}(\mathfrak{g})$, see [**KW92**, 1.10]. By [**BBBR95**, Prop. 4.10.1],

$$\mathsf{Ad}(n(w^{-1})) = \mathsf{Ad}(\zeta(w)) \circ \mathsf{Ad}(n(w))^{-1} \tag{2.11}$$

for all $w \in W$, where $\zeta(w) \in \widetilde{H}$ is defined by

$$\zeta(w)(\lambda) = \prod_{\alpha \in \Phi^+ \cap w(-\Phi^+)} (-1)^{\lambda(\alpha^\vee)}, \qquad \lambda \in Q. \tag{2.12}$$

**2.6. Subdiagrams of finite type.** Given a subset $X \subseteq I$ we may consider the Lie subalgebra $\mathfrak{g}_X := \langle \{e_i, f_i\}_{i \in X} \rangle$, the Cartan subalgebra $\mathfrak{h}_X = \mathfrak{h} \cap \mathfrak{g}_X$, the parabolic Weyl group $W_X := \langle \{s_i\}_{i \in X} \rangle$ and the root subsystem $\Phi_X := \Phi \cap Q_X$ where $Q_X = \sum_{i \in X} \mathbb{Z}\alpha_i$. The following statements are equivalent: $A_X$ is of finite type; the restriction of $(\ ,\ )$ to $\mathfrak{h}_X \times \mathfrak{h}_X$ is positive definite; $W_X$ is finite; $\mathfrak{g}_X$ is finite-dimensional; $\Phi_X$ is finite; all elements of $\Phi_X$ are real. In this case $\mathfrak{g}_X$ is semisimple and we simply say that $X$ is of finite type, which we assume henceforth.

The unique longest element $w_X \in W_X$ is necessarily an involution and there exists an involutive diagram automorphism $\mathsf{oi}_X \in \mathsf{Aut}(A_X)$, called *opposition involution*, such that $w_X(\alpha_i) = -\alpha_{\mathsf{oi}_X(i)}$ for all $i \in X$. Immediately we obtain

$$w_X \cdot s_i = s_{\mathsf{oi}_X(i)} \cdot w_X \qquad \text{for all } i \in X \tag{2.13}$$

in $W_X$. The opposition involution of a connected subset $X \subseteq I$ of finite type is the trivial diagram automorphism except in the case where $X$ is of type $\mathsf{A}_r$ with

$r > 1$, $\mathsf{D}_r$ with $r > 3$ odd or $\mathsf{E}_6$; in these cases $\mathsf{oi}_X$ is the unique nontrivial diagram automorphism.

Because any $\tau \in \mathsf{Aut}(A)$ preserves the length function, we obtain

$$\tau \circ w_X = w_X \circ \tau \qquad \Longleftrightarrow \qquad \tau \in \mathsf{Aut}_X(A) \tag{2.14}$$

where $\mathsf{Aut}_X(A)$ is the subgroup $\{\tau \in \mathsf{Aut}(A) : \tau(X) \subseteq X\}$. We denote $n_X := n(w_X)$ and obtain that $\tau \in \mathsf{Aut}(A)$ commutes with $\mathsf{Ad}(n_X)$ in $\mathsf{Aut}(\mathfrak{g})$ if and only if $\tau \in \mathsf{Aut}_X(A)$. By [**BBBR95**, Cor. 4.10.3] we have

$$\mathsf{Ad}(n_X^2) = \mathsf{Ad}(\zeta_X) \qquad \text{where} \qquad \zeta_X := \zeta(w_X) \in \widetilde{H}. \tag{2.15}$$

It follows that $\zeta_X(\lambda) = (-1)^{\lambda(2\rho_X^\vee)}$ for all $\lambda \in Q$ with $\rho_X^\vee$ half the sum of positive coroots of $\mathfrak{g}_X$ (the sum of the fundamental coweights of $\mathfrak{g}_X$).

**2.7. Automorphisms of $\mathfrak{g}$.** For any subset $\mathfrak{t} \subseteq \mathfrak{g}$ we denote by $\mathsf{Aut}(\mathfrak{g}, \mathfrak{t})$ the subgroup of $\mathsf{Aut}(\mathfrak{g})$ of automorphisms stabilizing $\mathfrak{t}$ and by $\mathsf{Aut}(\mathfrak{g}; \mathfrak{t})$ the subgroup of $\mathsf{Aut}(\mathfrak{g}, \mathfrak{t})$ of automorphisms fixing $\mathfrak{t}$ pointwise. Furthermore, if $K$ is a subgroup of $\mathsf{Aut}(\mathfrak{g})$ and $\theta, \theta' \in \mathsf{Aut}(\mathfrak{g})$ are such that $k \circ \theta = \theta' \circ k$ for some $k \in K$ we say that $\theta$ and $\theta'$ are *$K$-conjugate* and write $\theta \sim_K \theta'$.

The subgroup $\mathsf{Aut}(\mathfrak{g}; \mathfrak{g}')$ is the group of *transvections* of $\mathfrak{g}$ described in [**KW92**, 4.20]; its elements are of the form $\sigma_f := \mathsf{id} + f \circ \pi''$ where $f$ is an arbitrary $\mathbb{F}$-linear map: $\mathfrak{h}'' \to \mathfrak{c}$ and $\pi''$ is the projection from $\mathfrak{g}$ onto $\mathfrak{h}''$ with respect to the decomposition $\mathfrak{g} = \mathfrak{g}' \oplus \mathfrak{h}''$. Such elements commute with $\mathsf{Ad}(\widetilde{H} \ltimes G)$ and satisfy $\tau \circ \sigma_f \circ \tau^{-1} = \sigma_{\tau \circ f \circ \tau^{-1}}$ for all $\tau \in \mathsf{Aut}(A)$. The following then defines a subgroup of $\mathsf{Aut}(\mathfrak{g})$:

$$\mathsf{Inn}(\mathfrak{g}) := \mathsf{Aut}(\mathfrak{g}; \mathfrak{g}') \times \mathsf{Ad}(\widetilde{H} \ltimes G). \tag{2.16}$$

The *Chevalley involution* is the unique involutive Lie algebra automorphism $\omega$ such that $\omega(e_i) = -f_i$ and $\omega|_\mathfrak{h} = -\mathsf{id}_\mathfrak{h}$. Note that $\omega$ centralizes the subgroups $\mathsf{Aut}(A)$, $\mathsf{Aut}(\mathfrak{g}; \mathfrak{g}')$ and $\mathsf{Ad}(N)$ of $\mathsf{Aut}(\mathfrak{g})$. For all $X$ of finite type we have

$$\mathsf{Ad}(n_X)|_{\mathfrak{g}_X} = \omega \circ \mathsf{oi}_X|_{\mathfrak{g}_X}, \tag{2.17}$$

see e.g. [**BBBR95**, Lem. 4.9]; in particular $\omega \in \mathsf{Aut}(A) \ltimes \mathsf{Inn}(\mathfrak{g})$ if $A$ is of finite type. We set $\mathsf{Out}(A) = \mathsf{Aut}(A)$ if $A$ is of finite type and $\mathsf{Out}(A) = \{\mathsf{id}, \omega\} \times \mathsf{Aut}(A)$ otherwise. The following decomposition of $\mathsf{Aut}(\mathfrak{g})$ is established in [**KW92**, 4.23]:

$$\mathsf{Aut}(\mathfrak{g}) = \mathsf{Out}(A) \ltimes \mathsf{Inn}(\mathfrak{g}). \tag{2.18}$$

If $\theta \in \mathsf{Aut}(\mathfrak{g}, \mathfrak{h})$ then we denote the dual linear map $(\theta|_\mathfrak{h})^*$ on $\mathfrak{h}^*$ simply by $\theta^*$. From the definition of root space it follows that $\theta(\mathfrak{g}_\alpha) = \mathfrak{g}_{(\theta^*)^{-1}(\alpha)}$ for all $\alpha \in \Phi$. In particular, $\theta^*$ stabilizes $\Phi$ and $Q$. Furthermore, $\theta$ acts on $\widetilde{H}$ via $(\theta * \chi)(\lambda) = \chi((\theta^*)^{-1}(\lambda))$ for all $\lambda \in Q$, $\chi \in \widetilde{H}$, see [**KW92**, 4.23]. We denote by $\widetilde{H}^\theta$ the subgroup of $\widetilde{H}$ consisting of those $\chi \in \widetilde{H}$ such that $\theta * \chi = \chi$. Note that, since the field $\mathbb{F}$ is closed under taking square roots, one has

$$\forall (\theta, \chi) \in \mathsf{Aut}(\mathfrak{g}, \mathfrak{h}) \times \widetilde{H} \quad \exists \chi' \in \widetilde{H}^\theta : \ \mathsf{Ad}(\chi) \circ \theta \sim_{\mathsf{Ad}(\widetilde{H})} \mathsf{Ad}(\chi') \circ \theta. \tag{2.19}$$

By [**KW92**, 1.16 (i)], $\mathsf{Aut}(\mathfrak{g}, \mathfrak{h}) \cap \mathsf{Ad}(G) = \mathsf{Ad}(N)$; since elements of $\mathsf{Out}(A)$, $\mathsf{Aut}(\mathfrak{g}; \mathfrak{g}')$ and $\mathsf{Ad}(\widetilde{H})$ all stabilize $\mathfrak{h}$ we obtain

$$\mathsf{Aut}(\mathfrak{g}, \mathfrak{h}) = \mathsf{Out}(A) \ltimes \mathsf{Inn}(\mathfrak{g}, \mathfrak{h}), \qquad \mathsf{Inn}(\mathfrak{g}, \mathfrak{h}) := \mathsf{Aut}(\mathfrak{g}; \mathfrak{g}') \times \mathsf{Ad}(\widetilde{H} \ltimes N). \tag{2.20}$$

Note that elements of $\mathsf{Aut}(\mathfrak{g}, \mathfrak{h})$ send real root spaces to real root spaces.

By [**PK83**, Thm. 3], every Borel subalgebra of $\mathfrak{g}$ is $\mathsf{Ad}(G)$-conjugate to $\mathfrak{b}^+$ or $\mathfrak{b}^-$, where $\mathfrak{b}^\pm := \langle \mathfrak{h}, \mathfrak{n}^\pm \rangle$. In particular, given $\theta \in \mathsf{Aut}(\mathfrak{g})$, $\theta(\mathfrak{b}^+)$ is $\mathsf{Ad}(G)$-conjugate to $\mathfrak{b}^+$ or $\mathfrak{b}^-$ and we say that $\theta$ is *of the first* or *second kind*, respectively. According to [**KW92**, 4.6], $\theta$ is of the second kind if and only if $\theta(\mathfrak{b}^+) \cap \mathfrak{b}^+$ is finite-dimensional. The set of automorphisms of the first kind is a subgroup of $\mathsf{Aut}(\mathfrak{g})$ and its coset with respect to $\omega$ is the set of automorphisms of the second kind. By (2.17), if $\mathfrak{g}$ is finite-dimensional then these two subsets coincide; otherwise they define a partition of $\mathsf{Aut}(\mathfrak{g})$.

**2.8. Twisted involutions and compatible decorations.** Let $\theta \in \mathsf{Aut}(\mathfrak{g}, \mathfrak{h})$ be of the second kind such that $\theta^2|_{\mathfrak{h}} = \mathsf{id}_{\mathfrak{h}}$. By (2.20) there exist a linear map $f : \mathfrak{h}'' \to \mathfrak{c}$, $\phi \in \widetilde{H}$, $w \in W$ and $\tau \in \mathsf{Aut}(A)$ such that

$$\theta = \sigma_f \circ \mathsf{Ad}(\phi \cdot n(w)) \circ \omega \circ \tau. \tag{2.21}$$

By $\mathsf{Aut}(\mathfrak{g}; \mathfrak{g}')$-conjugacy we may assume that $f$, and hence $\sigma_f$, commutes with $\tau$. From the condition that the restriction of $\theta$ to $\mathfrak{h}$ is involutive we obtain

$$f = 0, \qquad \tau(w) = w^{-1}, \qquad \tau^2 = \mathsf{id}_I. \tag{2.22}$$

In particular, $w \in W$ is a $\tau$-twisted involution. Invoking the result [**Sp85**, Prop. 3.3], we obtain that there exists $v \in W$ and $X \subseteq I$ of finite type such that $\tau|_X = \mathsf{oi}_X$ and

$$w = v \circ w_X \circ \tau(v)^{-1}. \tag{2.23}$$

Recalling (2.11) we obtain

$$\theta = \mathsf{Ad}(\phi \cdot n(v) \cdot n_X) \circ \omega \circ \tau \circ \mathsf{Ad}(\zeta(w) \cdot n(v)^{-1}). \tag{2.24}$$

Hence there exists $\chi \in \widetilde{H}$, $X \subseteq I$ of finite type and involutive $\tau \in \mathsf{Aut}_X(A)$ such that $\tau|_X = \mathsf{oi}_X$ and

$$\theta \sim_{\mathsf{Ad}(N)} \mathsf{Ad}(\chi \cdot n_X) \circ \omega \circ \tau \tag{2.25}$$

Now by (2.19) we may assume that $\chi \in \widetilde{H}^{\theta(X,\tau)}$ where

$$\theta(X, \tau) := \mathsf{Ad}(n_X) \circ \omega \circ \tau. \tag{2.26}$$

Note that $\theta(X, \tau)$ is a particular case of a *special semisimple automorphism of the second kind* as defined in [**KW92**, 4.38]. We observe that the three factors in (2.26) pairwise commute so that $\theta(X, \tau)^2 = \mathsf{Ad}(\zeta_X)$ by (2.15); in particular, the order of $\theta(X, \tau)$ divides 4. Also, by (2.17), the condition $\tau|_X = \mathsf{oi}_X$ implies that $\theta(X, \tau)|_{\mathfrak{g}_X} = \mathsf{id}_{\mathfrak{g}_X}$, i.e. $\theta(X, \tau)$ fixes pointwise all $\theta(X, \tau)$-stable root spaces.

We are led to the following definition, the natural Kac-Moody analogue of [**RV20**, Eqn. (2.8)].

DEFINITION 2.1. Let $X \subseteq I$ and $\tau \in \mathsf{Aut}_X(A)$. We call $(X, \tau)$ a *compatible decoration* if $X$ is of finite type, $\tau^2 = \mathsf{id}_I$ and $\tau|_X = \mathsf{oi}_X$. Furthermore, we call $(X, \tau, \chi)$ an *enriched compatible decoration* if $(X, \tau)$ is a compatible decoration and $\chi \in \widetilde{H}^{\theta(X,\tau)}$. If $(X, \tau, \chi)$ is an enriched compatible decoration, in a minor abuse of notation we will write $\theta(X, \tau, \chi) = \mathsf{Ad}(\chi) \circ \theta(X, \tau)$.

The following result summarizes the above discussion.

PROPOSITION 2.2. *Let $\theta \in \mathsf{Aut}(\mathfrak{g}, \mathfrak{h})$ be a pseudo-involution of the second kind. Then there exists an enriched compatible decoration $(X, \tau, \chi)$ such that $\theta \sim_{\mathsf{Inn}(\mathfrak{g},\mathfrak{h})} \theta(X, \tau, \chi)$.*

Compatible decorations are indicated diagrammatically by decorating the underlying Dynkin diagram as follows: fill the nodes corresponding to $X$ and indicate the nontrivial $\tau$-orbits by bidirectional single arrows. The following basic properties will be useful later.

LEMMA 2.3. *Let $(X, \tau)$ be a compatible decoration, let $i \in I\backslash X$ and let $Z$ be the union of connected components of $X \cup \{i, \tau(i)\}$ containing $i$ or $\tau(i)$. The following statements are true.*

(1) *The pair $(X \cap Z, \tau|_Z)$ is a compatible decoration of the generalized Cartan matrix $A_Z$, and either $Z$ is of type $\mathsf{A}_1 \times \mathsf{A}_1$ or $Z$ is connected.*
(2) *One has the following identity for linear maps on $\mathfrak{h}$ (or linear maps on $\mathfrak{h}^*$):*

$$(\theta(X, \tau) - \mathsf{id}) \circ (\tau - \mathsf{id}) = 0. \tag{2.27}$$

(3) *If $\{\kappa_j^\vee\}_{j \in X}$ denotes the set of fundamental coweights of $\mathfrak{g}_X$, then*

$$w_X(\alpha_i) = \alpha_i + \sum_{j \in X} v_{ij}\alpha_j, \qquad v_{ij} = -(\alpha_i + \alpha_{\tau(i)})(\kappa_j^\vee) \in \mathbb{Z}_{\geq 0}. \tag{2.28}$$

PROOF.

1 Suppose $i \notin X^\perp$ and let $j \in X$ such that $a_{ij} \neq 0$. Let $X'$ be the connected component of $X$ containing $j$, so that $i \notin (X')^\perp$. Then $\tau|_{X'} = \mathsf{oi}_X|_{X'} = \mathsf{oi}_{X'}$. Taking the union of all such $X'$, we obtain the first statement. Furthermore, from $a_{\tau(i)\tau(j)} = a_{ij}$ it follows that $\tau(i) \notin (X')^\perp$ so that $X' \cup \{i, \tau(i)\}$ is connected. Now suppose $i \in X^\perp$. By the above analysis, also $\tau(i) \in X^\perp$. Consider the full Dynkin subdiagram whose set of vertices is $Z = \{i, \tau(i)\}$. If $Z$ is not connected, it must be of type $\mathsf{A}_1 \times \mathsf{A}_1$.

2 The linear map $(\theta(X, \tau) - \mathsf{id}) \circ (\tau - \mathsf{id}) = (w_X - \mathsf{id}) \circ (\tau - \mathsf{id})$ maps into $\mathfrak{h}_X$; on the other hand its image lies in $\mathfrak{h}^{-\theta(X,\tau)}$. Since $\theta(X, \tau)|_{\mathfrak{h}_X} = \mathsf{id}_{\mathfrak{h}_X}$ it follows that $(\theta(X, \tau) - \mathsf{id}) \circ (\tau - \mathsf{id}) = 0$. A similar argument applies to the action on $\mathfrak{h}^*$.

3 It is clear that the first equality in (2.28) holds for some nonnegative integers $v_{ij}$. Now apply that equality to the fundamental coweight $\kappa_j^\vee$ for $j \in X$. It yields $v_{ij} = (w_X(\alpha_i))(\kappa_j^\vee) - \alpha_i(\kappa_j^\vee)$. Since $(X, \tau)$ is a compatible decoration, we readily obtain $(w_X(\alpha_i))(\kappa_j^\vee) = \alpha_i(w_X(\kappa_j^\vee)) = -\alpha_i(\tau(\kappa_j^\vee)) = -\alpha_{\tau(i)}(\kappa_j^\vee)$ and we arrive at the desired expression for $v_{ij}$ in terms of $\kappa_j^\vee$. □

Satake diagrams (of Kac-Moody type) are particular types of compatible decorations, associated to almost split real forms of Kac-Moody algebras or equivalently to involutive automorphisms of $\mathfrak{g}$ of the second kind (see [**BBBR95**, Def. 4.10 (b), Cor. 4.10.4] and [**Ko14**, Def. 2.3]). We can restate this definition as follows.

DEFINITION 2.4. We call $i \in I\backslash X$ an *odd node* if $\tau(i) = i$ and $\zeta_X(\alpha_i) = -1$. A compatible decoration is called a *Satake diagram* if there are no odd nodes.

**2.9. Classification of pseudo-involutions of the second kind.** Now let $\theta \in \mathsf{Aut}(\mathfrak{g})$ be an arbitrary pseudo-involution of the second kind and let $\mathfrak{t}$ be a $\theta$-stable Cartan subalgebra of $\mathfrak{g}$ such that $\theta|_{\mathfrak{t}}$ is an involution. By [**PK83**, Thm. 2(a)], $\mathfrak{h}$ is $\mathsf{Ad}(G)$-conjugate to $\mathfrak{t}$. We can now apply Proposition 2.2 and hence deduce the following main result of this section.

THEOREM 2.5. *Any pseudo-involution of the second kind is* $\mathsf{Inn}(\mathfrak{g})$*-conjugate to* $\theta(X,\tau,\chi)$ *for some enriched compatible decoration* $(X,\tau,\chi)$.

Theorem 2.5 is a special case of [**KW92**, 4.39] with stronger constraints on the datum $(X,\tau)$. It can be restated as follows: the assignment $(X,\tau,\chi)\mapsto\theta(X,\tau,\chi)$ induces a surjection:

$$\{\text{enriched compatible decorations}\}\to\{\text{pseudo-involutions of the 2nd kind}\}/\mathsf{Inn}(\mathfrak{g}).$$

We can compare Theorem 2.5 with [**Ko14**, Prop. A.6], the analogous statement about involutions of the second kind in terms of Satake diagrams $(X,\tau)$. In that case without loss of generality we may let $\chi$ take values among fourth roots of unity, see [**Ko14**, (2.7)], in order to guarantee the involutiveness of $\theta(X,\tau,\chi)$.

Note that $\mathsf{Aut}(A)$ acts on enriched compatible decorations via

$$\psi\cdot(X,\tau,\chi)=(\psi(X),\psi\circ\tau\circ\psi^{-1},\psi*\chi),\qquad \psi\in\mathsf{Aut}(A). \tag{2.29}$$

Also, $\omega$ commutes with $\theta(X,\tau)$ for all compatible decorations $(X,\tau)$ and $\omega*\chi=\chi^{-1}\in\widetilde{H}^{\theta(X,\tau)}$ for all $\chi\in\widetilde{H}^{\theta(X,\tau)}$. Finally, note that all elements of $\mathsf{Out}(A)$ act by conjugation on the set of pseudo-involutions of the second kind. Hence we obtain from Theorem 2.5 the following result.

COROLLARY 2.6. *The assignment* $(X,\tau,\chi)\mapsto\theta(X,\tau,\chi)$ *induces a surjection* $\overline{\theta}$ *from*

$$\{\textit{enriched compatible decorations}\}/\mathsf{Aut}(A)$$

*to*

$$\{\textit{pseudo-involutions of the 2nd kind}\}/\mathsf{Aut}(\mathfrak{g}).$$

In the same way, [**Ko14**, Prop. A.6] can be promoted to the existence of a surjection from the set of $\mathsf{Aut}(A)$-orbits of Satake diagrams to the set of $\mathsf{Aut}(\mathfrak{g})$-conjugacy classes of involutions of the second kind. Owing to results in [**KW92**, Ch. 5] about involutive automorphisms of $\mathfrak{g}$, this surjection is in fact a bijection for any $A$, see [**Ko14**, Thm. 2.7].

However, the map $\overline{\theta}$ in Corollary 2.6 is not always injective. Indeed, if $A$ is such that there exists a compatible diagram $(X,\tau)$ with at least one odd node, then there always exists a choice of $\chi$ such that $\theta(X,\tau,\chi)$ is an involution. For instance, fix a square root $\sqrt{-1}\in\mathbb{F}$ of $-1$ and let $\chi\in\widetilde{H}$ be defined by $\chi(\alpha_i)=(\sqrt{-1})^{2\alpha_i(\rho^\vee_X)}$ for all $i\in I$. It can be checked using the formula (2.28) and the relation $\rho^\vee_X=\sum_{j\in X}\kappa^\vee_j$ that $\chi\in\widetilde{H}^{\theta(X,\tau)}$. Moreover, we have $\theta(X,\tau,\chi)^2=\mathsf{Ad}(\chi^2\cdot\zeta_X)=\mathsf{id}_{\mathfrak{g}}$, as claimed. By [**Ko14**, Thm. 2.7], $\theta(X,\tau,\chi)$ is $\mathsf{Aut}(\mathfrak{g})$-conjugate with some $\theta(X',\tau',\chi')$ where $(X',\tau')$ is a Satake diagram and $\chi'$ is the character $s(X,\tau)$ defined in [**Ko14**, (2.7)]. Since the $\mathsf{Aut}(A)$-action on the set of compatible decorations preserves the set of Satake diagrams, injectivity does not hold.

Conversely, if $\chi\in\widetilde{H}$ is such that $\theta(X,\tau,\chi)^2=\mathsf{id}_{\mathfrak{g}}$ and $i$ is an odd node then $\chi(\alpha_i-w_X(\alpha_i))=-1$, so $\chi(\alpha_j)\neq1$ for some $j\in X$. Hence, any involution of the form $(X,\tau,\chi)$ where $(X,\tau)$ is not a Satake diagram does not fix pointwise $\mathfrak{g}_X$. It is therefore natural to focus on pseudo-involutions of the second kind which fix pointwise all stable root spaces. Straightforwardly one obtains the following refinement of Corollary 2.6.

COROLLARY 2.7. *Any pseudo-involution of the second kind which fixes pointwise all stable root spaces is* $\mathsf{Inn}(\mathfrak{g})$*-conjugate to* $\theta(X,\tau,\chi)$ *for some enriched compatible decoration* $(X,\tau,\chi)$ *with* $\chi|_{Q_X}=1$.

The set of enriched compatible decorations $(X,\tau,\chi)$ such that $\chi|_{Q_X} = 1$ is $\mathsf{Aut}(A)$-stable. We conjecture that only allowing maps which fix pointwise all stable root spaces resolves the failure of the surjection $\overline{\theta}$ in Corollary 2.5 to be a bijection.

CONJECTURE 2.8. *The assignment* $(X,\tau,\chi) \mapsto \theta(X,\tau,\chi)$ *induces a bijection* $\overline{\theta}_{\mathsf{fix}}$ *from*

$$\{\textit{enriched compatible decorations } (X,\tau,\chi) \textit{ such that } \chi|_{Q_X} = 1\}/\mathsf{Aut}(A)$$

*to*

$$\{\textit{pseudo-involutions of the 2nd kind fixing pointwise all stable root spaces}\}/\mathsf{Aut}(\mathfrak{g}).$$

Note that if $\psi \in \mathsf{Aut}(A)$ and $(X,\tau,\chi)$ is a compatible decoration such that $\chi|_{Q_X} = 1$ then $\theta(\psi\cdot(X,\tau,\chi)) = \psi\circ\theta(X,\tau,\chi)\circ\psi^{-1}$. Considering Corollary 2.7 we see that in order to prove Conjecture 2.8, it suffices to establish the injectivity of $\overline{\theta}_{\mathsf{fix}}$. Note that pseudo-involutions $\theta$ of the second kind which fix pointwise stable root spaces automatically have a split pair, see [**KW92**, 5.15], so that a proof along the lines of the last part of [**Ko14**, App. A] may be possible. On the other hand, the crucial result [**KW92**, Thm. 5.31] does not directly extend to pseudo-involutions: it is no longer true that $\mathfrak{g} = \mathfrak{b}^+ + \mathfrak{g}^\theta$ if $\theta = \theta(X,\tau,\chi)$ is not an involution. Nevertheless, it is useful to follow this approach in first instance.

Consider a pair of enriched compatible decorations $(X,\tau,\chi)$ and $(X',\tau',\chi')$ such that $\chi|_{Q_X} = \chi'|_{Q_{X'}} = 1$ and

$$\theta\circ\phi = \phi\circ\theta' \tag{2.30}$$

for some $\phi \in \mathsf{Aut}(\mathfrak{g})$, where $\theta := \theta(X,\tau,\chi)$ and $\theta' := \theta(X',\tau',\chi')$. We wish to show that there exists $\psi \in \mathsf{Aut}(A)$ such that

$$\psi\cdot(X',\tau',\chi') = (X,\tau,\chi). \tag{2.31}$$

Using the decomposition (2.18) and observing that $\theta$ preserves $\mathfrak{g}'$ and satisfies $\theta\circ\sigma_f = \sigma_{\tau\circ f\circ\tau^{-1}}\circ\theta$ for all linear $f : \mathfrak{h}'' \to \mathfrak{c}$, we may assume that $\phi \in \mathsf{Aut}(\mathfrak{g}') = \mathsf{Out}(A) \ltimes \mathsf{Ad}(\widetilde{H}\ltimes G)$. Furthermore, by replacing $\phi$ by $\theta^{-1}\circ\phi$ if necessary, we may assume that $\phi$ is of the first kind, i.e. $\phi$ lies in

$$\mathsf{Aut}_{\mathrm{I}}(\mathfrak{g}') := \mathsf{Aut}(A) \ltimes \mathsf{Ad}(\widetilde{H}\ltimes G). \tag{2.32}$$

Denoting the factor of $\phi$ in $\mathsf{Aut}(A)$ by $\psi$, from (2.18) we deduce

$$\tau\circ\psi = \psi\circ\tau' \tag{2.33}$$

as desired. It would now be sufficient to prove that $\psi(X') = X$, $\psi * \chi' = \chi$ and $\phi\circ\psi^{-1} \in \mathsf{Ad}(\widetilde{H}\ltimes G)$ commutes with $\theta$.

We make one more general remark before proving the conjecture in a special case. Consider an enriched compatible decoration $(X,\tau,\chi)$ such that $\chi|_{Q_X} = 1$. Fix $i \in I\backslash X$. The conditions on $\chi$ imply that $\chi(\alpha_i + \alpha_{\tau(i)}) = 1$. Suppose in addition (if $\tau(i) = i$ this is automatic) that $\chi(\alpha_i) = \chi(\alpha_{\tau(i)})$, necessarily equal to $\pm1$. Then there exists $\chi_+ \in \widetilde{H}^{\theta(X,\tau)}$ such that

$$\theta(X,\tau,\chi) \sim_{\mathsf{Ad}(\widetilde{H})} \theta(X,\tau,\chi_+), \qquad \chi_+(\alpha_i) = \chi_+(\alpha_{\tau(i)}) = 1. \tag{2.34}$$

Indeed if $\chi(\alpha_i) = \chi(\alpha_{\tau(i)}) = -1$, one can conjugate by $\mathsf{Ad}(\xi)$ where $\xi \in \widetilde{H}$ satisfies $\xi(\alpha_i) = \xi(\alpha_{\tau(i)}) = \sqrt{-1}$ and $\xi(\alpha_j) = 1$ if $j \neq \{i,\tau(i)\}$.

To provide evidence for the conjecture, we now give an elementary proof in the crucial case $\mathfrak{g} = \mathfrak{sl}_3$ (the smallest indecomposable Kac-Moody algebra for which there exist non-involutive pseudo-involutions). Similar proofs, all based on the fact that conjugation of an automorphism preserves its order and the dimension of its eigenspaces, can straightforwardly be given for other low-rank cases.

Let $I = \{1,2\}$ and $A = (a_{ij})_{i,j\in I}$ be such that $a_{11} = a_{22} = 2$ and $a_{12} = a_{21} = -1$. In addition to the Satake diagrams $(\emptyset, \mathsf{id})$, $(\emptyset, \mathsf{oi}_I)$, $(I, \mathsf{oi}_I)$, where $\mathsf{oi}_I$ is the nontrivial diagram automorphism, we have the compatible decorations $(\{1\}, \mathsf{id})$ and $\mathsf{oi}_I \cdot (\{1\}, \mathsf{id}) = (\{2\}, \mathsf{id})$. Consider two enriched compatible decorations $(X, \tau, \chi)$ and $(X', \tau', \chi')$ such that $\chi|_{Q_X} = \chi'|_{Q_{X'}} = 1$. Suppose that $\theta(X, \tau, \chi) \sim_{\mathsf{Aut}_1(\mathfrak{g}')} \theta(X', \tau', \chi')$. We must show that $(X, \tau, \chi)$ and $(X', \tau', \chi')$ lie in the same $\mathsf{Aut}(A)$-orbit. By (2.33), we only need to show $(X, \chi) = (X', \chi')$ or $(X, \chi) = \mathsf{oi}_I \cdot (X', \chi')$.

$\tau = \tau' = \mathsf{id}$: By (2.34) we may assume $\chi = \chi' = 1$. Hence it suffices to study the conjugacy problem

$$\theta(X, \mathsf{id}) \sim_{\mathsf{Aut}_1(\mathfrak{g}')} \theta(X', \mathsf{id}). \tag{2.35}$$

Recall that $\theta(X, \tau)^2 = \mathsf{Ad}(\zeta_X)$ and note that $X$ is either empty or a singleton. Hence $\theta(X, \mathsf{id})$ is an involution if and only if $X = \emptyset$. As a consequence, either $X = \emptyset = X'$, and we are done, or $X \neq \emptyset \neq X'$. In the second case both $X$ and $X'$ are singletons and either $X = X'$, in which case $(X, \tau, \chi) = (X', \tau', \chi')$, or $X = \mathsf{oi}_I(X')$, in which case $(X, \tau, \chi) = \mathsf{oi}_I \cdot (X, \tau', \chi')$, as required.

$\tau = \tau' = \mathsf{oi}_I$: Note that in this case each of $X$ and $X'$ are equal to $I$ or $\emptyset$, and it is necessary to show that $X = X'$. On the one hand, $\theta(I, \mathsf{oi}_I) = \mathsf{id}$ so that $\theta(I, \mathsf{oi}_I, \chi) = \mathsf{id}$ for any $\chi \in \widetilde{H}^{\theta(I,\mathsf{oi}_I)}$ such that $\chi|_{Q_I} = 1$ (i.e. for $\chi = 1$). On the other hand, $\theta(\emptyset, \mathsf{oi}_I, \chi)$ sends $h_1$ to $-h_2$, and hence is not the identity, for any $\chi \in \widetilde{H}$. Therefore $\theta(X, \mathsf{oi}_I, \chi) \sim_{\mathsf{Aut}_1(\mathfrak{g}')} \theta(X', \mathsf{oi}_I, \chi')$ implies $X = X'$. If $X = X' = I$, we automatically have $\chi = \chi'(= 1)$, as required.

To deal with the case $X = X' = \emptyset$, note that $\theta(\emptyset, \mathsf{oi}_I, \chi)$ is an involution if and only if $\chi^2 = 1$, which is equivalent to $\chi(\alpha_1) = \chi(\alpha_2) \in \{\pm 1\}$. Again by (2.34), we may assume that both $\chi$ and $\chi'$ are equal to 1; we obtain $(X, \tau, \chi) = (X', \tau', \chi')$ as required.

It remains to deal with the case that $\chi^2 \neq 1 \neq (\chi')^2$. It suffices to prove that $\chi(\alpha_1) = \chi'(\alpha_1)$. We observe that $\mathsf{Ad}(x)$ sends $\mathfrak{g}^{\theta'}$ to $\mathfrak{g}^{\theta}$, as a consequence of (2.30). In this case $\mathfrak{g}^{\theta} = \mathfrak{g}^{\theta'}$ is the abelian subalgebra $\mathbb{F}(h_1 - h_2) \oplus \mathbb{F}b_{12}$ where $b_{12} := [f_1, f_2] - [e_1, e_2]$. Now we identify $\mathfrak{sl}_3$ with its faithful 3-dimensional representation according to

$$\begin{aligned} e_1 &= \begin{pmatrix} 0 & 1 & 0 \\ 0 & 0 & 0 \\ 0 & 0 & 0 \end{pmatrix}, \quad f_1 = \begin{pmatrix} 0 & 0 & 0 \\ 1 & 0 & 0 \\ 0 & 0 & 0 \end{pmatrix}, \\ e_2 &= \begin{pmatrix} 0 & 0 & 0 \\ 0 & 0 & 1 \\ 0 & 0 & 0 \end{pmatrix}, \quad f_2 = \begin{pmatrix} 0 & 0 & 0 \\ 0 & 0 & 0 \\ 0 & 1 & 0 \end{pmatrix}. \end{aligned} \tag{2.36}$$

Hence $x$ is identified with an element of $\mathsf{GL}(3, \mathbb{F})$ (well-defined up to a scalar multiple). The condition on $x$ that $\mathsf{Ad}(x)$ preserves $\mathfrak{g}^{\theta}$ has two solution classes.

(1) In the first class of solutions we have

$$\mathsf{Ad}(x)(h_1 - h_2) = h_1 - h_2, \qquad \mathsf{Ad}(x)(b_{12}) = \epsilon b_{12},$$

$$x = \begin{pmatrix} A & 0 & B \\ 0 & 1 & 0 \\ \epsilon B & 0 & \epsilon A \end{pmatrix}, \tag{2.37}$$

where $\epsilon \in \{\pm 1\}$ and $A, B \in \mathbb{F}$ are such that $A^2 \neq B^2$. Now we obtain

$$\begin{aligned} \mathsf{Ad}(x)(f_1) &= (A^2 - B^2)^{-1}(Af_1 - \epsilon B e_2), \\ \mathsf{Ad}(x)(e_2) &= (A^2 - B^2)^{-1}(-Bf_1 + \epsilon A e_2). \end{aligned} \tag{2.38}$$

In the case $\psi = \mathsf{id}$, applying (2.30) to $f_1$ gives $\chi(\alpha_2) = \chi'(\alpha_2)$, as required. If $\psi = \mathsf{oi}_I$, applying (2.30) to $f_2$ yields a contradiction, so (2.30) does not hold in this case and there is nothing to prove.

(2) In the second class, we find

$$\begin{aligned} \mathsf{Ad}(x)(h_1 - h_2) &= -\frac{1}{2}(h_1 - h_2 + 3\zeta b_{12}), \\ \mathsf{Ad}(x)(b_{12}) &= -\frac{\eta}{2}(h_1 - h_2 - \zeta b_{12}), \\ x &= \begin{pmatrix} 1 & C & \eta \\ -\eta D & 0 & D \\ \zeta & -\zeta C & \zeta\eta \end{pmatrix} \end{aligned} \tag{2.39}$$

where $\zeta, \eta \in \{\pm 1\}$ and $C, D \in \mathbb{F}^\times$. Now we obtain

$$\begin{aligned} \mathsf{Ad}(x)(f_1) &= \frac{C}{4}\Big(h_1 + h_2 + \zeta([e_1, e_2] + [f_1, f_2]) + 2\eta D^{-1}(\zeta f_2 - e_1)\Big), \\ \mathsf{Ad}(x)(e_2) &= \frac{\eta C}{4}\Big(h_1 + h_2 + \zeta([e_1, e_2] + [f_1, f_2]) - 2\eta D^{-1}(\zeta f_2 - e_1)\Big). \end{aligned} \tag{2.40}$$

Applying (2.30) to $f_{\psi(1)}$ and considering the projections to $\mathfrak{g}_{\alpha_1}$ and $\mathfrak{g}_{-\alpha_2}$ results in a contradiction. Hence (2.30) does not hold and there is nothing to prove.

## 3. Pseudo-fixed-point subalgebras in terms of generalized Satake diagrams

In this section we discuss constructions of pseudo-fixed-point subalgebras in the Kac-Moody setting, associated to pseudo-involutions of the second kind. In [**RV20**, Sec. 3] we considered a Lie subalgebra $\mathfrak{k} = \mathfrak{k}(X, \tau, \chi) \subseteq \mathfrak{g}$ of finite type which is a pseudo-fixed-point subalgebra if and only if the enriched compatible decoration $(X, \tau, \chi)$ satisfies certain additional conditions. Part of this earlier work immediately generalizes to the Kac-Moody setting, see [**RV20**, Rmk. 3 (ii)]. Here we give a brief synopsis of it for completeness. This is complemented with a statement of the Iwasawa decomposition for a pseudo-symmetric pair.

**3.1. The subalgebra $\mathfrak{k}$.** Let $\mathfrak{t}$ be an arbitrary Cartan subalgebra of the Kac-Moody algebra $\mathfrak{g}$. Denote by $\mathfrak{g}_\alpha^{(\mathfrak{t})}$ the $\mathfrak{t}$-root space associated to $\alpha \in \mathfrak{t}^*$,

$$\mathfrak{g}_\alpha^{(\mathfrak{t})} := \{x \in \mathfrak{g} : \forall t \in \mathfrak{t}\ [t, x] = \alpha(t)x\}. \tag{3.1}$$

and by $\Phi^{(\mathfrak{t})}$ the corresponding root system,

$$\Phi^{(\mathfrak{t})} := \{\alpha \in \mathfrak{t}^* : \mathfrak{g}_\alpha^{(\mathfrak{t})} \neq \{0\}\} \backslash \{0\}. \tag{3.2}$$

Choose a subset $\Pi^{(\mathfrak{t})} \subset \Phi^{(\mathfrak{t})}$ such that $\Phi^{(\mathfrak{t})} = \mathsf{Sp}_{\mathbb{Z}_{\geq 0}}\Pi^{(\mathfrak{t})} \cup \mathsf{Sp}_{\mathbb{Z}_{\leq 0}}\Pi^{(\mathfrak{t})}$. Moreover, for every $\alpha \in \Pi^{(\mathfrak{t})}$ and every choice of sign, choose nonzero elements $x_{\pm\alpha} \in \mathfrak{g}^{(\mathfrak{t})}_{\pm\alpha}$ (simple root vectors). Then $\mathfrak{g}$ is generated by $\mathfrak{t}$ and $\{x_\alpha, x_{-\alpha}\}_{\alpha\in\Pi^{(\mathfrak{t})}}$.

DEFINITION 3.1. With the notation defined above, let $\theta \in \mathsf{Aut}(\mathfrak{g})$ be a pseudo-involution of the second kind stabilizing $\mathfrak{t}$ such that $\theta|_{\mathfrak{t}}$ is an involution. To this datum we associate a subalgebra $\mathfrak{k}$, defined to be the subalgebra generated by $\mathfrak{t}^\theta$, all $x_\alpha$ and $x_{-\alpha}$ such that $\alpha \in \Pi^{\theta^*}$, and the elements $x_\alpha + \theta(x_\alpha)$ for $\alpha \in \Pi\backslash\Pi^{\theta^*}$.

This generating set is natural in view of Definition 1.2 and shortly we will prove that under an additional mild assumption on $(X, \tau, \chi)$ the subalgebra $\mathfrak{k}$ is indeed a pseudo-fixed-point subalgebra.

REMARK 3.2. Definition 3.1 also makes sense if $\theta$ is of the first kind, but then $\mathfrak{k}$ will typically not be a pseudo-fixed-point subalgebra in the sense of Definition 1.2. This can be easily checked if $\mathfrak{g}$ is of type $\widehat{\mathsf{A}}_1$ and $\theta$ is the nontrivial diagram automorphism, cf. [**KW92**, Example 5.11].

If $\theta$ is a pseudo-involution of the second kind then Theorem 2.5 implies that by $\mathsf{Aut}(\mathfrak{g})$-conjugacy we may assume

$$\theta = \mathsf{Ad}(\chi) \circ \theta(X, \tau) \tag{3.3}$$

for some enriched compatible decoration $(X, \tau, \chi)$. Then the corresponding subalgebra $\mathfrak{k}$ is generated by $\mathfrak{h}^\theta$, $\mathfrak{n}^+_X$ and the elements

$$b_i := \begin{cases} f_i & \text{if } i \in X, \\ f_i + \theta(f_i) & \text{otherwise,} \end{cases} \tag{3.4}$$

which recovers the description of a fixed-point subalgebra $\mathfrak{g}^\theta$ in the case that $\theta$ is involutive, see [**Ko14**, Lem. 2.8].

REMARK 3.3. Equally, we could have interchanged the role of $\mathfrak{n}^+$ and $\mathfrak{n}^-$ in the definition of $\mathfrak{k}$. The current choice is customary in recent work on q-deformations of $U\mathfrak{k}$, see e.g. [**Ko14, BK19, DK19, RV20, AV20**].

**3.2. Generalized Satake diagrams.** We now consider a particular type of (enriched) compatible decoration; we will see that, if and only if it is associated to such datum, the subalgebra $\mathfrak{k} = \langle \mathfrak{h}^\theta, \mathfrak{n}^+_X, \{b_i\}_{i\in I}\rangle$ is a pseudo-fixed-point subalgebra. For all $i \in I\backslash X$ we consider the set

$$X[i] := X \cup \{i, \tau(i)\}, \tag{3.5}$$

which we may indentify with the corresponding full subdiagram of $I$.

DEFINITION 3.4.

(1) Let $(X, \tau)$ be a compatible decoration. We call $(X, \tau)$ a *generalized Satake diagram* if for all $i \in I\backslash X$ such that $\tau(i) = i$ the connected component of the subdiagram $X[i]$ containing $i$ is not of type $\mathsf{A}_2$ (i.e. $X[i]$ does not have ○—● as a connected component).

(2) Let $(X, \tau, \chi)$ be an enriched compatible decoration. If $(X, \tau)$ is a generalized Satake diagram and $\chi(\alpha_{\tau(i)}) = \chi(\alpha_i)$ for all $i \in X^\perp$ such that $a_{i\tau(i)} = 0$ then we call $(X, \tau, \chi)$ an *enriched generalized Satake diagram*.

REMARK 3.5.

(1) Recall the notion of a Satake diagram given in Definition 2.4. If a compatible decoration $(X,\tau)$ is not a generalized Satake diagram, then there exists $(i,j) \in (I\backslash X)\times X$ such that $i,j \in (X\backslash\{j\})^{\perp}$, $\tau(i)=i$ and $a_{ij}=a_{ji}=-1$. Hence $\zeta_X(\alpha_i)=(-1)^{\alpha_i(h_j)}=-1$, so that $i$ is an odd node. It follows that all Satake diagrams are generalized Satake diagrams.

(2) The action of $\mathsf{Aut}(A)$ on the set of (enriched) compatible decoration given in (2.29) stabilizes the set of (enriched) generalized Satake diagrams.

We give the classification of generalized Satake diagrams when $A$ is of finite or affine type in Appendices A.3 and A.4. Here we make some remarks to support this classification in the case that $A$ is of classical Lie type. Depending on the Lie type, the corresponding generalized Satake diagrams naturally organize themselves into 2, 3 or 4 families.

To describe these families, it is convenient to say that a connected component of $X$ is $(X,\tau)$-*simple* if it is a singleton $\{j\}$ and there exists $i \in I\backslash X$ such that $a_{ij}a_{ji}=1$ and $\tau(i)=i$. Suppose that $\tau$ is trivial except possibly on subsets of type $\mathsf{A}_1\times\mathsf{A}_1$ (these occur when $A$ is of type $\mathsf{D}_n$, $\widehat{\mathsf{B}}_n$, $\widehat{\mathsf{B}}_n^\vee$ or $\widehat{\mathsf{D}}_n$). Then the condition that $\tau|_X=\mathsf{oi}_X$ requires that $X$ has at most two connected components which are not $(X,\tau)$-simple, which must be of type $\mathsf{B}_p$, $\mathsf{C}_p$ or $\mathsf{D}_p$ (with $p\le n$). More precisely, if $A$ is of type $\mathsf{A}_n$ or $\widehat{\mathsf{A}}_n$, then all connected components of $X$ are $(X,\tau)$-simple; if $A$ is of type $\mathsf{B}_n$, $\mathsf{C}_n$ or $\mathsf{D}_n$ then $X$ can have at most one connected component which is not $(X,\tau)$-simple; in all other cases $X$ can have two connected components which are not $(X,\tau)$-simple. The defining condition of generalized Satake diagrams now require that either there are no $(X,\tau)$-simple components, in which case we call $(X,\tau)$ *plain*, or, due to the definining condition of generalized Satake diagram, there are as many as fit in the complement of the other connected components of $X$, in which case we call $(X,\tau)$ *alternating*.

If $A$ is of type $\widehat{\mathsf{A}}_n$, $\widehat{\mathsf{C}}_n$, $\widehat{\mathsf{C}}_n^\vee$ or $\widehat{\mathsf{D}}_n$ then rotation by a half-turn is a diagram automorphism $\tau$. To extend $\tau$ to a generalized Satake diagram $(X,\tau)$, in the first case $X$ must be empty and otherwise of type $\mathsf{A}_p$. In this case we call $(X,\tau)$ *rotational*. Finally, if $A$ is of type $\mathsf{A}_n$ or $\widehat{\mathsf{A}}_n$ then $\tau$ can be a reflection of the Dynkin diagram, in which case $X$ is of type $\mathsf{A}_p$ or $\mathsf{A}_{p_1}\times\mathsf{A}_{p_2}$, respectively; we call such generalized Satake diagrams *reflecting*.

**3.3. Basic properties of $\mathfrak{k}$.** In this section we analyse the subalgebra $\mathfrak{k}$. In order to do that, for $\boldsymbol{i}=(i_1,\dots,i_\ell)\in I^\ell$ ($\ell\in\mathbb{Z}_{>0}$) we set

(3.6) $$e_{\boldsymbol{i}}=[e_{i_1},[e_{i_2},\dots,[e_{i_{\ell-1}},e_{i_\ell}]\cdots]]\in\mathfrak{n}^+,$$

(3.7) $$f_{\boldsymbol{i}}=[f_{i_1},[f_{i_2},\dots,[f_{i_{\ell-1}},f_{i_\ell}]\cdots]]\in\mathfrak{n}^-,$$

(3.8) $$b_{\boldsymbol{i}}=[b_{i_1},[b_{i_2},\dots,[b_{i_{\ell-1}},b_{i_\ell}]\cdots]],$$

(3.9) $$\alpha_{\boldsymbol{i}}=\alpha_{i_1}+\ldots+\alpha_{i_\ell}\in Q^+.$$

Fix a subset $\mathcal{J}\subset\cup_{\ell\in\mathbb{Z}_{>0}}I^\ell$ such that $\{f_{\boldsymbol{i}}:\boldsymbol{i}\in\mathcal{J}\}$ is a basis for $\mathfrak{n}^-$. In particular $I\subseteq\mathcal{J}$.

Recall the standard order on $Q$, defined for any $\lambda,\mu\in Q$ by $\lambda\ge\mu$ if and only if $\lambda-\mu\in Q^+$. Finally, for $i,j\in I$ such that $i\ne j$ we set $M_{ij}:=1-a_{ij}\in\mathbb{Z}_{>0}$ and denote

(3.10) $$\lambda_{ij}:=\alpha_i+M_{ij}\alpha_j\in Q^+,$$

which is not a root (this statement is equivalent to the Serre relations in $\mathfrak{g}$).

LEMMA 3.6. *Let $(X,\tau,\chi)$ be an enriched compatible decoration and let $\theta = \mathsf{Ad}(\chi)\circ\theta(X,\tau)$. Then*

$$b_{\boldsymbol{i}} - f_{\boldsymbol{i}} \in \bigoplus_{\substack{\alpha\in\Phi \\ \alpha > -\alpha_{\boldsymbol{i}}}} \mathfrak{g}_\alpha \tag{3.11}$$

*for all $\boldsymbol{i} \in I^\ell$, $\ell \in \mathbb{Z}_{>0}$. Hence, the projection of $b_{\boldsymbol{i}}$ on $\mathfrak{g}_{-\alpha_{\boldsymbol{i}}}$ with respect to the root space decomposition* (2.7) *is $f_{\boldsymbol{i}}$.*

PROOF. A straightforward induction with respect to $\ell$ and the explicit formula (3.4). □

The generators $b_i$ satisfy Serre relations which, unlike those for $f_i$, may have additional lower order terms.

LEMMA 3.7. [**RV20**, Eq. (3.7)] *Let $(X,\tau,\chi)$ be an enriched compatible decoration. Let $i,j\in I$ be such that $i\neq j$.*

(1) *Suppose $\theta^*(\alpha_i)+\alpha_i+\alpha_j \in -\Phi^+\cup\{0\}$. Then $i\in I\backslash X$, $\tau(i)=i$, $j\in X$, $\theta^*(\alpha_i)+\alpha_i+\alpha_j\in -\Phi^+_X\cup\{0\}$ and*

$$\mathsf{ad}(b_i)^{M_{ij}}(b_j) = \begin{cases} (1+\zeta_X(\alpha_i))[\theta(f_i),[f_i,f_j]] \in \mathfrak{n}^+_X & \\ & \text{if } \theta^*(\alpha_i)+\alpha_i+\alpha_j<0,\ a_{ij}=-1, \\ -18\chi(\alpha_i)^{-2}e_j & \text{if } \theta^*(\alpha_i)+\alpha_i+\alpha_j=0,\ a_{ij}=-3, \\ -\chi(\alpha_i)^{-1}(2h_i+h_j) & \text{if } \theta^*(\alpha_i)+\alpha_i+\alpha_j=0,\ a_{ij}=-1, \\ 0 & \text{otherwise.} \end{cases}$$

(2) *Suppose $\theta^*(\alpha_i)+\alpha_j\in-\Phi^+\cup\{0\}$ and $j\in I\backslash X$. Then $i\in I\backslash X$, $\tau(i)=j$, $\theta^*(\alpha_i)+\alpha_j\in -\Phi^+_X\cup\{0\}$ and*

$$\mathsf{ad}(b_i)^{M_{ij}}(b_j) = \begin{cases} (1+\zeta_X(\alpha_i)\chi(\alpha_i-\alpha_j))[\theta(f_i),f_j] \in \mathfrak{n}^+_X & \\ & \text{if } \theta^*(\alpha_i)+\alpha_j<0,\ a_{ij}=0, \\ \chi(\alpha_j)^{-1}h_i - \chi(\alpha_i)^{-1}h_j & \text{if } \theta^*(\alpha_i)+\alpha_j=0,\ a_{ij}=0, \\ 2(\chi(\alpha_i)^{-1}+\chi(\alpha_j)^{-1})b_i & \text{if } \theta^*(\alpha_i)+\alpha_j=0,\ a_{ij}=-1, \\ 0 & \text{otherwise.} \end{cases}$$

(3) *Suppose $\theta^*(\alpha_i)=-\alpha_i$ and $j\in I\backslash X$. Then $i\in X^\perp$, $\tau(i)=i$ and*

$$\mathsf{ad}(b_i)^{M_{ij}}(b_j) = \sum_{\substack{r\in\mathbb{Z}_{>0} \\ 2r\leq M_{ij}}} p_{ij}^{(r,M_{ij})}\chi(\alpha_i)^{-r}\mathsf{ad}(b_i)^{M_{ij}-2r}(b_j) \tag{3.12}$$

*where $p_{ij}^{(r,m)}$ for $m,r\in\mathbb{Z}_{\geq 0}$ such that $0\leq 2r\leq m\leq M_{ij}$ are negative integers defined as follows. We set $p_{ij}^{(0,m)}=-1$ and $p_{ij}^{(r,m)}=0$ if $2r>m$ and impose the recursion*

$$p_{ij}^{(r,m)} = p_{ij}^{(r,m-1)} + (m-1)(M+1-m)p_{ij}^{(r-1,m-2)} \qquad \text{if } 0<r\leq \lfloor \tfrac{m}{2}\rfloor. \tag{3.13}$$

(4) *If none of the three conditions above are satisfied then $\mathsf{ad}(b_i)^{M_{ij}}(b_j)=0$.*

PROOF. By a careful root space analysis, see [**RV20**, App. A]. □

The relation (3.12) and the coefficients $p_{ij}^{(r,m)}$ also appeared in [**St19**, Def. 2.3] which deals with the case $\theta = \omega$ and $\mathfrak{k} = \mathfrak{g}^\omega$. Such $\mathfrak{k}$ are also called (embedded) *generalized Onsager algebras*.

The following result is a key motivation for introducing generalized Satake diagrams.

THEOREM 3.8. [**RV20**, Thm. 3.3] *Let $(X, \tau, \chi)$ be an enriched compatible decoration and let $\theta = \mathsf{Ad}(\chi) \circ \theta(X, \tau)$. The following statements are equivalent:*

(1) *$(X, \tau, \chi)$ is an enriched generalized Satake diagram;*
(2) *for all $i, j \in I$ such that $i \neq j$ one has*

$$\mathsf{ad}(b_i)^{M_{ij}}(b_j) \in \mathfrak{n}_X^+ \oplus \mathfrak{h}^\theta \oplus \bigoplus_{\ell \in \mathbb{Z}_{>0}} \bigoplus_{\substack{\boldsymbol{i} \in I^\ell \\ \alpha_{\boldsymbol{i}} < \lambda_{ij}}} \mathbb{F} b_{\boldsymbol{i}}; \tag{3.14}$$

(3) *the following identity of $\mathfrak{h}^\theta$-modules holds:*

$$\mathfrak{k} = \mathfrak{n}_X^+ \oplus \mathfrak{h}^\theta \oplus \bigoplus_{\boldsymbol{i} \in \mathcal{J}} \mathbb{F} b_{\boldsymbol{i}}; \tag{3.15}$$

(4) *$\mathfrak{k}$ is a pseudo-fixed-point subalgebra.*

PROOF. We refer to [**RV20**] for the proof of the equivalence of the first 3 statements and condition (1.2) in the definition of pseudo-fixed-point subalgebra. It suffices to prove that 2 implies condition (1.3) in the definition of pseudo-fixed-point subalgebra. In fact, since $\mathfrak{g}_\alpha \subset \mathfrak{k}$ if $\alpha \in \Phi_X$ and $\theta^*$ interchanges $\Phi^+ \backslash \Phi_X$ and $-\Phi^+ \backslash \Phi_X$, it suffices to prove (1.3) with $\alpha \in -\Phi^+ \backslash \Phi_X$.

First of all, by an induction argument it follows from 2 that, for all $\boldsymbol{i} \in \mathcal{J}$,

$$b_{\boldsymbol{i}} - f_{\boldsymbol{i}} - \theta(f_{\boldsymbol{i}}) \in \mathfrak{n}_X^+ \oplus \mathfrak{h}^\theta \oplus \bigoplus_{\substack{\boldsymbol{j} \in \mathcal{J} \\ \alpha_{\boldsymbol{j}} < \alpha_{\boldsymbol{i}}}} \mathbb{F} b_{\boldsymbol{j}}, \tag{3.16}$$

see [**RV20**, Eq. (3.20)]. Hence, $f_{\boldsymbol{i}} + \theta(f_{\boldsymbol{i}})$ lies in $\mathfrak{k}$. On the other hand, $f_{\boldsymbol{i}} + \theta(f_{\boldsymbol{i}})$ lies in $\mathfrak{g}_{-\alpha_{\boldsymbol{i}}} + \mathfrak{g}_{-\theta^*(\alpha_{\boldsymbol{i}})}$. Now let $\alpha \in \Phi^+$ be arbitrary. Since $\{f_{\boldsymbol{i}} : \boldsymbol{i} \in \mathcal{J}, \alpha_{\boldsymbol{i}} = \alpha\}$ is a basis for $\mathfrak{g}_{-\alpha}$, Lemma 3.6 implies that $\{f_{\boldsymbol{i}} + \theta(f_{\boldsymbol{i}}) : \boldsymbol{i} \in \mathcal{J}, \alpha_{\boldsymbol{i}} = \alpha\}$ is a basis for $\mathfrak{g}_{-\alpha} + \mathfrak{g}_{-\theta^*(\alpha)}$, as required. □

REMARK 3.9. In [**RV20**, Thm. 3.3] we use the notation $\gamma_i$ for the scalar $\chi(\alpha_i)^{-1}$, $i \in I \backslash X$; note that the values $\chi(\alpha_j)$ for $j \in X$ are irrelevant for Theorem 3.8 and we may as well assume that $\chi|_{Q_X} = 1$, cf. Corollary 2.7.

**3.4. Iwasawa decomposition for pseudo-symmetric pairs.** Consider the subalgebra

$$\mathfrak{n}_\theta^+ := \mathfrak{n}^+ \cap \theta(\mathfrak{n}^-) \subseteq \mathfrak{n}^+. \tag{3.17}$$

Since $\mathfrak{n}^+ \cap \theta(\mathfrak{n}^+) = \mathfrak{n}_X^+$ and $\mathfrak{h}$ is $\theta$-stable, immediately we obtain $\mathfrak{n}^+ = \mathfrak{n}_X^+ \oplus \mathfrak{n}_\theta^+$. Hence the triangular decomposition of $\mathfrak{g}$ implies

$$\mathfrak{g} = \mathfrak{n}_X^+ \oplus \mathfrak{h}^\theta \oplus \mathfrak{n}^- \oplus \mathfrak{h}^{-\theta} \oplus \mathfrak{n}_\theta^+. \tag{3.18}$$

Next, $\mathfrak{n}^- = \bigoplus_{\boldsymbol{i} \in \mathcal{J}} \mathbb{F} f_{\boldsymbol{i}}$ by definition of $\mathcal{J}$. By Lemma 3.6 we may write

$$\mathfrak{g} = \mathfrak{n}_X^+ \oplus \mathfrak{h}^\theta \oplus \bigoplus_{\boldsymbol{i} \in \mathcal{J}} \mathbb{F} b_{\boldsymbol{i}} \oplus \mathfrak{h}^{-\theta} \oplus \mathfrak{n}_\theta^+. \tag{3.19}$$

which a priori is a decomposition of linear spaces, since $\bigoplus_{i\in\mathcal{J}} \mathbb{F}b_i$ is not a subalgebra of $\mathfrak{g}$. However Theorem 3.8 (3) now implies the following decomposition for $\mathfrak{g}$ in terms of subalgebras.

COROLLARY 3.10 (Iwasawa decomposition for the pseudo-symmetric pair $(\mathfrak{g},\mathfrak{k})$). *Let $(X,\tau,\chi)$ be an enriched compatible decoration and let $\theta = \mathsf{Ad}(\chi)\circ\theta(X,\tau)$. Then $\mathfrak{g}$ has the following decomposition in terms of subalgebras:*

$$\mathfrak{g} = \mathfrak{k} \oplus \mathfrak{h}^{-\theta} \oplus \mathfrak{n}_\theta^+ \tag{3.20}$$

*if and only if $(X,\tau,\chi)$ is an enriched generalized Satake diagram.*

**3.5. A combinatorial description of $\mathfrak{k}'$.** We complete this section with a description of the derived subalgebra $\mathfrak{k}'$, which indicates how the universal enveloping algebra $U\mathfrak{k}$ can be modified by scalar terms. This directly extends the discussion in [**RV20**, Sec. 3.3] to the Kac-Moody setting. Let $(X,\tau,\chi)$ be an enriched generalized Satake diagram and fix a subset $I^* \subset I\backslash X$ intersecting each $\tau$-orbit in a singleton. Set

$$\mathcal{J}_X := \mathcal{J} \cap \bigcup_{\ell>0} X^\ell. \tag{3.21}$$

From (2.27) it follows that $\{h_i\}_{i\in X} \cup \{h_i - h_{\tau(i)}\}_{i\in I^*,\tau(i)\neq i}$ is a basis for $\mathfrak{h}^\theta$. Hence, Theorem 3.8 implies that

$$\{e_i\}_{i\in\mathcal{J}_X} \cup \{h_i\}_{i\in X} \cup \{h_i - h_{\tau(i)}\}_{i\in I^*,\tau(i)\neq i} \cup \{b_i\}_{i\in\mathcal{J}} \tag{3.22}$$

is an $\mathbb{F}$-basis for $\mathfrak{k}$. Now consider the following subsets of $I^*$:

$$I_{\mathsf{diff}} = \{i \in I^* : i \notin (X\cup\tau(i))^\perp \wedge \tau(i)\neq i\}, \tag{3.23}$$

$$I_{\mathsf{ns}} = \{i \in I^* : i \in X^\perp \wedge \tau(i) = i\}, \tag{3.24}$$

$$I_{\mathsf{nsf}} = \{j \in I_{\mathsf{ns}} : \forall i \in I_{\mathsf{ns}}\ a_{ij} \in 2\mathbb{Z}\}. \tag{3.25}$$

Elements of $I_{\mathsf{diff}}$ and $I_{\mathsf{nsf}}$ are called *special $\tau$-orbits.*

PROPOSITION 3.11. [**RV20**, Prop. 3.2] *Let $(X,\tau,\chi)$ be an enriched generalized Satake diagram. Then the set*

$$\{e_i\}_{i\in\mathcal{J}_X} \cup \{h_i\}_{i\in X} \cup \{h_i - h_{\tau(i)}\}_{i\in I^*\backslash I_{\mathsf{diff}},\tau(i)\neq i} \cup \{b_i\}_{i\in\mathcal{J}\backslash I_{\mathsf{nsf}}} \tag{3.26}$$

*is an $\mathbb{F}$-basis for $\mathfrak{k}'$. Moreover,*

$$\mathfrak{k}\cap\mathfrak{g}' = \mathfrak{k}' \rtimes \left( \bigoplus_{i\in I_{\mathsf{diff}}} \mathbb{F}(h_i - h_{\tau(i)}) \oplus \bigoplus_{j\in I_{\mathsf{nsf}}} \mathbb{F}b_j \right). \tag{3.27}$$

REMARK 3.12.

(1) In (3.27), the intersection with $\mathfrak{g}'$ is only necessary since $\mathfrak{h}$ may contain additional $\theta$-fixed elements if $\mathsf{corank}(A) > 1$. In other words, $\mathfrak{k} \subseteq \mathfrak{g}'$ if $\mathsf{corank}(A) \leq 1$.
(2) By Definition 3.4 (2), if $(X,\tau,\chi)$ is an enriched generalized Satake diagram and $i \in I^*$ then $\chi(\alpha_i)$ and $\chi(\alpha_{\tau(i)})$ are allowed to be *different* precisely if $i \in I_{\mathsf{diff}}$. This explains the notation.
(3) The set $I_{\mathsf{ns}}$ is implicit in Lemma 3.7 (3). Its elements are called *non-standard*; this nomenclature goes back to [**Le02**, Sec. 7, Variation 2] and [**Ko14**, Sec. 5] where it was pointed out that if $i \in I_{\mathsf{nsf}}$ then a *free* parameter appears in the extra Cartan term in the quantum analogues of the

generators $b_i$ in $U_q\mathfrak{k}$. The non-deformed version of this statement is the result [**RV20**, Prop. 3.13], which also extends directly to the Kac-Moody setting. It characterizes, in terms of the set $I_{nsf}$, the essential freedom in the enveloping algebra $U\mathfrak{k}$ with respect to the additional scalar terms.

## 4. The restricted Weyl group and restricted root system

In this section we give a survey on results on the restricted Weyl group and restricted root system associated to a root system involution of a Kac-Moody algebra, thereby generalizing some of Heck's work [**He84**] for finite root systems and drawing on works by Lusztig [**Lu95**] (also see [**Lu03**, Ch. 25]) and Geck and Iancu [**GI14**].

For finite root systems $\Phi$, Heck [**He84**] studied involutive automorphisms of finite root systems $\Phi$. He obtained a diagrammatic constraint on involutions of finite root systems such that the associated restricted root system and restricted Weyl group have natural properties, thereby essentially obtaining the notion of a generalized Satake diagram (although the terminology in [**He84**] is different). Heck's work was a simplification of Schattschneider's work [**Sch69**] in the still more general case of an arbitrary group of automorphisms of a finite root system $\Phi$.

Note that the Weyl group $W$ is a Coxeter group (see e.g. [**Ka90**, §3.13], [**MT72**] and [**Lo80**]), acting by reflections in the root system of $\mathfrak{g}$. A particular obstacle in the restricted setting is that, given a compatible decoration $(X,\tau)$ and the associated root system involution $w_X \circ \tau$, there are no fewer than three different natural definitions of the restricted Weyl group, which we will denote by $W(\overline{\Phi})$, $\overline{W}$ and $\widetilde{W}$. We will see that, precisely if $(X,\tau)$ is a generalized Satake diagram, these three groups are isomorphic. Moreover in this case this group is a Coxeter group and the restricted root system is stable under its action.

**4.1. The $\mathbb{Q}$-span of the root system.** By (2.20), the subgroup of $\mathsf{GL}(\mathfrak{h}')$ of invertible linear maps on $\mathfrak{h}'$ that extend to Lie algebra automorphisms of $\mathfrak{g}$ is of the form $\mathsf{Out}(A) \ltimes W$. The definition of root space implies that the duals of such maps on $\mathfrak{h}'$ necessarily stabilize $\Phi$. Conversely, an invertible linear map on the dual of $\mathfrak{h}'$ which stabilizes $\Phi$ can be dualized and extended to a Lie algebra automorphism stabilizing $\mathfrak{h}$.

It is convenient here to work over the ordered subfield $\mathbb{Q} \subset \mathbb{F}$ in order to facilitate certain (geometric) constructions, so we consider the following $|I|$-dimensional vector space over $\mathbb{Q}$:

$$V := \mathsf{Sp}_{\mathbb{Q}}\Phi = \mathsf{Sp}_{\mathbb{Q}}\Pi. \tag{4.1}$$

The group of *root system automorphisms* $\mathsf{Aut}(\Phi) \cong \mathsf{Out}(A) \ltimes W$ is the subgroup of those $\sigma \in \mathsf{GL}(V)$ which stabilize $\Phi$. Note that the bilinear form $(\ ,\ )$ restricts to an $\mathsf{Aut}(\Phi)$-invariant $\mathbb{Q}$-valued bilinear form on $V$.

**4.2. Root system involutions and the corresponding orthogonal decompositions.** By the results of Section 2, any pseudo-involution of the second kind is $\mathsf{Inn}(\mathfrak{g})$-conjugate to $\mathsf{Ad}(\chi) \circ \theta(X,\tau)$ for some extended compatible decoration $(X,\tau,\chi)$. The automorphism $\mathsf{Ad}(\chi) \circ \theta(X,\tau)$ restricts to the linear involution $-w_X \circ \tau$ on $\mathfrak{h}$, stabilizing the subspace $\mathfrak{h}'$. The dual map on $(\mathfrak{h}')^* = \mathsf{Sp}_{\mathbb{F}}\Phi$ is given by the same formula. It restricts to the $\mathbb{Q}$-span $V$ of $\Phi$ and is an example of an involutive root system automorphism, which we call *root system involution.*

Our focus will be on the map

$$\sigma := w_X \circ \tau = -\theta(X,\tau)^*|_V \in \mathsf{Aut}(A) \ltimes W, \tag{4.2}$$

which is also the convention used in [**He84**]. Since $\sigma$ is an involution the following orthogonal linear decomposition holds:

$$V = V^\sigma \oplus V^{-\sigma}. \tag{4.3}$$

We denote by $\overline{\phantom{x}}$ the corresponding projection: $V \to V^\sigma$, so that

$$\overline{\lambda} = \frac{\lambda + \sigma(\lambda)}{2} \qquad \text{for all } \lambda \in V. \tag{4.4}$$

Because $\sigma$ is an isometry, the following property holds, which we will repeatedly use in this section:

$$(\overline{\beta},\overline{\gamma}) = (\overline{\beta},\gamma) = (\beta,\overline{\gamma}) \qquad \text{for all } \beta,\gamma \in V. \tag{4.5}$$

For $J \subseteq I$ we denote $V_J := \mathsf{Sp}_{\mathbb{Q}}\Phi_J$. Since $(X,\tau)$ is a compatible decoration, the identity $\sigma|_{V_X} = w_X \circ \mathsf{oi}_X|_{V_X} = -\mathsf{id}_{V_X}$ holds and hence the projection $\overline{\phantom{x}}$ annihilates $V_X$. The linear map $w_X - \mathsf{id}$ maps $V$ to $V_X$; since $\sigma$ commutes with $w_X$, we obtain

$$w_X \circ \overline{\phantom{x}} = \overline{\phantom{x}} \circ w_X = \overline{\phantom{x}}. \tag{4.6}$$

By multiplying by $\sigma$ we also obtain

$$\tau \circ \overline{\phantom{x}} = \overline{\phantom{x}} \circ \tau = \overline{\phantom{x}}. \tag{4.7}$$

From (4.6)-4.7 it follows that $V^\sigma \subseteq V^{w_X}$ and $V^\sigma \subseteq V^\tau$. Since $V^{w_X} \cap V^\tau \subseteq V^\sigma$ is clear, we obtain

$$V^\sigma = V^{w_X} \cap V^\tau. \tag{4.8}$$

### 4.3. The restricted root system.

DEFINITION 4.1. Let $\sigma$ be a root system involution. For any subset $\Sigma \subseteq \Phi$ (not necessarily a root subsystem), we denote

$$\overline{\Sigma} := \{\overline{\alpha} : \alpha \in \Sigma\}\backslash\{0\}. \tag{4.9}$$

We call $\overline{\Phi}$ the *restricted* (or *relative*) *root system* associated to $\sigma$ and its elements *restricted roots*; the dimension of $\mathsf{Sp}_{\mathbb{Q}}(\overline{\Phi})$ is called the *restricted rank* of $\sigma$.

If $\sigma = w_X \circ \tau$ with $(X,\tau)$ a compatible decoration, then $\overline{\alpha} = 0$ if and only if $\alpha \in \Phi_X$, so that

$$\overline{\Sigma} = \{\overline{\alpha} : \alpha \in \Sigma\backslash\Phi_X\} \tag{4.10}$$

for any $\Sigma \subseteq \Phi$. Equivalently, $\overline{\Phi}$ may be defined as $\{\alpha|_{\mathfrak{h}^{-\theta(X,\tau)}}\}\backslash\{0\}$, see e.g. [**DK19**, Sec. 2.3].

The restricted root system can be empty, non-reduced and non-crystallographic. More problematically, it is not necessarily stable under reflections in hyperplanes orthogonal to its elements. We will return to this question shortly, indicating constraints on $(X,\tau)$ to guarantee that $\overline{\Phi}$ is reflection-stable.

LEMMA 4.2. *Let $\sigma$ be a root system involution and let $J \subseteq I$ be such that $\sigma$ stabilizes $V_J$.*

(1) *If $J$ is of finite type, then $\overline{\Phi_J}$ is finite and, since $(V_J)^\sigma \subseteq V_J$, the restriction of $(\ ,\ )$ to $(V_J)^\sigma$ is positive definite.*

(2) *If $J$ is of infinite type we denote by $\Phi^+_{\mathsf{im},J}$ the set of positive imaginary roots in $\Phi_J$. Then $\overline{\Phi^+_{\mathsf{im},J}}$ is infinite and contained in $\overline{\Phi^+_J}$. Furthermore, if $(V_J)^\sigma$ is 1-dimensional then $(\beta,\beta)\le 0$ for all $\beta\in\overline{\Phi_J}$.*

PROOF.

1 If $J$ is of finite type, then $\Phi_J$ is finite and hence $\overline{\Phi_J}$ is finite. Since the restriction of the bilinear form $(\ ,\ )$ to $V_J=\mathsf{Sp}_{\mathbb{Q}}(\Phi_J)$ is positive definite and $\sigma$ is an isometry, the bilinear form $(\ ,\ )$ restricts to $(V_J)^\sigma$ and the restriction is positive definite.

2 If $J$ is of infinite type then by [**Ka90**, Thm. 5.6], $\Phi^-_{\mathsf{im},J}$ is nonempty. Note that $\mathsf{Aut}_J(A)\ltimes W_J$ stabilizes $\Phi^+_{\mathsf{im},J}$ as a consequence of [**Ka90**, Thm. 5.4]. In particular, $\sigma$ stabilizes $\Phi^+_{\mathsf{im},J}$ and therefore $\overline{\lambda}\neq 0$ if $\lambda\in\Phi^+_{\mathsf{im},J}$. It follows that $\overline{\Phi^+_{\mathsf{im},J}}\subseteq\overline{\Phi}$. By [**Ka90**, Prop. 5.5], $\Phi^+_{\mathsf{im},J}$ is stable under multiplication by $\mathbb{Z}_{\geq 0}$, so that $\overline{\Phi^+_{\mathsf{im},J}}$ is infinite. Finally, by [**Ka90**, Thm. 5.6] there exists $\lambda\in\Phi^+_{\mathsf{im},J}$ such that $(\lambda,\mu)\le 0$ for all $\mu\in\Phi^+$. Hence $(\lambda,\lambda)\le 0$ and moreover, since $\sigma(\lambda)\in\Phi^+$, also $(\lambda,\sigma(\lambda))\le 0$. Therefore $(\overline{\lambda},\overline{\lambda})=(\lambda,\overline{\lambda})\le 0$ as required. If $(V_J)^\sigma$ is 1-dimensional, it is spanned over $\mathbb{Q}$ by $\overline{\lambda}$ and we obtain $(\beta,\beta)\le 0$ for all $\beta\in\overline{\Phi_J}$. □

REMARK 4.3. If $J$ is of affine type then we can strengthen Lemma 4.2 (2) as a consequence of [**Ka90**, Theorem 5.6 (b)]. Namely, the kernel of $A_J$ is spanned by a unique tuple of setwise coprime positive integers $(a_j)_{j\in J}$ and $\Phi^+_{\mathsf{im},J}=\mathsf{Sp}_{\mathbb{Z}_{>0}}\delta_J$ where $\delta_J=\sum_{j\in J}a_j\alpha_j$ is the *basic imaginary root*. For all $j\in J$ it follows that $\delta_J(h_j)=0$ and hence $s_j(\delta_J)=\delta_j$. Since $\mathsf{Aut}(A_J)$ preserves the height function on $Q_J$, we obtain $\tau(\delta_J)=\delta_J$. It follows that $\overline{\delta_J}=\delta_J$ and hence $\overline{\Phi^+_{\mathsf{im},J}}=\Phi^+_{\mathsf{im},J}$.

**4.4. Combinatorial bases for $V^\sigma$ and $V^{-\sigma}$.** Recall that we have fixed a subset $I^*\subset I\backslash X$ which intersects each $\tau$-orbit in a singleton. Using (4.7) and (4.10), we obtain

$$\overline{\Pi}=\{\overline{\alpha_i}\}_{i\in I}\backslash\{0\}=\{\overline{\alpha_i}\}_{i\in I^*}. \tag{4.11}$$

Note that the set $\overline{\Pi}$ is linearly independent (for all $i\in I^*$, $\overline{\alpha_i}$ is the only element of $\overline{\Pi}$ such that when expanding with respect to the basis $\Pi$ of $V$, the coefficient of $\alpha_i$ is nonzero).

LEMMA 4.4. *Given a compatible decoration $(X,\tau)$, let $\sigma=w_X\circ\tau$ be the corresponding root system involution. Then*

$$V^\sigma=\mathsf{Sp}_{\mathbb{Q}}(\overline{\Pi})=\mathsf{Sp}_{\mathbb{Q}}(\overline{\Phi}). \tag{4.12}$$

*Moreover, $\overline{\Phi}=\overline{\Phi^+}\cup(-\overline{\Phi^+})$ with $\overline{\Phi^+}$ contained in $\mathsf{Sp}_{\mathbb{Z}_{\geq 0}}(\overline{\Pi})$.*

PROOF. Let $\beta=\sum_{i\in I}m_i\alpha_i\in V^\sigma$ with $m_i\in\mathbb{Q}$. Since $\beta$ is fixed by $\sigma$ it equals $\overline{\beta}$ given by

$$\overline{\beta}=\sum_{i\in I}m_i\overline{\alpha_i}=\sum_{\substack{i\in I^*\\ \tau(i)=i}}m_i\overline{\alpha_i}+\sum_{\substack{i\in I^*\\ \tau(i)\neq i}}(m_i+m_{\tau(i)})\overline{\alpha_i}. \tag{4.13}$$

Therefore $V^\sigma\subseteq\mathsf{Sp}_{\mathbb{Q}}(\overline{\Pi})$. On the other hand, one has the natural inclusions $\mathsf{Sp}_{\mathbb{Q}}(\overline{\Pi})\subseteq\mathsf{Sp}_{\mathbb{Q}}(\overline{\Phi})\subseteq V^\sigma$ and the first statement follows. To establish the second statement, suppose $\beta\in\overline{\Phi}$. Then $\beta=\overline{\pm\lambda}$ for some $\lambda\in\Phi^+$. Hence either $\beta$ or

$-\beta$ lies in $\sum_{i\in I} m_i\alpha_i$ with $m_i \in \mathbb{Z}_{\geq 0}$. Now (4.13) implies that $\pm\beta$ is a $\mathbb{Z}_{\geq 0}$-linear combination of the $\overline{\alpha_i}$. □

REMARK 4.5. If $B \subseteq R \subseteq V$ is an inclusion of sets with $V$ a $\mathbb{Q}$-linear space spanned by $R$, $B$ is called a *base* for $R$ if $B$ is a $\mathbb{Q}$-basis for $V$ and $R \subseteq \mathsf{Sp}_{\mathbb{Z}_{\geq 0}} B \cup \mathsf{Sp}_{\mathbb{Z}_{\leq 0}} B$. The root space decomposition (2.7) implies that $\Pi$ is a base for $\Phi$. Owing to Lemma 4.4, $\overline{\Pi}$ is a base for $\overline{\Phi}$.

It follows from Lemma 4.4 that the restricted rank of $\sigma$ is the number of $\tau$-orbits outside $X$; hence we will also call this number the restricted rank of $(X,\tau)$. We refer to the elements $\overline{\alpha_i}$ as *simple restricted roots.* For all $\beta \in \overline{\Phi}^+$ there exist unique $m_i \in \mathbb{Z}_{\geq 0}$ such that $\beta = \sum_{i\in I^*} m_i\overline{\alpha_i}$ and we call $\sum_{i\in I^*} m_i \in \mathbb{Z}_{\geq 0}$ the *restricted height* of $\beta$. Note that $\beta$ has restricted height 1 if and only if $\beta = \overline{\alpha_i}$ for some $i \in I^*$.

Knowledge of the signs of the inner products between simple restricted roots will allow us to study the action of the reflections with respect to the $\overline{\alpha_i}$ on $\overline{\Phi}$. For this purpose, recall the notation $X[i] = X \cup \{i,\tau(i)\}$ for $i \in I\backslash X$. It suffices to take $i \in I^*$. It will turn out to matter greatly whether $X[i]$ is of finite type, so we define

$$\widetilde{I} = \{i \in I^* \,:\, X[i] \text{ is of finite type}\}. \tag{4.14}$$

EXAMPLE 4.6. Let $(X,\tau)$ be a compatible decoration. If $I$ is of finite type then $\widetilde{I} = I^*$. If $I$ is of affine type then $I^*$ is nonempty (since $X \neq I$) and there are two possibilities: if $|I^*| = 1$ then $X[i] = I$ for $i \in I^*$ and hence $\widetilde{I} = \emptyset$; if $|I^*| > 1$ then $\widetilde{I} = I^*$.

LEMMA 4.7. *Let $i,j \in I^*$. Then $(\overline{\alpha_i},\overline{\alpha_j}) > 0$ if and only if $i = j \in \widetilde{I}$.*

PROOF. We split up the proof into some casework. We will repeatedly use that $\sigma$ stabilizes $V_{X[i]}$ for any $i \in I^*$ and is of restricted rank 1 as a root system involution of $V_{X[i]}$.

$i = j \in \widetilde{I}$: In this case $X[i]$ is of finite type and by Lemma 4.2 (1) we obtain $(\overline{\alpha_i},\overline{\alpha_i}) > 0$.

$i = j \notin \widetilde{I}$: Now $X[i]$ is of infinite type and $(\overline{\alpha_i},\overline{\alpha_j}) \leq 0$ by Lemma 4.2 (2).

$i \neq j$: In this case $i \notin X[j]$. Recalling Lemma 2.3 (3) one has $\sigma(\alpha_j) = \alpha_{\tau(j)} + \sum_{j\in X} v_{ij}\alpha_j$ with $v_{ij} \in \mathbb{Z}_{\geq 0}$. Hence

$$2(\overline{\alpha_i},\overline{\alpha_j}) = (\alpha_i,\alpha_j + \sigma(\alpha_j)) = (\alpha_i,\alpha_j) + (\alpha_i,\alpha_{\tau(j)}) + \sum_{k\in X} v_{ij}(\alpha_i,\alpha_k) \leq 0, \tag{4.15}$$

as required. □

It follows that, if $\widetilde{I}$ is a proper subset of $I^*$, there are imaginary simple restricted roots, as there are for Borcherds' generalized Kac-Moody algebras [**Bo88**].

We close this subsection with a description of the orthogonal complement $V^{-\sigma}$.

LEMMA 4.8. *Given a compatible decoration $(X,\tau)$, let $\sigma = w_X \circ \tau$ be the corresponding root system involution. Then*

$$V^{-\sigma} = \mathsf{Sp}_{\mathbb{Q}}(\{\alpha_i\}_{i\in X} \cup \{\alpha_i - \alpha_{\tau(i)} \,:\, \tau(i) \neq i\}_{i\in I^*}) \tag{4.16}$$

*and $\Phi \cap V^{-\sigma} = \Phi_X$.*

PROOF. Let $\beta \in V^{-\sigma}$. Then there exist $m_i \in \mathbb{Q}$ such that $\beta = \sum_{i\in I} m_i\alpha_i$. It follows that $\beta$ equals

$$\frac{\beta - \sigma(\beta)}{2} = \sum_{i\in I} m_i \frac{\alpha_i - \sigma(\alpha_i)}{2} = \sum_{i\in X} m_i\alpha_i + \sum_{i\in I\backslash X} \frac{m_i}{2}(\alpha_i - w_X(\alpha_{\tau(i)})). \tag{4.17}$$

Here, the element $\alpha_i - w_X(\alpha_{\tau(i)})$ equals $\alpha_i - \alpha_{\tau(i)}$ plus a term in $\mathsf{Sp}_{\mathbb{Q}}\{\alpha_i\}_{i\in X}$, so we obtain the first statement. The second statement follows from the fact that $\Phi = \Phi^+ \cup (-\Phi^+)$. □

**4.5. The Weyl group of the restricted root system.** Let $\beta \in V$ such that $(\beta, \beta) > 0$. Consider the idempotent linear map $p_\beta : V \to V$ defined by $p_\beta(\gamma) = \frac{(\beta,\gamma)}{(\beta,\beta)}\beta$ for all $\gamma \in V$; it is idempotent. Consider the involution

$$s_\beta := \mathsf{id} - 2p_\beta. \tag{4.18}$$

Note that the orthogonal decomposition $V = \mathbb{Q}\beta \oplus \{\lambda \in V : (\lambda, \beta) = 0\}$ diagonalizes $s_\beta$ and we refer to $s_\beta$ as the *orthogonal reflection* associated to $\beta$, respectively. Because $p_\beta$ is self-adjoint, so is $s_\beta$; we conclude that $s_\beta$ is an isometry. Finally, $s_\gamma = s_\beta$ for all $\gamma \in \mathbb{Q}_{\neq 0}\beta$.

If in addition $\beta \in V^\sigma$ for some root system involution $\sigma$ then one easily checks that $p_\beta$ commutes with $\sigma$, so that $s_\beta$ commutes with $\sigma$. Hence $s_\beta$ stabilizes $V^\sigma$ and is therefore an isometry of $V^\sigma$. If $f$ is any isometry of $V^\sigma$ then

$$f \circ s_\beta \circ f^{-1} = s_{f(\beta)}. \tag{4.19}$$

DEFINITION 4.9. The *Weyl group of* $\overline{\Phi}$ is the following group:

$$W(\overline{\Phi}) := \langle\{s_\beta|_{V^\sigma} : \beta \in \overline{\Phi}, (\beta, \beta) > 0\}\rangle \leq \mathsf{GL}(V^\sigma). \tag{4.20}$$

We call the elements $\overline{s_i} := s_{\overline{\alpha_i}}|_{V^\sigma}$ with $i \in \widetilde{I}$ *simple restricted reflections*.

The following basic property will later help us show that the set $\{\overline{s_i}\}_{i\in\widetilde{I}}$ generates $W(\overline{\Phi})$.

LEMMA 4.10. *For all* $i \in \widetilde{I}$, $\overline{s_i}$ *stabilizes* $\overline{\Phi^+}\backslash\mathbb{Q}\overline{\alpha_i}$.

PROOF. This is a variation on a standard argument, see e.g. [**Hu90**, Prop. 1.4]. Let $i \in \widetilde{I}$ be arbitrary and let $\beta \in \overline{\Phi^+}\backslash\mathbb{Q}\overline{\alpha_i}$. By Lemma 4.4, there exist $m_j \in \mathbb{Z}_{\geq 0}$ such that

$$\beta = \sum_{j\in I^*} m_j\overline{\alpha_j}. \tag{4.21}$$

Since $\beta \notin \mathbb{Q}\overline{\alpha_i}$ there exists $j \in I^*\backslash\{i\}$ such that $m_j \in \mathbb{Z}_{>0}$. At the same time $\overline{s_i}(\beta) = \beta - 2\frac{(\overline{\alpha_i},\beta)}{(\overline{\alpha_i},\overline{\alpha_i})}\overline{\alpha_i}$ with the same coefficient $m_j$. By the second statement of Lemma 4.4, $\overline{s_i}(\beta) \in \overline{\Phi^+}$. Finally, if $\overline{s_i}(\beta) \in \mathbb{Q}\overline{\alpha_i}$ then applying $\overline{s_i}$ produces a contradiction with $\beta \notin \mathbb{Q}\overline{\alpha_i}$. □

**4.6. The group $W^\sigma$ and the restricted Weyl group $\overline{W}$.** For any linear map $\psi : V \to V$ and any subset $S \subseteq W$ we may consider the following subset of $W$:

$$S^\psi = \{w \in S : w \circ \psi = \psi \circ w \text{ as linear maps on } V\}. \tag{4.22}$$

Given a root system automorphism $\sigma$, for all $w \in W^\sigma$, $w$ stabilizes $V^\sigma$ and we may consider $w|_{V^\sigma}$.

Definition 4.11 ([**Sch69, He84**]). Given a root system involution $\sigma$ of $\Phi$, we define the *restricted Weyl group* as

$$\overline{W} := \{w|_{V^\sigma} : w \in W^\sigma\}.$$

We may consider the groups $W(\overline{\Phi})$ and $\overline{W}$ for any root system involution. It is natural to expect that for suitable $\sigma$ they are isomorphic. In the remainder of this section we will always assume

$$\sigma = w_X \circ \tau \tag{4.23}$$

for some compatible decoration $(X, \tau)$. We will show in the remainder of this section that for certain compatible decorations both $W(\overline{\Phi})$ and $\overline{W}$ are isomorphic to a third combinatorially defined group $\widetilde{W} \leq W^\sigma$ (and hence to each other, as desired). In order to do this we develop some more properties of $W^\sigma$ and $\overline{W}$. Note that

$$W^\sigma = \{w \in W : w_X w = \tau(w) w_X\} \tag{4.24}$$

and that any $w \in W_X$ satisfies $w_X w w_X = \mathrm{oi}_X(w) = \tau(w)$, so that $W_X \leq W^\sigma$.

Lemma 4.12. *The group $W_X$ is the kernel of the homomorphism $\cdot|_{V_\sigma} : W^\sigma \to \overline{W}$.*

Proof. The proof of [**He84**, Prop. 3.1], which relies on [**Bo68**, Ch. V, §3.3, Prop. 1] and Lemma 4.8, immediately generalizes to the Kac-Moody setting. Also cf. [**Lo80**, Thm. 1.3]. □

As a consequence of Lemma 4.12, $W_X$ is a normal subgroup of $W^\sigma$ and

$$W^\sigma \cong W_X \rtimes \overline{W}. \tag{4.25}$$

Consider the set of *minimal left coset representatives* of $W_X$:

$$W^X = \{w \in W : \forall j \in X\ \ell(s_j w) > \ell(w)\}. \tag{4.26}$$

Because $\tau$ permutes $\{s_j\}_{j \in X}$ and preserves the length function on $W$, the set $W^X$ is also $\tau$-stable. By the natural generalization of the arguments in e.g. [**Hu90**, Prop. 1.10] to infinite Coxeter groups, one has

$$W = W_X \cdot W^X, \qquad W_X \cap W^X = \{1\}, \tag{4.27}$$

i.e. for all $w \in W$ there exists a unique $(u, v) \in W_X \times W^X$ such that $w = uv$. Since $\sigma$ fixes $W_X$ pointwise, we deduce

$$W^\sigma = W_X \cdot (W^X)^\sigma, \qquad W_X \cap (W^X)^\sigma = \{1\}. \tag{4.28}$$

For two subsets $S, S' \subseteq W$, we write $N_S(S') = \{w \in S | wS' = S'w\}$ for the normalizer of $S'$ in $S$. We will now show that $(W^X)^\sigma$ can be identified with the $\tau$-fixed point subset in

$$\mathcal{W} := N_{W^X}(W_X). \tag{4.29}$$

Lemma 4.13. $(W^X)^\sigma = \mathcal{W}^\tau$.

Proof. Let $w \in (W^X)^\sigma$ be arbitrary. Because $w$ commutes with $\sigma$ and $W_X$ is normal,

$$w_X \cdot \tau(w) = w \cdot w_X = u \cdot w \tag{4.30}$$

for some $u \in W_X$. Since decompositions in $W = W_X \cdot W^X$ are unique, it follows that $\tau(w) = w$ so that $(W^X)^\sigma \subseteq W^\tau$. Since $W_X$ is a normal subgroup of $W^\sigma$ we obtain $(W^X)^\sigma \subseteq \mathcal{W}^\tau$.

To prove the reverse inclusion, assume that $w \in \mathcal{W}^\tau$ so that $wW_Xw^{-1} = W_X$. For all $u \in W_X$ there exists a unique $v \in W_X$ such that $wu = vw$; since $w \in W^X$ we obtain $\ell(v) \le \ell(u)$. On the other hand, $W_X$ is finite so conjugation by $w$ is surjective. It follows that $ww_X = w_Xw$. Since $\tau(w) = w$, (4.24) now implies that $w \in W^\sigma$. □

By a statement in [**Lu03**, 25.1], which relies on [**Lu03**, Prop. 1.5, Lemma 2.2], the set $\mathcal{W}$ is a group. Comparing (4.25) and (4.28), making use of Lemma 4.13 we obtain

PROPOSITION 4.14. $\overline{W} \cong \mathcal{W}^\tau$.

**4.7. A combinatorial prescription of the simple restricted reflections: the group $\widetilde{W}$.** Recall that $W(\overline{\Phi})$ was not defined as a subgroup of $W$. We now discuss a particular subgroup $\widetilde{W} \le W$ defined in terms of explicit generators $\widetilde{s}_i$, see e.g. [**Lu76**, §5], [**Lu03**, Ch. 25] and [**GI14**], with the aim of having an isomorphic copy of $W(\overline{\Phi})$ in $W$. In particular we will strive to choose the $\widetilde{s}_i \in W$ so that $\widetilde{s}_i|_{V^\sigma} = \overline{s_i}$. Having achieved this, we will obtain a straightforward proof that $W(\overline{\Phi})$ is generated by the $\overline{s_i}$.

Note that the simple restricted reflection $\overline{s_i}$ fixes pointwise $\{\beta \in V^\sigma : (\beta, \overline{\alpha_i}) = 0\}$. The following lemma justifies selecting $\widetilde{s}_i$ from $W_{X[i]}$.

LEMMA 4.15. *Let $i \in I^*$. If $w \in W_{X[i]}$ stabilizes $V^\sigma$ then $w$ fixes pointwise $\{\beta \in V^\sigma : (\beta, \overline{\alpha_i}) = 0\}$.*

PROOF. Recall that $W_{X[i]}$ is generated by $\{s_j\}_{j\in X} \cup \{s_i, s_{\tau(i)}\}$. If $j \in X$ then $\overline{\alpha_j} = 0$ so that the orthogonality of the decomposition (4.3) implies that $s_j$ fixes $V^\sigma$ pointwise (and hence also the relevant subspace). Now let $\beta \in V^\sigma$ such that $(\beta, \overline{\alpha_i}) = 0$. It remains to prove that $s_i$ and $s_{\tau(i)}$ fix $\beta$. From (4.8) it follows that $\tau(\beta) = \beta$. Hence $(\beta, \overline{\alpha_i}) = 0$ implies $(\beta, \alpha_i) = (\beta, \alpha_{\tau(i)}) = 0$ so that both $s_i$ and $s_{\tau(i)}$ fix $\beta$. □

Recall that for $i \in I^*\backslash\widetilde{I}$ we have not defined a simple restricted reflection $\overline{s_i}$. Indeed, the following lemma shows that there are no involutions in $W$ that negate $\overline{\alpha_i}$.

LEMMA 4.16. *Suppose that $i \in I^*$ is such that $X[i]$ is of infinite type. No involutive element of $W$ sends $\overline{\alpha_i}$ to $-\overline{\alpha_i}$.*

PROOF. Lemma 4.4 implies that $(V_{X[i]})^\sigma$ is 1-dimensional, so that $(\overline{\alpha_i}, \overline{\alpha_i}) \le 0$ by Lemma 4.2 (2). Now suppose there is an involution $w \in W$ such that $w(\overline{\alpha_i}) = -\overline{\alpha_i}$. By [**Ri82**, Thm. A(a)][4] there exists $v \in W$ and $Z \subseteq I$ of finite type such that $w = v \circ w_Z \circ v^{-1}$. Hence $v^{-1}(\overline{\alpha_i})$ is fixed by $-w_Z$. Since $W_Z$ maps $Q^+\backslash Q_Z$ to $Q^+$, it follows that $v^{-1}(\overline{\alpha_i}) \in V_Z$. Since $Z$ is of finite type, the bilinear form $(\ ,\ )$ is positive definite on $V_Z$, so that $(\overline{\alpha_i}, \overline{\alpha_i}) = (v^{-1}(\overline{\alpha_i}), v^{-1}(\overline{\alpha_i})) > 0$, contradicting the first statement of the Lemma. □

Given $i \in I^*$, we now engineer the desired element $\widetilde{s}_i$. There are two clear desirable properties: $\widetilde{s}_i$ should commute with $\sigma$ (so that it stabilizes $V^\sigma$) and should send $\overline{\alpha_i} \in V_{X[i]}$ to $-\overline{\alpha_i}$ (hence $\widetilde{s}_i$ will be involutive). By Lemma 4.15, it is natural to choose $\widetilde{s}_i \in W_{X[i]}$. Since the longest element acts as multiplication by

[4]This is a special case of the result [**Sp85**, Prop. 3.3] which we used in Section 2.8.

$-1$ on $V_{X[i]}$ up to a diagram automorphism, it is natural to set $\widetilde{s}_i = u \cdot w_{X[i]}$ for some $u \in W_{X[i]}$. The two required properties are now seen to be equivalent to

$$(4.31) \qquad \alpha_i + \sigma(\alpha_i) \in (Q_{X[i]})^{u \circ \mathsf{oi}_{X[i]}}, \qquad u \cdot w_{\mathsf{oi}_{X[i]}(X)} = w_X \cdot \tau(u).$$

If we choose $u \in W_X$ then $u \circ \mathsf{oi}_{X[i]}$ maps $\alpha_i + \sigma(\alpha_i) \in Q^+ \backslash Q_X$ to $Q^+ \backslash Q_X$, which supports the first condition in (4.31). A natural choice for $u$ is now $u = w_X$, since in this case $\widetilde{s}_i$ has minimal length. For this choice the last condition in (4.31) amounts to $\mathsf{oi}_{X[i]}(w_X) = w_X$. Assuming this and recalling Lemma 2.28, we see that the first condition in (4.31) is satisfied. Hence we are led to the following definition.

DEFINITION 4.17 (See e.g. [**Lu03**, 25.1] and [**GI14**, Rmk. 8]). We define $\widetilde{W} := \langle \{\widetilde{s}_i\}_{i \in \widetilde{I}} \rangle \leq W$ where

$$(4.32) \qquad \widetilde{s}_i := w_X \cdot w_{X[i]} \in W_{X[i]}, \qquad i \in \widetilde{I}.$$

Since the $\widetilde{s}_i$ are fixed by $\tau$ in $\mathsf{Aut}(A) \ltimes W$, we immediately obtain

$$(4.33) \qquad \widetilde{W} \leq W^\tau.$$

EXAMPLE 4.18. It is instructive to spell out the generator $\widetilde{s}_i$ in certain typical configurations. Let $i \in \widetilde{I}$.

(1) If $i \in X^\perp$ then $\tau(i) \in X^\perp$ by Lemma 2.3 (1). In this case $w_{X[i]} = w_X \cdot w_{\{i,\tau(i)\}}$ so that $\widetilde{s}_i = w_{\{i,\tau(i)\}}$. There are now three possibilities: $\tau(i) = i$ (i.e. $i \in I_{\mathsf{ns}}$), so that $\{i, \tau(i)\}$ is of type $\mathsf{A}_1$ and $\widetilde{s}_i = s_i$; $a_{i\tau(i)} = 0$ (i.e. $\tau(i) \neq i$ and $i \notin I_{\mathsf{diff}}$), so that $\{i, \tau(i)\}$ is of type $\mathsf{A}_1 \times \mathsf{A}_1$ and $\widetilde{s}_i = s_i s_{\tau(i)}$; $a_{i\tau(i)} = 1$, so that $\{i, \tau(i)\}$ is of type $\mathsf{A}_2$ and $\widetilde{s}_i = s_i s_{\tau(i)} s_i$.
(2) Suppose that $\tau(i) = i$ and that $X[i]$ is of rank 2. Hence $X = \{j\}$ and $a_{ij}a_{ji} \in \{0,1,2,3\}$. It follows that

$$\widetilde{s}_i = w_{\{j\}} \cdot w_{\{i,j\}} \in \{s_i, s_i s_j, s_i s_j s_i, s_i s_j s_i s_j s_i\},$$

respectively. Note that $\widetilde{s}_i$ is an involution except if $a_{ij}a_{ji} = 1$.

The following key result shows that, given the definition (4.32), the desirable properties of $\widetilde{s}_i$ are equivalent *to each other* and to the condition that $(X, \tau)$ is a generalized Satake diagram. It extends the result [**RV20**, Thm. 2.1] which itself combined and extended results in [**He84**] and [**Lu03**].

PROPOSITION 4.19. *The following statements are equivalent:*

(1) $\widetilde{W} \leq W^\sigma$;
(2) $\widetilde{s}_i$ *is an involution for all* $i \in \widetilde{I}$;
(3) $\{s_j\}_{j \in X}$ *is stable under conjugation by elements of* $\widetilde{W}$;
(4) $\Phi_X^+$ *is* $\widetilde{W}$*-stable;*
(5) $(X, \tau)$ *is a generalized Satake diagram;*
(6) $\widetilde{W}$ *stabilizes* $V^\sigma$;
(7) $\widetilde{s}_i(\overline{\alpha_i}) = -\overline{\alpha_i}$ *for all* $i \in \widetilde{I}$;
(8) $\widetilde{W} \leq N_W(W_X)$.

PROOF. Note that each statement except (5) is a statement about the finite group $W_{X[i]}$; for instance, (6) is equivalent to the statement that for all $i \in \widetilde{I}$, $\widetilde{s}_i$ stabilizes $V^\sigma$. Furthermore, recall Definition 3.4 (1). Since the excluded connected component ○—● is of finite type, it follows that $(X, \tau)$ is a generalized Satake diagram if and only if for all $i \in \widetilde{I}$ such that $\tau(i) = i$ the connected component of the subdiagram $X[i]$ containing $i$ is not of type $\mathsf{A}_2$. Hence all statements in Proposition 4.19, including (5), can be written in the form "For all $i \in \widetilde{I}$, ...". It therefore suffices to prove the equivalence of the statements for a given fixed $i \in \widetilde{I}$. We can refer to the proof of the finite-type result [**RV20**, Thm. 2.1] for the equivalence of statements (1), (5) and (7), as well as the statement

$$w_X \cdot w_{X[i]} = w_{X[i]} \cdot w_X, \tag{4.34}$$

which is easily seen to be equivalent to (2). By (2.13) (with $X$ replaced by $X[i]$), (4.34) can be rephrased as the condition

$$\mathsf{oi}_{X[i]}(w_X) = w_X \tag{4.35}$$

which we encountered above, and which in turn is equivalent to (3). By (2.14), (4.35) is equivalent to

$$\mathsf{oi}_{X[i]}(X) = X. \tag{4.36}$$

In turn, this is equivalent to

$$w_{X[i]}(\Phi_X^+) = -\Phi_X^+. \tag{4.37}$$

Since also $w_X(\Phi_X^+) = -(\Phi_X^+)$, (4.37) is equivalent to (4).

In order prove that the remaining statements (6) and (8) are equivalent to the others, it remains to establish some implications.

(1) $\Rightarrow$ (6): This is clear.

(3) $\Rightarrow$ (8): This is clear.

(6) $\Rightarrow$ (5) and (8) $\Rightarrow$ (5): We prove the contrapositives. Suppose ○–● is a connected component of $X[i]$. Then $\tau(i) = i$ and there exists $j \in X$ such that $w_{X[i]}|_{\mathsf{Sp}_{\mathbb{Q}}\{\alpha_i,\alpha_j\}}$ equals $s_i s_j s_i = s_j s_i s_j$. Hence $\widetilde{s}_i = w_{\{i,j\}} w_{\{i\}} = s_i s_j$. Note that $\sigma|_{\mathsf{Sp}_{\mathbb{Q}}\{\alpha_i,\alpha_j\}} = s_j$. We obtain $\overline{\alpha_i} = \alpha_i + \frac{1}{2}\alpha_j \in V^\sigma$. It follows that $\widetilde{s}_i(\overline{\alpha_i}) = s_i(\alpha_i + \frac{1}{2}\alpha_j) = \frac{1}{2}(\alpha_j - \alpha_i)$ which is not fixed by $\sigma$, as required. Furthermore, conjugation by $w_{X[i]}$ maps $s_j$ to $s_i$, so that $w_{X[i]}$ does not normalize $W_X$. □

REMARK 4.20. Suppose $(X, \tau)$ is *not* a generalized Satake diagram. Let $i \in I^*$ be such that $X[i]$ has ○–● as a connected component.

(1) Note that $\widetilde{s}_i$ does not stabilize $V^\sigma$, but it does stabilize the subspace $\{\beta \in V^\sigma : (\beta, \overline{\alpha_i}) = 0\}$, so there is no contradiction between Lemma 4.15 and Proposition 4.19 (6).
(2) The generator $\widetilde{s}_i$ is of order 3. Hence in this case $\widetilde{W}$ is a (possibly infinite) complex reflection group. We will not pursue this further here.

**4.8. The group $W(\overline{\Phi})$ revisited.** Note that Lemma 4.15 and Proposition 4.19 (6)-7 combine to imply that, for all $i \in \widetilde{I}$,

$$\widetilde{s}_i|_{V_\sigma} = \overline{s_i}. \tag{4.38}$$

As a consequence we obtain a group map $\phi : \widetilde{W} \to W(\overline{\Phi})$. We now proceed as in [**DK19**, Sec. 2.3] to show that $\phi$ is an isomorphism. Denote by $W'(\overline{\Phi})$ the subgroup of $W(\overline{\Phi})$ generated by $\{\overline{s_i}\}_{i \in \widetilde{I}}$. The following lemma is similar to a key technical step in the standard proof that Weyl groups of root systems are generated by the simple reflections, see e.g. [**Hu90**, Thm. 1.5, Step (1) of proof], but we feel it is beneficial to go through it carefully in the restricted Kac-Moody setting.

LEMMA 4.21. *Let $\beta \in \overline{\Phi^+}$ be and assume that $(\beta, \beta) > 0$. Then the $W'(\overline{\Phi})$-orbit through $\beta$ intersects $\mathbb{Z}_{\geq 0}\overline{\alpha_i}$ for some $i \in \widetilde{I}$.*

PROOF. Denote the orbit under consideration by $[\beta]$. By Lemma 4.7, if $\overline{\alpha_i} \in [\beta]$ then $i \in \widetilde{I}$. Hence it is sufficient to prove that $[\beta]$ intersects $\mathbb{Z}_{\geq 0}\overline{\alpha_i}$ for some $i \in I^*$.

Since $[\beta] \cap \overline{\Phi^+}$ contains $\beta$, it is nonempty. Now pick any $\mu \in \Phi^+$ such that $\overline{\mu}$ is an element of $[\beta] \cap \overline{\Phi^+}$. By Lemma 4.4, $\overline{\mu} = \sum_{j \in I^*} m_j \overline{\alpha_j}$ with $m_j \in \mathbb{Z}_{\geq 0}$. Since $(\overline{\mu}, \overline{\mu}) = (\beta, \beta) > 0$, there exists $i \in I^*$ such that $(\overline{\mu}, \overline{\alpha_i}) > 0$. If $\overline{\mu} \in \mathbb{Z}_{\geq 0}\overline{\alpha_i}$, we are done. Otherwise consider $\overline{s_i}(\overline{\mu})$ which lies in $\overline{\Phi^+} \backslash \mathbb{Q}\overline{\alpha_i}$ by Lemma 4.10. We obtain

$$\overline{s_i}(\overline{\mu}) - \overline{\mu} = -2\frac{(\overline{\alpha_i}, \overline{\mu})}{(\overline{\alpha_i}, \overline{\alpha_i})}\overline{\alpha_i} \in \mathbb{Q}_{<0}\overline{\alpha_i}. \tag{4.39}$$

On the other hand,

$$\overline{s_i}(\overline{\mu}) - \overline{\mu} = \widetilde{s}_i(\overline{\mu}) - \overline{\mu} = \overline{\widetilde{s}_i(\mu) - \mu} \tag{4.40}$$

as a consequence of (4.38) and Proposition 4.19 (1). Since $\widetilde{s}_i(\mu) - \mu \in \mathsf{Sp}_{\mathbb{Z}}\{\alpha_j\}_{j \in X[i]}$ it follows that $\overline{s_i}(\overline{\mu}) - \overline{\mu} \in \mathsf{Sp}_{\mathbb{Z}}\{\overline{\alpha_j}\}_{j \in X[i]} = \mathbb{Z}\overline{\alpha_i}$. Now (4.39) implies that $\overline{s_i}(\overline{\mu}) - \overline{\mu}$ lies in $\mathbb{Z}_{<0}\overline{\alpha_i}$, so that $\overline{s_i}(\overline{\mu}) \in \overline{\Phi^+}$ has smaller restricted height than $\overline{\mu}$.

We repeat the above steps with $\overline{\mu}$ replaced by $\overline{s_i}(\overline{\mu})$. This process terminates, since the restricted simple roots are precisely the elements of $\overline{\Phi^+}$ with restricted height 1 so that at some point an element of $\mathbb{Z}_{>0}\overline{\alpha_i}$ is obtained for some $i \in I^*$. □

PROPOSITION 4.22. *The group $W(\overline{\Phi})$ is generated by $\overline{s_i}$ for $i \in \widetilde{I}$.*

PROOF. We need to prove that $W'(\overline{\Phi})$ equals $W(\overline{\Phi})$. From Lemma 4.21 it follows that

$$\{\beta \in \overline{\Phi^+} : (\beta, \beta) > 0\} \tag{4.41}$$

lies in the $W'(\overline{\Phi})$-image of $\cup_{i \in \widetilde{I}} \mathbb{Z}_{>0}\overline{\alpha_i}$. Since elements of $W(\overline{\Phi})$ are isometries, by (4.19) we obtain that for all $\beta \in \overline{\Phi^+}$ such that $(\beta, \beta) > 0$ there exist $w \in W'(\overline{\Phi})$, $i \in \widetilde{I}$ and $m \in \mathbb{Z}_{>0}$ such that

$$s_\beta = w s_{m\overline{\alpha_i}} w^{-1}. \tag{4.42}$$

We complete the proof by observing that $s_{-\beta} = s_\beta$ and $s_{m\overline{\alpha_i}} = \overline{s_i}$. □

Proposition 4.22 shows that $W(\overline{\Phi})$ is generated by $\{\widetilde{s}_i\}_{i \in \widetilde{I}}$ so that the map $\phi$ is surjective. The following result proves that the map $\phi$ is injective.

LEMMA 4.23. *If $(X, \tau)$ is a generalized Satake diagram, the action of $\widetilde{W}$ on $\overline{\Phi}$ is faithful.*

PROOF. We give a simplified version of the proof of [**DK19**, Lem. 2.8]. Suppose that $w \in \widetilde{W}$ fixes $\overline{\Phi}$ pointwise. We need to prove that $w = 1$. In particular, $w(\overline{\alpha_i}) = \overline{\alpha_i}$ for all $i \in I^*$, i.e.

$$w(\alpha_i + \sigma(\alpha_i)) = \alpha_i + \sigma(\alpha_i) \qquad \text{for all } i \in I^*. \tag{4.43}$$

Recalling that $\sigma = w_X \circ \tau$ and using Lemma 2.3 (3) we obtain

$$\sigma(\alpha_i) = \alpha_{\tau(i)} + \sum_{j \in X} v_{ij}\alpha_j \tag{4.44}$$

with $v_{ij} \in \mathbb{Z}_{\geq 0}$. Since the generators of $\widetilde{W}$ commute with $\tau$ and act on $\Pi_X := \{\alpha_j\}_{j\in X}$ as a diagram automorphism, we deduce from (4.43) the relation

$$w(\alpha_i) + \tau(w(\alpha_i)) + \sum_{j\in X} v_{ij} w(\alpha_j) = \alpha_i + \alpha_{\tau(i)} + \sum_{j\in X} v_{ij}\alpha_j. \tag{4.45}$$

Now applying $^-$ and using (4.6-4.7), we obtain $w(\alpha_i) = \alpha_i$ for all $i \in I^*$. Using once again that $w$ commutes with $\tau$ and acts on $\Pi_X$ as a diagram automorphism, it follows that $w$ stabilizes $\Pi$. From $\mathsf{Aut}(\Phi) = \mathsf{Out}(A) \ltimes W$ we deduce $w = 1$. $\square$

We can now identify the two groups $\widetilde{W}$ and $W(\overline{\Phi})$.

THEOREM 4.24. *$\widetilde{W} \cong W(\overline{\Phi})$ if and only if $(X, \tau)$ is a generalized Satake diagram.*

PROOF. If $(X, \tau)$ is a generalized Satake diagram, by Proposition 4.22 and Lemma 4.23, the map $\phi$ is a group isomorphism. It only remains to prove the "only if" part of the first statement, which can be done as before via the contrapositive. Suppose $(X, \tau)$ is not a generalized Satake diagram. By Proposition 4.19, there exists $i \in \widetilde{I}$ and $j \in X$ such that $\widetilde{s}_i = s_i s_j$ is of order 3. It suffices to prove that $\widetilde{W} \cap W_{X[i]} = \langle \widetilde{s}_i \rangle$ is not isomorphic to $W(\overline{\Phi_{\{i,j\}}})$. This is clear since by Lemma 4.4 $W(\overline{\Phi_{\{i,j\}}}) = W(\{\pm\overline{\alpha_i}\})$ is a group of two elements. $\square$

As a consequence, we obtain that $\overline{\Phi}$ is a root system in the following sense:

COROLLARY 4.25. *If $(X, \tau)$ is a generalized Satake diagram, then $\overline{\Phi}$ is $W(\overline{\Phi})$-stable.*

PROOF. Let $i \in \widetilde{I}$ be arbitrary. In view of Theorem 4.24 it suffices to prove that $\overline{\Phi}$ is stable under the action of $\widetilde{s}_i|_{V^\sigma}$. Note that an arbitrary element of $\overline{\Phi}$ is of the form $\overline{\lambda}$ with $\lambda \in \Phi\backslash\Phi_X$. Then $\widetilde{s}_i(\overline{\lambda}) = \overline{\widetilde{s}_i(\lambda)}$ by Proposition 4.19 (1). Combining $\lambda \in \Phi\backslash\Phi_X$ with Proposition 4.19 (4), we obtain $\widetilde{s}_i(\lambda) \in \Phi\backslash\Phi_X$ so that $\overline{\widetilde{s}_i(\lambda)} \in \overline{\Phi}$ as required. $\square$

The converse to Corollary 4.25 is not true: $\overline{\Phi}$ is stable under $W(\overline{\Phi})$ also if $(X, \tau)$ equals ○—● (cf. [**He84**, Thm. 6.1], which is a result for the group $\overline{W}$).

**4.9. The restricted Weyl group as a Coxeter group.** It turns out that the condition that $(\widetilde{W}, \{\widetilde{s}_i\})_{i\in\widetilde{I}}$ is a Coxeter system is also equivalent to $(X, \tau)$ being a generalized Satake diagram. We give a brief synopsis of the proof of this in the approaches of [**Lu03**, Appendix] and [**GI14**]. In particular, we highlight the following property as it identifies where in this approach the condition that $(X, \tau)$ is a generalized Satake diagram is necessary.

LEMMA 4.26. *$\widetilde{W} \leq \mathcal{W}^\tau$ if and only if $(X, \tau)$ is a generalized Satake diagram.*

PROOF. By Proposition 4.19, $\widetilde{W} \leq N_W(W_X)$ if and only if $(X, \tau)$ is a generalized Satake diagram. Combining this with (4.33), it remains to prove that $\ell(s_j\widetilde{s}_i) > \ell(\widetilde{s}_i)$ for all $i \in \widetilde{I}$ and $j \in X$, assuming that $(X, \tau)$ is a generalized Satake diagram. For all $j \in X$, $s_j\widetilde{s}_i = \widetilde{s}_i s_{(\mathsf{oi}_X \mathsf{oi}_{X[i]})(j)}$ with $(\mathsf{oi}_X\mathsf{oi}_{X[i]})(j) \in X$. Therefore it suffices to prove that $\ell(\widetilde{s}_i w_j) > \ell(\widetilde{s}_i)$ for all $j \in X$, which by a basic property of Coxeter groups is equivalent to $\widetilde{s}_i$ sending $\{\alpha_i\}_{i\in X}$ into $\Phi^+_{X[i]}$. This follows directly from $\widetilde{s}_i(\alpha_j) = \alpha_{(\mathsf{oi}_X\mathsf{oi}_{X[i]})(j)}$ for all $j \in X$. $\square$

The inductive proof of [**GI14**, Lem. 2] can be directly generalized to the current setting, which yields the following statement.

LEMMA 4.27. *For all $w \in \mathcal{W}^\tau$ there exists $(i_1, \ldots, i_{\ell(w)}) \in \widetilde{I}^{\ell(w)}$ such that*

$$w = \widetilde{s}_{i_1} \cdots \widetilde{s}_{i_{\ell(w)}}. \tag{4.46}$$

*Furthermore, if $i \in I$ satisfies $\ell(s_i w) < \ell(w)$ then we can choose $(i_1, \ldots, i_{\ell(w)})$ so that $i \in X[i_1]$.*

We obtain from Lemmas 4.26 and 4.27 the following result.

COROLLARY 4.28. *$\widetilde{W} \cong \mathcal{W}^\tau$ if and only if $(X, \tau)$ is a generalized Satake diagram.*

The following desirable property of $\widetilde{W}$ can now be obtained.

THEOREM 4.29. *The group $\widetilde{W}$ is a Coxeter group if and only if $(X, \tau)$ is a generalized Satake diagram. In this case, with $\widetilde{\ell}$ denoting the length function for this Coxeter system, we have*

$$\widetilde{\ell}(ww') = \widetilde{\ell}(w) + \widetilde{\ell}(w') \quad \Longleftrightarrow \quad \ell(ww') = \ell(w) + \ell(w') \qquad \text{for all } w, w' \in \widetilde{W}.$$

PROOF. We may proceed as in [**GI14**, Lem. 4 - Prop. 7] (also cf. [**Hé91**] and [**Lu03**, Appendix]) by replacing $W^\tau$ by $\mathcal{W}^\tau$ according to [**GI14**, Rmk. 8]. In this approach, $(\widetilde{W}, \{\widetilde{s}_i\}_{i \in \widetilde{I}})$ is shown to be a Coxeter system via the characterization that the exchange condition is satisfied, see e.g. [**Bo68**, Ch. IV, §1.6, Thm. 1]. We leave the details to the reader.

To complete the proof, we also need to show that $(\widetilde{W}, \{\widetilde{s}_i\}_{i \in \widetilde{I}})$ is not a Coxeter system if $(X, \tau)$ is not a generalized Satake diagram, which follows immediately from Proposition 4.19. □

Note that Proposition 4.14 and Corollary 4.28 combine to yield the following final identification.

THEOREM 4.30. *$\widetilde{W} \cong \overline{W}$ if and only if $(X, \tau)$ is a generalized Satake diagram.*

REMARK 4.31.

(1) We leave it to the reader to check that the results in this section extend to non-crystallographic Coxeter groups, if we make the following adjustment: call a compatible decoration $(X, \tau)$ a generalized Satake diagram if for all $i \in I \backslash X$ such that $\tau(i) = i$ the connected component of $X[i]$ containing $i$ is not a nontrivial odd dihedral group, i.e. not of type $\mathsf{I}_2(m)$ with $m > 1$ odd. In the crystallographic case only $\mathsf{I}_2(3) = \mathsf{A}_2$ can occur, which recovers Definition 3.4 (1).

(2) In the case $(X, \tau) =$ ○—● the three groups $\overline{W}$, $W(\overline{\Phi})$ and $\widetilde{W}$ are all distinct: their orders are 1, 2 and 3, respectively.

**4.10. Non-reduced and non-crystallographic root systems.** In Appendices A.3 and A.4 we also indicate the restricted root system, which can be non-reduced. In Table 1 below we recall the standard notation and Dynkin diagrams for non-reduced crystallographic root systems of finite and affine type. The notation is of the form $(X, Y)_n$ for $n \geq 1$; here $X_n$ is the type of the underlying reduced crystallographic root system which is obtained by deleting $2\beta$ for all roots $\beta$ such that

TABLE 1. Dynkin diagrams for non-reduced crystallographic finite and affine root systems. If and only if a node in such a Dynkin diagram is marked by x, the root system contains in addition to the corresponding simple root $\beta$ also the root $2\beta$.

| Type | Diagram | Constraints | Special low-rank case | |
|---|---|---|---|---|
| $(\mathsf{B},\mathsf{C})_n$ | 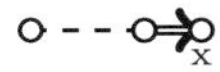 | $n\geq 1$ |  | $n=1$ |
| $(\widehat{\mathsf{B}},\widehat{\mathsf{B}}^\vee)_n$ | 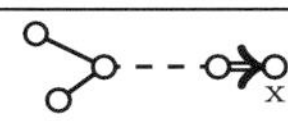 | $n\geq 2$ |  | $n=2$ |
| $(\widehat{\mathsf{C}}',\widehat{\mathsf{C}})_n$ |  | $n\geq 1$ |  | $n=1$ |
| $(\widehat{\mathsf{C}}^\vee,\widehat{\mathsf{C}})_n$ |  | $n\geq 1$ | 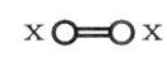 | $n=1$ |
| $(\widehat{\mathsf{C}}^\vee,\widehat{\mathsf{C}}')_n$ | 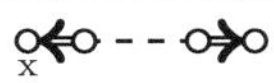 | $n\geq 1$ | 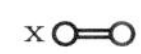 | $n=1$ |

$(\beta,\beta)>0$ and $Y_n$ is the type of the underlying reduced crystallographic obtained by deleting all roots $\beta$ such that $(\beta,\beta)>0$ and $2\beta$ is also a root.

Only in one finite-type and one affine-type case a generalized Satake diagram yields a non-crystallographic root system (which is also non-reduced). We consider these in detail now, also as examples of the computation of the restricted root system via subdiagrams of restricted rank 1.

EXAMPLE 4.32.

(1) Suppose $\mathfrak{g}$ is of type $\mathsf{G}_2$. Label the Dynkin diagram as follows: . Let $\sigma=s_1$. From

$$\Phi=\pm\{\alpha_1,\alpha_2,\alpha_1+\alpha_2,\alpha_1+2\alpha_2,\alpha_1+3\alpha_2,2\alpha_1+3\alpha_2\} \tag{4.47}$$

and

$$\begin{aligned}\overline{\alpha_2}=\overline{\alpha_1+\alpha_2}&=\tfrac{1}{2}\alpha_1+\alpha_2,\\ \overline{\alpha_1+2\alpha_2}&=\alpha_1+2\alpha_2,\\ \overline{\alpha_1+3\alpha_2}=\overline{2\alpha_1+3\alpha_2}&=\tfrac{3}{2}\alpha_1+3\alpha_2\end{aligned} \tag{4.48}$$

it follows that $\overline{\Phi}=\pm\{1,2,3\}\overline{\alpha_2}$, a non-reduced non-crystallographic root system of rank 1. We denote it[5] by $(\mathsf{B},\mathsf{C})_1^+$, since we may obtain it by adjoining $3\beta$ for all short roots $\beta$ to the non-reduced root system of type $(\mathsf{B},\mathsf{C})_1$.

(2) By adjoining a node corresponding to minus the highest short root to the Dynkin diagram of the previous example we obtain the affine Lie algebra of type $\widehat{\mathsf{G}}_2^\vee$. More precisely, we may label the nodes in the Dynkin diagram as follows: 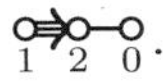.

Again, let $\sigma=s_1$. As before, $\overline{\alpha_2}=\frac{1}{2}\alpha_1+\alpha_2$ and straightforwardly we find $\overline{\alpha_0}=\alpha_0$. We may set $\epsilon_0=\epsilon_1=1$ and $\epsilon_2=3$. We compute the

[5]This case is exceptional in another way: the corresponding subalgebra $\mathfrak{k}$ is the only reductive one with $(X,\tau,\gamma)$ an enriched generalized Satake diagram of finite type such that $(X,\tau)$ is *not* a Satake diagram, see [**RV20**, Ex. 2 (ii)].

inner products:

$$(\overline{\alpha_0},\overline{\alpha_0}) = (\alpha_0,\alpha_0) = 2,$$
$$(4.49)\qquad (\overline{\alpha_0},\overline{\alpha_2}) = (\alpha_0,\alpha_2) = -1,$$
$$(\overline{\alpha_2},\overline{\alpha_2}) = (\alpha_2,\tfrac{1}{2}\alpha_1+\alpha_2) = \tfrac{1}{2}.$$

Hence $\overline{\alpha_2}$ and $\overline{\alpha_0}$ generate a crystallographic root system $\overline{\Phi}_{red}$ of type $\widehat{\mathsf{C}}'_1$. It is reduced in the sense that $\mathbb{Z}\beta\cap\overline{\Phi}_{red} = \{\pm\beta\}$ for all real roots $\beta \in \overline{\Phi}_{red}$. Integer multiples of $\overline{\alpha_2}$ arise in this root system: $\overline{\alpha_1+2\alpha_2} = 2\overline{\alpha_2}$ and $\overline{2\alpha_1+3\alpha_2} = 3\overline{\alpha_2}$. One verifies that $\overline{\Phi}$ is obtained from $\overline{\Phi}_{red}$ by adjoining $2\alpha$ and $3\alpha$ for all real short roots $\alpha$. Similar to the previous example, we denote this non-reduced non-crystallographic root system by $(\widehat{\mathsf{C}}',\widehat{\mathsf{C}})^+_1$.

## Appendix A. Classification of generalized Satake diagrams

This Appendix provides a classification of $\mathsf{Aut}(A)$-orbits of generalized Satake diagrams whose underlying Dynkin diagram is of finite or affine type[6]. We refer the reader to Definition 3.4 (1) for the definition of generalized Satake diagram and to (2.29) for the definition of the $\mathsf{Aut}(A)$-action on the set of generalized Satake diagrams. For $A$ of finite type we recover the classification obtained in [**He84**, Table I]. We have also indicated the special $\tau$-orbits (see Section 3.5) and odd nodes (see Section 2.8) by marking the corresponding orbits in the diagram $\mathsf{s}$ and $\mathsf{o}$, respectively. The absence of odd nodes in a diagram allows the reader easily to recognize Satake diagrams and to recover the classification obtained in [**Ar62**, §4 and §5] for $A$ of finite type (also see [**NS95**, Sec. 3], [**Le03**, Sec. 7] and [**He78**, page 2, page 34]) and in [**BBBR95**, Sec. 6] for $A$ of affine type.

For each diagram we indicate the type of the restricted root system $\overline{\Phi}$ (see Section 4). We also give the constraints on the discrete parameters determining the type and the description of $(X,\tau)$ in terms of these parameters. Finally, we also enumerate the special cases where the nature of the diagram is different from the general case (e.g. due to a different number of special $\tau$-orbits or odd nodes or a low rank case), in order of decreasing restricted rank.

**A.1. Notation.** We use the following descriptive notation for the generalized Satake diagrams:

$$(\mathrm{T})^{\text{type of }(X,\tau)}_{\text{description of }X}.$$

Here T denotes the type of the underlying Dynkin diagram. The other details are as follows:

- If T is of a classical Lie type, the type of $(X,\tau)$ can be plain (indicated by $\mathsf{pl}$), alternating ($\mathsf{alt}$), reflecting ($\mathsf{rfl}$) or rotational ($\mathsf{rot}$), as explained in Section 3.2. The description of $X$ is a listing of up to two nonnegative integers, except for copies of $\mathsf{A}_1$ for diagrams of alternating type. Each such integer $p$ indicates a connected component of type $\mathsf{A}_{p-1}$ (for reflecting and rotational types) or $\mathsf{B}_p$, $\mathsf{C}_p$ or $\mathsf{D}_p$ (for plain and alternating types).
- If T is of an exceptional Lie type, the type of $(X,\tau)$ amounts to specifying whether $\tau = \mathsf{id}$ (indicated by an absence of a symbol) or $\tau$ is a nontrivial involution (indicated by an apostrophe). The latter is only relevant if T is

[6] For completeness we always include the split and compact Satake diagrams $(\emptyset,\mathsf{id})$ and $(I,\mathsf{oi}_I)$.

$\mathsf{E}_6$, $\widehat{\mathsf{E}}_6$ or $\widehat{\mathsf{E}}_7$; in each of these cases there is a unique diagram automorphism of order 2 up to $\mathsf{Aut}(A)$-conjugacy. The description of $X$ is simply its Lie type. When $\mathrm{T} = \mathsf{G}_2$, it is supplemented by a superscript $\mathsf{lo}$ or $\mathsf{sho}$ indicating whether $X$, when it has a single element, corresponds to the long or short simple root of $\mathsf{G}_2$.

We hope this notation will allow the reader to identify the diagram more easily than the roman numeral notation commonly used for finite type and the notation in [BBBR95] used for affine type.

**A.2. Low-rank coincidences.** We will allow, in the infinite families of classical Lie type, the rank to go below the usual lower bounds with the following natural interpretations:

| Rank | Diagrams |
|---|---|
| 0 | $\mathsf{A}_{-1} = \mathsf{A}_0 = \mathsf{B}_0 = \mathsf{C}_0 = \mathsf{D}_0 = \mathsf{D}_1 = (\mathsf{B},\mathsf{C})_0 = \mathsf{Z}_0$, |
| 1 | $\mathsf{A}_1 = \mathsf{B}_1 = \mathsf{C}_1$, |
| 2 | $\mathsf{A}_1 \times \mathsf{A}_1 = \mathsf{D}_2$, $\quad \mathsf{B}_2 = \mathsf{C}_2$ |
| 3 | $\mathsf{A}_3 = \mathsf{D}_3$ |

Here $\mathsf{Z}_0$ denotes the type of the empty set (a finite root system) when considered in isolation (i.e. not as part of an infinite family).

Similarly, the set of nonzero integer multiples of a fixed nonzero vector is an affine root system, called the *basic imaginary root system*. We denote the type of this root system by $\widehat{\mathsf{Z}}_0$ if considered in isolation. In the infinite families of classical Lie type we will allow the rank to go below the usual lower bounds with the following natural interpretations:

| Rank | Diagrams |
|---|---|
| 0 | $\widehat{\mathsf{A}}_0 = \widehat{\mathsf{B}}_0 = \widehat{\mathsf{B}}_0^\vee = \widehat{\mathsf{C}}_0 = \widehat{\mathsf{C}}_0^\vee = \widehat{\mathsf{C}}_0' =$ |
| | $= (\widehat{\mathsf{B}},\widehat{\mathsf{B}}^\vee)_0 = (\widehat{\mathsf{C}}',\widehat{\mathsf{C}}^\vee)_0 = (\widehat{\mathsf{C}}^\vee,\widehat{\mathsf{C}}^\vee)_0 = (\widehat{\mathsf{C}}^\vee,\widehat{\mathsf{C}}')_0 = \widehat{\mathsf{Z}}_0$ |
| 1 | $\widehat{\mathsf{A}}_1 = \widehat{\mathsf{B}}_1 = \widehat{\mathsf{B}}_1^\vee = \widehat{\mathsf{C}}_1 = \widehat{\mathsf{C}}_1^\vee = (\widehat{\mathsf{B}},\widehat{\mathsf{B}}^\vee)_1$ |
| 2 | $\widehat{\mathsf{B}}_2 = \widehat{\mathsf{C}}_2$, $\quad \widehat{\mathsf{B}}_2^\vee = \widehat{\mathsf{C}}_2^\vee$ |
| 3 | $\widehat{\mathsf{A}}_3 = \widehat{\mathsf{D}}_3$ |

**A.3. Finite type.** Tables 2, 3 and 4 list the generalized Satake diagrams. For each Satake diagram we also give the standard notation in terms of roman numerals. Note that types $\mathsf{AIV}$ and $\mathsf{BDII}$ are distinguished from the larger families $\mathsf{AIII}$ and $\mathsf{BDI}$, respectively, by the condition $|I^*| = 1$.

Table 2: Generalized Satake diagrams of finite Lie type $\mathsf{A}_n$. We set $N := n + 1$.

| Type | | Diagram | Type of $\overline{\Phi}$ | Constraints |
|---|---|---|---|---|
| $(\mathsf{A}_n)^{\mathsf{pl}}$ | $\mathsf{AI}$ | 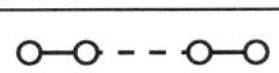 | $\mathsf{A}_n$ | $(n > 1)$ |
| | | $X = \emptyset$ | $\tau = \mathrm{id}$ | |

| Type | Diagram | Type of $\overline{\Phi}$ | Constraints |
|---|---|---|---|
| $(\mathsf{A}_n)^{\mathsf{alt}}$ AII |  | $\mathsf{A}_{\frac{n-1}{2}}$ | $n$ odd ($n > 1$) |
| | $X$ is of type $\mathsf{A}_1^{N/2}$ $\quad \tau = \mathsf{id}$ | | |
| $(\mathsf{A}_n)^{\mathsf{rfl}}_p$ AIII AIV | 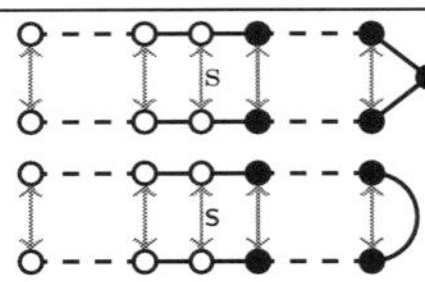 | $(\mathsf{B},\mathsf{C})_{\frac{N-p}{2}}$ | $0 \leq p \leq N$, $N-p$ even |
| | $X$ is of type $\mathsf{A}_{p-1}$ $\quad \tau$ has $\mathsf{mod}(p-1,2)$ fixed points | | |
| Special cases: | 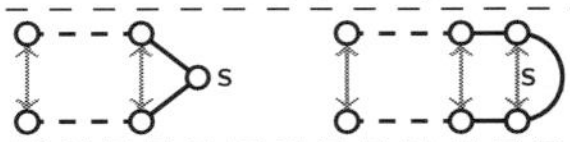 | $\mathsf{C}_{\frac{N}{2}}$, $(\mathsf{B},\mathsf{C})_{\frac{N-1}{2}}$ | $p < 1$ |
| |  | $\mathsf{Z}_0$ | $p = N$ |
| |  | $\mathsf{A}_1$, $\mathsf{Z}_0$ | $n = 1$, $p \in \{0,2\}$ |

Table 3: Generalized Satake diagrams of finite classical Lie type other than $\mathsf{A}_n$.

| Type | Diagram | Type of $\overline{\Phi}$ | Constraints |
|---|---|---|---|
| $(\mathsf{B}_n)^{\mathsf{pl}}_p$ BI BII |  | $\mathsf{B}_{n-p}$ | $0 \leq p \leq n$ |
| | $X$ is of type $\mathsf{B}_p$ $\quad \tau = \mathsf{id}$ | | |
| Special cases: |  | $\mathsf{C}_2$ | $p = n-2$ |
| $(\mathsf{B}_n)^{\mathsf{alt}}_p$ | 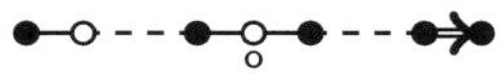 | $(\mathsf{B},\mathsf{C})_{\frac{n-p}{2}}$ | $0 \leq p < n$, $n-p$ even |
| | $X$ is of type $\mathsf{A}_1^{\frac{n-p}{2}} \times \mathsf{B}_p$ $\quad \tau = \mathsf{id}$ | | |
| Special cases: |  | $(\mathsf{B},\mathsf{C})_{\frac{n}{2}}$ | $p = 0$ |
| $(\mathsf{C}_n)^{\mathsf{pl}}_p$ |  | $(\mathsf{B},\mathsf{C})_{n-p}$ | $0 \leq p < n$ |
| | $X$ is of type $\mathsf{C}_p$ $\quad \tau = \mathsf{id}$ | | |
| Special cases: CI | 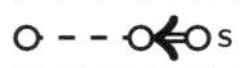 | $\mathsf{C}_n$ | $p = 0$ |
| |  | $(\mathsf{B},\mathsf{C})_2$ | $p = n-2$ |
| $(\mathsf{C}_n)^{\mathsf{alt}}_p$ CII | 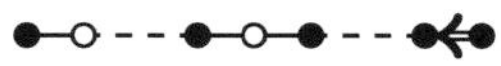 | $(\mathsf{B},\mathsf{C})_{\frac{n-p}{2}}$ | $0 \leq p \leq n$, $n-p$ even |
| | $X$ is of type $\mathsf{A}_1^{\frac{n-p}{2}} \times \mathsf{C}_p$ $\quad \tau = \mathsf{id}$ | | |
| Special cases: |  | $\mathsf{C}_{n/2}$ | $p = 0$ |
| |  | $\mathsf{Z}_0$ | $p = n$ |
| $(\mathsf{D}_n)^{\mathsf{pl}}_p$ DI DII | 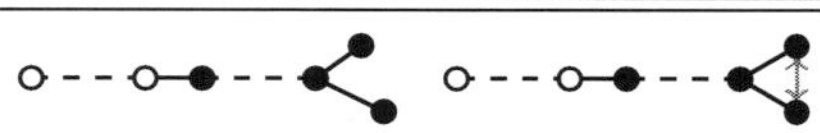 | $\mathsf{B}_{n-p}$ | $0 \leq p \leq n$ |
| | $X$ is of type $\mathsf{D}_p$ $\quad \tau$ has $\mathsf{mod}(p,2)$ nontrivial orbits | | |

| Type | Diagram | Type of $\overline{\Phi}$ | Constraints |
|---|---|---|---|
| Special cases: |  | $\mathsf{D}_n$, $\mathsf{B}_{n-1}$ | $p \in \{0, 1\}$ |
| |  | $\mathsf{C}_2$ | $p = n - 2$ |
| | 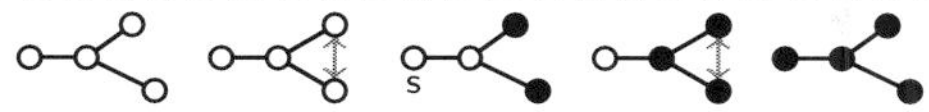 | $\mathsf{D}_4$, $\mathsf{B}_3$, $\mathsf{C}_2$, $\mathsf{A}_1$, $\mathsf{Z}_0$ | $n = 4$, $0 \le p \le 4$ |
| $(\mathsf{D}_n)^{\mathsf{alt}}_p$ | 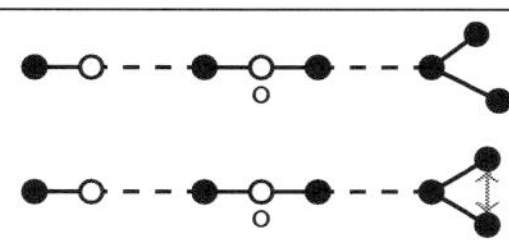 | $(\mathsf{B}, \mathsf{C})_{\frac{n-p}{2}}$ | $0 \le p < n$, $n - p$ even |
| | $X$ is of type $\mathsf{A}_1^{\frac{n-p}{2}} \times \mathsf{D}_p$ $\quad\tau$ has $\mathsf{mod}(p, 2)$ nontrivial orbits | | |
| Special cases: DIII |  | $\mathsf{C}_{\frac{n}{2}}$, $(\mathsf{B}, \mathsf{C})_{\frac{n-1}{2}}$ | $p \le 1$ $(n > 4)$ |
| |  | $(\mathsf{B}, \mathsf{C})_1$ | $n = 4$, $p = 2$ |

Table 4: Generalized Satake diagrams of finite exceptional Lie type.

| Type | | Diagram | Type of $\overline{\Phi}$ | Type | | Diagram | Type of $\overline{\Phi}$ |
|---|---|---|---|---|---|---|---|
| $(\mathsf{E}_6)_\emptyset$ | EI | 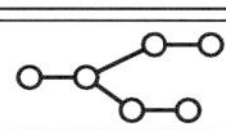 | $\mathsf{E}_6$ | $(\mathsf{E}_6)_{\mathsf{D}_4}$ | EIV | 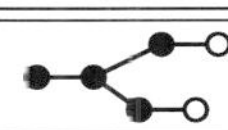 | $\mathsf{A}_2$ |
| $(\mathsf{E}_6)'_{\mathsf{A}_3}$ | EIII | 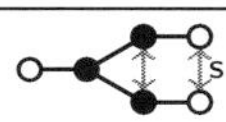 | $(\mathsf{B}, \mathsf{C})_2$ | $(\mathsf{E}_6)'_{\mathsf{A}_5}$ | | 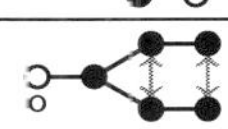 | $(\mathsf{B}, \mathsf{C})_1$ |
| $(\mathsf{E}_6)'_\emptyset$ | EII |  | $\mathsf{F}_4$ | $(\mathsf{E}_6)'_{\mathsf{E}_6}$ | | 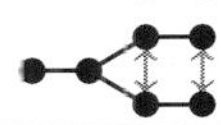 | $\mathsf{Z}_0$ |
| $(\mathsf{E}_7)_\emptyset$ | EV | 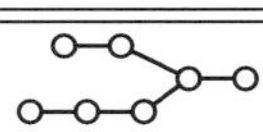 | $\mathsf{E}_7$ | $(\mathsf{E}_7)_{\mathsf{D}_4 \times \mathsf{A}_1}$ | | 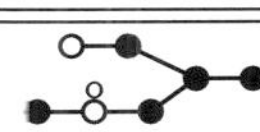 | $(\mathsf{B}, \mathsf{C})_2$ |
| $(\mathsf{E}_7)_{\mathsf{A}_1^3}$ | EVI | 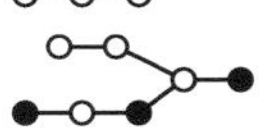 | $\mathsf{F}_4$ | $(\mathsf{E}_7)_{\mathsf{D}_6}$ | | 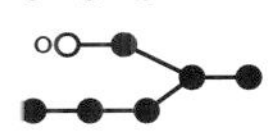 | $(\mathsf{B}, \mathsf{C})_1$ |
| $(\mathsf{E}_7)_{\mathsf{D}_4}$ | EVII | 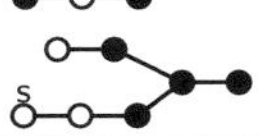 | $\mathsf{C}_3$ | $(\mathsf{E}_7)_{\mathsf{E}_7}$ | | 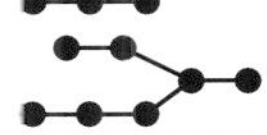 | $\mathsf{Z}_0$ |
| $(\mathsf{E}_8)_\emptyset$ | EVIII | 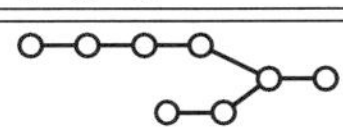 | $\mathsf{E}_8$ | $(\mathsf{E}_8)_{\mathsf{E}_7}$ | | 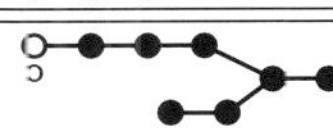 | $(\mathsf{B}, \mathsf{C})_1$ |
| $(\mathsf{E}_8)_{\mathsf{D}_4}$ | EIX | 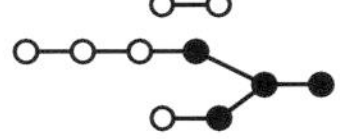 | $\mathsf{F}_4$ | $(\mathsf{E}_8)_{\mathsf{E}_8}$ | | 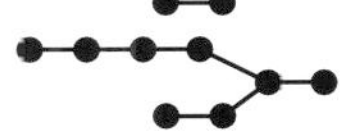 | $\mathsf{Z}_0$ |
| $(\mathsf{E}_8)_{\mathsf{D}_6}$ | | 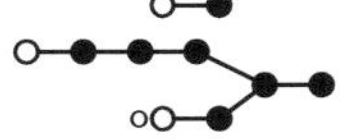 | $(\mathsf{B}, \mathsf{C})_2$ | | | | |
| $(\mathsf{F}_4)_\emptyset$ | FI | 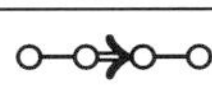 | $\mathsf{F}_4$ | $(\mathsf{F}_4)_{\mathsf{C}_3}$ | | 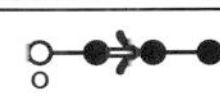 | $(\mathsf{B}, \mathsf{C})_1$ |
| $(\mathsf{F}_4)_{\mathsf{C}_2}$ | |  | $(\mathsf{B}, \mathsf{C})_2$ | $(\mathsf{F}_4)_{\mathsf{F}_4}$ | |  | $\mathsf{Z}_0$ |
| $(\mathsf{F}_4)_{\mathsf{B}_3}$ | FII |  | $(\mathsf{B}, \mathsf{C})_1$ | | | | |

| Type | Diagram | Type of $\overline{\Phi}$ | Type | Diagram | Type of $\overline{\Phi}$ |
|---|---|---|---|---|---|
| $(\mathsf{G}_2)_{\emptyset}$   G |  | $\mathsf{G}_2$ | $(\mathsf{G}_2)^{\mathsf{lo}}_{\mathsf{A}_1}$ |  | $(\mathsf{B},\mathsf{C})_1^+$ |
| $(\mathsf{G}_2)^{\mathsf{sho}}_{\mathsf{A}_1}$ |  | $(\mathsf{B},\mathsf{C})_1$ | $(\mathsf{G}_2)_{\mathsf{G}_2}$ |  | $\mathsf{Z}_0$ |

**A.4. Affine type.** Tables 5, 6 and 7 list the generalized Satake diagrams.

Table 5: Generalized Satake diagrams of affine Lie type $\widehat{\mathsf{A}}_n$. We set $N := n+1$.

| Type | Diagram | | Type of $\overline{\Phi}$ | Constraints |
|---|---|---|---|---|
| $(\widehat{\mathsf{A}}_n)^{\mathsf{pl}}$ | 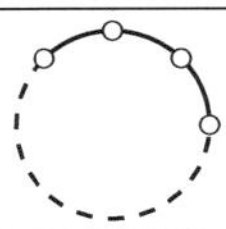 | $X = \emptyset$<br>$\tau = \mathsf{id}$ | $\widehat{\mathsf{A}}_n$ | $(n>1)$ |
| $(\widehat{\mathsf{A}}_n)^{\mathsf{alt}}$ | 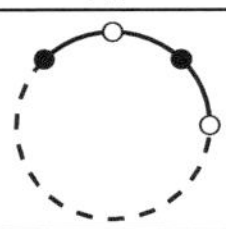 | $X$ is of type $\mathsf{A}_1^{N/2}$<br>$\tau = \mathsf{id}$ | $\widehat{\mathsf{A}}_{\frac{n-1}{2}}$ | $n$ odd $(n>1)$ |
| $(\widehat{\mathsf{A}}_n)^{\mathsf{rot}}$ | 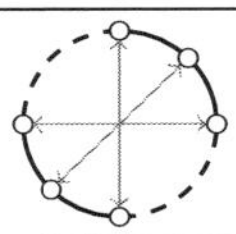 | $X = \emptyset$<br>$\tau =$ half-turn | $\widehat{\mathsf{A}}_{\frac{n-1}{2}}$ | $n$ odd $(n>1)$ |
| $(\widehat{\mathsf{A}}_n)^{\mathsf{rfl}}_{p_1,p_2}$ | 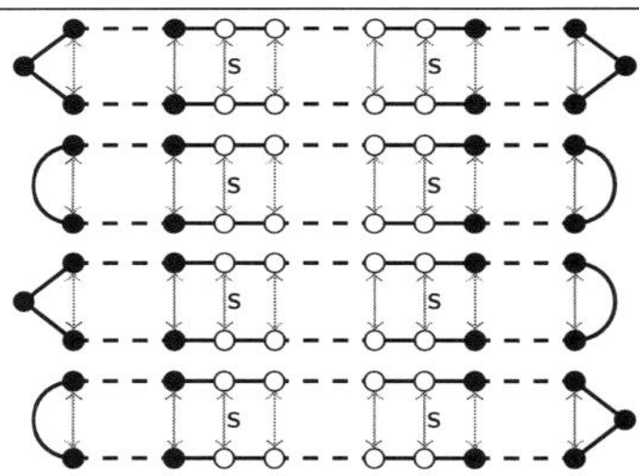 | | $(\widehat{\mathsf{C}}^\vee,\widehat{\mathsf{C}})_{\frac{N-p_1-p_2}{2}}$ | $0 \le p_1 \le p_2$<br>$p_1+p_2 \le N$<br>$N-p_1-p_2$ even |
| | $X$ is of type $\mathsf{A}_{p_1-1}\times\mathsf{A}_{p_2-1}$ | $\tau$ has $\mathsf{mod}(p_1-1,2)+\mathsf{mod}(p_2-1,2)$ fixed points | | |
| | 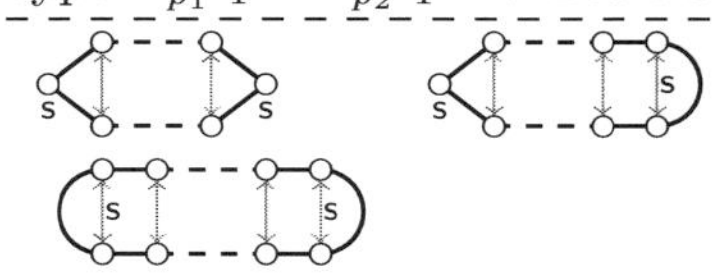 | | $\widehat{\mathsf{C}}_{\frac{N}{2}}$, $(\widehat{\mathsf{C}}',\widehat{\mathsf{C}})_{\frac{N-1}{2}}$<br>$(\widehat{\mathsf{C}}^\vee,\widehat{\mathsf{C}})_{\frac{N-2}{2}}$ | $(p_1,p_2)\in$<br>$\{(0,0),(0,1),(1,1)\}$ |
| | 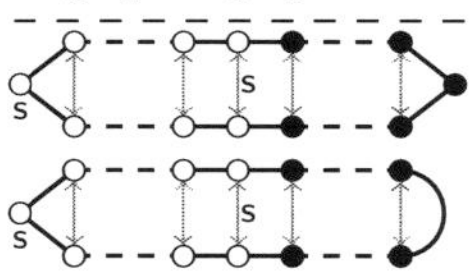 | | $(\widehat{\mathsf{C}}',\widehat{\mathsf{C}})_{\frac{N-p_2}{2}}$ | $p_1=0$, $p_2\ge 2$ |
| Special cases: | 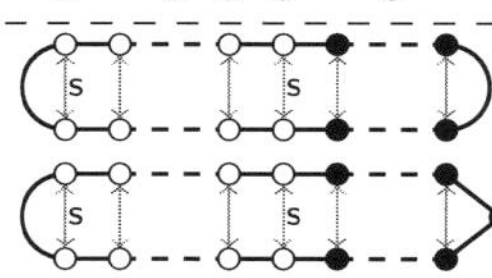 | | $(\widehat{\mathsf{C}}^\vee,\widehat{\mathsf{C}})_{\frac{N+1-p_2}{2}}$ | $p_1=1$, $p_2\ge 2$ |

| Type | Diagram | Type of $\overline{\Phi}$ | Constraints |
|---|---|---|---|
| | 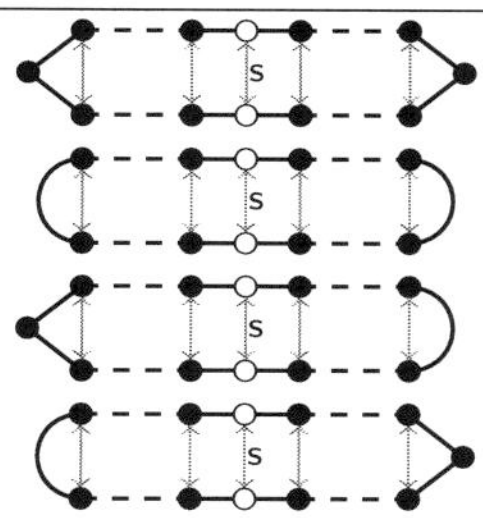 | $\widehat{\mathsf{Z}}_0$ | $p_1 + p_2 = N$<br>$p_1 \geq 2$ |
| | 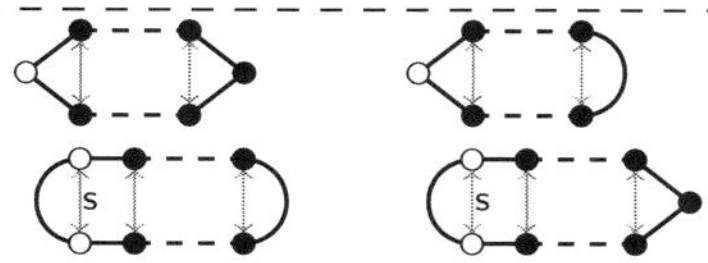 | $\widehat{\mathsf{Z}}_0$ | $(p_1, p_2) \in$<br>$\{(0, N), (1, N-1)\}$ |
| |  | $\widehat{\mathsf{A}}_1$, $\widehat{\mathsf{Z}}_0$, $\widehat{\mathsf{Z}}_0$ | $n = 1$, $(p_1, p_2) \in$<br>$\{(0,0), (0,2), (1,1)\}$ |

Table 6: Generalized Satake diagrams of affine classical Lie type other than $\widehat{\mathsf{A}}_n$.

| Type | Diagram | Type of $\overline{\Phi}$ | Constraints |
|---|---|---|---|
| $(\widehat{\mathsf{B}}_n)^{\mathsf{pl}}_{p_1,p_2}$ | 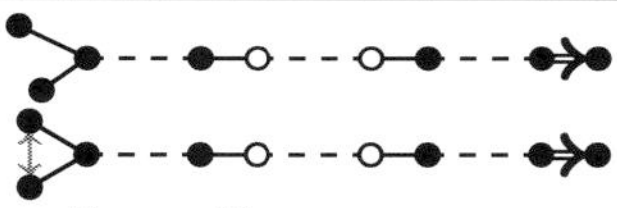 | $\widehat{\mathsf{C}}^{\vee}_{n-p_1-p_2}$ | $p_1, p_2 \geq 0$<br>$p_1 + p_2 \leq n$ |
| $X$ is of type $\mathsf{D}_{p_1} \times \mathsf{B}_{p_2}$ | | $\tau$ has $\mathsf{mod}(p_1, 2)$ nontrivial orbits | |
| Special cases: | 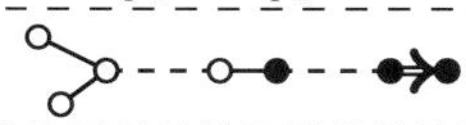 | $\widehat{\mathsf{B}}_{n-p_2}$ | $p_1 = 0$ |
| | 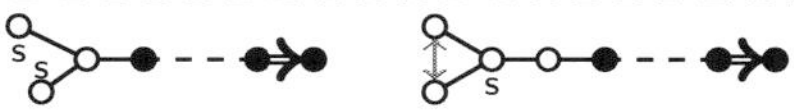 | $\widehat{\mathsf{C}}_2$, $\widehat{\mathsf{C}}^{\vee}_2$ | $p_1 + p_2 = n - 2$<br>$p_1 < 2$ |
| | 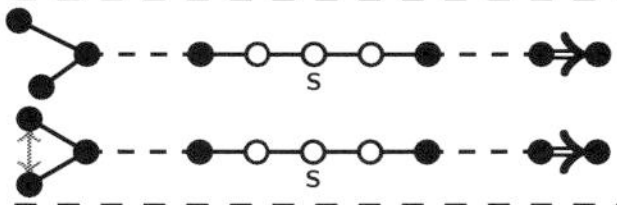 | $\widehat{\mathsf{C}}^{\vee}_2$ | $p_1 + p_2 = n - 2$<br>$p_1 \geq 2$ |
| | 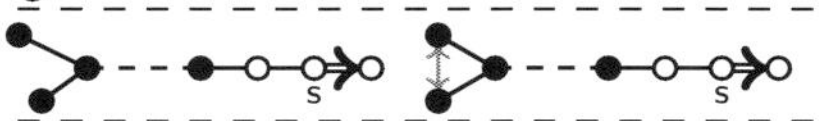 | $\widehat{\mathsf{C}}^{\vee}_2$ | $(p_1, p_2) = (n-2, 0)$ |
| |  | $\widehat{\mathsf{A}}_1$ | $(p_1, p_2) = (n-1, 0)$ |
| |  | $\widehat{\mathsf{Z}}_0$ | $(p_1, p_2) \in$<br>$\{(0, n), (1, n-1)\}$ |
| $(\widehat{\mathsf{B}}_n)^{\mathsf{alt}}_{p_1,p_2}$ | 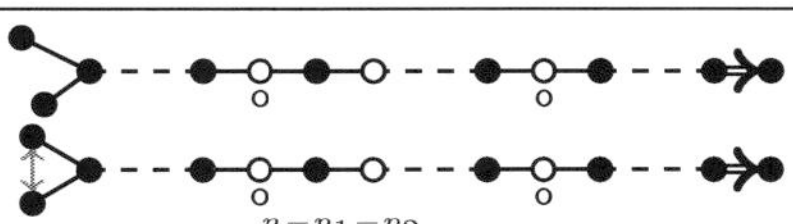 | $(\widehat{\mathsf{C}}^{\vee}, \widehat{\mathsf{C}})_{\frac{n-p_1-p_2}{2}}$ | $p_1, p_2 \geq 0$<br>$p_1 + p_2 < n$<br>$n - p_1 - p_2$ even |
| $X$ is of type $\mathsf{D}_{p_1} \times \mathsf{A}_1^{\frac{n-p_1-p_2}{2}} \times \mathsf{B}_{p_2}$ | | $\tau$ has $\mathsf{mod}(p_1, 2)$ nontrivial orbits | |
| Special cases: | 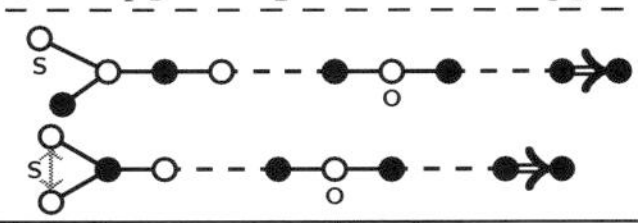 | $(\widehat{\mathsf{C}}', \widehat{\mathsf{C}})_{\frac{n-p_2}{2}}$<br>$(\widehat{\mathsf{C}}^{\vee}, \widehat{\mathsf{C}})_{\frac{n-1-p_2}{2}}$ | $p_1 < 2$ |

| Type | Diagram | Type of $\overline{\Phi}$ | Constraints |
|---|---|---|---|
| $(\widehat{\mathsf{B}}^\vee_n)^{\mathsf{pl}}_{p_1,p_2}$ | 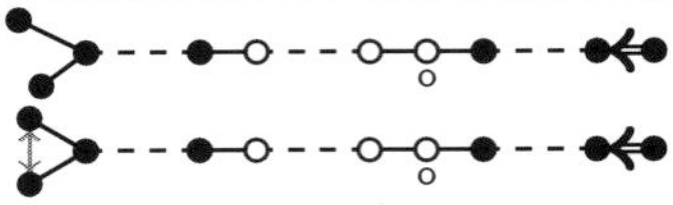 | $(\widehat{\mathsf{C}}^\vee,\widehat{\mathsf{C}}')_{n-p_1-p_2}$ | $p_1,p_2\geq 0$<br>$p_1+p_2<n$ |
| $X$ is of type $\mathsf{D}_{p_1}\times\mathsf{C}_{p_2}$ | | $\tau$ has $\mathsf{mod}(p_1,2)$ nontrivial orbits | |
| Special cases: | 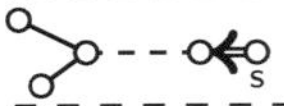 | $\widehat{\mathsf{B}}^\vee_n$ | $(p_1,p_2)=(0,0)$ |
| | 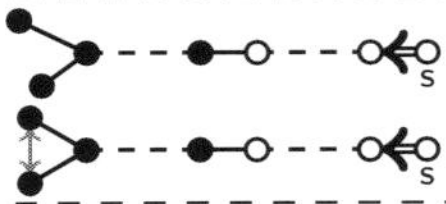 | $\widehat{\mathsf{C}}'_{n-p_1}$ | $p_1>0$<br>$p_2=0$ |
| |  | $(\widehat{\mathsf{B}},\widehat{\mathsf{B}}^\vee)_2$, $(\widehat{\mathsf{C}}^\vee,\widehat{\mathsf{C}}')_2$ | $p_1+p_2=n-2$<br>$p_1<2$ |
| | 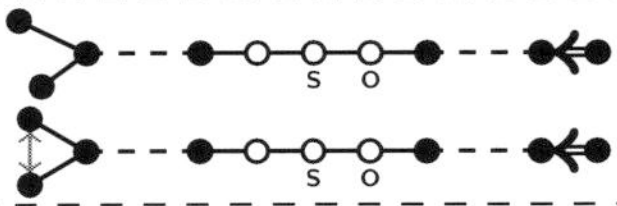 | $(\widehat{\mathsf{C}}^\vee,\widehat{\mathsf{C}}')_2$ | $p_1+p_2=n-2$<br>$p_1\geq 2$ |
| |  | $(\widehat{\mathsf{C}}^\vee,\widehat{\mathsf{C}})_1$ | $(p_1,p_2)=(0,n-1)$ |
| |  | $\widehat{\mathsf{Z}}_0$ | $p_1+p_2=n$, $p_1<2$ |
| $(\widehat{\mathsf{B}}^\vee_n)^{\mathsf{alt}}_{p_1,p_2}$ | 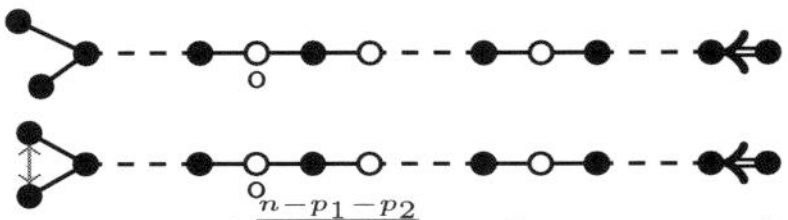 | $(\widehat{\mathsf{C}}^\vee,\widehat{\mathsf{C}})_{\frac{n-p_1-p_2}{2}}$ | $p_1,p_2\geq 0$<br>$p_1+p_2<n$<br>$n-p_1-p_2$ even |
| $X$ is of type $\mathsf{D}_{p_1}\times\mathsf{A}_1^{\frac{n-p_1-p_2}{2}}\times\mathsf{C}_{p_2}$ | | $\tau$ has $\mathsf{mod}(p_1,2)$ nontrivial orbits | |
| Special cases: |  | $\widehat{\mathsf{C}}_{\frac{n}{2}}$, $(\widehat{\mathsf{C}}',\widehat{\mathsf{C}})_{\frac{n-1}{2}}$ | $(p_1,p_2)\in\{(0,0),(1,0)\}$ |
| | 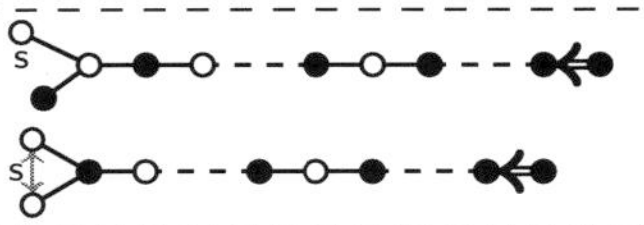 | $(\widehat{\mathsf{C}}',\widehat{\mathsf{C}})_{\frac{n-p_2}{2}}$<br>$(\widehat{\mathsf{C}}^\vee,\widehat{\mathsf{C}})_{\frac{n-1-p_2}{2}}$ | $p_1<2$, $p_2\geq 1$ |
| | 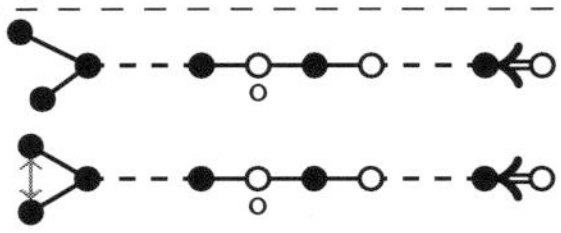 | $(\widehat{\mathsf{C}}',\widehat{\mathsf{C}})_{\frac{n-p_1}{2}}$ | $p_1>0$, $p_2=0$ |
| $(\widehat{\mathsf{C}}_n)^{\mathsf{pl}}_{p_1,p_2}$ |  | $(\widehat{\mathsf{C}}^\vee,\widehat{\mathsf{C}})_{n-p_1-p_2}$ | $0\leq p_1\leq p_2$<br>$p_1+p_2<n$ |
| $X$ is of type $\mathsf{C}_{p_1}\times\mathsf{C}_{p_2}$ | | $\tau=\mathsf{id}$ | |
| Special cases: | 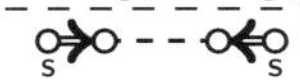 | $\widehat{\mathsf{C}}_n$ | $(p_1,p_2)=(0,0)$ |
| | 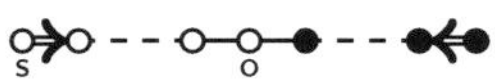 | $(\widehat{\mathsf{C}}',\widehat{\mathsf{C}})_{n-p_2}$ | $p_1=0$, $p_2>0$ |
| | 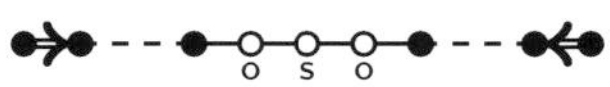 | $(\widehat{\mathsf{C}}^\vee,\widehat{\mathsf{C}})_2$ | $p_1+p_2=n-2$<br>$p_1>0$ |
| $(\widehat{\mathsf{C}}_n)^{\mathsf{alt}}_{p_1,p_2}$ | 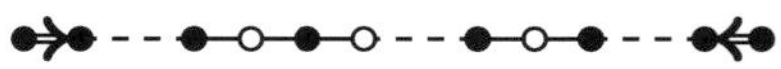 | $(\widehat{\mathsf{C}}^\vee,\widehat{\mathsf{C}})_{\frac{n-p_1-p_2}{2}}$ | $0\leq p_1\leq p_2$<br>$p_1+p_2\leq n$<br>$n-p_1-p_2$ even |
| $X$ is of type $\mathsf{C}_{p_1}\times\mathsf{A}_1^{\frac{n-p_1-p_2}{2}}\times\mathsf{C}_{p_2}$ | | $\tau=\mathsf{id}$ | |

| Type | Diagram | Type of $\overline{\Phi}$ | Constraints |
|---|---|---|---|
| Special cases: | 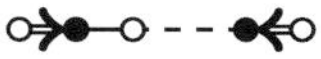 | $\widehat{\mathsf{C}}_{\frac{n}{2}}$ | $(p_1, p_2) = (0, 0)$ |
| |  | $(\widehat{\mathsf{C}}', \widehat{\mathsf{C}})_{\frac{n-p_2}{2}}$ | $p_1 = 0$, $p_2 > 0$ |
| $(\widehat{\mathsf{C}}_n)^{\mathsf{rot}}_p$ | 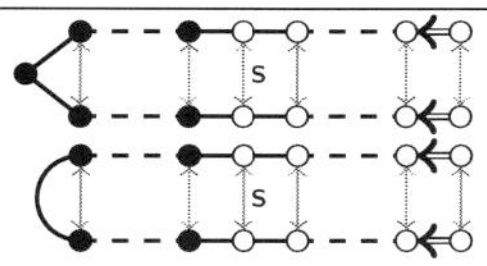 | $(\widehat{\mathsf{C}}', \widehat{\mathsf{C}})_{\frac{n-p}{2}}$ | $0 \leq p \leq n$<br>$n - p$ even |
| $X$ is of type $\mathsf{A}_{p-1}$ | | $\tau =$ half-turn with $\mathsf{mod}(p-1, 2)$ fixed points | |
| Special cases: | 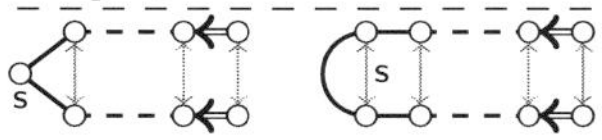 | $\widehat{\mathsf{C}}_{\frac{n}{2}}$, $(\widehat{\mathsf{C}}', \widehat{\mathsf{C}})_{\frac{n-1}{2}}$ | $p < 2$ |
| $(\widehat{\mathsf{C}}^{\vee}_n)^{\mathsf{pl}}_{p_1,p_2}$ |  | $\widehat{\mathsf{C}}^{\vee}_{n-p_1-p_2}$ | $0 \leq p_1 \leq p_2$<br>$p_1 + p_2 \leq n$ |
| $X$ is of type $\mathsf{B}_{p_1} \times \mathsf{B}_{p_2}$ | | $\tau = \mathsf{id}$ | |
| Special cases: | 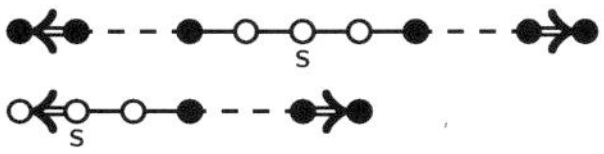 | $\widehat{\mathsf{C}}^{\vee}_2$ | $p_1 + p_2 = n - 2$ |
| |  | $\widehat{\mathsf{A}}_1$ | $(p_1, p_2) = (0, n-1)$ |
| $(\widehat{\mathsf{C}}^{\vee}_n)^{\mathsf{alt}}_{p_1,p_2}$ | 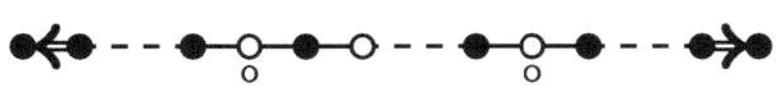 | $(\widehat{\mathsf{C}}^{\vee}, \widehat{\mathsf{C}})_{\frac{n-p_1-p_2}{2}}$ | $0 \leq p_1 \leq p_2$<br>$p_1 + p_2 < n$<br>$n - p_1 - p_2$ even |
| $X$ is of type $\mathsf{B}_{p_1} \times \mathsf{A}_1^{\frac{n-p_1-p_2}{2}} \times \mathsf{B}_{p_2}$ | | $\tau = \mathsf{id}$ | |
| $(\widehat{\mathsf{C}}^{\vee}_n)^{\mathsf{rot}}_p$ | 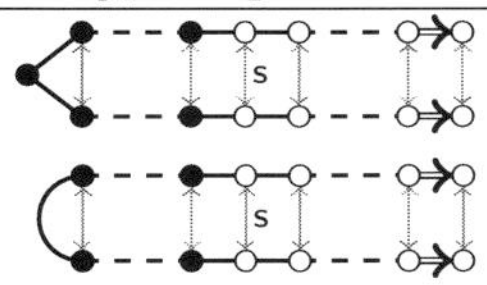 | $(\widehat{\mathsf{C}}^{\vee}, \widehat{\mathsf{C}}')_{\frac{n-p}{2}}$ | $0 \leq p \leq n$<br>$n - p$ even |
| $X$ is of type $\mathsf{A}_{p-1}$ | | $\tau =$ half-turn with $\mathsf{mod}(p-1, 2)$ fixed points | |
| Special cases: | 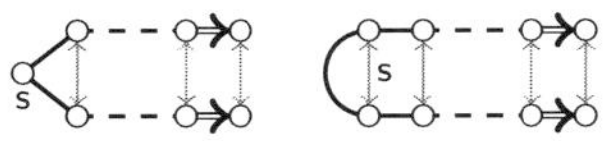 | $\widehat{\mathsf{C}}'_{\frac{n}{2}}$<br>$(\widehat{\mathsf{C}}^{\vee}, \widehat{\mathsf{C}}')_{\frac{n-1}{2}}$ | $p < 2$ |
| $(\widehat{\mathsf{C}}'_n)^{\mathsf{pl}}_{p_1,p_2}$ | 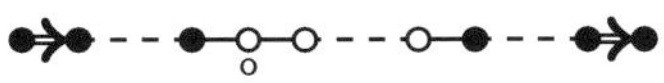 | $(\widehat{\mathsf{C}}^{\vee}, \widehat{\mathsf{C}}')_{n-p_1-p_2}$ | $p_1, p_2 \geq 0$<br>$p_1 + p_2 \leq n$ |
| $X$ is of type $\mathsf{C}_{p_1} \times \mathsf{B}_{p_2}$ | | $\tau = \mathsf{id}$ | |
| Special cases: | 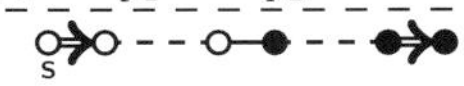 | $\widehat{\mathsf{C}}'_{n-p_2}$ | $p_1 = 0$, $p_2 < n$ |
| | 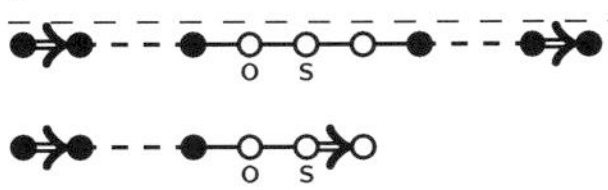 | $(\widehat{\mathsf{C}}^{\vee}, \widehat{\mathsf{C}}')_2$ | $p_1 + p_2 = n - 2$<br>$p_1 > 0$ |
| | 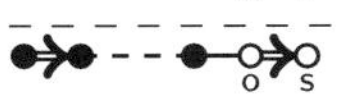 | $(\widehat{\mathsf{C}}^{\vee}, \widehat{\mathsf{C}}')_1$ | $(p_1, p_2) = (n-1, 0)$ |
| | 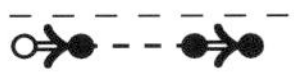 | $\widehat{\mathsf{Z}}_0$ | $(p_1, p_2) = (0, n)$ |
| |  | $(\widehat{\mathsf{C}}^{\vee}, \widehat{\mathsf{C}}')_1$, $\widehat{\mathsf{Z}}_0$, $\widehat{\mathsf{Z}}_0$ | $n = 1$, $(p_1, p_2) \in \{(0,0), (0,1), (1,0)\}$ |

| Type | Diagram | Type of $\overline{\Phi}$ | Constraints |
|---|---|---|---|
| $\left(\widehat{\mathsf{C}}'_n\right)^{\mathsf{alt}}_{p_1,p_2}$ | 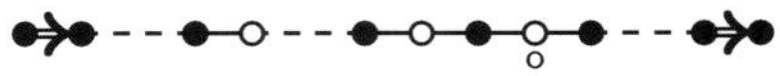 | $(\widehat{\mathsf{C}}^\vee,\widehat{\mathsf{C}})_{\frac{n-p_1-p_2}{2}}$ | $p_1,p_2 \geq 0$<br>$p_1+p_2<n$<br>$n-p_1-p_2$ even |
| $X$ is of type $\mathsf{C}_{p_1} \times \mathsf{A}_1^{\frac{n-p_1-p_2}{2}} \times \mathsf{B}_{p_2}$ | $\tau = \mathsf{id}$ | | |
| Special cases: |  | $(\widehat{\mathsf{C}}',\widehat{\mathsf{C}})_{\frac{n-p_2}{2}}$ | $p_1=0$ |
| $\left(\widehat{\mathsf{D}}_n\right)^{\mathsf{pl}}_{p_1,p_2}$ | 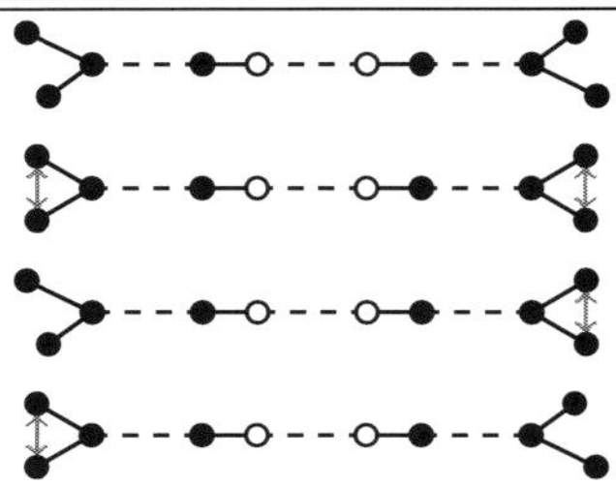 | $\widehat{\mathsf{C}}^\vee_{n-p_1-p_2}$ | $0 \leq p_1 \leq p_2$<br>$p_1+p_2 \leq n$ |
| $X$ is of type $\mathsf{D}_{p_1} \times \mathsf{D}_{p_2}$ | $\tau$ has $\mathsf{mod}(p_1,2)+\mathsf{mod}(p_2,2)$ nontriv. orbits | | |
| Special cases: | 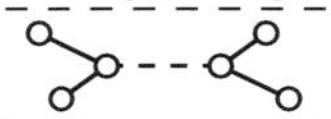 | $\widehat{\mathsf{D}}_n$ | $(p_1,p_2)=(0,0)$ |
| | 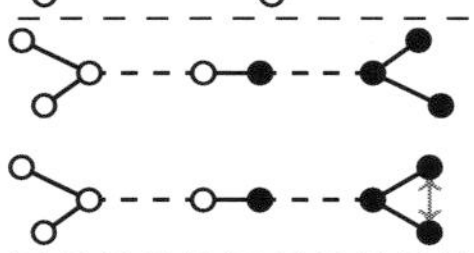 | $\widehat{\mathsf{B}}_{n-p_2}$ | $p_1=0$, $p_2>0$ |
| | 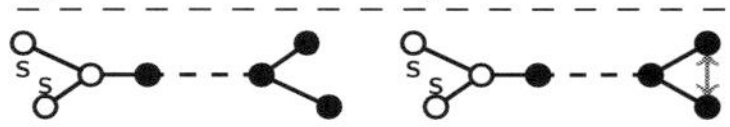 | $\widehat{\mathsf{C}}_2$ | $p_1+p_2=n-2$<br>$p_1=0$ |
| | 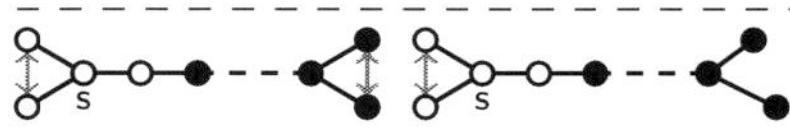 | $\widehat{\mathsf{C}}^\vee_2$ | $p_1+p_2=n-2$<br>$p_1=1$ |
| | 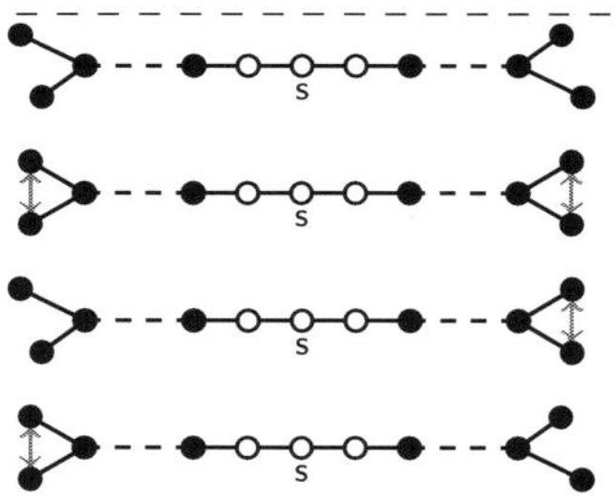 | $\widehat{\mathsf{C}}^\vee_2$ | $p_1+p_2=n-2$<br>$p_1 \geq 2$ |
| | 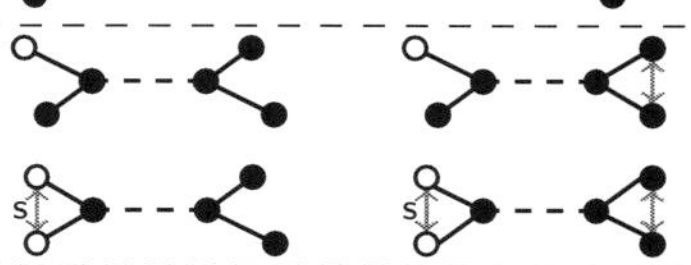 | $\widehat{\mathsf{Z}}_0$ | $p_1+p_2=n$<br>$p_1<2$ |
| | 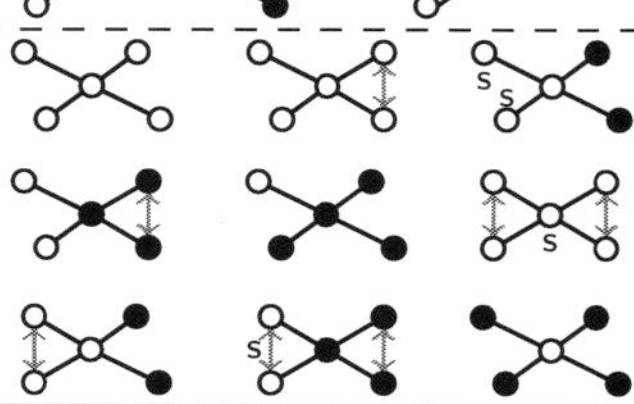 | $\widehat{\mathsf{D}}_4$, $\widehat{\mathsf{B}}_3$, $\widehat{\mathsf{C}}_2$,<br>$\widehat{\mathsf{A}}_1$, $\widehat{\mathsf{Z}}_0$, $\widehat{\mathsf{C}}^\vee_2$,<br>$\widehat{\mathsf{A}}_1$, $\widehat{\mathsf{Z}}_0$, $\widehat{\mathsf{Z}}_0$ | $n=4$, $(p_1,p_2) \in$<br>$\{(0,0),(0,1),(0,2),$<br>$(0,3),(0,4),(1,1),$<br>$(1,2),(1,3),(2,2)\}$ |

| Type | Diagram | Type of $\overline{\Phi}$ | Constraints |
|---|---|---|---|
| $(\widehat{\mathsf{D}}_n)^{\mathsf{alt}}_{p_1,p_2}$ | 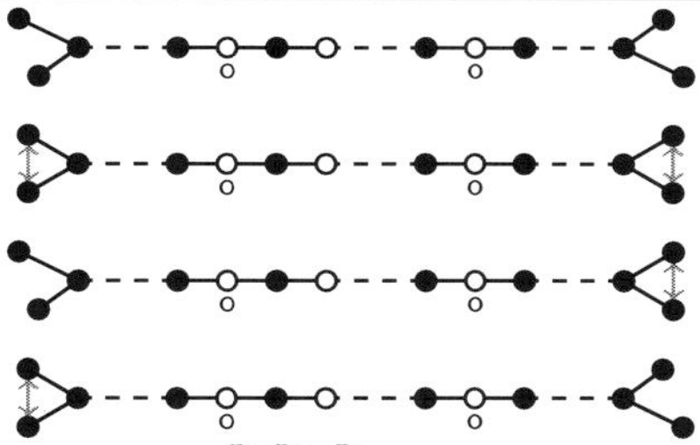 | $(\widehat{\mathsf{C}}^\vee,\widehat{\mathsf{C}})_{\frac{n-p_1-p_2}{2}}$ | $p_1,p_2 \geq 0$<br>$p_1+p_2<n$<br>$n-p_1-p_2$ even<br>($n>4$ if $p_1+p_2\leq 1$) |
| $X$ is of type $\mathsf{D}_{p_1}\times\mathsf{A}_1^{\frac{n-p_1-p_2}{2}}\times\mathsf{D}_{p_2}$, | | $\tau$ has $\mathsf{mod}(p_1,2)+\mathsf{mod}(p_2,2)$ nontriv. orbits | |
| | 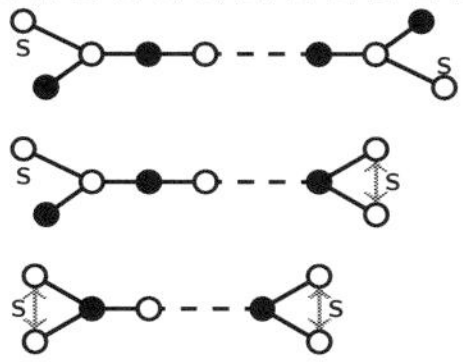 | $\widehat{\mathsf{C}}_{\frac{n}{2}}$<br>$(\widehat{\mathsf{C}}',\widehat{\mathsf{C}})_{\frac{n-1}{2}}$<br>$(\widehat{\mathsf{C}}^\vee,\widehat{\mathsf{C}})_{\frac{n-2}{2}}$ | $(p_1,p_2)\in$<br>$\{(0,0),(0,1),(1,1)\}$ |
| Special cases: | 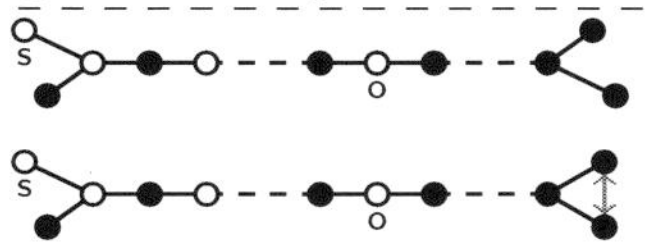 | $(\widehat{\mathsf{C}}',\widehat{\mathsf{C}})_{\frac{n-p_2}{2}}$ | $p_1=0$, $p_2\geq 2$ |
| | 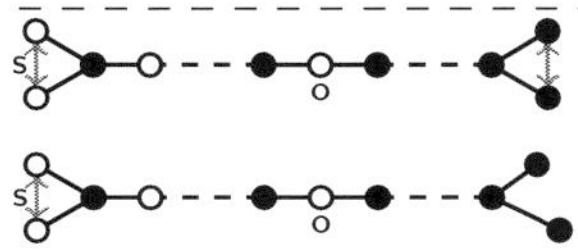 | $(\widehat{\mathsf{C}}^\vee,\widehat{\mathsf{C}})_{\frac{n-1-p_2}{2}}$ | $p_1=1$, $p_2\geq 2$ |
| |  | $(\widehat{\mathsf{C}}',\widehat{\mathsf{C}})_1$, $(\widehat{\mathsf{C}}^\vee,\widehat{\mathsf{C}})_1$ | $n=4$, $(p_1,p_2)\in$<br>$\{(0,2),(1,1)\}$ |
| $(\widehat{\mathsf{D}}_n)^{\mathsf{rot}}_{p}$ | 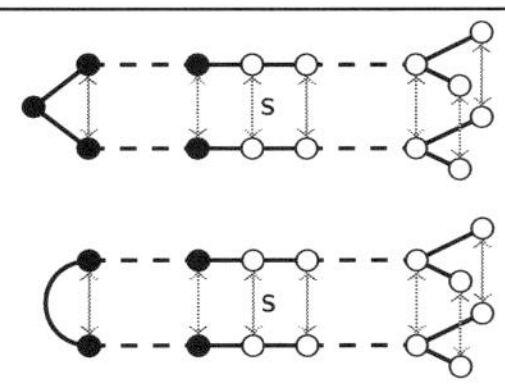 | $(\widehat{\mathsf{B}},\widehat{\mathsf{B}}^\vee)_{\frac{n-p}{2}}$ | $0\leq p\leq n$<br>$n-p$ even |
| $X$ is of type $\mathsf{A}_{p-1}$ | | $\tau=$ half-turn with $\mathsf{mod}(p-1,2)$ fixed points | |
| | 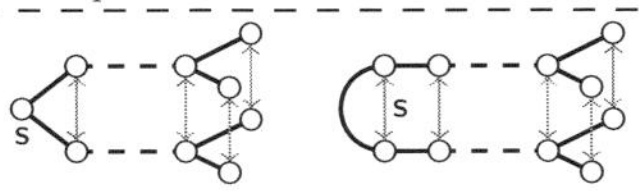 | $\widehat{\mathsf{B}}^\vee_{\frac{n}{2}}$, $(\widehat{\mathsf{B}},\widehat{\mathsf{B}}^\vee)_{\frac{n-1}{2}}$ | $p<2$ |
| Special cases: | 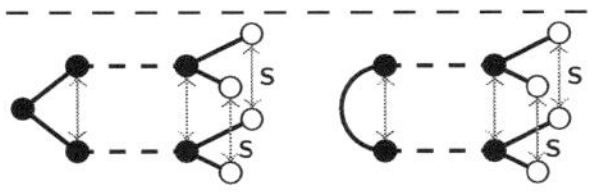 | $\widehat{\mathsf{A}}_1$ | $p=n-2$ |
| | 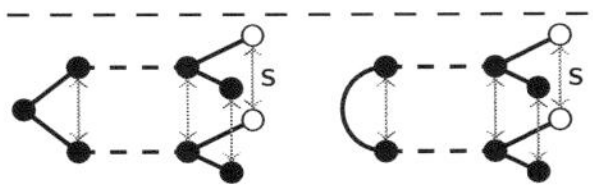 | $\widehat{\mathsf{Z}}_0$ | $p=n$ |

Table 7: Generalized Satake diagrams of affine exceptional Lie type.

| Type | Diagram | Type of $\overline{\Phi}$ | Type | Diagram | Type of $\overline{\Phi}$ |
|---|---|---|---|---|---|
| $(\widehat{E}_6)_{\emptyset}$ | 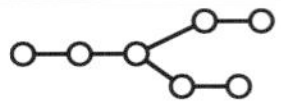 | $\widehat{E}_6$ | $(\widehat{E}_6)_{D_4}$ |  | $\widehat{A}_2$ |
| $(\widehat{E}_6)_{A_1^4}$ | 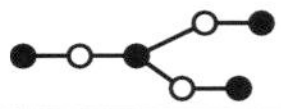 | $\widehat{A}_2$ | | | |
| $(\widehat{E}_6)'_{\emptyset}$ | 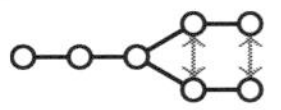 | $\widehat{F}_4$ | $(\widehat{E}_6)'_{A_5}$ | 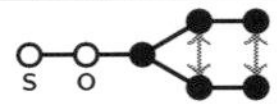 | $(\widehat{C}',\widehat{C})_1$ |
| $(\widehat{E}_6)'_{A_1^2}$ | 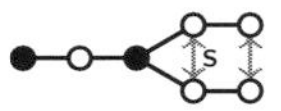 | $(\widehat{B},\widehat{B}^\vee)_2$ | $(\widehat{E}_6)'_{A_5\times A_1}$ | 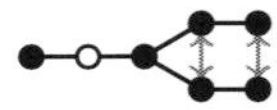 | $\widehat{Z}_0$ |
| $(\widehat{E}_6)'_{A_3}$ | 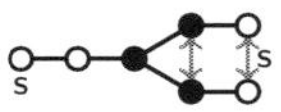 | $(\widehat{C}',\widehat{C})_2$ | $(\widehat{E}_6)'_{D_5}$ | 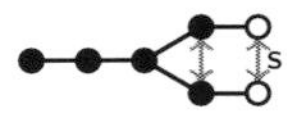 | $\widehat{Z}_0$ |
| $(\widehat{E}_6)'_{A_3\times A_1}$ | 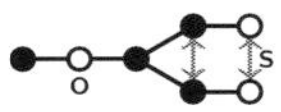 | $(\widehat{C}^\vee,\widehat{C}')_1$ | $(\widehat{E}_6)'_{E_6}$ | 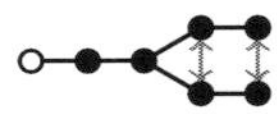 | $\widehat{Z}_0$ |
| $(\widehat{E}_7)_{\emptyset}$ | 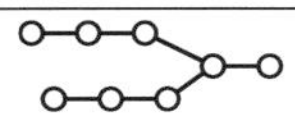 | $\widehat{E}_7$ | $(\widehat{E}_7)_{D_4\times A_1}$ | 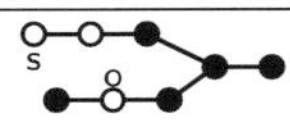 | $(\widehat{C}',\widehat{C})_2$ |
| $(\widehat{E}_7)_{A_1^3}$ | 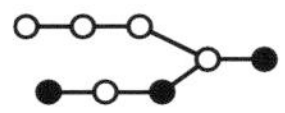 | $\widehat{F}_4$ | $(\widehat{E}_7)_{D_4\times A_1^2}$ | 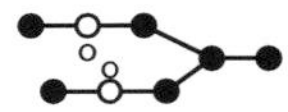 | $(\widehat{C}^\vee,\widehat{C})_1$ |
| $(\widehat{E}_7)_{A_1^4}$ | 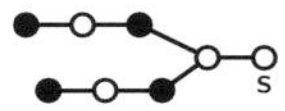 | $\widehat{B}_3^\vee$ | $(\widehat{E}_7)_{D_6}$ | 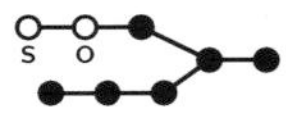 | $(\widehat{C}',\widehat{C})_1$ |
| $(\widehat{E}_7)_{D_4}$ | 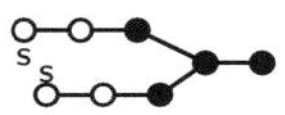 | $\widehat{C}_3$ | $(\widehat{E}_7)_{D_6\times A_1}$ | 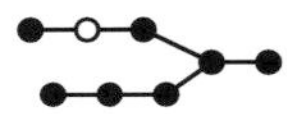 | $\widehat{Z}_0$ |
| $(\widehat{E}_7)_{A_1^5}$ | 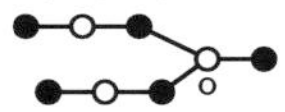 | $(\widehat{B},\widehat{B}^\vee)_2$ | $(\widehat{E}_7)_{E_7}$ | 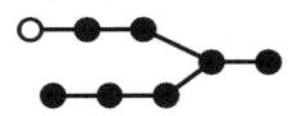 | $\widehat{Z}_0$ |
| $(\widehat{E}_7)'_{\emptyset}$ | 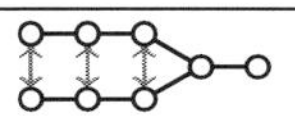 | $\widehat{F}_4^\vee$ | $(\widehat{E}_7)'_{E_6}$ | 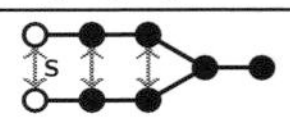 | $\widehat{Z}_0$ |
| $(\widehat{E}_7)'_{A_3}$ | 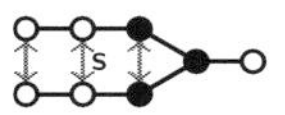 | $(\widehat{B},\widehat{B}^\vee)_2$ | $(\widehat{E}_7)'_{A_7}$ | 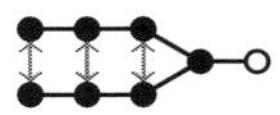 | $\widehat{Z}_0$ |
| $(\widehat{E}_7)'_{A_5}$ | 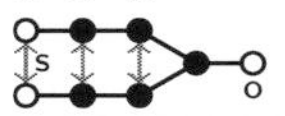 | $(\widehat{C}^\vee,\widehat{C})_1$ | | | |
| $(\widehat{E}_8)_{\emptyset}$ | 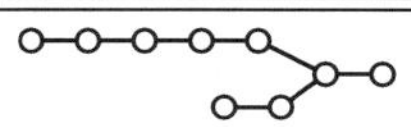 | $\widehat{E}_8$ | $(\widehat{E}_8)_{D_6\times A_1}$ | 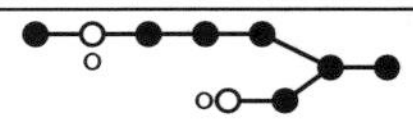 | $(\widehat{C}^\vee,\widehat{C})_1$ |
| $(\widehat{E}_8)_{A_1^4}$ | 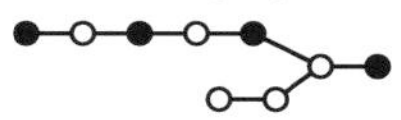 | $\widehat{F}_4^\vee$ | $(\widehat{E}_8)_{E_7}$ | 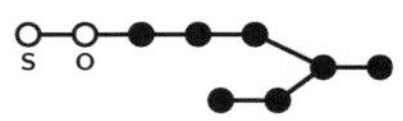 | $(\widehat{C}',\widehat{C})_1$ |
| $(\widehat{E}_8)_{D_4}$ | 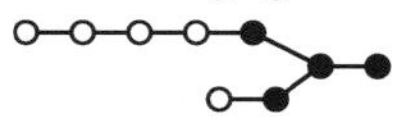 | $\widehat{F}_4$ | $(\widehat{E}_8)_{E_7\times A_1}$ | 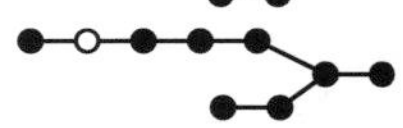 | $\widehat{Z}_0$ |
| $(\widehat{E}_8)_{D_4\times A_1^2}$ |  | $(\widehat{B},\widehat{B}^\vee)_2$ | $(\widehat{E}_8)_{D_8}$ | 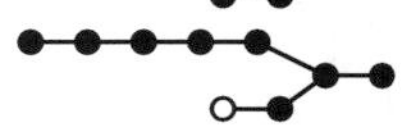 | $\widehat{Z}_0$ |
| $(\widehat{E}_8)_{D_6}$ | 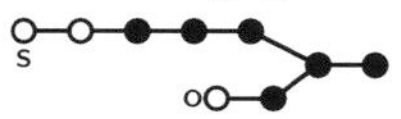 | $(\widehat{C}',\widehat{C})_2$ | $(\widehat{E}_8)_{E_8}$ | 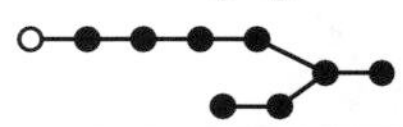 | $\widehat{Z}_0$ |

| Type | Diagram | Type of $\overline{\Phi}$ | Type | Diagram | Type of $\overline{\Phi}$ |
|---|---|---|---|---|---|
| $(\widehat{F}_4)_\emptyset$ | 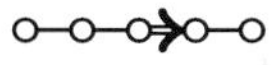 | $\widehat{F}_4$ | $(\widehat{F}_4)_{B_3}$ |  | $(\widehat{C}',\widehat{C})_1$ |
| $(\widehat{F}_4)_{A_1^2}$ |  | $(\widehat{B},\widehat{B}^\vee)_2$ | $(\widehat{F}_4)_{C_3}$ |  | $(\widehat{C}',\widehat{C})_1$ |
| $(\widehat{F}_4)_{C_2}$ |  | $(\widehat{C}',\widehat{C})_2$ | $(\widehat{F}_4)_{C_3\times A_1}$ |  | $\widehat{Z}_0$ |
| $(\widehat{F}_4)_{A_1^3}$ |  | $(\widehat{C}',\widehat{C})_1$ | $(\widehat{F}_4)_{B_4}$ |  | $\widehat{Z}_0$ |
| $(\widehat{F}_4)_{C_2\times A_1}$ |  | $(\widehat{C}^\vee,\widehat{C})_1$ | $(\widehat{F}_4)_{F_4}$ |  | $\widehat{Z}_0$ |
| $(\widehat{F}_4^\vee)_\emptyset$ | 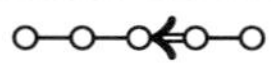 | $\widehat{F}_4^\vee$ | $(\widehat{F}_4^\vee)_{C_3}$ |  | $(\widehat{C}^\vee,\widehat{C})_1$ |
| $(\widehat{F}_4^\vee)_{A_1^2}$ |  | $\widehat{C}_2'$ | $(\widehat{F}_4^\vee)_{B_3}$ |  | $(\widehat{C}',\widehat{C})_1$ |
| $(\widehat{F}_4^\vee)_{C_2}$ |  | $(\widehat{B},\widehat{B}^\vee)_2$ | $(\widehat{F}_4^\vee)_{B_3\times A_1}$ |  | $\widehat{Z}_0$ |
| $(\widehat{F}_4^\vee)_{A_1^3}$ |  | $(\widehat{C}',\widehat{C})_1$ | $(\widehat{F}_4^\vee)_{C_4}$ |  | $\widehat{Z}_0$ |
| $(\widehat{F}_4^\vee)_{C_2\times A_1}$ |  | $(\widehat{C}',\widehat{C})_1$ | $(\widehat{F}_4^\vee)_{F_4}$ |  | $\widehat{Z}_0$ |
| $(\widehat{G}_2)_\emptyset$ | 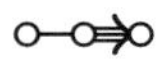 | $\widehat{G}_2$ | $(\widehat{G}_2)_{A_1^2}$ | 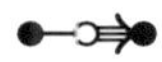 | $\widehat{Z}_0$ |
| $(\widehat{G}_2)_{A_1}$ | 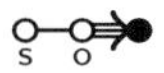 | $(\widehat{C}',\widehat{C})_1$ | $(\widehat{G}_2)_{G_2}$ |  | $\widehat{Z}_0$ |
| $(\widehat{G}_2^\vee)_\emptyset$ | 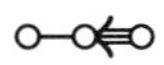 | $\widehat{G}_2^\vee$ | $(\widehat{G}_2^\vee)_{A_1^2}$ | 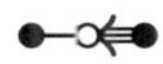 | $\widehat{Z}_0$ |
| $(\widehat{G}_2^\vee)_{A_1}$ | 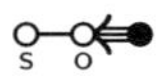 | $(\widehat{C}',\widehat{C})_1^+$ | $(\widehat{G}_2^\vee)_{G_2}$ |  | $\widehat{Z}_0$ |

## Acknowledgments

The authors are grateful to the referees for valuable suggestions.

## References

[Ar62] Shôrô Araki, *On root systems and an infinitesimal classification of irreducible symmetric spaces*, J. Math. Osaka City Univ. **13** (1962), 1–34. MR153782

[AV20] Andrea Appel, Bart Vlaar, *Universal k-matrices for quantum Kac-Moody algebras*. To appear in Representation Theory (2022).

[Ba86] Jean Bausch, *Automorphismes des algèbres de Kac-Moody affines* (French, with English summary), C. R. Acad. Sci. Paris Sér. I Math. **302** (1986), no. 11, 409–412. MR838590

[BBBR95] Valérie Back-Valente, Nicole Bardy-Panse, Hechmi Ben Messaoud, and Guy Rousseau, *Formes presque-déployées des algèbres de Kac-Moody: classification et racines relatives* (French), J. Algebra **171** (1995), no. 1, 43–96, DOI 10.1006/jabr.1995.1004. MR1314093

[Be89] Stephen Berman, *On generators and relations for certain involutory subalgebras of Kac-Moody Lie algebras*, Comm. Algebra **17** (1989), no. 12, 3165–3185, DOI 10.1080/00927878908823899. MR1030614

[BGG71] Joseph N. Bernšteĭn, I. M. Gel′fand, and Sergei I. Gel′fand, *Structure of representations that are generated by vectors of highest weight* (Russian), Funkcional. Anal. i Priložen. **5** (1971), no. 1, 1–9. MR0291204

[BK19] Martina Balagović and Stefan Kolb, *Universal K-matrix for quantum symmetric pairs*, J. Reine Angew. Math. **747** (2019), 299–353, DOI 10.1515/crelle-2016-0012. MR3905136

[Bo68] Nicolas Bourbaki, *Éléments de mathématique, Groupes et algèbres de Lie. Chapitres IV, V & VI.* Hermann, Paris (1968).

[Bo88] Richard Borcherds, *Generalized Kac-Moody algebras*, J. Algebra **115** (1988), no. 2, 501–512, DOI 10.1016/0021-8693(88)90275-X. MR943273

[BR03] Hechmi Ben Messaoud and Guy Rousseau, *Classification des formes réelles presque compactes des algèbres de Kac-Moody affines* (French, with French summary), J. Algebra **267** (2003), no. 2, 443–513, DOI 10.1016/S0021-8693(03)00345-4. MR2003338

[BR06] Hechmi Ben Messaoud and Guy Rousseau, *Sous-algèbres de Cartan des algèbres de Kac-Moody réelles presque déployées* (French, with English summary), J. Math. Soc. Japan **58** (2006), no. 4, 1009–1030. MR2276179

[BR07] Hechmi Ben Messaoud and Guy Rousseau, *Sous-algèbres de Cartan des algèbres de Kac-Moody affines réelles presque compactes* (French, with English summary), J. Lie Theory **17** (2007), no. 1, 1–25. MR2286878

[BW18] Huanchen Bao and Weiqiang Wang, *A new approach to Kazhdan-Lusztig theory of type B via quantum symmetric pairs* (English, with English and French summaries), Astérisque **402** (2018), vii+134. MR3864017

[Ca05] Roger W. Carter, *Lie algebras of finite and affine type*, Cambridge Studies in Advanced Mathematics, vol. 96, Cambridge University Press, Cambridge, 2005, DOI 10.1017/CBO9780511614910. MR2188930

[Ch84] Ivan V. Cherednik, *Factorizing particles on a half line, and root systems* (Russian, with English summary), Teoret. Mat. Fiz. **61** (1984), no. 1, 35–44. MR774205

[Ch92] Ivan Cherednik, *Quantum Knizhnik-Zamolodchikov equations and affine root systems*, Comm. Math. Phys. **150** (1992), no. 1, 109–136. MR1188499

[CH04] Meng-Kiat Chuah and Chu-Chin Hu, *Equivalence classes of Vogan diagrams*, J. Algebra **279** (2004), no. 1, 22–37, DOI 10.1016/j.jalgebra.2003.10.011. MR2078384

[CZ17] Meng-Kiat Chuah and Mingjing Zhang, *Outer automorphism groups of simple Lie algebras and symmetries of painted diagrams*, Forum Math. **29** (2017), no. 3, 555–562, DOI 10.1515/forum-2016-0023. MR3641665

[DK19] Liam Dobson and Stefan Kolb, *Factorisation of quasi $K$-matrices for quantum symmetric pairs*, Selecta Math. (N.S.) **25** (2019), no. 4, 1–55, DOI 10.1007/s00029-019-0508-5. MR4021849

[DM03] Gustav W. Delius and Niall J. MacKay, *Quantum group symmetry in sine-Gordon and affine Toda field theories on the half-line*, Comm. Math. Phys. **233** (2003), no. 1, 173–190, DOI 10.1007/s00220-002-0758-4. MR1958056

[Dr87] Vladimir G. Drinfelʹd, *Quantum groups*, Proceedings of the International Congress of Mathematicians, Vol. 1, 2 (Berkeley, Calif., 1986), Amer. Math. Soc., Providence, RI, 1987, pp. 798–820. MR934283

[ES18] Michael Ehrig and Catharina Stroppel, *Nazarov-Wenzl algebras, coideal subalgebras and categorified skew Howe duality*, Adv. Math. **331** (2018), 58–142, DOI 10.1016/j.aim.2018.01.013. MR3804673

[Ge70] I.M. Gel'fand, *The cohomology of infinite dimensional Lie algebras: some questions of integral geometry* (in Russian). Actes du Congres Intern. Math. (1970), tome 1: 95–111.

[GI14] Meinolf Geck, Lacrimioara Iancu, *Coxeter groups and automorphisms.* Preprint (2014), `arXiv:1412.5428`.

[Gü10] Hasan Gündoğan, *The component group of the automorphism group of a simple Lie algebra and the splitting of the corresponding short exact sequence*, J. Lie Theory **20** (2010), no. 4, 709–737. MR2778233

[He78] Sigurdur Helgason, *Differential geometry, Lie groups, and symmetric spaces*, Pure and Applied Mathematics, vol. 80, Academic Press, Inc. [Harcourt Brace Jovanovich, Publishers], New York-London, 1978. MR514561

[He84] André Heck, *Involutive automorphisms of root systems*, J. Math. Soc. Japan **36** (1984), no. 4, 643–658, DOI 10.2969/jmsj/03640643. MR759421

[Hé91] Jean-Yves Hée, *Système de racines sur un anneau commutatif totalement ordonné* (French, with English summary), Geom. Dedicata **37** (1991), no. 1, 65–102, DOI 10.1007/BF00150405. MR1094228

[HG09] Ernst Heintze and Christian Groß, *Finite order automorphisms and real forms of affine Kac-Moody algebras in the smooth and algebraic category*, Mem. Amer. Math. Soc. **219** (2012), no. 1030, viii+66, DOI 10.1090/S0065-9266-2012-00650-2. MR2985521

[Hu90] James E. Humphreys, *Reflection groups and Coxeter groups*, Cambridge Studies in Advanced Mathematics, vol. 29, Cambridge University Press, Cambridge, 1990, DOI 10.1017/CBO9780511623646. MR1066460

[Jim86] Michio Jimbo, *A $q$-analogue of $U(\mathfrak{gl}(N+1))$, Hecke algebra, and the Yang-Baxter equation*, Lett. Math. Phys. **11** (1986), no. 3, 247–252, DOI 10.1007/BF00400222. MR841713

[JZh01] Quanqin Jin and Zhixue Zhang, *On automorphisms of affine Kac-Moody algebras*, Comm. Algebra **29** (2001), no. 7, 2827–2858, DOI 10.1081/AGB-100104989. MR1848384

[Ka68] Victor G. Kac, *Simple irreducible graded Lie algebras of finite growth* (Russian), Izv. Akad. Nauk SSSR Ser. Mat. **32** (1968), 1323–1367. MR0259961

[Ka69] Victor G. Kac, *Automorphisms of finite order of semisimple Lie algebras* (Russian), Funkcional. Anal. i Priložen. **3** (1969), no. 3, 94–96. MR0251091

[Ka74] Victor G. Kac, *Infinite-dimensional Lie algebras, and the Dedekind $\eta$-function* (Russian), Funkcional. Anal. i Priložen. **8** (1974), no. 1, 77–78. MR0374210

[Ka90] Victor G. Kac, *Infinite-dimensional Lie algebras*, 3rd ed., Cambridge University Press, Cambridge, 1990, DOI 10.1017/CBO9780511626234. MR1104219

[Ko14] Stefan Kolb, *Quantum symmetric Kac-Moody pairs*, Adv. Math. **267** (2014), 395–469, DOI 10.1016/j.aim.2014.08.010. MR3269184

[Ko21] Stefan Kolb, *The bar involution for quantum symmetric pairs - hidden in plain sight*. Preprint (2021), `arXiv:2104.06120`.

[KS92] Petr P. Kulish and Evgeny K. Sklyanin, *Algebraic structures related to reflection equations*, J. Phys. A **25** (1992), no. 22, 5963–5975. MR1193836

[KS93] Petr P. Kulish and Ryu Sasaki, *Covariance properties of reflection equation algebras*, Progr. Theoret. Phys. **89** (1993), no. 3, 741–761, DOI 10.1143/PTP.89.741. MR1215910

[KS09] Stefan Kolb and Jasper V. Stokman, *Reflection equation algebras, coideal subalgebras, and their centres*, Selecta Math. (N.S.) **15** (2009), no. 4, 621–664, DOI 10.1007/s00029-009-0007-1. MR2565052

[KW92] Victor G. Kac and Shu Ping Wang, *On automorphisms of Kac-Moody algebras and groups*, Adv. Math. **92** (1992), no. 2, 129–195, DOI 10.1016/0001-8708(92)90063-Q. MR1155464

[Le88] Fernando Levstein, *A classification of involutive automorphisms of an affine Kac-Moody Lie algebra*, J. Algebra **114** (1988), no. 2, 489–518, DOI 10.1016/0021-8693(88)90308-0. MR936987

[Le99] Gail Letzter, *Symmetric pairs for quantized enveloping algebras*, J. Algebra **220** (1999), no. 2, 729–767, DOI 10.1006/jabr.1999.8015. MR1717368

[Le02] Gail Letzter, *Coideal Subalgebras and Quantum Symmetric Pairs*. In: New Directions in Hopf Algebras, MSRI publications 43, CUP (2002): 117–166.

[Le03] Gail Letzter, *Quantum symmetric pairs and their zonal spherical functions*, Transform. Groups **8** (2003), no. 3, 261–292, DOI 10.1007/s00031-003-0719-9. MR1996417

[Lo80] Eduard Looijenga, *Invariant theory for generalized root systems*, Invent. Math. **61** (1980), no. 1, 1–32, DOI 10.1007/BF01389892. MR587331

[Lu76] George Lusztig, *Coxeter orbits and eigenspaces of Frobenius*, Invent. Math. **38** (1976/77), no. 2, 101–159, DOI 10.1007/BF01408569. MR453885

[Lu94] George Lusztig, *Introduction to quantum groups*. Birkhäuser, Boston, 1994.

[Lu95] George Lusztig, *Classification of unipotent representations of simple p-adic groups*. Int. Math. Res. Not. (1995): 517–589.

[Lu03] George Lusztig, *Hecke algebras with unequal parameters*, CRM Monograph Series, vol. 18, American Mathematical Society, Providence, RI, 2003, DOI 10.1090/crmm/018. MR1974442

[LW20] Ming Lu and Weiqiang Wang, *Hall algebras and quantum symmetric pairs II: reflection functors*, Comm. Math. Phys. **381** (2021), no. 3, 799–855, DOI 10.1007/s00220-021-03965-8. MR4218672

[Mo68] Robert V. Moody, *A new class of Lie algebras*, J. Algebra **10** (1968), 211–230, DOI 10.1016/0021-8693(68)90096-3. MR229687

[MT72] Robert V. Moody and K. L. Teo, *Tits' systems with crystallographic Weyl groups*, J. Algebra **21** (1972), 178–190, DOI 10.1016/0021-8693(72)90016-6. MR320165

[Na92] Satoshi Naito, *Embedding into Kac-Moody algebras and construction of folding subalgebras for generalized Kac-Moody algebras*, Japan. J. Math. (N.S.) **18** (1992), no. 1, 155–171, DOI 10.4099/math1924.18.155. MR1173834

[NDS97] Masatoshi Noumi, Mathijs S. Dijkhuizen, and Tetsuya Sugitani, *Multivariable Askey-Wilson polynomials and quantum complex Grassmannians*, Special functions, $q$-series and related topics (Toronto, ON, 1995), Fields Inst. Commun., vol. 14, Amer. Math. Soc., Providence, RI, 1997, pp. 167–177, DOI 10.1090/s0002-9947-98-01971-0. MR1448686

[NS95] Masatoshi Noumi and Tetsuya Sugitani, *Quantum symmetric spaces and related q-orthogonal polynomials*, Group theoretical methods in physics (Toyonaka, 1994), World Sci. Publ., River Edge, NJ, 1995, pp. 28–40. MR1413733

[OV94] Arkady L. Onishchik, Ernest B. Vinberg, *Lie Groups and Lie Algebras III: Structure of Lie Groups and Lie Algebras.* Encyclopaedia of Mathematical Sciences **41**, Springer (1994).

[PK83] Dale H. Peterson and Victor G. Kac, *Infinite flag varieties and conjugacy theorems*, Proc. Nat. Acad. Sci. U.S.A. **80** (1983), no. 6, i, 1778–1782, DOI 10.1073/pnas.80.6.1778. MR699439

[Ri82] Roger W. Richardson, *Conjugacy classes of involutions in Coxeter groups*, Bull. Austral. Math. Soc. **26** (1982), no. 1, 1–15, DOI 10.1017/S0004972700005554. MR679916

[RS20] Nicolai Reshetikhin, Jasper Stokman, *Asymptotic boundary KZB operators and quantum Calogero-Moser spin chains.* Preprint (2020), `arXiv:2012.13497`.

[RSV15a] Nicolai Reshetikhin, Jasper Stokman, and Bart Vlaar, *Boundary quantum Knizhnik-Zamolodchikov equations and Bethe vectors*, Comm. Math. Phys. **336** (2015), no. 2, 953–986, DOI 10.1007/s00220-014-2227-2. MR3322392

[RSV15b] Nicolai Reshetikhin, Jasper Stokman, Bart Vlaar, *Boundary quantum Knizhnik-Zamolodchikov equations and fusion.* Ann. Henri Poincaré (2015): 1–41.

[RSV18] Nicolai Reshetikhin, Jasper Stokman, and Bart Vlaar, *Integral solutions to boundary quantum Knizhnik-Zamolodchikov equations*, Adv. Math. **323** (2018), 486–528, DOI 10.1016/j.aim.2017.10.041. MR3725884

[RV16] Vidas Regelskis, Bart Vlaar, *Reflection matrices, coideal subalgebras and generalized Satake diagrams of affine type.* Preprint (2016), `arXiv:1602.08471`.

[RV20] Vidas Regelskis, Bart Vlaar, *Quasitriangular coideal subalgebras of $U_q(\mathfrak{g})$ in terms of generalized Satake diagrams.* Bull. London Math. Soc. **54** (2020), no. 4: 693–715.

[Sa60] Ichirô Satake, *On representations and compactifications of symmetric Riemannian spaces*, Ann. of Math. (2) **71** (1960), 77–110, DOI 10.2307/1969880. MR118775

[Sch69] Doris J. Schattschneider, *On restricted roots of semi-simple algebraic groups*, J. Math. Soc. Japan **21** (1969), 94–115, DOI 10.2969/jmsj/02110094. MR236187

[Sk88] Evgeny K. Sklyanin, *Boundary conditions for integrable quantum systems*, J. Phys. A **21** (1988), no. 10, 2375–2389. MR953215

[Sp85] Tonny A. Springer, *Some results on algebraic groups with involutions*, Algebraic groups and related topics (Kyoto/Nagoya, 1983), Adv. Stud. Pure Math., vol. 6, North-Holland, Amsterdam, 1985, pp. 525–543, DOI 10.2969/aspm/00610525. MR803346

[SR20] Jasper V. Stokman, Nicolai Reshetikhin, *N-point spherical functions and asymptotic boundary KZB equations.* Inventiones mathematicae 229 (2022), 1–86. DOI 10.1007/s00222-022-01102-3.

[St19] Jasper V. Stokman, *Generalized Onsager algebras*, Algebr. Represent. Theory **23** (2020), no. 4, 1523–1541, DOI 10.1007/s10468-019-09903-6. MR4125590

[St21] Jasper V. Stokman, *Folded and contracted solutions of coupled classical dynamical Yang-Baxter and reflection equations*, Indag. Math. (N.S.) **32** (2021), no. 6, 1372–1411, DOI 10.1016/j.indag.2021.07.003. MR4334172

[SV15] Jasper Stokman and Bart Vlaar, *Koornwinder polynomials and the $XXZ$ spin chain*, J. Approx. Theory **197** (2015), 69–100, DOI 10.1016/j.jat.2014.03.003. MR3351541

Department of Physics, Astronomy and Mathematics, University of Hertfordshire, Hatfield AL10 9AB, UK, and Institute of Theoretical Physics and Astronomy, Vilnius University, Saulėtekio av. 3, Vilnius 10257, Lithuania.

*Email address*: vidas.regelskis@gmail.com

Department of Mathematics, Heriot-Watt University, Edinburgh, EH14 4AS, UK, and Max Planck Institute for Mathematics, Vivatsgasse 7, 53111 Bonn, Germany

*Email address*: B.Vlaar@hw.ac.uk

Contemporary Mathematics
Volume **780**, 2022
https://doi.org/10.1090/conm/780/15691

# Asymptotic boundary KZB operators and quantum Calogero-Moser spin chains

N. Reshetikhin and J. V. Stokman

ABSTRACT. Asymptotic boundary KZB equations describe the consistency conditions of degenerations of correlation functions for boundary Wess-Zumino-Witten-Novikov conformal field theory on a cylinder. In the first part of the paper we define asymptotic boundary KZB operators for connected real semisimple Lie groups $G$ with finite center. We prove their main properties algebraically using coordinate versions of Harish-Chandra's radial component map. We show that their commutativity is governed by a system of equations involving coupled versions of classical dynamical Yang-Baxter equations and reflection equations.

We use the coordinate radial components maps to introduce a new class of quantum superintegrable systems, called quantum Calogero-Moser spin chains. A quantum Calogero-Moser spin chain is a mixture of a quantum spin Calogero-Moser system associated to the restricted root system of $G$ and a one-dimensional spin chain with two-sided reflecting boundaries. The asymptotic boundary KZB operators provide explicit expressions for its first-order quantum Hamiltonians. We also explicitly describe the Schrödinger operator.

## 1. Introduction

This paper is part two of a series aimed at connecting harmonic analysis on affine symmetric spaces to boundary Wess-Zumino-Witten-Novikov (WZWN) conformal field theory (see [**20**] for part one).

**1.1.** Let $G$ be a noncompact real connected semisimple Lie group of real rank $r$ with finite center, $K = G^{\Theta}$ the compact subgroup of fixed points of a Cartan involution $\Theta$ of $G$, and $\mathfrak{h}_{\mathbb{R}}$ a corresponding maximally noncompact Cartan subalgebra of the Lie algebra $\mathfrak{g}_{\mathbb{R}}$ of $G$. We write $\mathfrak{g}$ and $\mathfrak{h}$ for the complexifications of $\mathfrak{g}_{\mathbb{R}}$ and $\mathfrak{h}_{\mathbb{R}}$. Let $\mathcal{M}_{1,n}$ be the moduli space of elliptic curves with $n$ marked points.

In [**10**] Felder and Wieczerkowski introduced the twisted WZWN conformal block bundle over $\mathcal{M}_{1,n} \times \mathfrak{h}$ with flat connection. The explicit realisation of the flat connection leads to explicit commuting first-order differential operators, called the Knizhnik-Zamolodchikov-Bernard (KZB) operators [**1, 10**]. In the limit when the period and marked points go to infinity the KZB flat connection over $\mathcal{M}_{1,n} \times \mathfrak{h}$

2020 *Mathematics Subject Classification.* 17B80, 43A90.

Both authors are supported by the Netherlands Organization for Scientific Research (NWO). The first author was partially supported by the grants NSF DMS-1902226 and RFBR No. 18-01-00916.

becomes a flat connection over $\mathfrak{h}$. We call the corresponding differential operators the *asymptotic KZB operators*. They were studied in [**6**].

In a similar way WZWN conformal field theory on a disc (or on a half plane, or on a strip) with $n$ marked points and Cardy type conformal boundary conditions [**2, 24**] leads to commuting *asymptotic boundary KZB operators* on $\mathfrak{h}$, see [**20**]. In this case $G$ is real split, and $K = G^{\Theta}$ is the symmetry group at the boundary. One of the main goals of this paper is to introduce the *asymptotic boundary KZB operators for all noncompact real connected semisimple Lie groups $G$ with finite center.*

The resulting asymptotic boundary KZB operators (see Section 6) are first-order commuting differential operators on $\mathfrak{a}$, where $\mathfrak{a}$ is the complexification of the $(-1)$-eigenspace of $\mathfrak{h}_{\mathbb{R}}$ with respect to the Cartan involution $\theta_{\mathbb{R}}$ of $\mathfrak{g}_{\mathbb{R}}$ associated to $\Theta$. We construct these asymptotic boundary KZB operators and derive their commutativity using generalised Harish-Chandra [**3, 11**] radial component maps with respect to the Cartan decomposition $G = KAK$ of $G$, where $A := \exp(\mathfrak{a}_{\mathbb{R}})$ (see Section 4 for the description of the radial component maps). Furthermore, we show that the asymptotic boundary KZB operators are particular first-order quantum Hamiltonians for a quantum superintegrable system whose algebra of Hamiltonians is isomorphic to a quotient of $Z(U(\mathfrak{g}))^{\otimes(n+1)}$, where $Z(U(\mathfrak{g}))$ denotes the center of the universal enveloping algebra $U(\mathfrak{g})$ (see, e.g., [**16**] for a survey on classical and quantum integrability). These quantum superintegrable systems are parametrized by a choice of $n$ representations of $G$ and two representations of $K$. We call them *quantum Calogero-Moser spin chains* because they can be regarded both as a system of interacting "spin" particles and as a version of a spin chain of Gaudin type with reflecting boundary conditions.

One of the surprises is the fact that the commutativity of the asymptotic boundary KZB operators is equivalent to a hierarchy of *coupled* classical dynamical Yang-Baxter and reflection type equations for the local factors of the operators. These local factors are folded and contracted versions of a natural generalisation (5.10) of Felder's [**9**] trigonometric dynamical $r$-matrix.

For $G$ real split the coupled integrability equations decouple, reducing them to the mixed classical dynamical Yang-Baxter equations and associated classical dynamical reflection equations from [**20**]. This decoupling is caused by the fact that the centraliser $M = Z_K(A)$ of $A$ in $K$ is finite and discrete for real split $G$. In this case the mixed classical dynamical Yang-Baxter equations and the associated classical dynamical reflection equation are a direct consequence of the well-known fact that Felder's trigonometric $r$-matrix satisfies the classical dynamical Yang-Baxter equation (this will be explained in [**21**]).

**1.2.** We now describe the results in a bit more detail from the viewpoint of harmonic analysis. Harmonic analysis on symmetric spaces provides a natural representation theoretical context for a class of quantum superintegrable one-dimensional particle systems called quantum spin Calogero-Moser systems [**12, 13, 17, 18, 20**]. The main observation is as follows. For two finite-dimensional $K$-representations $(\sigma_\ell, V_\ell)$, $(\sigma_r, V_r)$ the space $C^\infty_{\sigma_\ell,\sigma_r}(G)$ of spherical functions consists of the smooth functions $f : G \to \mathrm{Hom}(V_r, V_\ell)$ satisfying the equivariance property

$$f(k_\ell g k_r^{-1}) = \sigma_\ell(k_\ell) f(g) \sigma_r(k_r^{-1}) \qquad \forall\, g \in G,\ \forall\, k_\ell, k_r \in K$$

(see, e.g., [**3, 11, 23**]). A spherical function $f \in C^\infty_{\sigma_\ell,\sigma_r}(G)$ is uniquely determined by its radial component $f|_A$. The radial components of the biinvariant differential operators acting on $C^\infty_{\sigma_\ell,\sigma_r}(G)$ become vector-valued differential operators on $A$. They form an algebra of commuting differential operators acting on the space $C^\infty_{\sigma_\ell,\sigma_r}(G)|_A$. This algebra is the algebra of quantum Hamiltonians of the associated quantum spin Calogero-Moser system. The Schrödinger operator of this system corresponds to the quadratic Casimir element $\Omega \in Z(U(\mathfrak{g}))$, regarded as a biinvariant differential operator on $C^\infty_{\sigma_\ell,\sigma_r}(G)$. The quantum integrals are the radial components of the action of $U(\mathfrak{g})^{K,\mathrm{opp}} \otimes_{Z(U(\mathfrak{g}))} U(\mathfrak{g})^K$ on $C^\infty_{\sigma_\ell,\sigma_r}(G)$ by left and right $G$-invariant differential operators, where $U(\mathfrak{g})^K$ is the algebra of $\mathrm{Ad}(K)$-invariant elements in $U(\mathfrak{g})$ and $U(\mathfrak{g})^{K,\mathrm{opp}}$ is the associated opposite algebra.

The connection to asymptotic boundary WZWN conformal field theory arises when the left $K$-representation $\sigma_\ell$ is of the form $(\sigma_{\ell;n}, V_\ell \otimes \underline{U})$ with $\underline{U} = U_1 \otimes \cdots \otimes U_n$ the tensor product of $n$ finite-dimensional $G$-representations $(\tau_i, U_i)$ and $\sigma_{\ell;n}$ the diagonal $K$-action on $V_\ell \otimes \underline{U}$. The analytic Weyl group $W := N_K(A)/M$ naturally acts on $A$, as well as on the space $(V_\ell \otimes \underline{U} \otimes V_r^*)^M$ of $M$-invariant vectors in $V_\ell \otimes \underline{U} \otimes V_r^*$. Restriction to $A$ now defines a linear isomorphism

$$|_A : C^\infty_{\sigma_{\ell;n},\sigma_r}(G) \xrightarrow{\sim} C^\infty\big(A; (V_\ell \otimes \underline{U} \otimes V_r^*)^M\big)^W. \tag{1.1}$$

The space $C^\infty\big(A; (V_\ell \otimes \underline{U} \otimes V_r^*)^M\big)^W$ is the smooth analogue of the space of asymptotic twisted conformal blocks for boundary WZWN conformal field theory on a cylinder [**2, 6, 10, 20, 24**], asymptotic in the sense that the insertion points have been sent to infinity within an appropriate asymptotic sector. Typical examples of spherical functions in $C^\infty_{\sigma_{\ell;n},\sigma_r}(G)$ are

$$g \mapsto (\phi_\ell \otimes \mathrm{id}_{\underline{U}})(\Psi_1 \otimes \mathrm{id}_{U_2 \otimes \cdots \otimes U_n}) \cdots (\Psi_{n-1} \otimes \mathrm{id}_{U_n})\Psi_n \pi_n(g)\phi_r, \tag{1.2}$$

where $(\pi_i, \mathcal{H}_i)$ $(0 \leq i \leq n)$ are smooth $G$-representations, $\Psi_i \in \mathrm{Hom}_G(\mathcal{H}_i, \mathcal{H}_{i-1} \otimes U_i)$ $(1 \leq i \leq n)$ are $G$-intertwiners (asymptotic vertex operators) and $\phi_\ell \in \mathrm{Hom}_K(\mathcal{H}_0, V_\ell)$, $\phi_r \in \mathrm{Hom}_K(V_r, \mathcal{H}_n)$ are $K$-intertwiners (asymptotic boundary states). Functions of the form (1.2) appear naturally as limits of $n$-point functions in twisted conformal blocks for boundary WZWN conformal field theory on a cylinder, cf. [**6, 10**]. The twisting corresponds to the insertion of $\pi_n(g)$ in (1.2).

**1.3.** Replacing the $\Psi_i$ in (1.2) by secondary asymptotic field operators $(\pi_{i-1}(g_{i-1}) \otimes \mathrm{id}_{U_i})\Psi_i$ is providing smooth $V_\ell \otimes \underline{U} \otimes V_r^*$-valued functions on $G^{\times(n+1)}$, equivariant with respect to the $K \times G^{\times n} \times K$-action (2.1) on $G^{\times(n+1)}$. The corresponding vector space $C_{\sigma_\ell,\underline{\tau},\sigma_r}(G^{\times(n+1)})$ of $K \times G^{\times n} \times K$-equivariant functions is naturally isomorphic to $C^\infty_{\sigma_{\ell;n},\sigma_r}(G)$,

$$C^\infty_{\sigma_\ell,\underline{\tau},\sigma_r}(G^{\times(n+1)}) \xrightarrow{\sim} C^\infty_{\sigma_{\ell;n},\sigma_r}(G), \qquad f \mapsto f^\flat \tag{1.3}$$

with $f^\flat(g) := f(1, \ldots, 1, g)$. The upshot of this observation is that $C^\infty_{\sigma_\ell,\underline{\tau},\sigma_r}(G^{\times(n+1)})$ admits a natural $U(\mathfrak{g})^{K,\mathrm{opp}} \otimes Z(U(\mathfrak{g}))^{\otimes(n-1)} \otimes U(\mathfrak{g})^K$-action in terms of coordinate-wise invariant differential operators. For the commutative subalgebra $Z(U(\mathfrak{g}))^{\otimes(n+1)}$ the action is given in terms of coordinate-wise biinvariant differential operators. Pushing this action through the two isomorphisms (1.3) and (1.1) provides algebras $A^{\sigma_\ell,\underline{\tau},\sigma_r} \subseteq B^{\sigma_\ell,\underline{\tau},\sigma_r}$ of $W$-invariant differential operators on $A$ with coefficients in $\mathrm{End}((V_\ell \otimes \underline{U} \otimes V_r^*)^M)$, with $A^{\sigma_\ell,\underline{\tau},\sigma_r}$ commutative. They serve, after an appropriate gauge, as the algebras of quantum Hamiltonians and quantum integrals for a quantum superintegrable system, which we call the quantum Calogero-Moser spin chain.

Gauged radial components of the action of the Casimir element $\Omega \in Z(U(\mathfrak{g}))$ as biinvariant differential operator on the $j$th coordinate of $C^{\infty}_{\sigma_\ell,\underline{\tau},\sigma_r}(G^{\times(n+1)})$ $(0 \leq j \leq n)$ are providing commuting second-order quantum Hamiltonians $\widetilde{D}^{\sigma_\ell,\underline{\tau},\sigma_r}_{\Omega,j;M}$ $(0 \leq j \leq n)$. We will derive explicit expressions for $\widetilde{D}^{\sigma_\ell,\underline{\tau},\sigma_r}_{\Omega,j;M}$ by computing $\widetilde{D}^{\sigma_\ell,\underline{\tau},\sigma_r}_{\Omega,n;M}$, which serves as the Schrödinger operator of the quantum Calogero-Moser spin chain, and the differences

$$\widetilde{\mathcal{D}}^{\sigma_\ell,\underline{\tau},\sigma_r}_{i;M} := \widetilde{D}^{\sigma_\ell,\underline{\tau},\sigma_r}_{\Omega,i;M} - \widetilde{D}^{\sigma_\ell,\underline{\tau},\sigma_r}_{\Omega,i-1;M} \qquad (1 \leq i \leq n).$$

The vector-valued potential $V^{\sigma_\ell,\underline{\tau},\sigma_r}$ (see (5.16)) of the Schrödinger operator $\widetilde{D}^{\sigma_\ell,\underline{\tau},\sigma_r}_{\Omega,n;M}$, which we will compute explicitly using standard techniques [**3, 11, 12**] from harmonic analysis on symmetric spaces, has the typical pairwise $\sinh^{-2}$-interaction terms associated to the roots of the restricted root system of $G$. The first-order commuting differential operators $\widetilde{\mathcal{D}}^{\sigma_\ell,\underline{\tau},\sigma_r}_{i;M}$ are the asymptotic boundary KZB operators. They are of the form

$$\widetilde{\mathcal{D}}^{\sigma_\ell,\underline{\tau},\sigma_r}_{i;M} = \sum_{j=1}^{r} \tau_i(x_j)\partial_{x_j} - \widehat{\kappa}_i - \sum_{k=1}^{i-1} r^+_{ki} - \sum_{k=i+1}^{N} r^-_{ik}$$

with $\{x_i\}_{i=1}^r$ an orthonormal basis of $\mathfrak{a}_{\mathbb{R}}$. The local factors $\kappa_i, r^+_{ki}, r^-_{ik}$ are $\mathrm{End}((V_\ell \otimes \underline{U} \otimes V_r^*)^M)$-valued functions on $\mathfrak{a}_{\mathbb{R}}$, acting nontrivially on the tensor components $V_\ell \otimes U_i \otimes V_r^*$, $U_k \otimes U_i$ and $U_i \otimes U_k$ within $(V_\ell \otimes \underline{U} \otimes V_r^*)^M$, respectively. We will show that the operators $r^{\pm}_{ik}(a)$ are given by the action on $U_i \otimes U_k$ of $\theta$-(anti)symmetrised versions of the non-split analogue of Felder's [**9**] classical dynamical $r$-matrix $r(a)$, and the core component $\widehat{\kappa}_i^{\mathrm{core}}(a)$ of $\widehat{\kappa}_i(a)$ is the action on $U_i$ of a $\theta$-twisted contraction of $r(a)$ (see Proposition 6.10 and (6.8)).

As mentioned earlier, the operators $\widehat{\kappa}_i, r^+_{ki}$ and $r^-_{ik}$ are solutions of a hierarchy of integrability equations (see Theorem 6.12). It contains a classical dynamical reflection type equation for $\widehat{\kappa}$ relative to $(r^+, r^-)$ (see (6.15)) and three coupled classical dynamical Yang-Baxter-reflection type equations for the triple $(r^+, r^-, \widehat{\kappa})$ (see (6.16)).

In [**20**] formal $n$-point spherical functions were introduced when $G$ is real split. They provide common eigenfunctions for all the quantum Hamiltonians after a suitable gauge. In [**20**] this approach was used to arrive at the explicit form of the asymptotic boundary KZB operators, and to prove their commutativity. The theory of formal $n$-point spherical functions for non-split $G$ is yet to be developed, but we do shortly discuss global $n$-point spherical functions.

The results and techniques in this paper are closely related to the Etingof-Schiffmann-Varchenko theory on generalised trace functions, see, e.g., [**5–8**] (our current trigonometric, non-affine level of the theory was discussed for generalised trace functions in [**6**]). From the harmonic analysis point of view the Etingof-Schiffmann-Varchenko theory is related to the symmetric pair $(G \times G, \mathrm{diag}(G))$, with $\mathrm{diag}(G)$ the diagonal embedding of $G$ into $G \times G$. The extension of the results in the current paper to the level of (quantum) affine symmetric pairs is currently under investigation.

The content of the paper is as follows.

In Section 2 we introduce the space of global $n$-point spherical functions and study the coordinate-wise action of (bi)invariant differential operators on $C^{\infty}_{\sigma_\ell,\underline{\tau},\sigma_r}(G^{\times(n+1)}) \simeq C^{\infty}_{\sigma_{\ell;n},\sigma_r}(G)$. This part of the paper, which does not yet require radial component maps, will be developed for the more general context involving an arbitrary real connected Lie group $G$ and two closed Lie subgroups $K_\ell$ and $K_r$.

Section 3 is a short section in which we recall some basic facts about the structure theory of real semisimple Lie groups. This section is mainly meant to fix notations.

In Section 4 we introduce the coordinate radial component maps and prove their main algebraic properties.

In Section 5 we use the coordinate radial component maps to define the algebras of quantum Hamiltonians and quantum integrals for the quantum Calogero-Moser spin chain. We furthermore derive the explicit expression for the Schrödinger operator of the quantum Calogero-Moser spin chain.

In Section 6 we derive the explicit expressions for the asymptotic boundary KZB operators and we show that the local factors of the asymptotic boundary KZB operators satisfy coupled classical dynamical Yang-Baxter-reflection equations.

Finally, in Section 7 we explicitly compute the main structural ingredients in case $G = \mathrm{SU}(p,r)$ with $1 \leq r \leq p$ (in this case the underlying restricted root system is of type $\mathrm{BC}_r$).

**Conventions.** We write $\otimes_{\mathcal{A}}$ for the tensor product over a complex associative algebra $\mathcal{A}$. For $\mathcal{A} = \mathbb{C}$ we simply denote it by $\otimes$. A similar convention is used for hom-spaces; $\mathrm{Hom}_{\mathcal{A}}(E,F)$ denotes the space of $\mathcal{A}$-linear maps, and $\mathrm{Hom}(E,F)$ the space of complex linear maps.

## 2. $n$-Point spherical functions

Let $G$ be a real connected Lie group, and $K_\ell, K_r \subseteq G$ two closed Lie subgroups. Write for $n \in \mathbb{Z}_{\geq 1}$,

$$\mathcal{G}_n := K_\ell \times G^{\times n} \times K_r.$$

Elements in $G^{\times n}$ will be denoted by $\underline{h} = (h_1, \ldots, h_n)$, and elements in $G^{\times(n+1)}$ by $\mathbf{g} = (g_0, \ldots, g_n)$. Consider the left $\mathcal{G}_n$-action on $G^{\times(n+1)}$ defined by

$$(k_\ell, \underline{h}, k_r) \cdot \mathbf{g} := (k_\ell g_0 h_1^{-1}, h_1 g_1 h_2^{-1}, \ldots, h_n g_n k_r^{-1}). \tag{2.1}$$

For finite-dimensional smooth $G$-representations $(\tau_i, U_i)$ $(1 \leq i \leq n)$ write $\underline{\tau} = \tau_1 \otimes \cdots \otimes \tau_n$ for the tensor product representation of $G^{\times n}$ on $\underline{U} := U_1 \otimes \cdots \otimes U_n$. Furthermore, fix finite-dimensional smooth representations $(\sigma_\ell, V_\ell)$, $(\sigma_r, V_r)$ of $K_\ell$ and $K_r$ respectively, and write $(\sigma_r^*, V_r^*)$ for the $K_r$-representation dual to $(\sigma_r, V_r)$.

Definition 2.1. Let $C^{\infty}_{\sigma_\ell,\underline{\tau},\sigma_r}(G^{\times(n+1)})$ be the space of smooth functions

$$f : G^{\times(n+1)} \to V_\ell \otimes \underline{U} \otimes V_r^*$$

satisfying

$$f((k_\ell, \underline{h}, k_r)\cdot \mathbf{g}) = (\sigma_\ell(k_\ell) \otimes \underline{\tau}(\underline{h}) \otimes \sigma_r^*(k_r)) f(\mathbf{g}) \qquad \forall\, (k_\ell, \underline{h}, k_r) \in \mathcal{G}_n,\ \forall\, \mathbf{g} \in G^{\times(n+1)}.$$

Remark 2.2. For $n = 0$ the space $C^{\infty}_{\sigma_\ell,\sigma_r}(G)$ consists of the smooth functions $f : G \to V_\ell \otimes V_r^*$ satisfying

$$f(k_\ell g k_r^{-1}) = (\sigma_\ell(k_\ell) \otimes \sigma_r^*(k_r)) f(g) \qquad \forall\, k_\ell \in K_\ell,\ \forall\, k_r \in K_r,\ \forall\, g \in G.$$

When $K_\ell = K_r$, the space $C^\infty_{\sigma_\ell,\sigma_r}(G)$ is called the space of $\sigma_\ell\otimes\sigma_r^*$-spherical functions on $G$ with respect to the Lie subgroup $K_\ell$ of $G$.

REMARK 2.3. We will use standard tensor-leg notations. For example, for $v_\ell\in V_\ell$, $\underline{u}\in\underline{U}$, $\phi_r\in V_r^*$ and $g_1,g_2\in G$ the tensor

$$(\mathrm{id}_{V_\ell}\otimes\tau_1(g_1)\otimes\tau_2(g_2)\otimes\mathrm{id}_{U_3}\otimes\cdots\otimes\mathrm{id}_{U_n}\otimes\mathrm{id}_{V_r^*})(v_\ell\otimes\underline{u}\otimes\phi_r)$$

in $V_\ell\otimes\underline{U}\otimes V_r^*$ will be denoted by $(\tau_1(g_1)\otimes\tau_2(g_2))(v_\ell\otimes\underline{u}\otimes\phi_r)$, or $\tau_1(g_1)\tau_2(g_2)(v_\ell\otimes\underline{u}\otimes\phi_r)$. We will write omitted components in vectors with a cap. For instance, $(g_0,\ldots,\hat{g}_j,\ldots,g_n)$ stands for the $n$-vector $(g_0,\ldots,g_{j-1},g_{j+1},\ldots,g_n)$ in $G^{\times n}$.

We write $(\sigma_{\ell;n},V_\ell\otimes\underline{U})$ for the finite-dimensional $K_\ell$-representation defined by

$$\sigma_{\ell;n}(k_\ell):=\sigma_\ell(k_\ell)\otimes\tau_1(k_\ell)\otimes\cdots\otimes\tau_n(k_\ell)\qquad\forall\,k_\ell\in K_\ell.$$

For a smooth function $f: G^{\times(n+1)}\to V_\ell\otimes\underline{U}\otimes V_r^*$ we define $f^\flat: G\to V_\ell\otimes\underline{U}\otimes V_r^*$ by

$$f^\flat(g):=f(1,\ldots,1,g)\qquad(g\in G).\tag{2.2}$$

LEMMA 2.4. *The linear map $f\mapsto f^\flat$ restricts to a linear isomorphism*

$$C^\infty_{\sigma_\ell,\underline{\tau},\sigma_r}(G^{\times(n+1)})\xrightarrow{\sim}C^\infty_{\sigma_{\ell;n},\sigma_r}(G)$$

*The preimage of $f^\flat\in C^\infty_{\sigma_{\ell;n},\sigma_r}(G)$ is the function $f\in C^\infty_{\sigma_\ell,\underline{\tau},\sigma_r}(G^{\times(n+1)})$ defined by*

$$f(\mathbf{g}):=(\tau_1(g_0^{-1})\otimes\tau_2(g_1^{-1}g_0^{-1})\otimes\cdots\otimes\tau_n(g_{n-1}^{-1}\cdots g_0^{-1}))f^\flat(g_0g_1\cdots g_n)\tag{2.3}$$

*for $\mathbf{g}\in G^{\times(n+1)}$.*

PROOF. Note that $f^\flat\in C^\infty_{\sigma_{\ell;n},\sigma_r}(G)$ for $f\in C^\infty_{\sigma_\ell,\underline{\tau},\sigma_r}(G^{\times(n+1)})$ since

$$(k_\ell,(k_\ell,\ldots,k_\ell),k_r)\cdot(1,\ldots,1,g)=(1,\ldots,1,k_\ell gk_r^{-1})$$

in $G^{\times(n+1)}$ for $k_\ell\in K_\ell$, $k_r\in K_r$ and $g\in G$. Conversely, for $f^\flat\in C^\infty_{\sigma_{\ell;n},\sigma_r}(G)$, it follows from the formula

$$(1,g_0,g_0g_1,\ldots,g_0g_1\cdots g_{n-1},1)\cdot\mathbf{g}=(1,\ldots,1,g_0g_1\cdots g_n)\qquad(\mathbf{g}\in G^{\times(n+1)})$$

that $f$, defined by (2.3), lies in $C^\infty_{\sigma_\ell,\underline{\tau},\sigma_r}(G^{\times(n+1)})$. The lemma now follows easily. □

For a complex Lie algebra $L$ we write $U(L)$ for its universal enveloping algebra. It is a Hopf algebra, with counit $\epsilon: U(L)\to\mathbb{C}$ the unital algebra homomorphism satisfying $\epsilon(y)=0$ $(y\in L)$, comultiplication $\Delta: U(L)\to U(L)\otimes U(L)$ the unital algebra homomorphism satisfying $\Delta(y)=y\otimes 1+1\otimes y$ $(y\in L)$, and with antipode $S: U(L)\to U(L)$ the unital algebra anti-isomorphism satisfying $S(y)=-y$ $(y\in L)$. For $k\in\mathbb{Z}_{>0}$ let $\Delta^{(k)}: U(L)\to U(L)^{\otimes(k+1)}$ be the $k$th iterated comultiplication, defined inductively by $\Delta^{(1)}=\Delta$ and $\Delta^{(k)}=(\Delta\otimes\mathrm{id}_{U(L)}^{\otimes(k-1)})\Delta^{(k-1)}$ for $k>1$. We will use Sweedler's notation for the expression of $\Delta^{(k)}(u)$ $(u\in U(L))$ as sum of pure tensors,

$$\Delta^{(k)}(u)=\sum_{(u)}u_{(1)}\otimes\cdots\otimes u_{(k+1)}.$$

For a finite-dimensional $U(L)$-module $(\sigma,V)$ we write $(\sigma^*,V^*)$ for the dual $U(L)$-module, defined by $(\sigma^*(u)f)(v)=-f(\sigma(u)v)$ for $u\in U(L)$, $v\in V$ and $f\in V^*$. We write $V^L$ for the space of $L$-invariant elements in $V$.

Write $\mathfrak{g}_{\mathbb{R}}$, $\mathfrak{k}_{\ell,\mathbb{R}}$ and $\mathfrak{k}_{r,\mathbb{R}}$ for the Lie algebras of $G$, $K_\ell$ and $K_r$ respectively. Their complexifications are denoted by $\mathfrak{g}$, $\mathfrak{k}_\ell$ and $\mathfrak{k}_r$. Differentiating the representations $(\sigma_\ell, V_\ell)$, $(\sigma_r, V_r)$ and $(\tau_i, U_i)$ turns $V_\ell$ into a left $U(\mathfrak{k}_\ell)$-module, $V_r$ into a left $U(\mathfrak{k}_r)$-module and the $U_i$ into left $U(\mathfrak{g})$-modules. The corresponding representation maps will again be denoted by $\sigma_\ell, \sigma_r$ and $\tau_i$.

Let $V$ be a finite-dimensional complex vector space. We denote by $C^\infty(G^{\times(n+1)}; V)$ the space of smooth $V$-valued functions on $G^{\times(n+1)}$. We write $C^\infty(G^{\times(n+1)})$ when $V = \mathbb{C}$.

We have a left $U(\mathfrak{g})^{\otimes(n+1)}$-action on $C^\infty(G^{\times(n+1)}; V)$, defined by

$$u_0 \otimes \cdots \otimes u_n \mapsto u_0[0]u_1[1]\cdots u_n[n] \qquad (u_j \in U(\mathfrak{g}))$$

with $u \mapsto u[j]$ $(u \in U(\mathfrak{g}))$ the action of $U(\mathfrak{g})$ on the $j^{th}$ coordinate of $f \in C^\infty(G^{\times(n+1)}; V)$ by left $G$-invariant differential operators. Concretely, the action $u \mapsto u[j]$ of $U(\mathfrak{g})$ on $C^\infty(G^{\times(n+1)}; V)$ is determined by

$$(y[j]f)(\mathbf{g}) := \frac{d}{dt}\bigg|_{t=0} f(g_0, \ldots, g_{j-1}, g_j \exp(ty), g_{j+1}, \ldots, g_n) \qquad (y \in \mathfrak{g}_{\mathbb{R}}),$$

where $\exp : \mathfrak{g}_{\mathbb{R}} \to G$ is the exponential map.

REMARK 2.5. We write $u \mapsto u\langle j\rangle$ for the coordinate-wise action of $U(\mathfrak{g})$ on $C^\infty(G^{\times(n+1)}; V)$ by right $G$-invariant differential operators, so that

$$(y\langle j\rangle f)(\mathbf{g}) := \frac{d}{dt}\bigg|_{t=0} f(g_0, \ldots, g_{j-1}, \exp(-ty)g_j, g_{j+1}, \ldots, g_n) \qquad (y \in \mathfrak{g}_{\mathbb{R}}).$$

Note that

$$(u\langle j\rangle f)(\mathbf{g}) = \big(\mathrm{Ad}_{g_j^{-1}}(S(u))[j]f\big)(\mathbf{g})$$

for $u \in U(\mathfrak{g})$. In particular, $u\langle j\rangle f = S(u)[j]f$ for $u \in Z(U(\mathfrak{g}))$, with $Z(U(\mathfrak{g}))$ the center of $U(\mathfrak{g})$.

Let $C_g$ be the conjugation action of $g \in G$ on $G$, defined by $C_g(h) := ghg^{-1}$ for $g, h \in G$. Differentiating $C_g$ defines the adjoint action $\mathrm{Ad}_g \in \mathrm{Aut}(\mathfrak{g}_{\mathbb{R}})$ of $g \in G$ on $\mathfrak{g}_{\mathbb{R}}$. It extends to an action of $G$ on $U(\mathfrak{g})$ by unital algebra automorphisms, which will also be denoted by $g \mapsto \mathrm{Ad}_g$.

PROPOSITION 2.6. *For $f \in C^\infty_{\sigma_\ell,\underline{\tau},\sigma_r}(G^{\times(n+1)})$ and $0 \le j < n$ we have*

$$\begin{aligned}&\big(u[j]f\big)(\mathbf{g})\\ &\quad= \sum_{(u)} A_{j,(u)}(\mathbf{g})\big(\mathrm{Ad}_{g_n^{-1}\cdots g_{j+1}^{-1}}(u_{(n-j+1)})[n]f\big)(g_0, \ldots, \hat{g}_j, \ldots, g_n, C_{g_n^{-1}\cdots g_{j+1}^{-1}}(g_j))\end{aligned}$$

*for $\mathbf{g} \in G^{\times(n+1)}$ and $u \in U(\mathfrak{g})$, with $A_{j,(u)}(\mathbf{g}) \in \mathrm{End}(V_\ell \otimes \underline{U} \otimes V_r^*)$ given by*

$$\begin{aligned}&A_{j,(u)}(\mathbf{g})\\ &\quad:= \tau_{j+1}(S(u_{(1)}))\tau_{j+1}(g_j^{-1}g_{j+1}) \otimes \tau_{j+2}(g_{j+1}^{-1})\tau_{j+2}(S(u_{(2)}))\tau_{j+2}(g_j^{-1}g_{j+1}g_{j+2}) \otimes \cdots\\ &\qquad\cdots \otimes \tau_n(g_{n-1}^{-1}\cdots g_{j+1}^{-1})\tau_n(S(u_{(n-j)}))\tau_n(g_j^{-1}g_{j+1}\cdots g_n).\end{aligned}$$

PROOF. For the duration of the proof we use the shorthand notation

$$F^{(j)}(\mathbf{g}) := F(g_0, \ldots, \hat{g}_j, \ldots, g_n, C_{g_n^{-1}\cdots g_{j-1}^{-1}}(g_j))$$

for $F \in C^\infty(G^{\times(n+1)}; V_\ell \otimes \underline{U} \otimes V_r^*)$ and $0 \leq j < n$. One then easily checks that for $y \in \mathfrak{g}_\mathbb{R}$,

$$\big(y[j]F^{(j)}\big)(\mathbf{g}) = \big(\mathrm{Ad}_{g_n^{-1}\cdots g_{j+1}^{-1}}(y)[n]F\big)^{(j)}(\mathbf{g}). \tag{2.4}$$

Consider now the standard filtration $U(\mathfrak{g}) = \bigcup_{k\in\mathbb{Z}_{\geq 0}} U_k(\mathfrak{g})$ of $U(\mathfrak{g})$. We prove the proposition for $u \in U_k(\mathfrak{g})$ by induction in $k$. For $k = 0$ the result follows from the formula

$$f(\mathbf{g}) = \big(\tau_j(g_j^{-1}g_{j+1}) \otimes \cdots \otimes \tau_n(g_{n-1}^{-1}\cdots g_j^{-1}g_{j+1}\cdots g_n)\big)f^{(j)}(\mathbf{g}),$$

which holds true since $f \in C^\infty_{\sigma_\ell,\underline{\tau},\sigma_r}(G^{\times(n+1)})$. For the induction step, suppose the proposition is correct for $u \in U_k(\mathfrak{g})$, so

$$\big(u[j]f\big)(\mathbf{g}) = \sum_{(u)} A_{j,(u)}(\mathbf{g})\big(\mathrm{Ad}_{g_n^{-1}\cdots g_{j+1}^{-1}}(u_{(n-j+1)})[n]f\big)^{(j)}(\mathbf{g}).$$

Then for $y \in \mathfrak{g}_\mathbb{R}$ we have by (2.4),

$$\begin{aligned}\big((yu)[j]f\big)(\mathbf{g}) &= \sum_{(u)} \big(y[j]A_{j,(u)}\big)(\mathbf{g})\big(\mathrm{Ad}_{g_n^{-1}\cdots g_{j+1}^{-1}}(u_{(n-j+1)})[n]f\big)^{(j)}(\mathbf{g})\\ &\quad+ \sum_{(u)} A_{j,(u)}(\mathbf{g})\big(\mathrm{Ad}_{g_n^{-1}\cdots g_{j+1}^{-1}}(yu_{(n-j+1)})[n]f\big)^{(j)}(\mathbf{g}).\end{aligned} \tag{2.5}$$

By the explicit expression for $A_{j,(u)}(\mathbf{g})$ and the product rule we have

$$\big(y[j]A_{j,(u)}\big)(\mathbf{g}) = \sum_{s=j+1}^{n} \Big(\cdots\otimes\tau_s(g_{s-1}^{-1}\cdots g_{j+1}^{-1})\tau_s(S(yu_{(s-j)}))\tau_s(g_j^{-1}g_{j+1}\cdots g_s)\otimes\cdots\Big)$$

where we specified only the tensor component that differs from the corresponding tensor component of $A_{j,(u)}(\mathbf{g})$. Substituting into (2.5) proves the proposition for $yu \in U_{k+1}(\mathfrak{g})$. This completes the proof. □

COROLLARY 2.7. *For $f \in C^\infty_{\sigma_\ell,\underline{\tau},\sigma_r}(G^{\times(n+1)})$ and $0 \leq j < n$ we have*

$$\big(u[j]f\big)^\flat(g) = \sum_{(u)} \tau_{j+1}(S(u_{(1)}))\cdots\tau_n(S(u_{(n-j)}))\big(S(u_{(n-j+1)})\langle n\rangle f\big)^\flat(g)$$

*for $g \in G$ and $u \in U(\mathfrak{g})$.*

PROOF. By Proposition 2.6,

$$\begin{aligned}&\big(u[j]f\big)^\flat(g)\\ &= \sum_{(u)} \tau_{j+1}(S(u_{(1)}))\cdots\tau_n(S(u_{(n-j)}))\tau_n(g)\big(\mathrm{Ad}_{g^{-1}}(u_{(n-j+1)})[n]f\big)(1,\ldots,1,g,1).\end{aligned}$$

The result now follows from Remark 2.5 and the observation that

$$\tau_n(g)f(g_0,\ldots,g_{n-2},g,g_n) = f(g_0,\ldots,g_{n-2},1,gg_n)$$

since $f \in C^\infty_{\sigma_\ell,\underline{\tau},\sigma_r}(G^{\times(n+1)})$. □

REMARK 2.8. A similar analysis can be done for the coordinate-wise action of $U(\mathfrak{g})$ by right $G$-invariant differential operators. It leads to the following analogue of Corollary 2.7. Suppose that $f \in C^\infty_{\sigma_\ell,\underline{\tau},\sigma_r}(G^{\times(n+1)})$ and $0 \leq j < n$, then

$$(u\langle j\rangle f)^\flat(g) = \sum_{(u)} \tau_{j+1}(u_{(1)}) \cdots \tau_n(u_{(n-j)})\big(u_{(n-j+1)}\langle n\rangle f\big)^\flat(g)$$

for $g \in G$ and $u \in U(\mathfrak{g})$.

For a closed Lie subgroup $K \subseteq G$ write $U(\mathfrak{g})^K$ for the subalgebra of $\mathrm{Ad}(K)$-invariant elements in $U(\mathfrak{g})$, and $U(\mathfrak{g})^{K,\mathrm{opp}}$ for the opposite algebra. Note that $S : U(\mathfrak{g})^K \to U(\mathfrak{g})^{K,\mathrm{opp}}$ is an isomorphism of algebras.

LEMMA 2.9. *The space* $C^\infty_{\sigma_\ell,\underline{\tau},\sigma_r}(G^{\times(n+1)})$ *is a* $U(\mathfrak{g})^{K_\ell,\mathrm{opp}} \otimes Z(U(\mathfrak{g}))^{\otimes(n-1)} \otimes U(\mathfrak{g})^{K_r}$*-module, with the action defined by*

$$(v_\ell \otimes (u_1 \otimes \cdots \otimes u_{n-1}) \otimes v_r) * f := S(v_\ell)\langle 0\rangle u_1[1] \cdots u_{n-1}[n-1]v_r[n]f \tag{2.6}$$

*for* $v_\ell \in U(\mathfrak{g})^{K_\ell}$, $u_i \in Z(U(\mathfrak{g}))$ $(1 \leq i < n)$, $v_r \in U(\mathfrak{g})^{K_r}$ *and* $f \in C^\infty_{\sigma_\ell,\underline{\tau},\sigma_r}(G^{\times(n+1)})$. *Furthermore, for* $u_0 \otimes \cdots \otimes u_n \in Z(U(\mathfrak{g}))^{\otimes(n+1)}$ *we have the action*

$$(u_0 \otimes \cdots \otimes u_n) * f = u_0[0]u_1[1] \cdots u_n[n]f \qquad (f \in C^\infty_{\sigma_\ell,\underline{\tau},\sigma_r}(G^{\times(n+1)}))$$

*by coordinate-wise biinvariant differential operators.*

PROOF. The straightforward proof is left to the reader. □

REMARK 2.10. For $n = 0$ the appropriate analogue of Lemma 2.9 is the statement that $U(\mathfrak{g})^{K_\ell,\mathrm{opp}} \otimes_{Z(U(\mathfrak{g}))} U(\mathfrak{g})^{K_r}$ acts on $C^\infty_{\sigma_\ell,\sigma_r}(G)$ by

$$(v_\ell \otimes_{Z(U(\mathfrak{g}))} v_r) * f = S(v_\ell)\langle 1\rangle v_r[1]f \qquad (v_\ell \in U(\mathfrak{g})^{K_\ell}, v_r \in U(\mathfrak{g})^{K_r})$$

for $f \in C^\infty_{\sigma_\ell,\sigma_r}(G)$, where $u\langle 1\rangle f$ and $u[1]f$ now denote the action of $u \in U(\mathfrak{g})$ on $f \in C^\infty(G)$ as right and left $G$-invariant differential operator, respectively. This is a well-defined action since $S(u)\langle 1\rangle f = u[1]f$ for $u \in Z(U(\mathfrak{g}))$, cf. Remark 2.5 (it is for this reason that we have taken the opposite algebra $U(\mathfrak{g})^{K_\ell,\mathrm{opp}}$ in Lemma 2.9).

We can now extend the definition of $n$-point spherical functions from [**20**] to the current context.

DEFINITION 2.11. We call $f \in C^\infty_{\sigma_\ell,\underline{\tau},\sigma_r}(G^{\times(n+1)})$ an $n$-point spherical function if there exist algebra homomorphisms $\chi_j : Z(U(\mathfrak{g})) \to \mathbb{C}$ $(0 \leq j \leq n)$ such that

$$u[j]f = \chi_j(u)f \qquad \forall\, u \in Z(U(\mathfrak{g})),\ \forall\, j \in \{0, \ldots, n\}. \tag{2.7}$$

We write

$$C^\infty_{\sigma_\ell,\underline{\tau},\sigma_r}(G^{\times(n+1)};\chi) \subseteq C^\infty_{\sigma_\ell,\underline{\tau},\sigma_r}(G^{\times(n+1)})$$

for the $U(\mathfrak{g})^{K_\ell} \otimes Z(U(\mathfrak{g}))^{\otimes(n-1)} \otimes U(\mathfrak{g})^{K_r}$-module of $n$-point spherical functions satisfying (2.7) with respect to the $(n+1)$-tuple $\chi := (\chi_0, \ldots, \chi_n)$ of characters of $Z(U(\mathfrak{g}))$.

Examples of $n$-point spherical functions $f \in C^\infty_{\sigma_\ell,\underline{\tau},\sigma_r}(G^{\times(n+1)};\chi)$ can be constructed using $G$-intertwiners $\Psi_i : \mathcal{H}_i \to \mathcal{H}_{i-1} \otimes U_i$ $(1 \leq i \leq n)$, with $(\pi_j, \mathcal{H}_j)$

($0 \leq j \leq n$) quasi-simple smooth $G$-representations with central character $\chi_j$, together with a $K_\ell$-intertwiner $\phi_\ell : \mathcal{H}_0 \to V_\ell$ and a $K_r$-intertwiner $\phi_r : V_r \to \mathcal{H}_n$, via the formula

$$f(\mathbf{g}) := (\phi_\ell \pi_0(g_0) \otimes \mathrm{id}_{\underline{U}})(\Psi_1 \pi_1(g_1) \otimes \mathrm{id}_{U_2 \otimes \cdots \otimes U_n}) \\ \cdots (\Psi_{n-1}\pi_{n-1}(g_{n-1}) \otimes \mathrm{id}_{U_n})\Psi_n \pi_n(g_n)\phi_r.$$

See [**20**] for important explicit classes of such examples.

For special choices of $K_\ell$ and $K_r$, the action of the commuting differential operators $x[j]$ ($x \in Z(U(\mathfrak{g}))$, $0 \leq j \leq N$) on $C^\infty_{\sigma_\ell, \underline{\tau}, \sigma_r}(G^{\times(n+1)})$ gives rise to commuting vector-valued differential operators on a suitable abelian Lie subgroup $A \subseteq G$ by pulling the invariant differential operators through the restriction map $f \mapsto f^\flat|_A$ ($f \in C^\infty_{\sigma_\ell, \underline{\tau}, \sigma_r}(G^{\times(n+1)})$). Typically, one needs a natural parametrisation of the double $(K_\ell, K_r)$-cosets in $G$ in terms of orbits in $A$ with respect to a discrete group action on $A$. In the remainder of this paper we give a detailed analysis when $K := K_\ell = K_r$ is a maximal compact Lie subgroup of a real connected semisimple Lie group $G$ with finite center.

## 3. Structure theory of real semisimple Lie groups

We introduce standard facts and notations regarding the structure theory of real semisimple Lie groups. For more information see, e.g., [**15**, Chpt. VI], [**23**, Chpt. 9] and [**3**]. As a concrete example we will work out the structure theory in detail for $G = \mathrm{SU}(r,p)$ in Section 7.

We fix from now on a real connected semisimple Lie group $G$ with finite center and Lie algebra $\mathfrak{g}_\mathbb{R}$. Fix a Cartan involution $\theta_\mathbb{R} \in \mathrm{Aut}(\mathfrak{g}_\mathbb{R})$ and write

$$\mathfrak{g}_\mathbb{R} = \mathfrak{k}_\mathbb{R} \oplus \mathfrak{p}_\mathbb{R}$$

for the resulting Cartan decomposition, with $\mathfrak{k}_\mathbb{R} = \mathfrak{g}_\mathbb{R}^{\theta_\mathbb{R}}$ the fix-point Lie subalgebra with respect to $\theta_\mathbb{R}$ and $\mathfrak{p}_\mathbb{R}$ the $(-1)$-eigenspace of $\theta_\mathbb{R}$. We fix the scalar product

$$(x,y) := -K_{\mathfrak{g}_\mathbb{R}}(x, \theta_\mathbb{R}(y)), \qquad x,y \in \mathfrak{g}_\mathbb{R}$$

on $\mathfrak{g}_\mathbb{R}$, with $K_{\mathfrak{g}_\mathbb{R}}$ the Killing form of $\mathfrak{g}_\mathbb{R}$.

Choose a maximal abelian subspace $\mathfrak{a}_\mathbb{R}$ of $\mathfrak{p}_\mathbb{R}$, which we view as Euclidean space with scalar product inherited from $(\cdot,\cdot)$. We endow the linear dual $\mathfrak{a}_\mathbb{R}^*$ with the scalar product such that the linear isomorphism $\mathfrak{a}_\mathbb{R} \xrightarrow{\sim} \mathfrak{a}_\mathbb{R}^*$, $h \mapsto (h,\cdot)$, becomes an isomorphism of Euclidean spaces. We denote the scalar product on $\mathfrak{a}_\mathbb{R}^*$ again by $(\cdot,\cdot)$.

For $\lambda \in \mathfrak{a}_\mathbb{R}^*$ set

$$\mathfrak{g}_\mathbb{R}^\lambda := \{x \in \mathfrak{g}_\mathbb{R} \mid [h,x] = \lambda(h)x \quad \forall\, h \in \mathfrak{a}_\mathbb{R}\}, \\ \Sigma := \{\lambda \in \mathfrak{a}_\mathbb{R}^* \setminus \{0\} \mid \mathfrak{g}_\mathbb{R}^\lambda \neq \{0\}\}.$$

Then $\Sigma \subset \mathfrak{a}_\mathbb{R}^*$ is a (possibly non-reduced) root system, called the restricted root system. Furthermore,

$$\mathfrak{g}_\mathbb{R} = \mathfrak{g}_\mathbb{R}^0 \oplus \bigoplus_{\lambda \in \Sigma} \mathfrak{g}_\mathbb{R}^\lambda \tag{3.1}$$

and $\mathfrak{g}_\mathbb{R}^0 = Z_{\mathfrak{g}_\mathbb{R}}(\mathfrak{a}_\mathbb{R}) = \mathfrak{m}_\mathbb{R} \oplus \mathfrak{a}_\mathbb{R}$, with $\mathfrak{m}_\mathbb{R} := Z_{\mathfrak{k}_\mathbb{R}}(\mathfrak{a}_\mathbb{R})$ the centraliser of $\mathfrak{a}_\mathbb{R}$ in $\mathfrak{k}_\mathbb{R}$. The decomposition (3.1) is orthogonal with respect to $(\cdot,\cdot)$, and $\theta(\mathfrak{g}_\mathbb{R}^\lambda) = \mathfrak{g}_\mathbb{R}^{-\lambda}$.

Fix a maximal abelian subspace $\mathfrak{t}_{\mathbb{R}}$ of $\mathfrak{m}_{\mathbb{R}}$. Then

$$\mathfrak{h}_{\mathbb{R}} := \mathfrak{t}_{\mathbb{R}} \oplus \mathfrak{a}_{\mathbb{R}}$$

is a maximally noncompact $\theta_{\mathbb{R}}$-stable Cartan subalgebra of $\mathfrak{g}_{\mathbb{R}}$. Write $\mathfrak{g}, \mathfrak{g}^{\lambda}, \mathfrak{k}, \mathfrak{m}, \mathfrak{h}, \mathfrak{a}, \mathfrak{t}, \ldots$ for the complexifications of the Lie algebras $\mathfrak{g}_{\mathbb{R}}, \mathfrak{g}_{\mathbb{R}}^{\lambda}, \mathfrak{k}_{\mathbb{R}}, \mathfrak{m}_{\mathbb{R}}, \mathfrak{h}_{\mathbb{R}}, \mathfrak{a}_{\mathbb{R}}, \mathfrak{t}_{\mathbb{R}}, \ldots$. The root space decomposition of $\mathfrak{g}$ with respect to $\mathfrak{h}$ is denoted by

$$\mathfrak{g} = \mathfrak{h} \oplus \bigoplus_{\alpha \in R} \mathfrak{g}_{\alpha}$$

with

$$\mathfrak{g}_{\beta} := \{x \in \mathfrak{g} \mid [h, x] = \beta(h)x \quad \forall\, h \in \mathfrak{h}\} \qquad (\beta \in \mathfrak{h}^*)$$

and root system $R := \{\alpha \in \mathfrak{h}^* \setminus \{0\} \mid \mathfrak{g}_{\alpha} \neq \{0\}\}$. We write $\theta \in \mathrm{Aut}(\mathfrak{g})$ for the complex linear extension of $\theta_{\mathbb{R}}$ to an involution of $\mathfrak{g}$. Then $\mathfrak{g} = \mathfrak{k} \oplus \mathfrak{p}$ is the decomposition of $\mathfrak{g}$ in $(+1)$–and $(-1)$-eigenspaces of $\theta$. The involution $\theta$ restricts to an automorphism of $\mathfrak{h}$. The transpose of $\theta|_{\mathfrak{h}}$ is a linear involution of $\mathfrak{h}^*$, which we will also denote by $\theta$. It restricts to an involution of $R$ satisfying $\theta\alpha|_{\mathfrak{a}_{\mathbb{R}}} = -\alpha|_{\mathfrak{a}_{\mathbb{R}}}$ for all $\alpha \in R$. We then have $\theta(\mathfrak{g}_{\alpha}) = \mathfrak{g}_{\theta\alpha}$.

For $\lambda \in \Sigma \cup \{0\}$ we write

$$R_{\lambda} := \{\alpha \in R \mid \alpha|_{\mathfrak{a}_{\mathbb{R}}} = \lambda\}.$$

Roots in $R_0$ are called imaginary roots (they are the roots in $R$ that take purely imaginary values on $\mathfrak{h}_{\mathbb{R}}$). Note that $\theta$ fixes $R_0$ point-wise and maps $R_{\lambda}$ to $R_{-\lambda}$ for $\lambda \in \Sigma$. Furthermore,

$$\mathfrak{m} = \mathfrak{t} \oplus \bigoplus_{\alpha \in R_0} \mathfrak{g}_{\alpha}, \qquad \mathfrak{g}^{\lambda} = \bigoplus_{\alpha \in R_{\lambda}} \mathfrak{g}_{\alpha} \quad (\lambda \in \Sigma).$$

In particular, $R$ is the disjoint union of the $R_{\lambda}$ ($\lambda \in \Sigma \cup \{0\}$), $R_{\lambda}$ is nonempty for $\lambda \in \Sigma$, and the restriction map

$$R \setminus R_0 \twoheadrightarrow \Sigma, \quad \alpha \mapsto \alpha|_{\mathfrak{a}_{\mathbb{R}}} \tag{3.2}$$

is surjective. The strictly positive numbers

$$\mathrm{mtp}(\lambda) := \#R_{\lambda}, \qquad \lambda \in \Sigma$$

are called the restricted root multiplicities. The Lie subalgebra $\mathfrak{m}$ of $\mathfrak{g}$ is reductive. The root system of the semisimple part $[\mathfrak{m}, \mathfrak{m}]$ of $\mathfrak{m}$ with respect to the Cartan subalgebra $\mathfrak{t}_{\mathrm{ss}} := \mathfrak{t} \cap [\mathfrak{m}, \mathfrak{m}]$ is $\{\alpha|_{\mathfrak{t}_{\mathrm{ss}}}\}_{\alpha \in R_0}$.

Choose a decomposition $R = R^+ \cup R^-$ of the root system $R$ into positive and negative roots such that

(1) $R_0 = R_0^+ \cup R_0^-$ with $R_0^{\pm} := (R^{\pm} \cap R_0)$ is a decomposition of $R_0$ in positive and negative roots,
(2) $\Sigma = \Sigma^+ \cup \Sigma^-$ with $\Sigma^{\pm} := \{\alpha|_{\mathfrak{a}_{\mathbb{R}}} \mid \alpha \in R^{\pm} \setminus R_0^{\pm}\}$ is a decomposition of $\Sigma$ in positive and negative roots

(see, e.g., [**4**, 11.1.16]).

REMARK 3.1. Restricted root systems appear more generally for so-called normal $\sigma$-systems of roots, see, e.g., [**22**, §1.3]. A normal $\sigma$-system of roots consists of a root system $R$ and an involutive isometry $\sigma$ of the ambient Euclidean space, satisfying the additional properties that $\sigma$ stabilises $R$ and $\sigma\alpha - \alpha \notin R$ for all $\alpha \in R$.

In our current setup, $R$ is a normal $(-\theta)$-system of roots. Indeed, suppose that $\alpha \in R$ satisfies $\alpha + \theta\alpha \in R$. Then $[e_{\alpha}, \theta(e_{\alpha})] \in \mathfrak{p} \setminus \{0\}$. On the other hand,

$\alpha+\theta\alpha \in R_0$ since the root $\alpha+\theta\alpha$ vanishes on $\mathfrak{a}_{\mathbb{R}}$, and hence $[e_\alpha, \theta(e_\alpha)] \in \mathfrak{k}$. This is a contradiction.

Consider the nilpotent Lie subalgebra

$$\mathfrak{n}_{\mathbb{R}} := \bigoplus_{\lambda\in\Sigma^+} \mathfrak{g}_{\mathbb{R}}^{\lambda}$$

of $\mathfrak{g}_{\mathbb{R}}$. Its complexification within $\mathfrak{g}$ is the nilpotent Lie subalgebra $\mathfrak{n}=\bigoplus_{\alpha\in R^+\setminus R_0^+}\mathfrak{g}_\alpha$. We then have the direct sum decompositions

$$\mathfrak{g}_{\mathbb{R}} = \mathfrak{k}_{\mathbb{R}} \oplus \mathfrak{a}_{\mathbb{R}} \oplus \mathfrak{n}_{\mathbb{R}}, \qquad \mathfrak{g} = \mathfrak{k} \oplus \mathfrak{a} \oplus \mathfrak{n}$$

as vector spaces (infinitesimal Iwasawa decomposition).

Let $K, A, N \subset G$ be the connected Lie subgroups with Lie algebra $\mathfrak{k}_{\mathbb{R}}$, $\mathfrak{a}_{\mathbb{R}}$ and $\mathfrak{n}_{\mathbb{R}}$, respectively. Then $K \subset G$ is maximal compact, the multiplication map

$$K \times A \times N \to G, \qquad (k,a,n) \mapsto kan$$

is a diffeomorphism (the Iwasawa decomposition), and the exponential map exp of $G$ restricts to an isomorphism $\exp : \mathfrak{a}_{\mathbb{R}} \xrightarrow{\sim} A$ of Lie groups. We write $\log : A \to \mathfrak{a}_{\mathbb{R}}$ for its inverse.

Write $M := Z_K(A)$ for the centraliser of $A$ in $K$. It is a closed, but not necessarily connected, Lie subgroup of $K$ with Lie algebra $\mathfrak{m}_{\mathbb{R}}$. Note that $M = Z_K(\mathfrak{a})$ (the centraliser of $\mathfrak{a}$ in $K$ with respect to the adjoint action). Note that $\theta_{\mathbb{R}}$ and $\theta$ are $\mathrm{Ad}(K)$-linear, and $\mathfrak{g}^\lambda$ is $\mathrm{Ad}(M)$-stable ($\lambda \in \Sigma \cup \{0\}$). In particular, $\mathfrak{n}_{\mathbb{R}}$ and $\mathfrak{n}$ are $\mathrm{Ad}(M)$-stable.

The analytic Weyl group of $G$ is $W := N_K(A)/M$, with $N_K(A)$ the normaliser of $A$ in $K$. It acts naturally on $A$, as well as on the set $\Sigma$ of restricted roots. Denote by $W\backslash A$ the set of $W$-orbits in $A$, and write $K\backslash G/K$ for the double $(K,K)$-coset space of $G$. The map

$$W\backslash A \xrightarrow{\sim} K\backslash G/K, \qquad Wa \mapsto KaK$$

is a bijection (see, e.g., [**15**, §VII.3]). The decomposition $G = KAK$ is sometimes called the Cartan decomposition of $G$ with respect to $K$.

For $\nu \in \mathfrak{h}^*$ write $h_\nu \in \mathfrak{h}$ for the unique element such that $K_{\mathfrak{g}}(h, h_\nu) = \nu(h)$ for all $h \in \mathfrak{h}$, with $K_{\mathfrak{g}}$ the Killing form of $\mathfrak{g}$. Choose root vectors $e_\alpha \in \mathfrak{g}_\alpha$ such that $[e_\alpha, e_{-\alpha}] = h_\alpha$ ($\alpha \in R$). For $\alpha \in R$ we have

$$\theta(e_\alpha) = c_\alpha e_{\theta\alpha}$$

for a unique nonzero scalar $c_\alpha \in \mathbb{C}^*$. Note that $c_\alpha = 1$ for $\alpha \in R_0$. Define for $\alpha \in R \setminus R_0$,

$$y_\alpha := e_\alpha + \theta(e_\alpha) = e_\alpha + c_\alpha e_{\theta\alpha} \in \mathfrak{k}.$$

LEMMA 3.2. *For $\alpha \in R \setminus R_0$ we have $c_{\theta\alpha} = c_\alpha^{-1} = c_{-\alpha}$. Furthermore, $y_\alpha = c_\alpha y_{\theta\alpha}$.*

PROOF. We have $e_\alpha = \theta^2(e_\alpha) = c_\alpha c_{\theta\alpha} e_\alpha$, hence $c_{\theta\alpha} = c_\alpha^{-1}$. Furthermore,

$$\theta(h_\alpha) = \theta([e_\alpha, e_{-\alpha}]) = [\theta(e_\alpha), \theta(e_{-\alpha})] = c_\alpha c_{-\alpha}[e_{\theta\alpha}, e_{-\theta\alpha}] = c_\alpha c_{-\alpha} h_{\theta\alpha}.$$

But $\theta(h_\alpha) = h_{\theta\alpha}$ since $K_{\mathfrak{g}}(\theta(x), \theta(y)) = K_{\mathfrak{g}}(x,y)$ for $x, y \in \mathfrak{g}$, hence $c_{-\alpha} = c_\alpha^{-1}$. Finally,

$$y_\alpha = e_\alpha + \theta(e_\alpha) = \theta^2(e_\alpha) + c_\alpha e_{\theta\alpha} = c_\alpha(e_{\theta\alpha} + \theta(e_{\theta\alpha})) = c_\alpha y_{\theta\alpha}.$$

$\square$

Consider the direct sum decomposition

$$\mathfrak{k} = \mathfrak{m} \oplus \mathfrak{q}, \qquad \mathfrak{q} := \bigoplus_{\alpha \in R^+ \setminus R_0^+} \mathbb{C} y_\alpha$$

as vector spaces. Note that $\mathfrak{q}$ is the complex linear span of the orthogonal complement $\mathfrak{q}_{\mathbb{R}}$ of $\mathfrak{m}_{\mathbb{R}}$ in $\mathfrak{k}_{\mathbb{R}}$ with respect to $(\cdot,\cdot)$. Another description of $\mathfrak{q}$ is as follows. Let $\mathrm{pr}_{\mathfrak{k}} : \mathfrak{g} \twoheadrightarrow \mathfrak{k}$ and $\mathrm{pr}_{\mathfrak{p}} : \mathfrak{g} \twoheadrightarrow \mathfrak{p}$ be the projections along the direct sum decomposition $\mathfrak{g} = \mathfrak{k} \oplus \mathfrak{p}$. Then $\mathrm{pr}_{\mathfrak{k}} = \frac{1}{2}(\mathrm{id}_{\mathfrak{g}} + \theta)$ and $\mathrm{pr}_{\mathfrak{p}} = \frac{1}{2}(\mathrm{id}_{\mathfrak{g}} - \theta)$, both projections are $\mathrm{Ad}(K)$-equivariant, and $\mathfrak{q}$ is the image of $\mathfrak{n}$ under the projection $\mathrm{pr}_{\mathfrak{k}}$. In particular, $\mathfrak{q}$ is $\mathrm{Ad}(M)$-stable.

For $\lambda \in \mathfrak{a}^*$ the map $\xi_\lambda : A \to \mathbb{C}^*$, $a \mapsto a^\lambda := e^{\lambda(\log a)}$, defines a complex-valued multiplicative character of $A$. Note that $\xi_\lambda \xi_\nu = \xi_{\lambda+\nu}$ $(\lambda, \nu \in \mathfrak{a}^*)$ and $\xi_0 \equiv 1$. Furthermore, for $\lambda \in \Sigma \cup \{0\}$ we have

$$\mathrm{Ad}_a(x) = a^\lambda x \qquad (x \in \mathfrak{g}^\lambda, a \in A).$$

Write $A_{\mathrm{reg}} := \{a \in A \mid a^\lambda \neq 1 \ \ \forall \lambda \in \Sigma\}$. For $a \in A_{\mathrm{reg}}$ and $\alpha \in R_\lambda$ $(\lambda \in \Sigma)$ a direct computation shows that

$$\begin{aligned} e_\alpha &= \frac{a^\lambda \,\mathrm{Ad}_{a^{-1}}(y_\alpha) - a^{2\lambda} y_\alpha}{1 - a^{2\lambda}}, \\ \theta(e_\alpha) &= \frac{a^\lambda \,\mathrm{Ad}_{a^{-1}}(y_\alpha) - y_\alpha}{a^{2\lambda} - 1}. \end{aligned} \tag{3.3}$$

This leads to the direct sum decomposition

$$\mathfrak{g} = \mathfrak{a} \oplus \mathrm{Ad}_{a^{-1}}(\mathfrak{q}) \oplus \mathfrak{k} \qquad (a \in A_{\mathrm{reg}}) \tag{3.4}$$

as vector spaces. It is this infinitesimal analogue of the Cartan decomposition that plays a crucial role in computing radial components of invariant differential operators, see [**3, 11**] and Section 4.

## 4. Generalised radial component maps

**4.1. Differential operators.** The conjugation action of $N_K(A)$ on $A$ descends to an action of the analytic Weyl group $W = N_K(A)/M$. Its contragredient action on $C^\infty(A)$ is explicitly given by $(wf)(a) := f(g^{-1}ag)$ for $w = gM \in W$, $f \in C^\infty(A)$ and $a \in A$. Note that the $W$-action on $A$ preserves $A_{\mathrm{reg}}$, and hence $C^\infty(A_{\mathrm{reg}})$ admits the same $W$-module algebra structure.

Note furthermore that $N_K(A)$ acts canonically on $\mathfrak{a}^*$, and the action descends to an action of $W$. Then $w\xi_\lambda := \xi_{w\lambda}$ for $w \in W$ and $\lambda \in \mathfrak{a}^*$.

DEFINITION 4.1. We write $\mathcal{R}$ for the unital subalgebra of $C^\infty(A_{\mathrm{reg}})$ generated by $(1 \pm \xi_\lambda)^{-1}$ for $\lambda \in \Sigma$.

Note that $\mathcal{R}$ is a $W$-module subalgebra of $C^\infty(A_{\mathrm{reg}})$.

The action of $N_K(A)$ and $W$ on $C^\infty(A)$ has the following twisted vector-valued analogue. Let $(\sigma, V)$ be a finite-dimensional $N_K(A)$-representation. The space $C^\infty(A; V)$ of smooth $V$-valued functions on $A$ is a left $N_K(A)$-module with action

$$(g \cdot f)(a) := \sigma(g) f(g^{-1}ag) \qquad (g \in N_K(A)) \tag{4.1}$$

for $f \in C^\infty(A; V)$ and $a \in A$. Let $V^M \subseteq V$ be the subspace of $M$-invariant vectors in $V$. It is a $N_K(A)$-subrepresentation of $V$, and hence $C^\infty(A; V^M)$ is a $N_K(A)$-submodule of $C^\infty(A; V)$. The action of $N_K(A)$ on $C^\infty(A; V^M)$ descends to an action of $W$. We write $C^\infty(A;V^M)^W$ for the associated subspace of $W$-invariant functions.

The results from the previous paragraph also hold true for smooth $V$-valued functions on $A_{\mathrm{reg}}$. Then $\mathcal{R}\otimes V$ is a $N_K(A)$-submodule of $C^\infty(A_{\mathrm{reg}};V)$, and $\mathcal{R}\otimes V^M$ is a $W$-submodule of $C^\infty(A_{\mathrm{reg}};V^M)$.

We write $h\mapsto \partial(h)$ ($h\in U(\mathfrak{a})$) for the action of $U(\mathfrak{a})$ on $C^\infty(A)$ by constant coefficient differential operators. Concretely, for $h\in\mathfrak{a}_{\mathbb{R}}$ the differential operator $\partial(h)$ is the directional derivative $\partial_h$ in direction $h$,

$$(\partial_h f)(a) := \frac{d}{dt}\Big|_{t=0} f(a\exp(th)) \qquad (f\in C^\infty(A),\ a\in A).$$

We fix an orthonormal basis $\{x_1,\ldots,x_r\}$ of $\mathfrak{a}_{\mathbb{R}}$ with respect to $(\cdot,\cdot)$ once and for all, and we write

$$\partial^{\mathbf{m}} := \partial(x_1^{m_1}\cdots x_r^{m_r}) = \partial_{x_1}^{m_1}\cdots\partial_{x_r}^{m_r}$$

for $\mathbf{m}=(m_1,\ldots,m_r)\in\mathbb{Z}_{\geq 0}^r$.

We now take $V=\mathrm{Hom}(E,F)$ with $E$ and $F$ two finite-dimensional $N_K(A)$-representations, viewed as $N_K(A)$-representation with respect to the conjugation action. Then $V$ is canonically isomorphic to the tensor product $N_K(A)$-representation $F\otimes E^*$.

Definition 4.2. We write $\mathbb{D}(E,F)$ for the $N_K(A)$-module consisting of differential operators $D=\sum_{\mathbf{m}} q_{\mathbf{m}}\partial^{\mathbf{m}}$ on $A_{\mathrm{reg}}$ with coefficients $q_{\mathbf{m}}\in\mathcal{R}\otimes\mathrm{Hom}(E,F)$.

Concretely, the $N_K(A)$-module structure on $\mathbb{D}(E,F)$ is given by

$$g\cdot\Big(\sum_i q_i\partial(h_i)\Big) = \sum_i (g\cdot q_i)\partial(\mathrm{Ad}_g(h_i)) \tag{4.2}$$

for $g\in N_K(A)$, $q_i\in\mathcal{R}\otimes\mathrm{Hom}(E,F)$ and $h_i\in U(\mathfrak{a})$. The $N_K(A)$-action on differential operators $D\in\mathbb{D}(E,F)$, viewed as linear operators $D: C^\infty(A_{\mathrm{reg}};E)\to C^\infty(A_{\mathrm{reg}};F)$, is compatible with the $N_K(A)$-actions on $C^\infty(A_{\mathrm{reg}};E)$ and $C^\infty(A_{\mathrm{reg}};F)$,

$$g\cdot(Df) = (g\cdot D)(g\cdot f)$$

for $g\in N_K(A)$, $D\in\mathbb{D}(E,F)$ and $f\in C^\infty(A_{\mathrm{reg}};E)$.

Note that the $M$-action on $\mathbb{D}(E,F)$ is the trivial extension to $\mathbb{D}(E,F)$ of the conjugation $M$-action on $\mathrm{Hom}(E,F)$.

Definition 4.3. We write $\mathbb{D}_M(E,F)$ for the subspace of $M$-invariant differential operators in $\mathbb{D}(E,F)$. Concretely, it is the $N_K(A)$-module of differential operators $D=\sum_{\mathbf{m}} q_{\mathbf{m}}\partial^{\mathbf{m}}$ with $q_{\mathbf{m}}\in\mathcal{R}\otimes\mathrm{Hom}_M(E,F)$.

The $N_K(A)$-action on $\mathbb{D}_M(E,F)$ descends to a $W$-action. We write $\mathbb{D}_M(E,F)^W$ for the subspace of $W$-invariant differential operators in $\mathbb{D}_M(E,F)$ (equivalently, it is the subspace of $N_K(A)$-invariant differential operators in $\mathbb{D}(E,F)$).

If $E=F$ then we write $\mathbb{D}(E)$, $\mathbb{D}_M(E)$ and $\mathbb{D}_M(E)^W$ for the algebras $\mathbb{D}(E,E)$, $\mathbb{D}_M(E,E)$ and $\mathbb{D}_M(E,E)^W$, respectively. Furthermore, for $D=\sum_{\mathbf{m}} q_{\mathbf{m}}\partial^{\mathbf{m}}\in\mathbb{D}(E,F)$ we call

$$D_M := \sum_{\mathbf{m}} q_{\mathbf{m}}(\cdot)|_{E^M}\partial^{\mathbf{m}}\in\mathbb{D}(E^M,F)$$

the $M$-restriction of $D$. It satisfies

$$Df = D_M f \qquad (D\in\mathbb{D}(E,F),\ f\in C^\infty(A_{\mathrm{reg}};E^M)). \tag{4.3}$$

Note that the assignment $\mathbb{D}(E,F)\to\mathbb{D}(E^M,F)$, $D\mapsto D_M$ restricts to a linear map

$$\mathbb{D}_M(E,F)\to\mathbb{D}(E^M,F^M),$$

which is an algebra map $\mathbb{D}_M(E)\to\mathbb{D}(E^M)$ when $F=E$.

Typically $E$ and $F$ are going to be $N_K(A)$-subrepresentations of the tensor product representation $V_\ell\otimes V_r^*$, with $(\sigma_\ell,V_\ell)$ and $(\sigma_r,V_r)$ finite-dimensional $K$-representations.

**4.2. Radial component maps.** This section recalls some well-known facts about Harish-Chandra's radial component maps. We follow closely the reference [**3**]. By [**3**, Thm. 2.4] there exist for $u\in U(\mathfrak{g})$ unique elements

$$\begin{aligned}\Pi(u)&=\sum_i f_i\otimes h_i\otimes(u_i\otimes_{U(\mathfrak{m})}v_i)\in\mathcal{R}\otimes U(\mathfrak{a})\otimes(U(\mathfrak{k})\otimes_{U(\mathfrak{m})}U(\mathfrak{k})),\\ \widetilde{\Pi}(u)&=\sum_i \widetilde{f}_i\otimes \widetilde{h}_i\otimes(\widetilde{u}_i\otimes_{U(\mathfrak{m})}\widetilde{v}_i)\in\mathcal{R}\otimes U(\mathfrak{a})\otimes(U(\mathfrak{k})\otimes_{U(\mathfrak{m})}U(\mathfrak{k})),\end{aligned}\tag{4.4}$$

such that

$$\sum_i f_i(a)\,\mathrm{Ad}_{a^{-1}}(u_i)h_iv_i=u=\sum_i\widetilde{f}_i(a)\widetilde{u}_i\widetilde{h}_i\,\mathrm{Ad}_a(\widetilde{v}_i)\qquad\forall\,a\in A_{\mathrm{reg}}.\tag{4.5}$$

The two decompositions (4.5) of $u\in U(\mathfrak{g})$ are called infinitesimal Cartan decompositions of $u$ relative to $a\in A_{\mathrm{reg}}$. The maps $\Pi,\widetilde{\Pi}:U(\mathfrak{g})\to\mathcal{R}\otimes U(\mathfrak{a})\otimes(U(\mathfrak{k})\otimes_{U(\mathfrak{m})}U(\mathfrak{k}))$ are called radial component maps. They provide the algebraic description of the radial components of the action of $u\in U(\mathfrak{g})$ as left and right $G$-invariant differential operators on $C^\infty(G)$ (see [**3**, **11**] and Theorem 4.7).

REMARK 4.4. Our present choice of coefficient algebra $\mathcal{R}$ (see Definition 4.1), which is different from the one used in [**3**], is more convenient for our purposes since $\mathcal{R}\subset C^\infty(A_{\mathrm{reg}})$ is $W$-stable. The results in [**3**, §2–3] are also valid in the present setup, with the same proofs.

EXAMPLE 4.5. For $\alpha\in R_\lambda$ with $\lambda\in\Sigma$,

$$\begin{aligned}\Pi(e_\alpha)&=\frac{\xi_\lambda}{1-\xi_{2\lambda}}\otimes1\otimes(y_\alpha\otimes_{U(\mathfrak{m})}1)-\frac{\xi_{2\lambda}}{1-\xi_{2\lambda}}\otimes1\otimes(1\otimes_{U(\mathfrak{m})}y_\alpha),\\ \widetilde{\Pi}(e_\alpha)&=\frac{1}{1-\xi_{2\lambda}}\otimes1\otimes(y_\alpha\otimes_{U(\mathfrak{m})}1)-\frac{\xi_\lambda}{1-\xi_{2\lambda}}\otimes1\otimes(1\otimes_{U(\mathfrak{m})}y_\alpha)\end{aligned}\tag{4.6}$$

in view of the first line of (3.3).

Note that

$$\Pi(u)=\widetilde{\Pi}(u)\qquad\forall\,u\in U(\mathfrak{g})^A,\tag{4.7}$$

with $U(\mathfrak{g})^A\subseteq U(\mathfrak{g})$ the subalgebra of $\mathrm{Ad}(A)$-invariant elements in $U(\mathfrak{g})$.

Fix two finite-dimensional $K$-representations $(\sigma_\ell,V_\ell)$ and $(\sigma_r,V_r)$. Write $(V_\ell\otimes V_r^*)^{\mathfrak{m}}$ for the subspace of $\mathfrak{m}$-invariant elements in the tensor product $\mathfrak{k}$-module $V_\ell\otimes V_r^*$, i.e.,

$$\big(V_\ell\otimes V_r^*\big)^{\mathfrak{m}}:=\{T\in V_\ell\otimes V_r^*\mid(\sigma_\ell(y)\otimes\mathrm{id}_{V_r^*})T+(\mathrm{id}_{V_\ell}\otimes\sigma_r^*(y))T=0\quad\forall\,y\in\mathfrak{m}\}.$$

Under the natural vector space identification $V_\ell\otimes V_r^*\simeq\mathrm{Hom}(V_r,V_\ell)$, the invariant subspace $(V_\ell\otimes V_r^*)^{\mathfrak{m}}$ corresponds to the subspace $\mathrm{Hom}_{\mathfrak{m}}(V_r,V_\ell)$ of $\mathfrak{m}$-intertwiners $V_r\to V_\ell$. Note that $(V_\ell\otimes V_r^*)^{\mathfrak{m}}$ is an $N_K(A)$-subrepresentation of $V_\ell\otimes V_r^*$ containing $(V_\ell\otimes V_r^*)^M$. It may be a strict inclusion since $M$ is not necessarily connected.

Consider now the linear map

$$\zeta_{\sigma_\ell,\sigma_r}: U(\mathfrak{k}) \otimes_{U(\mathfrak{m})} U(\mathfrak{k}) \to \mathrm{Hom}\big((V_\ell \otimes V_r^*)^{\mathfrak{m}}, V_\ell \otimes V_r^*\big)$$

defined by

$$\zeta_{\sigma_\ell,\sigma_r}(u \otimes_{U(\mathfrak{m})} v)T := (\sigma_\ell(u) \otimes \sigma_r^*(S(v)))T \qquad (u, v \in U(\mathfrak{k}), T \in (V_\ell \otimes V_r^*)^{\mathfrak{m}}).$$

The map is clearly well-defined.

DEFINITION 4.6. Let

$$\mathcal{R} \otimes U(\mathfrak{a}) \otimes (U(\mathfrak{k}) \otimes_{U(\mathfrak{m})} U(\mathfrak{k})) \to \mathbb{D}((V_\ell \otimes V_r^*)^{\mathfrak{m}}, V_\ell \otimes V_r^*), \qquad z \mapsto L_z^{\sigma_\ell,\sigma_r}$$

be the linear map such that

$$L_z^{\sigma_\ell,\sigma_r} := \zeta_{\sigma_\ell,\sigma_r}(u \otimes_{U(\mathfrak{m})} v) f \partial(h)$$

for a pure tensor $z = f \otimes h \otimes (u \otimes_{U(\mathfrak{m})} v) \in \mathcal{R} \otimes U(\mathfrak{a}) \otimes (U(\mathfrak{k}) \otimes_{U(\mathfrak{m})} U(\mathfrak{k}))$.

Write $f|_A \in C^\infty(A; V_\ell \otimes V_r^*)$ for the restriction of $f \in C^\infty(G; V_\ell \otimes V_r^*)$ to $A$. Note that $|_A$ restricts to an injective linear map

$$|_A : C^\infty_{\sigma_\ell,\sigma_r}(G) \hookrightarrow C^\infty(A; (V_\ell \otimes V_r^*)^M)^W.$$

Recall from Section 2 the actions $u[1]$ and $u\langle 1\rangle$ of $u \in U(\mathfrak{g})$ on $C^\infty(G; V)$ as $G$-invariant differential operators. Recall furthermore that

$$L_{z;M}^{\sigma_\ell,\sigma_r} \in \mathbb{D}((V_\ell \otimes V_r^*)^M, V_\ell \otimes V_r^*)$$

denotes the restriction of $L_z^{\sigma_\ell,\sigma_r}$ to $\mathbb{D}((V_\ell \otimes V_r^*)^M, V_\ell \otimes V_r^*)$. We now have the following theorem, which is essentially [**3**, Thm. 3.1].

THEOREM 4.7. *Let $u \in U(\mathfrak{g})$, $f \in C^\infty_{\sigma_\ell,\sigma_r}(G)$ and $a \in A_{\mathrm{reg}}$. Then*

$$\begin{aligned} (u[1]f)(a) &= \big(L^{\sigma_\ell,\sigma_r}_{\Pi(u);M} f|_A\big)(a), \\ (u\langle 1\rangle f)(a) &= \big(L^{\sigma_\ell,\sigma_r}_{\widetilde{\Pi}(S(u));M} f|_A\big)(a). \end{aligned} \tag{4.8}$$

PROOF. By Remark 2.5 the second equality of (4.8) is equivalent to

$$\big(\mathrm{Ad}_{a^{-1}}(u)[1]f\big)(a) = \big(L^{\sigma_\ell,\sigma_r}_{\widetilde{\Pi}(u);M} f|_A\big)(a). \tag{4.9}$$

By the definition of $\Pi(u)$ and $\widetilde{\Pi}(u)$ as well as the definition of $L_z^{\sigma_\ell,\sigma_r}$ (see Definition 4.6) it then suffices to prove that

$$\big((\mathrm{Ad}_{a^{-1}}(u)hv)f\big)(a) = \zeta_{\sigma_\ell,\sigma_r}(u \otimes_{U(\mathfrak{m})} v)\big(\partial(h)f\big)(a)$$

for $u, v \in U(\mathfrak{k})$, $h \in U(\mathfrak{a})$ and $a \in A_{\mathrm{reg}}$. This is shown in the proof of [**3**, Thm. 3.1]. □

Note that

$$L^{\sigma_\ell,\sigma_r}_{\widetilde{\Pi}(u)} = L^{\sigma_\ell,\sigma_r}_{\Pi(u)} \qquad (u \in U(\mathfrak{g})^A) \tag{4.10}$$

in view of (4.7).

The following proposition shows that the differential operators $L^{\sigma_\ell,\sigma_r}_{\Pi(u);M}$ and $L^{\sigma_\ell,\sigma_r}_{\widetilde{\Pi}(u);M}$ are determined by the property (4.8) for all $f \in C^\infty_{\sigma_\ell,\sigma_r}(G)$.

PROPOSITION 4.8. *Suppose that $D \in \mathbb{D}((V_\ell \otimes V_r^*)^M, V_\ell \otimes V_r^*)$ satisfies $D(f|_A) = 0$ for all $f \in C^\infty_{\sigma_\ell,\sigma_r}(G)$. Then $D = 0$.*

PROOF. This is shown as part of the proof of [**3**, Thm. 3.3]. □

The radial components satisfy the following equivariance property.

LEMMA 4.9. *For* $u \in U(\mathfrak{g})$ *and* $g \in N_K(A)$ *we have*

$$L^{\sigma_\ell,\sigma_r}_{\Pi(\mathrm{Ad}_g(u))} = g \cdot L^{\sigma_\ell,\sigma_r}_{\Pi(u)}, \qquad L^{\sigma_\ell,\sigma_r}_{\widetilde{\Pi}(\mathrm{Ad}_g(u))} = g \cdot L^{\sigma_\ell,\sigma_r}_{\widetilde{\Pi}(u)} \tag{4.11}$$

*in* $\mathbb{D}((V_\ell \otimes V_r^*)^{\mathfrak{m}}, V_\ell \otimes V_r^*)$.

PROOF. We only prove the first equality, the second is proved in a similar manner. Fix $g \in N_K(A)$ and write $w := gM \in W$. Using the notation (4.4) we have for $a \in A_{\mathrm{reg}}$,

$$\mathrm{Ad}_g(u) = \sum_i (w \cdot f_i)(wa)\, \mathrm{Ad}_{(wa)^{-1}}(\mathrm{Ad}_g(u_i))\, \mathrm{Ad}_g(h_i)\, \mathrm{Ad}_g(v_i). \tag{4.12}$$

Since $wf_i \in \mathcal{R}$, $\mathrm{Ad}_g(h_i) \in U(\mathfrak{a})$ and $\mathrm{Ad}_g(u_i), \mathrm{Ad}_g(v_i) \in U(\mathfrak{k})$, formula (4.12) is an infinitesimal Cartan decomposition of $\mathrm{Ad}_g(u)$ relative to $wa \in A_{\mathrm{reg}}$ for each $a \in A_{\mathrm{reg}}$. Hence

$$\Pi(\mathrm{Ad}_g(u)) = \sum_i w \cdot f_i \otimes \mathrm{Ad}_g(h_i) \otimes (\mathrm{Ad}_g(u_i) \otimes_{U(\mathfrak{m})} \mathrm{Ad}_g(v_i)).$$

The lemma then follows from the definition of $L^{\sigma_\ell,\sigma_r}_z$ (Definition 4.6) and the fact that

$$\zeta_{\sigma_\ell,\sigma_r}(\mathrm{Ad}_g(u_i) \otimes_{U(\mathfrak{m})} \mathrm{Ad}_g(v_i)) = (\sigma_\ell(g) \otimes \sigma_r^*(g)) \circ \zeta_{\sigma_\ell,\sigma_r}(u_i \otimes v_i) \circ (\sigma_\ell(g^{-1}) \otimes \sigma_r^*(g^{-1}))$$

in $\mathrm{Hom}((V_\ell \otimes V_r^*)^{\mathfrak{m}}, V_\ell \otimes V_r^*)$. □

**4.3. Coordinate radial component maps.** Fix finite-dimensional $G$-representations $(\tau_i, U_i)$ $(1 \leq i \leq n)$ and two finite-dimensional $K$-representations $(\sigma_\ell, V_\ell)$ and $(\sigma_r, V_r)$. Recall that $(\sigma_{\ell;n}, V_\ell \otimes \underline{U})$ is the $K$-representation with $K$ acting diagonally on $V_\ell \otimes \underline{U}$.

DEFINITION 4.10. For $u \in U(\mathfrak{g})$ and $0 \leq j \leq n$ define differential operators

$$D^{\sigma_\ell,\underline{\tau},\sigma_r}_{u,j} \in \mathbb{D}((V_\ell \otimes \underline{U} \otimes V_r^*)^{\mathfrak{m}}, V_\ell \otimes \underline{U} \otimes V_r^*)$$

by the formulas

$$D^{\sigma_\ell,\underline{\tau},\sigma_r}_{u,j} := \sum_{(u)} \tau_{j+1}(S(u_{(1)})) \cdots \tau_n(S(u_{(n-j)})) L^{\sigma_{\ell;n},\sigma_r}_{\widetilde{\Pi}(u_{(n-j+1)})} \qquad (0 \leq j < n),$$

$$D^{\sigma_\ell,\underline{\tau},\sigma_r}_{u,n} := L^{\sigma_{\ell;n},\sigma_r}_{\Pi(u)}.$$

The following lemma follows directly from Lemma 4.9.

LEMMA 4.11. *For* $u \in U(\mathfrak{g})$, $0 \leq j \leq n$ *and* $g \in N_K(A)$ *we have*

$$D^{\sigma_\ell,\underline{\tau},\sigma_r}_{\mathrm{Ad}_g(u),j} = g \cdot D^{\sigma_\ell,\underline{\tau},\sigma_r}_{u,j}$$

*in* $\mathbb{D}((V_\ell \otimes \underline{U} \otimes V_r^*)^{\mathfrak{m}}, V_\ell \otimes \underline{U} \otimes V_r^*)$.

Let $U(\mathfrak{g})^{\mathfrak{m}}$ be the centraliser of $\mathfrak{m}$ in $U(\mathfrak{g})$. It contains $U(\mathfrak{g})^M$ as a subalgebra. The following is a direct consequence of Lemma 4.11.

COROLLARY 4.12. *The differential operator* $D^{\sigma_\ell,\underline{\tau},\sigma_r}_{u,j} \in \mathbb{D}((V_\ell \otimes \underline{U} \otimes V_r^*)^{\mathfrak{m}}, V_\ell \otimes \underline{U} \otimes V_r^*)$ *lies in*

$$\begin{aligned} &\mathbb{D}((V_\ell \otimes \underline{U} \otimes V_r^*)^{\mathfrak{m}}) && \text{if } u \in U(\mathfrak{g})^{\mathfrak{m}}, \\ &\mathbb{D}_M((V_\ell \otimes \underline{U} \otimes V_r^*)^{\mathfrak{m}}) && \text{if } u \in U(\mathfrak{g})^M, \\ &\mathbb{D}_M((V_\ell \otimes \underline{U} \otimes V_r^*)^{\mathfrak{m}})^W && \text{if } u \in U(\mathfrak{g})^{N_K(A)}. \end{aligned}$$

*In particular,* $D^{\sigma_\ell,\underline{\mathcal{I}},\sigma_r}_{u,j;M} \in \mathbb{D}((V_\ell\otimes \underline{U}\otimes V_r^*)^M, V_\ell\otimes\underline{U}\otimes V_r^*)$ *lies in*

$$\begin{aligned}&\mathbb{D}((V_\ell\otimes \underline{U}\otimes V_r^*)^M) && \text{if } u\in U(\mathfrak{g})^M,\\ &\mathbb{D}((V_\ell\otimes \underline{U}\otimes V_r^*)^M)^W && \text{if } u\in U(\mathfrak{g})^{N_K(A)}.\end{aligned}$$

Next we show that the differential operator $D^{\sigma_\ell,\underline{\mathcal{I}},\sigma_r}_{u,j}$ arises as the radial component of the action $u[j]$ on $C^\infty_{\sigma_\ell,\underline{\mathcal{I}},\sigma_r}(G^{\times(n+1)})$. The relevant restriction map is defined as follows.

DEFINITION 4.13. Define the linear map

$$\mathrm{Res}^\flat : C^\infty(G^{\times(n+1)}; V_\ell\otimes\underline{U}\otimes V_r^*) \to C^\infty(A; V_\ell\otimes\underline{U}\otimes V_r^*)$$

by $\mathrm{Res}^\flat(f) := f^\flat|_A$. Concretely, for $f\in C^\infty(G^{\times(n+1)}; V_\ell\otimes\underline{U}\otimes V_r^*)$ and $a\in A$,

$$\mathrm{Res}^\flat(f)(a) := f(1,\ldots,1,a).$$

By Lemma 2.4 the restriction of $\mathrm{Res}^\flat$ to $C^\infty_{\sigma_\ell,\underline{\mathcal{I}},\sigma_r}(G^{\times(n+1)})$ is an injective linear map

$$\mathrm{Res}^\flat : C^\infty_{\sigma_\ell,\underline{\mathcal{I}},\sigma_r}(G^{\times(n+1)}) \hookrightarrow C^\infty(A; (V_\ell\otimes\underline{U}\otimes V_r^*)^M)^W.$$

Theorem 4.7 now generalises as follows.

THEOREM 4.14. *For* $u\in U(\mathfrak{g})$, $0\le j\le n$, $f\in C^\infty_{\sigma_\ell,\underline{\mathcal{I}},\sigma_r}(G^{\times(n+1)})$ *and* $a\in A_{\mathrm{reg}}$ *we have*

$$\mathrm{Res}^\flat\big(u[j]f\big)(a) = \big(D^{\sigma_\ell,\underline{\mathcal{I}},\sigma_r}_{u,j;M}\,\mathrm{Res}^\flat(f)\big)(a). \tag{4.13}$$

PROOF. Let $u\in U(\mathfrak{g})$, $f\in C^\infty_{\sigma_\ell,\underline{\mathcal{I}},\sigma_r}(G^{\times(n+1)})$ and $a\in A_{\mathrm{reg}}$. Then $f^\flat\in C^\infty_{\sigma_{\ell;n},\sigma_r}(G)$ by Lemma 2.4, and we have

$$\begin{aligned}\mathrm{Res}^\flat\big(u[n]f\big)(a) = \big(u[n]f\big)^\flat(a) = \big(u[1]f^\flat\big)(a) &= \big(L^{\sigma_{\ell;n},\sigma_r}_{\Pi(u);M} f^\flat|_A\big)(a)\\ &= \big(D^{\sigma_\ell,\underline{\mathcal{I}},\sigma_r}_{u,n:M}\,\mathrm{Res}^\flat(f)\big)(a),\end{aligned}$$

where the third equality follows from the first line of (4.8). This proves (4.13) for $j=n$.

Suppose that $0\le j<n$. By Corollary 2.7 we have

$$\big(u[j]f\big)^\flat(a) = \sum_{(u)} \tau_{j+1}(S(u_{(1)}))\cdots\tau_n(S(u_{(n-j)}))\big(S(u_{(n-j+1)})\langle n\rangle f\big)^\flat(a).$$

Formula (4.13) then follows from the fact that

$$\big(S(u_{(n-j+1)})\langle n\rangle f\big)^\flat(a) = \big(S(u_{(n-j+1)})\langle 1\rangle f^\flat\big)(a) = \big(L^{\sigma_{\ell;n},\sigma_r}_{\widetilde{\Pi}(u_{(n-j+1)});M}\,\mathrm{Res}^\flat(f)\big)(a),$$

where the last equality is due to the second equation of (4.8). □

REMARK 4.15. Replacing the role of left $G$-invariant differential operators by right $G$-invariant differential operators gives

$$\mathrm{Res}^\flat\big(u\langle j\rangle f\big)(a) = \big(\widehat{D}^{\sigma_\ell,\underline{\mathcal{I}},\sigma_r}_{S(u),j;M}\,\mathrm{Res}^\flat(f)\big)(a)$$

for $u\in U(\mathfrak{g})$, $0\le j\le n$, $f\in C^\infty_{\sigma_\ell,\underline{\mathcal{I}},\sigma_r}(G^{\times(n+1)})$ and $a\in A_{\mathrm{reg}}$, with the differential operators $\widehat{D}^{\sigma_\ell,\underline{\mathcal{I}},\sigma_r}_{u,j}\in\mathbb{D}((V_\ell\otimes\underline{U}\otimes V_r^*)^{\mathfrak{m}}, V_\ell\otimes\underline{U}\otimes V_r^*)$ given by

$$\widehat{D}^{\sigma_\ell,\underline{\mathcal{I}},\sigma_r}_{u,j} := \sum_{(u)} \tau_{j+1}(S(u_{(n-j+1)}))\cdots\tau_n(S(u_{(2)}))L^{\sigma_{\ell;n},\sigma_r}_{\widetilde{\Pi}(u_{(1)})}. \tag{4.14}$$

Lemma 4.11 and Corollary 4.12 also apply for $\widehat{D}^{\sigma_\ell,\underline{\tau},\sigma_r}_{u,j}$.

Define the linear map

$$(4.15)\qquad \big(U(\mathfrak{g})^M\big)^{\otimes(n+1)} \to \mathbb{D}_M((V_\ell\otimes\underline{U}\otimes V_r^*)^{\mathfrak{m}}),\qquad X\mapsto D^{\sigma_\ell,\underline{\tau},\sigma_r}_X$$

by

$$D^{\sigma_\ell,\underline{\tau},\sigma_r}_X := \widehat{D}^{\sigma_\ell,\underline{\tau},\sigma_r}_{u_0,0}\circ\Big(D^{\sigma_\ell,\underline{\tau},\sigma_r}_{u_1,1}\circ\cdots\circ D^{\sigma_\ell,\underline{\tau},\sigma_r}_{u_{n-1},n-1}\Big)\circ D^{\sigma_\ell,\underline{\tau},\sigma_r}_{u_n,n}\qquad (X=u_0\otimes\cdots\otimes u_n).$$

Note that its restriction $D^{\sigma_\ell,\underline{\tau},\sigma_r}_{X;M}$ to $\mathbb{D}((V_\ell\otimes\underline{U}\otimes V_r^*)^M)$ is the composition of the restricted coordinate-wise radial components.

THEOREM 4.16. *The assignment* $X\mapsto D^{\sigma_\ell,\underline{\tau},\sigma_r}_{X;M}$ *restricts to an algebra homomorphism*

$$U(\mathfrak{g})^{K,\mathrm{opp}}\otimes Z(U(\mathfrak{g}))^{\otimes(n-1)}\otimes U(\mathfrak{g})^K\to\mathbb{D}((V_\ell\otimes\underline{U}\otimes V_r^*)^M)^W.$$

PROOF. Fix an element $X\in U(\mathfrak{g})^{K,\mathrm{opp}}\otimes Z(U(\mathfrak{g}))^{\otimes(n-1)}\otimes U(\mathfrak{g})^K$ and $f\in C^\infty_{\sigma_\ell,\underline{\tau},\sigma_r}(G^{\times(n+1)})$. We have $D^{\sigma_\ell,\underline{\tau},\sigma_r}_{X;M}\in\mathbb{D}((V_\ell\otimes\underline{U}\otimes V_r^*)^M)^W$ by Corollary 4.12 and Remark 4.15. Furthermore, $X*f\in C^\infty_{\sigma_\ell,\underline{\tau},\sigma_r}(G^{\times(n+1)})$ by Lemma 2.9. By Theorem 4.14, Remark 4.15 and (2.6) we then have

$$\mathrm{Res}^\flat(X*f)=D^{\sigma_\ell,\underline{\tau},\sigma_r}_{X;M}(\mathrm{Res}^\flat f).$$

Choosing a second element $Y\in U(\mathfrak{g})^{K,\mathrm{opp}}\otimes Z(U(\mathfrak{g}))^{\otimes(n-1)}\otimes U(\mathfrak{g})^K$ we get

$$(4.16)\qquad D^{\sigma_\ell,\underline{\tau},\sigma_r}_{XY;M}(\mathrm{Res}^\flat f)=\mathrm{Res}^\flat(X*(Y*f))=D^{\sigma_\ell,\underline{\tau},\sigma_r}_{X;M}\big(D^{\sigma_\ell,\underline{\tau},\sigma_r}_{Y;M}(\mathrm{Res}^\flat f)\big).$$

Lemma 2.4 and (4.16) now imply that

$$D^{\sigma_\ell,\underline{\tau},\sigma_r}_{XY;M}(F|_A)=\big(D^{\sigma_\ell,\underline{\tau},\sigma_r}_{X;M}\circ D^{\sigma_\ell,\underline{\tau},\sigma_r}_{Y;M}\big)(F|_A)\qquad\forall F\in C^\infty_{\sigma_{\ell;n},\sigma_r}(G).$$

applying Proposition 4.8 with respect to the two $K$-representations $(\sigma_{\ell;n},V_\ell\otimes\underline{U})$ and $(\sigma_r,V_r)$ we conclude that

$$D^{\sigma_\ell,\underline{\tau},\sigma_r}_{XY;M}=D^{\sigma_\ell,\underline{\tau},\sigma_r}_{X;M}\circ D^{\sigma_\ell,\underline{\tau},\sigma_r}_{Y;M}$$

in $\mathbb{D}((V_\ell\otimes\underline{U}\otimes V_r^*)^M)^W$. □

REMARK 4.17. For $n=0$, Theorem 4.16 should be read as follows: the assignment

$$u\otimes_{Z(U(\mathfrak{g}))}v\mapsto L^{\sigma_\ell,\sigma_r}_{\widetilde{\Pi}(u);M}\circ L^{\sigma_\ell,\sigma_r}_{\Pi(v);M}\qquad(u,v\in U(\mathfrak{g})^K)$$

defines an algebra homomorphism $U(\mathfrak{g})^{K,\mathrm{opp}}\otimes_{Z(U(\mathfrak{g}))}U(\mathfrak{g})^K\to\mathbb{D}((V_\ell\otimes V_r^*)^M)^W$ (cf. [**3**, Thm. 3.3]). The balancing condition of the tensor product over the center is justified by (4.10), see also Remark 2.10.

By Theorem 4.16 we have the inclusion of algebras

$$(4.17)\qquad A^{\sigma_\ell,\underline{\tau},\sigma_r}\subseteq B^{\sigma_\ell,\underline{\tau},\sigma_r}\subseteq\mathbb{D}((V_\ell\otimes\underline{U}\otimes V_r^*)^M)^W$$

with

$$B^{\sigma_\ell,\underline{\tau},\sigma_r}:=\{D^{\sigma_\ell,\underline{\tau},\sigma_r}_{X;M}\mid X\in U(\mathfrak{g})^{K,\mathrm{opp}}\otimes Z(U(\mathfrak{g}))^{\otimes(n-1)}\otimes U(\mathfrak{g})^K\}$$

and $A^{\sigma_\ell,\underline{\tau},\sigma_r}$ the commutative subalgebra

$$A^{\sigma_\ell,\underline{\tau},\sigma_r}:=\{D^{\sigma_\ell,\underline{\tau},\sigma_r}_{X;M}\mid X\in Z(U(\mathfrak{g}))^{\otimes(n+1)}\}.$$

Furthermore, we have

$$D^{\sigma_\ell,\underline{\tau},\sigma_r}_{X;M}=D^{\sigma_\ell,\underline{\tau},\sigma_r}_{u_0,0;M}\circ\cdots\circ D^{\sigma_\ell,\underline{\tau},\sigma_r}_{u_{n-1},n-1;M}\circ D^{\sigma_\ell,\underline{\tau},\sigma_r}_{u_n,n;M}\quad(X=u_0\otimes\cdots\otimes u_n\in Z(U(\mathfrak{g}))^{\otimes(n+1)})$$

due to the following lemma.

LEMMA 4.18. *For $u \in Z(U(\mathfrak{g}))$ and $0 \leq j \leq n$ we have*

$$\widehat{D}^{\sigma_\ell,\underline{\tau},\sigma_r}_{u,j;M} = D^{\sigma_\ell,\underline{\tau},\sigma_r}_{u,j;M}$$

*in* $\mathbb{D}((V_\ell \otimes \underline{U} \otimes V_r^*)^M)^W$.

PROOF. Let $u \in Z(U(\mathfrak{g}))$ and $f \in C^\infty_{\sigma_\ell,\underline{\tau},\sigma_r}(G^{\times(n+1)})$. Then we have for $a \in A_{\mathrm{reg}}$,

$$\begin{aligned}\big(\widehat{D}^{\sigma_\ell,\underline{\tau},\sigma_r}_{u,j;M}\,\mathrm{Res}^\flat(f)\big)(a) &= \mathrm{Res}^\flat\big(S(u)\langle j\rangle f\big)(a)\\ &= \mathrm{Res}^\flat(u[j]f)(a) = \big(D^{\sigma_\ell,\underline{\tau},\sigma_r}_{u,j;M}\,\mathrm{Res}^\flat(f)\big)(a)\end{aligned}$$

with the first equality by Remark 4.15, the second by Remark 2.5, and the third by Theorem 4.14. Now the conclusion follows by Lemma 2.4 and Proposition 4.8, cf. the proof of Theorem 4.16. □

Suitable gauged versions of the subalgebras $B^{\sigma_\ell,\underline{\tau},\sigma_r}$ and $A^{\sigma_\ell,\underline{\tau},\sigma_r}$ will serve as the algebras of quantum integrals and quantum Hamiltonians of a quantum superintegrable system, see Subsection 5.

The following corollary of Theorem 4.16 and Lemma 4.18 is now immediate.

COROLLARY 4.19. *For $u, v \in Z(U(\mathfrak{g}))$ and $0 \leq i, j \leq n$ we have*

$$[D^{\sigma_\ell,\underline{\tau},\sigma_r}_{u,i;M}, D^{\sigma_\ell,\underline{\tau},\sigma_r}_{v,j;M}] = 0$$

*in* $\mathbb{D}((V_\ell \otimes \underline{U} \otimes V_r^*)^M)^W$.

REMARK 4.20. For finite-dimensional $\mathfrak{k}$-modules $(\sigma_\ell, V_\ell)$ and $(\sigma_r, V_r)$ we expect that

$$[D^{\sigma_\ell,\underline{\tau},\sigma_r}_{u,i}, D^{\sigma_\ell,\underline{\tau},\sigma_r}_{v,j}] = 0$$

in $\mathbb{D}_M((V_\ell \otimes \underline{U} \otimes V_r^*)^{\mathfrak{m}})$ for $u, v \in Z(U(\mathfrak{g}))$ and $0 \leq i, j \leq n$. This follows from the theory of formal $n$-point spherical functions when $G$ is real split, see [**20**].

## 5. The quantum Calogero-Moser spin chain

Fix throughout this section finite-dimensional smooth $K$-repesentations $(\sigma_\ell, V_\ell)$, $(\sigma_r, V_r)$ and finite-dimensional smooth $G$-representations $(\tau_i, U_i)$ $(1 \leq i \leq n)$.

**5.1. A dynamical factorisation of the Casimir element.** Denote the multiplication map of $U(\mathfrak{g})$ by $\mu \in \mathrm{Hom}(U(\mathfrak{g}) \otimes U(\mathfrak{g}), U(\mathfrak{g}))$.

DEFINITION 5.1. The Casimir element $\Omega \in Z(U(\mathfrak{g}))$ is

$$\Omega := \mu(\varpi)$$

with $\varpi \in S^2(\mathfrak{g})^G$ the $G$-invariant symmetric tensor associated to $K_{\mathfrak{g}}$.

Explicitly, for a basis $\{b_s\}_s$ of $\mathfrak{g}$ we have $\varpi = \sum_s b_s \otimes b_s'$ and $\Omega = \sum_s b_s b_s'$, where $\{b_s'\}_s$ is the basis of $\mathfrak{g}$ dual to $\{b_s\}_s$ with respect to $K_{\mathfrak{g}}$.

In the computations we will use a basis of $\mathfrak{g}$ compatible with the root space decomposition of $\mathfrak{g}$. It contains a basis $\{z_j\}_{j=1}^{\dim(\mathfrak{h})}$ of $\mathfrak{h}$ such that

(1) $K_{\mathfrak{g}}(z_j, z_{j'}) = \delta_{j,j'}$,
(2) $x_j := z_j$ $(1 \leq j \leq r)$ forms a basis of $\mathfrak{a}_{\mathbb{R}}$,
(3) $iz_j$ $(r+1 \leq j \leq \dim(\mathfrak{h}))$ forms a basis of $\mathfrak{t}_{\mathbb{R}}$.

Recall that $e_\alpha \in \mathfrak{g}_\alpha$ are root vectors such that $[e_\alpha, e_{-\alpha}] = h_\alpha$. Then $\{z_j\}_{j=1}^{\dim(\mathfrak{h})} \cup \{e_{-\alpha}\}_{\alpha\in R}$ is the basis of $\mathfrak{g}$ dual to $\{z_j\}_{j=1}^{\dim(\mathfrak{h})} \cup \{e_\alpha\}_{\alpha\in R}$ with respect to $K_\mathfrak{g}$ (cf. [**14**, Prop. 8.3 c]), hence

$$\varpi = \sum_{j=1}^{\dim(\mathfrak{h})} z_j \otimes z_j + \sum_{\alpha\in R} e_{-\alpha} \otimes e_\alpha \tag{5.1}$$

and

$$\Omega = \sum_{j=1}^{\dim(\mathfrak{h})} z_j^2 + \sum_{\alpha\in R} e_{-\alpha} e_\alpha. \tag{5.2}$$

Note that $\varpi$ can be recovered from $\Omega$ by

$$\varpi = \frac{1}{2}\big(\Delta(\Omega) - \Omega \otimes 1 - 1 \otimes \Omega\big).$$

Write $\Omega_\mathfrak{m} := \mu(\varpi_\mathfrak{m}) \in Z(U(\mathfrak{m}))$, with $\varpi_\mathfrak{m} \in S^2(\mathfrak{m})^M$ the symmetric tensor associated to the nondegenerate bilinear form $K_\mathfrak{g}(\cdot,\cdot)|_{\mathfrak{m}\times\mathfrak{m}}$,

$$\varpi_\mathfrak{m} := \sum_{j=r+1}^{\dim(\mathfrak{h})} z_j \otimes z_j + \sum_{\alpha\in R_0} e_{-\alpha} \otimes e_\alpha. \tag{5.3}$$

Finally, $\varpi' := \varpi - \varpi_\mathfrak{m} \in S^2(\mathfrak{g})^M$ and $\Omega' := \mu(\varpi') \in U(\mathfrak{g})^M$ admit the explicit expressions

$$\varpi' = \sum_{j=1}^{r} x_j \otimes x_j + \sum_{\lambda\in\Sigma} \varpi_\lambda, \qquad \Omega' = \sum_{j=1}^{r} x_j^2 + \sum_{\lambda\in\Sigma} \Omega_\lambda \tag{5.4}$$

with, for a restricted root $\lambda \in \Sigma$,

$$\varpi_\lambda := \sum_{\alpha\in R_\lambda} e_{-\alpha} \otimes e_\alpha \in \mathfrak{g}^{-\lambda} \otimes \mathfrak{g}^{\lambda}, \qquad \Omega_\lambda := \mu(\varpi_\lambda) = \sum_{\alpha\in R_\lambda} e_{-\alpha} e_\alpha \in U(\mathfrak{g}). \tag{5.5}$$

LEMMA 5.2. *We have $\varpi_\lambda \in (\mathfrak{g}^{-\lambda} \otimes \mathfrak{g}^\lambda)^M$ and $\Omega_\lambda \in U(\mathfrak{g})^M$.*

PROOF. The restricted root spaces $\mathfrak{g}^\lambda$ and $\mathfrak{g}^{-\lambda}$ are $\mathrm{Ad}(M)$-stable since $M = Z_K(\mathfrak{a})$. Let $g \in M$. The basis $\{\mathrm{Ad}_g(e_{-\alpha})\}_{\alpha\in R_\lambda}$ of $\mathfrak{g}^{-\lambda}$ is dual to the basis $\{\mathrm{Ad}_g(e_\alpha)\}_{\alpha\in R_\lambda}$ of $\mathfrak{g}^\lambda$ with respect to the perfect $\mathrm{Ad}(M)$-invariant pairing $K_\mathfrak{g}(\cdot,\cdot)|_{\mathfrak{g}^{-\lambda}\times\mathfrak{g}^\lambda} : \mathfrak{g}^{-\lambda} \times \mathfrak{g}^\lambda \to \mathbb{C}$. Hence the associated 2-tensor $\sum_{\alpha\in R_\lambda} \mathrm{Ad}_g(e_{-\alpha}) \otimes \mathrm{Ad}_g(e_\alpha)$ does not depend on $g \in M$. □

The adjoint action of $N_K(A)$ on $(\mathfrak{g}\otimes\mathfrak{g})^M$ and $U(\mathfrak{g})^M$ factorises to a $W$-action. We write $w \cdot x$ for the action of $w \in W$ on $x \in (\mathfrak{g}\otimes\mathfrak{g})^M$ or $x \in U(\mathfrak{g})^M$.

LEMMA 5.3. *We have*

(1) $w \cdot \varpi_\lambda = \varpi_{w\lambda}$ *and* $w \cdot \Omega_\lambda = \Omega_{w\lambda}$ *for* $w \in W$ *and* $\lambda \in \Sigma$.
(2) $\varpi_\mathfrak{m} \in S^2(\mathfrak{m})^{N_K(A)}$ *and* $\Omega_\mathfrak{m} \in U(\mathfrak{m})^{N_K(A)}$.
(3) $\sum_{i=1}^r x_i \otimes x_i \in S^2(\mathfrak{a})^{N_K(A)}$ *and* $\sum_{i=1}^r x_i^2 \in U(\mathfrak{a})^{N_K(A)}$.

PROOF. (1) For $w = gM \in W$ ($g \in N_K(A)$) the $\mathrm{Ad}_g$-invariance of $K_\mathfrak{g}$ implies that the basis $\{\mathrm{Ad}_g(e_{-\alpha})\}_{\alpha\in R_\lambda}$ of $\mathfrak{g}^{-w\lambda}$ is dual to the basis $\{\mathrm{Ad}_g(e_\alpha)\}_{\alpha\in R_\lambda}$ of $\mathfrak{g}^{w\lambda}$ with respect to the perfect pairing $K_\mathfrak{g}(\cdot,\cdot)|_{\mathfrak{g}^{-w\lambda}\times\mathfrak{g}^{w\lambda}}$. The result now follows immediately, cf. the proof of Lemma 5.2.
(2) This follows from the $\mathrm{Ad}(N_K(A))$-invariance of $\mathfrak{m}$ and $K_\mathfrak{g}(\cdot,\cdot)|_{\mathfrak{m}\times\mathfrak{m}}$.

(3) The proof is similar to the proof of (2), now considering the $\mathrm{Ad}(N_K(A))$-invariant nondegenerate symmetric bilinear form $K_{\mathfrak{g}}(\cdot,\cdot)|_{\mathfrak{a}\times\mathfrak{a}}$ on $\mathfrak{a}$. □

REMARK 5.4. For $\lambda\in\Sigma^+$ let $\{b_{\lambda,i}\}_{i=1}^{\mathrm{mtp}(\lambda)}$ be an orthonormal basis of $\mathfrak{g}_{\mathbb{R}}^{\lambda}$ with respect to the restriction of the scalar product $(\cdot,\cdot)$ to $\mathfrak{g}_{\mathbb{R}}^{\lambda}$. Define $b_{-\lambda,i}:=-\theta_{\mathbb{R}}(b_{\lambda,i})$ for $i=1,\ldots,\mathrm{mtp}(\lambda)$, then $\{b_{-\lambda,i}\}_{i=1}^{\mathrm{mtp}(\lambda)}$ is an orthonormal basis of $\mathfrak{g}_{\mathbb{R}}^{-\lambda}$ with respect to the restriction of $(\cdot,\cdot)$ to $\mathfrak{g}_{\mathbb{R}}^{-\lambda}$. In particular, $\{b_{-\lambda,i}\}_{i=1}^{\mathrm{mtp}(\lambda)}$ and $\{b_{\lambda,i}\}_{i=1}^{\mathrm{mtp}(\lambda)}$ are bases of $\mathfrak{g}^{-\lambda}$ and $\mathfrak{g}^{\lambda}$ respectively, dual with respect to $K_{\mathfrak{g}}(\cdot,\cdot)|_{\mathfrak{g}^{-\lambda}\times\mathfrak{g}^{\lambda}}$. We conclude that

$$\varpi_\lambda=\sum_{i=1}^{\mathrm{mtp}(\lambda)} b_{-\lambda,i}\otimes b_{\lambda,i} \tag{5.6}$$

for $\lambda\in\Sigma$. This expression of $\varpi_\lambda$ is commonly used in the radial component computation of the Casimir element, see, e.g., [**23**, §9.1.2]. Note that by (5.6),

$$(\theta\otimes\mathrm{id})(\varpi_\lambda)=-\sum_{i=1}^{\mathrm{mtp}(\lambda)} b_{\lambda,i}\otimes b_{\lambda,i}\in S^2(\mathfrak{g}^{\lambda})$$

and $(\theta\otimes\mathrm{id})(\varpi_\lambda)=(\mathrm{id}\otimes\theta)(\varpi_{-\lambda})$.

For $\nu\in\mathfrak{a}^*$ we denote by $t_\nu$ the unique element in $\mathfrak{a}$ such that $K_{\mathfrak{g}}(t_\nu,h)=\nu(h)$ for all $h\in\mathfrak{a}$. Note that $t_\lambda\in\mathfrak{a}_{\mathbb{R}}$ for $\lambda\in\Sigma$.

LEMMA 5.5. *Let $\lambda\in\Sigma$.*

(1) *If $\alpha\in R_0$ then $h_\alpha\in\mathfrak{t}$.*
(2) *If $\alpha\in R_\lambda$ then $\mathrm{pr}_{\mathfrak{p}}(h_\alpha)=t_\lambda$.*
(3) $\sum_{\alpha\in R_\lambda}h_\alpha\in\mathfrak{g}^M$.

PROOF. (1) This follows from the observation that $\mathfrak{t}=\{h\in\mathfrak{h}\mid K_{\mathfrak{g}}(h,\mathfrak{a})=0\}$.
(2) Let $\alpha\in R_\lambda$. Then $\mathrm{pr}_{\mathfrak{p}}(h_\alpha)\in\mathfrak{a}$ satisfies for all $h\in\mathfrak{a}$,

$$K_{\mathfrak{g}}(\mathrm{pr}_{\mathfrak{p}}(h_\alpha),h)=K_{\mathfrak{g}}(h_\alpha,h)=\alpha(h)=\lambda(h),$$

hence $\mathrm{pr}_{\mathfrak{p}}(h_\alpha)=t_\lambda$.
(3) Note that the $G$-equivariant linear map $\mathfrak{g}\otimes\mathfrak{g}\to\mathfrak{g}$, $x\otimes y\mapsto[x,y]$ maps $\varpi_\lambda$ to $-\sum_{\alpha\in R_\lambda}h_\alpha$. The result now follows from Lemma 5.2. □

We set for $\lambda\in\Sigma$,

$$z_\lambda:=\sum_{\alpha\in R_\lambda}\mathrm{pr}_{\mathfrak{k}}(h_\alpha)\in\mathfrak{t}.$$

COROLLARY 5.6. *We have $z_\lambda\in Z(\mathfrak{m})$, $z_{-\lambda}=-z_\lambda$ and*

$$\sum_{\alpha\in R_\lambda}h_\alpha=z_\lambda+\mathrm{mtp}(\lambda)t_\lambda. \tag{5.7}$$

*Furthermore, $w\cdot z_\lambda=z_{w\lambda}$ and $w\cdot t_\lambda=t_{w\lambda}$ for $w\in W$.*

PROOF. By Lemma 5.5(3) we have $z_\lambda\in\mathfrak{t}\cap\mathfrak{g}^M=Z(\mathfrak{m})$. The identity $z_{-\lambda}=-z_\lambda$ is obvious. Formula (5.7) is due to Lemma 5.5(2). It remains to prove the last identities.

Since $\mathrm{pr}_{\mathfrak{k}}$ and $\mathrm{pr}_{\mathfrak{p}}$ are $\mathrm{Ad}(K)$-linear, it suffices to show that $w\cdot(\sum_{\alpha\in R_\lambda}h_\alpha)=\sum_{\alpha\in R_{w\lambda}}h_\alpha$. This follows from Lemma 5.3(1) and the proof of Lemma 5.5(3). □

Consider the tensor product $W$-module algebra $\mathcal{R}\otimes U(\mathfrak{g})^M$. Its subalgebra $(\mathcal{R}\otimes U(\mathfrak{g})^M)^W$ of $W$-invariant elements contains $U(\mathfrak{g})^{N_K(A)}$ as a subalgebra. The Casimir $\Omega$ decomposes as $\Omega = \Omega_{\mathfrak{m}} + \Omega'$ with $\Omega_{\mathfrak{m}}, \Omega' \in U(\mathfrak{g})^{N_K(A)}$. The second term $\Omega' \in U(\mathfrak{g})^{N_K(A)}$ admits the following dynamical factorisation within $(\mathcal{R}\otimes U(\mathfrak{g})^M)^W$.

PROPOSITION 5.7. *We have*

$$\Omega' = \frac{1}{2}\sum_{\lambda\in\Sigma}\Big(\frac{1+\xi_{2\lambda}}{1-\xi_{2\lambda}}\Big)z_\lambda + \sum_{j=1}^r x_j^2 + \frac{1}{2}\sum_{\lambda\in\Sigma}\Big(\frac{1+\xi_{2\lambda}}{1-\xi_{2\lambda}}\Big)\,\mathrm{mtp}(\lambda)t_\lambda + 2\sum_{\lambda\in\Sigma}\frac{\Omega_\lambda}{1-\xi_{2\lambda}} \tag{5.8}$$

*in* $(\mathcal{R}\otimes U(\mathfrak{g})^M)^W$.

PROOF. By Lemma 5.3 and Corollary 5.6 it is clear that the right-hand side of (5.8) lies in $(\mathcal{R}\otimes U(\mathfrak{g})^M)^W$. By (5.7) we have

$$\Omega_{-\lambda} = \Omega_\lambda + z_\lambda + \mathrm{mtp}(\lambda)t_\lambda \tag{5.9}$$

for $\lambda\in\Sigma$. Hence

$$\begin{aligned}
\sum_{\lambda\in\Sigma}\frac{\Omega_\lambda}{1-\xi_{2\lambda}} &= \sum_{\lambda\in\Sigma^+}\frac{z_\lambda+\mathrm{mtp}(\lambda)t_\lambda}{1-\xi_{-2\lambda}} + \sum_{\lambda\in\Sigma^+}\Big(\frac{1}{1-\xi_{2\lambda}}+\frac{1}{1-\xi_{-2\lambda}}\Big)\Omega_\lambda\\
&= \sum_{\lambda\in\Sigma^+}\frac{z_\lambda+\mathrm{mtp}(\lambda)t_\lambda}{1-\xi_{-2\lambda}} + \sum_{\lambda\in\Sigma^+}\Omega_\lambda\\
&= \sum_{\lambda\in\Sigma^+}\Big(\frac{1}{1-\xi_{-2\lambda}}-\frac{1}{2}\Big)(z_\lambda+\mathrm{mtp}(\lambda)t_\lambda) + \frac{1}{2}\sum_{\lambda\in\Sigma}\Omega_\lambda\\
&= -\frac{1}{4}\sum_{\lambda\in\Sigma}\Big(\frac{1+\xi_{2\lambda}}{1-\xi_{2\lambda}}\Big)(z_\lambda+\mathrm{mtp}(\lambda)t_\lambda) + \frac{1}{2}\sum_{\lambda\in\Sigma}\Omega_\lambda,
\end{aligned}$$

where we have used (5.9) in the first and third equality. Substituting this identity in the right-hand side of the dynamical expression (5.8) reduces it to the expression (5.4) for $\Omega'$. □

REMARK 5.8. There is no dynamical expression for $\varpi'$ in $(\mathcal{R}\otimes(\mathfrak{g}\otimes\mathfrak{g})^M)^W$ that leads to (5.8) by applying the multiplication map $\mu$ of $U(\mathfrak{g})$. We can however write

$$\Omega' = \Big(\frac{1}{2}\sum_{\lambda\in\Sigma}\Big(\frac{1+\xi_{2\lambda}}{1-\xi_{2\lambda}}\Big)z_\lambda - \Omega_{\mathfrak{m}}\Big) + \frac{1}{2}\sum_{\lambda\in\Sigma}\Big(\frac{1+\xi_{2\lambda}}{1-\xi_{2\lambda}}\Big)\,\mathrm{mtp}(\lambda)t_\lambda - 2\mu(r_{21})$$

involving the dynamical element

$$r = -\frac{1}{2}\varpi_{\mathfrak{m}} - \frac{1}{2}\sum_{j=1}^r x_j\otimes x_j - \sum_{\lambda\in\Sigma}\frac{\varpi_\lambda}{1-\xi_{-2\lambda}} \in (\mathcal{R}\otimes(\mathfrak{g}\otimes\mathfrak{g})^M)^W \tag{5.10}$$

(here $r_{21}$ is obtained from $r$ by interchanging its tensor components). The 2-tensor $r$ will play an important role in the explicit description of the asymptotic boundary KZB operators, see Theorem 6.3 and Proposition 6.10. Note that $r$ satisfies the quasi-unitarity condition

$$r + r_{21} = -\varpi \in S^2(\mathfrak{g})^G.$$

In the real split case $r$ is Felder's **[9]** trigonometric solution of the classical dynamical Yang-Baxter equation. We will show in Subsection 5 that for arbitrary $G$, folded versions of $r$ satisfy coupled classical dynamical Yang-Baxter-reflection equations.

**5.2. The radial component of the action of the Casimir element.** We first define an explicit second order differential operator $H^{\sigma_\ell,\underline{\mathcal{I}},\sigma_r} \in \mathbb{D}_M((V_\ell \otimes \underline{U} \otimes V_r^*)^{\mathfrak{m}})^W$ and then show that the differential operator $D^{\sigma_\ell,\underline{\mathcal{I}},\sigma_r}_{\Omega,n}$ is equal to $H^{\sigma_\ell,\underline{\mathcal{I}},\sigma_r}$. The technique to compute $D^{\sigma_\ell,\underline{\mathcal{I}},\sigma_r}_{\Omega,n}$ goes back to Harish-Chandra, see, e.g., [**23**, §9.1.2].

Write for $\lambda \in \Sigma$,

$$\upsilon_\lambda := 4(\mathrm{pr}_{\mathfrak{k}} \otimes \mathrm{pr}_{\mathfrak{k}})(\varpi_\lambda) = \sum_{\alpha \in R_\lambda} y_{-\alpha} \otimes y_\alpha \in \mathfrak{q} \otimes \mathfrak{q}$$

and set $\Upsilon_\lambda := \mu(\upsilon_\lambda) = \sum_{\alpha \in R_\lambda} y_{-\alpha} y_\alpha \in U(\mathfrak{k})$.

LEMMA 5.9. *Let* $\lambda \in \Sigma$.

(1) $\upsilon_\lambda = \upsilon_{-\lambda}$ *and* $\Upsilon_\lambda = \Upsilon_{-\lambda}$.
(2) $\upsilon_\lambda \in (\mathfrak{q} \otimes \mathfrak{q})^M$ *and* $\Upsilon_\lambda \in U(\mathfrak{k})^M$.
(3) $w \cdot \upsilon_\lambda = \upsilon_{w\lambda}$ *and* $w \cdot \Upsilon_\lambda = \Upsilon_{w\lambda}$ *for* $w \in W$.

PROOF. (1) Observe that $(\mathrm{pr}_{\mathfrak{k}} \otimes \mathrm{pr}_{\mathfrak{k}})(\varpi_\lambda) = (\mathrm{pr}_{\mathfrak{k}} \otimes \mathrm{pr}_{\mathfrak{k}})((\theta \otimes \mathrm{id})(\varpi_\lambda))$ is a symmetric tensor by Remark 5.4. Hence $\upsilon_\lambda = \upsilon_{-\lambda}$, and consequently $\Upsilon_\lambda = \Upsilon_{-\lambda}$.
(2) The $\mathrm{Ad}(M)$-invariance of $\upsilon_\lambda$ follows from Lemma 5.2 and the fact that $\mathrm{pr}_{\mathfrak{k}}$ is $\mathrm{Ad}(K)$-equivariant. The $\mathrm{Ad}(M)$-invariance of $\Upsilon_\lambda \in U(\mathfrak{k})$ now follows immediately.
(3) This follows from Lemma 5.3(1). □

Note that $S(\Upsilon_\lambda) = \Upsilon_\lambda$ in view of Lemma 5.9(1).

DEFINITION 5.10. Define $H^{\sigma_\ell,\underline{\mathcal{I}},\sigma_r} \in \mathbb{D}_M((V_\ell \otimes \underline{U} \otimes V_r^*)^{\mathfrak{m}})^W$ by

$$\begin{aligned} H^{\sigma_\ell,\underline{\mathcal{I}},\sigma_r} := & \sum_{j=1}^r \partial_{x_j}^2 + \frac{1}{2}\sum_{\lambda \in \Sigma} \mathrm{mtp}(\lambda)\Big(\frac{\xi_\lambda + \xi_{-\lambda}}{\xi_\lambda - \xi_{-\lambda}}\Big)\partial_{t_\lambda} + \sigma_r^*(\Omega_{\mathfrak{m}}) \\ & + \frac{1}{2}\sum_{\lambda \in \Sigma}\Big(\frac{\xi_\lambda + \xi_{-\lambda}}{\xi_\lambda - \xi_{-\lambda}}\Big)\sigma_r^*(z_\lambda) \\ & - \sum_{\lambda \in \Sigma} \frac{\sigma_{\ell;n}(\Upsilon_\lambda) + \sigma_r^*(\Upsilon_\lambda) + (\xi_\lambda + \xi_{-\lambda})(\sigma_{\ell;n} \otimes \sigma_r^*)(\upsilon_\lambda)}{(\xi_\lambda - \xi_{-\lambda})^2}. \end{aligned}$$

The linear operators $\sigma_r^*(\Omega_{\mathfrak{m}})$, $\sigma_r^*(z_\lambda)$, $\sigma_{\ell;n}(\Upsilon_\lambda)$, $\sigma_r^*(\Upsilon_\lambda)$ and $(\sigma_{\ell;n} \otimes \sigma_r^*)(\upsilon_\lambda)$ occurring in the definition of $H^{\sigma_\ell,\underline{\mathcal{I}},\sigma_r}$ are viewed as linear operators on $(V_\ell \otimes \underline{U} \otimes V_r^*)^{\mathfrak{m}}$. Note that this is well-defined and that $H^{\sigma_\ell,\underline{\mathcal{I}},\sigma_r}$ lies in $\mathbb{D}_M((V_\ell \otimes \underline{U} \otimes V_r^*)^{\mathfrak{m}})^W$ in view of Lemma 5.9, Lemma 5.3 and Corollary 5.6. We now have the following result, cf. [**23**, Prop. 9.1.2.11].

THEOREM 5.11. *We have*

$$D^{\sigma_\ell,\underline{\mathcal{I}},\sigma_r}_{\Omega,n} = H^{\sigma_\ell,\underline{\mathcal{I}},\sigma_r}_{M_\circ} \tag{5.11}$$

*in* $\mathbb{D}_M((V_\ell \otimes \underline{U} \otimes V_r^*)^{\mathfrak{m}})^W$.

PROOF. By Corollary 4.12 the differential operator $D^{\sigma_\ell,\underline{\mathcal{I}},\sigma_r}_{\Omega,n}$ lies in $\mathbb{D}_M((V_\ell \otimes \underline{U} \otimes V_r^*)^{\mathfrak{m}})^W$. Furthermore,

$$D^{\sigma_\ell,\underline{\mathcal{I}},\sigma_r}_{\Omega,n} = L^{\sigma_{\ell;n},\sigma_r}_{\Pi(\Omega)} = L^{\sigma_{\ell;n},\sigma_r}_{\Pi(\Omega')} + \sigma_r^*(\Omega_{\mathfrak{m}}).$$

Hence it suffices to show that $L^{\sigma_{\ell;n},\sigma_r}_{\Pi(\Omega')}$ is the restriction to $C^\infty(A_{\mathrm{reg}};(V_\ell\otimes\underline{U}\otimes V_r^*)^{\mathfrak{m}})$ of the differential operator

$$\sum_{j=1}^r \partial_{x_j}^2+\frac{1}{2}\sum_{\lambda\in\Sigma}\mathrm{mtp}(\lambda)\Big(\frac{\xi_\lambda+\xi_{-\lambda}}{\xi_\lambda-\xi_{-\lambda}}\Big)\partial_{t_\lambda}+\frac{1}{2}\sum_{\lambda\in\Sigma}\Big(\frac{\xi_\lambda+\xi_{-\lambda}}{\xi_\lambda-\xi_{-\lambda}}\Big)\sigma_r^*(z_\lambda)$$
$$-\sum_{\lambda\in\Sigma}\frac{\sigma_{\ell;n}(\Upsilon_\lambda)+\sigma_r^*(\Upsilon_\lambda)+(\xi_\lambda+\xi_{-\lambda})(\sigma_{\ell;n}\otimes\sigma_r^*)(\upsilon_\lambda)}{(\xi_\lambda-\xi_{-\lambda})^2}.$$

For this it suffices to prove that

$$\begin{aligned}\Pi(\Omega')=&\sum_{j=1}^r 1\otimes x_j^2\otimes(1\otimes_{U(\mathfrak{m})}1)+\frac{1}{2}\sum_{\lambda\in\Sigma}\mathrm{mtp}(\lambda)\Big(\frac{\xi_\lambda+\xi_{-\lambda}}{\xi_\lambda-\xi_{-\lambda}}\Big)\otimes t_\lambda\otimes(1\otimes_{U(\mathfrak{m})}1)\\&-\frac{1}{2}\sum_{\lambda\in\Sigma}\Big(\frac{\xi_\lambda+\xi_{-\lambda}}{\xi_\lambda-\xi_{-\lambda}}\Big)\otimes 1\otimes(z_\lambda\otimes_{U(\mathfrak{m})}1)\;+\sum_{\lambda\in\Sigma}\frac{(\xi_\lambda+\xi_{-\lambda})}{(\xi_\lambda-\xi_{-\lambda})^2}\otimes 1\otimes\upsilon_\lambda\\&-\sum_{\lambda\in\Sigma}\frac{1}{(\xi_\lambda-\xi_{-\lambda})^2}\otimes 1\otimes\Big(\Upsilon_\lambda\otimes_{U(\mathfrak{m})}1+1\otimes_{U(\mathfrak{m})}\Upsilon_\lambda\Big),\end{aligned}$$

where, in the fourth term, $\upsilon_\lambda\in\mathfrak{q}\otimes\mathfrak{q}$ is viewed as element in $U(\mathfrak{k})\otimes_{U(\mathfrak{m})}U(\mathfrak{k})$ via the canonical injection $\mathfrak{q}\otimes\mathfrak{q}\hookrightarrow U(\mathfrak{k})\otimes_{U(\mathfrak{m})}U(\mathfrak{k})$, $x\otimes y\mapsto x\otimes_{U(\mathfrak{m})}y$.

This in turn will follow from the identity

$$\begin{aligned}\Omega'=&\sum_{j=1}^r x_j^2+\frac{1}{2}\sum_{\lambda\in\Sigma}\Big(\frac{a^\lambda+a^{-\lambda}}{a^\lambda-a^{-\lambda}}\Big)(\mathrm{mtp}(\lambda)t_\lambda-z_\lambda)\\&+\sum_{\lambda\in\Sigma}\frac{(a^\lambda+a^{-\lambda})\mu((\mathrm{Ad}_{a^{-1}}\otimes\mathrm{id})(\upsilon_\lambda))-\mathrm{Ad}_{a^{-1}}(\Upsilon_\lambda)-\Upsilon_\lambda}{(a^\lambda-a^{-\lambda})^2}\end{aligned}$$

in $U(\mathfrak{g})$ for all $a\in A_{\mathrm{reg}}$.

Fix $a\in A_{\mathrm{reg}}$ and $\lambda\in\Sigma$. By applying (3.3) to both $e_{-\alpha}$ and $e_\alpha$ and using Lemma 5.9(1) we get

$$\begin{aligned}\Omega_\lambda=\sum_{\alpha\in R_\lambda}e_{-\alpha}e_\alpha=&\frac{(a^\lambda+a^{-\lambda})\mu(\mathrm{Ad}_{a^{-1}}\otimes\mathrm{id})(\upsilon_\lambda)-\mathrm{Ad}_{a^{-1}}(\Upsilon_\lambda)-\Upsilon_\lambda}{(a^\lambda-a^{-\lambda})^2}\\&+\frac{a^{-\lambda}}{(a^\lambda-a^{-\lambda})^2}\sum_{\alpha\in R_\lambda}[y_\alpha,\mathrm{Ad}_{a^{-1}}(y_{-\alpha})].\end{aligned}$$

In view of (5.4) it thus suffices to show that

(5.12)
$$\sum_{\lambda\in\Sigma}\frac{a^{-\lambda}}{(a^\lambda-a^{-\lambda})^2}\sum_{\alpha\in R_\lambda}[y_\alpha,\mathrm{Ad}_{a^{-1}}(y_{-\alpha})]=\frac{1}{2}\sum_{\lambda\in\Sigma}\Big(\frac{a^\lambda+a^{-\lambda}}{a^\lambda-a^{-\lambda}}\Big)(\mathrm{mtp}(\lambda)t_\lambda-z_\lambda).$$

Note that for $\alpha\in R_\lambda$,

(5.13)
$$\begin{aligned}[y_\alpha,\mathrm{Ad}_{a^{-1}}(y_{-\alpha})]&=a^\lambda h_\alpha+a^{-\lambda}\theta(h_\alpha)+a^\lambda[\theta(e_\alpha),e_{-\alpha}]+a^{-\lambda}[e_\alpha,\theta(e_{-\alpha})]\\&=(a^\lambda+a^{-\lambda})\mathrm{pr}_{\mathfrak{k}}(h_\alpha)+(a^\lambda-a^{-\lambda})t_\lambda+a^\lambda[\theta(e_\alpha),e_{-\alpha}]+a^{-\lambda}[e_\alpha,\theta(e_{-\alpha})].\end{aligned}$$

When substituting into the left-hand side of (5.12), the first term of (5.13) contributes

$$\begin{aligned}\sum_{\lambda\in\Sigma}\frac{a^{-\lambda}(a^{\lambda}+a^{-\lambda})}{(a^{\lambda}-a^{-\lambda})^2}z_\lambda&=\sum_{\lambda\in\Sigma^+}\frac{a^{-\lambda}(a^{\lambda}+a^{-\lambda})-a^{\lambda}(a^{\lambda}+a^{-\lambda})}{(a^{\lambda}-a^{-\lambda})^2}z_\lambda\\&=\sum_{\lambda\in\Sigma^+}\Big(\frac{a^{-\lambda}+a^{\lambda}}{a^{-\lambda}-a^{\lambda}}\Big)z_\lambda=-\frac{1}{2}\sum_{\lambda\in\Sigma}\Big(\frac{a^{\lambda}+a^{-\lambda}}{a^{\lambda}-a^{-\lambda}}\Big)z_\lambda\end{aligned}$$

since $z_{-\lambda}=-z_\lambda$ (see Corollary 5.6), and the second term of (5.13) contributes

$$\begin{aligned}\sum_{\lambda\in\Sigma}\mathrm{mtp}(\lambda)\frac{a^{-\lambda}(a^{\lambda}-a^{-\lambda})}{(a^{\lambda}-a^{-\lambda})^2}t_\lambda&=\sum_{\lambda\in\Sigma^+}\mathrm{mtp}(\lambda)\Big(\frac{1}{a^{2\lambda}-1}-\frac{1}{a^{-2\lambda}-1}\Big)t_\lambda\\&=\sum_{\lambda\in\Sigma^+}\mathrm{mtp}(\lambda)\Big(\frac{a^{\lambda}+a^{-\lambda}}{a^{\lambda}-a^{-\lambda}}\Big)t_\lambda\\&=\frac{1}{2}\sum_{\lambda\in\Sigma}\mathrm{mtp}(\lambda)\Big(\frac{a^{\lambda}+a^{-\lambda}}{a^{\lambda}-a^{-\lambda}}\Big)t_\lambda\end{aligned}$$

since $t_{-\lambda}=-t_\lambda$ and $\mathrm{mtp}(-\lambda)=\mathrm{mtp}(\lambda)$. It thus suffices to show that

$$\sum_{\lambda\in\Sigma}\frac{a^{-\lambda}}{(a^{\lambda}-a^{-\lambda})^2}\sum_{\alpha\in R_\lambda}\big(a^{\lambda}[\theta(e_\alpha),e_{-\alpha}]+a^{-\lambda}[e_\alpha,\theta(e_{-\alpha})]\big)=0. \tag{5.14}$$

This follows from the anti-symmetry of the Lie bracket and the fact that

$$\sum_{\alpha\in R_\lambda}e_\alpha\otimes\theta(e_{-\alpha})\in S^2(\mathfrak{g}^{\lambda})$$

for all $\lambda\in\Sigma$, which was observed in Remark 5.4. □

REMARK 5.12. In the real split case one can choose the root vectors $e_\alpha$ in such a way that $\theta(e_\alpha)=-e_{-\alpha}$ for $\alpha\in R$. In that case one has $v_\alpha=-y_\alpha\otimes y_\alpha$ and $\Upsilon_\alpha=-y_\alpha^2$ for $\alpha\in R=\Sigma$, and Theorem 5.11 reduces to [**20**, Cor. 3.5].

**5.3. The Schrödinger operator.** The following explicit gauge will turn $H^{\sigma_\ell,\underline{\tau},\sigma_r}$ into a Schrödinger operator.

DEFINITION 5.13. We write $D\mapsto\widetilde{D}$ for the $W$-equivariant $\mathcal{R}\otimes\mathrm{End}_M((V_\ell\otimes\underline{U}\otimes V_r^*)^{\mathfrak{m}})$-linear algebra automorphism of $\mathbb{D}_M((V_\ell\otimes\underline{U}\otimes V_r^*)^{\mathfrak{m}})$ satisfying

$$\widetilde{\partial_h}=\frac{1}{4}\sum_{\lambda\in\Sigma}\lambda(h)\,\mathrm{mtp}(\lambda)\Big(\frac{1+\xi_{2\lambda}}{1-\xi_{2\lambda}}\Big)+\partial_h\qquad\forall\,h\in\mathfrak{a}_{\mathbb{R}}.$$

REMARK 5.14. Consider the positive Weyl chamber

$$A^+:=\{a\in A\mid a^{\lambda}>1\quad\forall\,\lambda\in\Sigma^+\}$$

and define an analytic function $\delta$ on $A^+$ by

$$\delta(a):=a^{\rho}\prod_{\lambda\in\Sigma^+}(1-a^{-2\lambda})^{\frac{\mathrm{mtp}(\lambda)}{2}},\qquad\rho:=\frac{1}{2}\sum_{\lambda\in\Sigma^+}\mathrm{mtp}(\lambda)\lambda. \tag{5.15}$$

A straightforward computation then shows that

$$\big(\widetilde{D}(\delta f)\big)(a)=\delta(a)\big(Df\big)(a)$$

for $D\in\mathbb{D}_M((V_\ell\otimes\underline{U}\otimes V_r^*)^{\mathfrak{m}})$, $f\in C^\infty(A^+;(V_\ell\otimes\underline{U}\otimes V_r^*)^{\mathfrak{m}})$ and $a\in A^+$.

Define the potential $V^{\sigma_\ell,\underline{\tau},\sigma_r} \in \big(\mathcal{R} \otimes \mathrm{End}_M((V_\ell \otimes \underline{U} \otimes V_r^*)^{\mathfrak{m}})\big)^W$ by

$$
\begin{aligned}
V^{\sigma_\ell,\underline{\tau},\sigma_r} := &-\sum_{\lambda\in\Sigma} \frac{k(\lambda)}{(\xi_\lambda - \xi_{-\lambda})^2} + \frac{1}{2}\sum_{\lambda\in\Sigma}\Big(\frac{\xi_\lambda+\xi_{-\lambda}}{\xi_\lambda-\xi_{-\lambda}}\Big)\sigma_r^*(z_\lambda)\\
&-\sum_{\lambda\in\Sigma}\frac{\sigma_{\ell;n}(\Upsilon_\lambda)+\sigma_r^*(\Upsilon_\lambda)+(\xi_\lambda+\xi_{-\lambda})(\sigma_{\ell;n}\otimes\sigma_r^*)(\upsilon_\lambda)}{(\xi_\lambda-\xi_{-\lambda})^2}
\end{aligned}
\tag{5.16}
$$

with $k(\lambda) := \mathrm{mtp}(\lambda)\big(\mathrm{mtp}(2\lambda) + \frac{\mathrm{mtp}(\lambda)}{2} - 1\big)(t_\lambda, t_\lambda)$.

PROPOSITION 5.15. *The gauged differential operator* $\widetilde{H}^{\sigma_\ell,\underline{\tau},\sigma_r} \in \mathbb{D}_M((V_\ell \otimes \underline{U} \otimes V_r^*)^{\mathfrak{m}})^W$ *is explicitly given by*

$$
\widetilde{H}^{\sigma_\ell,\underline{\tau},\sigma_r} = \sum_{j=1}^r \partial_{x_j}^2 + V^{\sigma_\ell,\underline{\tau},\sigma_r} + \sigma_r^*(\Omega_{\mathfrak{m}}) - (t_\rho, t_\rho). \tag{5.17}
$$

PROOF. The proof of the first statement is similar to the scalar case ($n = 0$, $\sigma_\ell = \sigma_r$ the trivial representation), cf. the proof of [**13**, Thm. 2.1.1]. □

Write $\widetilde{A}^{\sigma_\ell,\underline{\tau},\sigma_r}$ and $\widetilde{B}^{\sigma_\ell,\underline{\tau},\sigma_r}$ for the images of the algebras $A^{\sigma_\ell,\underline{\tau},\sigma_r}$ and $B^{\sigma_\ell,\underline{\tau},\sigma_r}$ (see (4.17)) under the isomorphism $D \mapsto \widetilde{D}$ of $\mathbb{D}((V_\ell \otimes \underline{U} \otimes V_r^*)^M)^W$. We have an inclusion of algebras

$$
\widetilde{A}^{\sigma_\ell,\underline{\tau},\sigma_r} \subseteq \widetilde{B}^{\sigma_\ell,\underline{\tau},\sigma_r} \subset \mathbb{D}((V_\ell \otimes \underline{U} \otimes V_r^*)^M)^W, \tag{5.18}
$$

with $\widetilde{A}^{\sigma_\ell,\underline{\tau},\sigma_r}$ commutative and containing $\widetilde{H}_M^{\sigma_\ell,\underline{\tau},\sigma_r} = \widetilde{D}_{\Omega,n;M}^{\sigma_\ell,\underline{\tau},\sigma_r}$.

DEFINITION 5.16. The quantum Calogero-Moser spin chain is the quantum superintegrable system with algebra of quantum Hamiltonians $\widetilde{A}^{\sigma_\ell,\underline{\tau},\sigma_r}$, algebra of quantum integrals $\widetilde{B}^{\sigma_\ell,\underline{\tau},\sigma_r}$ and Schrödinger operator $\widetilde{H}_M^{\sigma_\ell,\underline{\tau},\sigma_r} \in \widetilde{A}^{\sigma_\ell,\underline{\tau},\sigma_r}$.

The quantum Calogero-Moser spin chain is a mixture of a quantum spin Calogero-Moser system of rank $r$ and a one-dimensional spin chain with $n$ internal sites and two reflecting boundaries. For the bosonic theory, the quantum state space is a suitable completion of the space $C^\infty(A_{\mathrm{reg}}; (V_\ell \otimes \underline{U} \otimes V_r^*)^M)^W$.

COROLLARY 5.17. *Let* $f \in C^\infty_{\sigma_\ell,\underline{\tau},\sigma_r}(G^{\times(n+1)}; \chi)$ *be an* $n$*-point spherical function (see Definition* 2.11*). Write*

$$
\widetilde{f}^\flat \in C^\infty(A_{\mathrm{reg}}; (V_\ell \otimes \underline{U} \otimes V_r^*)^M)^W
$$

*for the* $W$*-invariant* $(V_\ell \otimes \underline{U} \otimes V_r^*)^M$*-valued function such that*

$$
\widetilde{f}^\flat(a) := \delta(a) f^\flat(a) \qquad \forall\, a \in A^+.
$$

*For all* $\mathbb{X} \in Z(U(\mathfrak{g}))^{\otimes(n+1)}$ *we have*

$$
\widetilde{D}^{\sigma_\ell,\underline{\tau},\sigma_r}_{\mathbb{X};M}(\widetilde{f}^\flat) = \chi(\mathbb{X})\widetilde{f}^\flat,
$$

*with* $\chi$ *viewed as the character* $\chi_0 \otimes \cdots \otimes \chi_n$ *of* $Z(U(\mathfrak{g}))^{\otimes(n+1)}$*. In particular,*

$$
\widetilde{H}_M^{\sigma_\ell,\underline{\tau},\sigma_r}(\widetilde{f}^\flat) = \chi_n(\Omega)\widetilde{f}^\flat.
$$

See [**20**] for explicit examples of classes of $n$-point spherical functions when $G$ is real split.

## 6. The asymptotic boundary KZB operators

Fix throughout this section finite-dimensional $K$-representations $(\sigma_\ell, V_\ell)$, $(\sigma_r, V_r)$ and finite-dimensional $G$-representations $(\tau_i, U_i)$ $(1 \leq i \leq n)$.

The quadratic Hamiltonians $\widetilde{D}^{\sigma_\ell,\underline{\tau},\sigma_r}_{\Omega,j;M} \in \widetilde{A}^{\sigma_\ell,\underline{\tau},\sigma_r}$ $(0 \leq j \leq n)$ of the quantum Calogero-Moser spin chain are related to the Schrödinger operator $\widetilde{H}^{\sigma_\ell,\underline{\tau},\sigma_r}_M$ by

$$\widetilde{D}^{\sigma_\ell,\underline{\tau},\sigma_r}_{\Omega,j;M} = \widetilde{H}^{\sigma_\ell,\underline{\tau},\sigma_r}_M - 2\sum_{i=j+1}^{n} \widetilde{\mathcal{D}}^{\sigma_\ell,\underline{\tau},\sigma_r}_{j;M} \in \widetilde{A}^{\sigma_\ell,\underline{\tau},\sigma_r} \qquad (0 \leq j \leq n),$$

with $\widetilde{\mathcal{D}}^{\sigma_\ell,\underline{\tau},\sigma_r}_{i;M} \in \widetilde{A}^{\sigma_\ell,\underline{\tau},\sigma_r}$ the quantum Hamiltonian obtained by gauging and $M$-restricting the differential operator

$$\mathcal{D}^{\sigma_\ell,\underline{\tau},\sigma_r}_i := \frac{1}{2}\big(D^{\sigma_\ell,\underline{\tau},\sigma_r}_{\Omega,i} - D^{\sigma_\ell,\underline{\tau},\sigma_r}_{\Omega,i-1}\big) \in \mathbb{D}_M((V_\ell \otimes \underline{U} \otimes V_r^*)^{\mathfrak{m}})^W \qquad (1 \leq i \leq n).$$

DEFINITION 6.1. The quantum Hamiltonians $\widetilde{\mathcal{D}}^{\sigma_\ell,\underline{\tau},\sigma_r}_{i;M} \in \widetilde{A}^{\sigma_\ell,\underline{\tau},\sigma_r}$ $(1 \leq i \leq n)$ are called the asymptotic boundary KZB operators.

We now first derive explicit expressions for the asymptotic boundary KZB operators.

**6.1. The derivation of the explicit expression.** The asymptotic boundary KZB operators turn out to be first-order differential operators, whose local terms we define now first. Write

$$r^+ := -\varpi_{\mathfrak{m}} - \sum_{\lambda\in\Sigma} \frac{2(\mathrm{pr}_{\mathfrak{k}} \otimes \mathrm{id})(\varpi_\lambda)}{1-\xi_{-2\lambda}} \in (\mathcal{R} \otimes (\mathfrak{k} \otimes \mathfrak{g})^M)^W,$$

$$r^- := \sum_{j=1}^{r} x_j \otimes x_j + \sum_{\lambda\in\Sigma} \frac{2(\mathrm{pr}_{\mathfrak{p}} \otimes \mathrm{id})(\varpi_\lambda)}{1-\xi_{-2\lambda}} \in (\mathcal{R} \otimes (\mathfrak{p} \otimes \mathfrak{g})^M)^W$$

(the $M$-invariance follows from Lemma 5.2, and the $W$-invariance from Lemma 5.3). Define $\kappa^{\mathrm{core}} \in (\mathcal{R} \otimes U(\mathfrak{g})^M)^W$ by

$$\kappa^{\mathrm{core}} := \frac{1}{2}\Omega + \mu(r^+_{21}) = -\frac{1}{2}\Omega_{\mathfrak{m}} + \frac{1}{2}\Omega' - \sum_{\lambda\in\Sigma}\sum_{\alpha\in R_\lambda} \frac{e_{-\alpha}y_\alpha}{1-\xi_{2\lambda}} \tag{6.1}$$

and $\kappa \in (\mathcal{R} \otimes (U(\mathfrak{k}) \otimes U(\mathfrak{g}) \otimes U(\mathfrak{k}))^M)^W$ by

$$\kappa := 1 \otimes \kappa^{\mathrm{core}} \otimes 1 - \Big(\varpi_{\mathfrak{m}} + \sum_{\lambda\in\Sigma} \frac{2(\mathrm{pr}_{\mathfrak{k}} \otimes \mathrm{id})(\varpi_\lambda)}{1-\xi_{-2\lambda}}\Big) \otimes 1 - 1 \otimes \Big(\sum_{\lambda\in\Sigma} \frac{2(\mathrm{id} \otimes \mathrm{pr}_{\mathfrak{k}})(\varpi_\lambda)}{\xi_{-\lambda}-\xi_\lambda}\Big).$$

For $1 \leq i < j \leq n$ and $1 \leq s \leq n$ we write

$$r^{\pm}_{ij} := (\tau_i \otimes \tau_j)(r^\pm), \qquad \kappa_s := (\sigma_\ell \otimes \tau_s \otimes \sigma_r^*)(\kappa),$$

which we view as elements in $(\mathcal{R} \otimes \mathrm{End}_M((V_\ell \otimes \underline{U} \otimes V_r^*)^{\mathfrak{m}}))^W$.

DEFINITION 6.2. Define first-order differential operators $\mathcal{D}^{\sigma_\ell,\underline{\tau},\sigma_r}_i \in \mathbb{D}_M((V_\ell \otimes \underline{U} \otimes V_r^*)^{\mathfrak{m}})^W$ $(1 \leq i \leq n)$ by

$$\mathcal{D}^{\sigma_\ell,\underline{\tau},\sigma_r}_i = \sum_{j=1}^{r} \tau_i(x_j)\partial_{x_j} - \kappa_i - \sum_{k=1}^{i-1} r^+_{ki} - \sum_{k=i+1}^{n} r^-_{ik}.$$

The following theorem is the main result of this section.

THEOREM 6.3. *For $1 \leq i \leq n$ we have the identity*

$$\frac{1}{2}\big(D^{\sigma_\ell,\underline{\tau},\sigma_r}_{\Omega,i} - D^{\sigma_\ell,\underline{\tau},\sigma_r}_{\Omega,i-1}\big) = \mathcal{D}^{\sigma_\ell,\underline{\tau},\sigma_r}_{i} \tag{6.2}$$

*in* $\mathbb{D}_M\big((V_\ell \otimes \underline{U} \otimes V_r^*)^{\mathfrak{m}}\big)^W$.

Before discussing the proof of Theorem 6.3, we note some immediate consequences first.

COROLLARY 6.4. (1) *The $M$-restricted gauged differential operators $\widetilde{\mathcal{D}}^{\sigma_\ell,\underline{\tau},\sigma_r}_{i;M} \in \widetilde{A}^{\sigma_\ell,\underline{\tau},\sigma_r}$ are first-order quantum Hamiltonians for the quantum Calogero-Moser spin chain. In particular,*

$$[\widetilde{\mathcal{D}}^{\sigma_\ell,\underline{\tau},\sigma_r}_{i;M}, \widetilde{\mathcal{D}}^{\sigma_\ell,\underline{\tau},\sigma_r}_{j;M}] = 0, \qquad [\widetilde{\mathcal{D}}^{\sigma_\ell,\underline{\tau},\sigma_r}_{i;M}, \widetilde{H}^{\sigma_\ell,\underline{\tau},\sigma_r}_{M}] = 0 \tag{6.3}$$

*for $1 \leq i,j \leq n$.*

(2) *The quadratic Hamiltonian $\widetilde{D}^{\sigma_\ell,\underline{\tau},\sigma_r}_{\Omega,j;M} \in \widetilde{A}^{\sigma_\ell,\underline{\tau},\sigma_r}$ for $j \in \{0,\ldots,n\}$ is explicitly given by*

$$\widetilde{D}^{\sigma_\ell,\underline{\tau},\sigma_r}_{\Omega,j;M} = \widetilde{H}^{\sigma_\ell,\underline{\tau},\sigma_r}_{M} - 2\sum_{i=j+1}^{n} \widetilde{\mathcal{D}}^{\sigma_\ell,\underline{\tau},\sigma_r}_{j;M}.$$

(3) *We have*

$$\widetilde{\mathcal{D}}^{\sigma_\ell,\underline{\tau},\sigma_r}_{i;M}(\widetilde{f}^\flat) = \frac{1}{2}\big(\chi_i(\Omega) - \chi_{i-1}(\Omega)\big)\widetilde{f}^\flat$$

*for $f \in C^\infty_{\sigma_\ell,\underline{\tau},\sigma_r}(G^{\times(n+1)};\chi)$ and $1 \leq i \leq n$.*

(4) *We have in $\mathbb{D}_M((V_\ell \otimes \underline{U} \otimes V_r^*)^{\mathfrak{m}})^W$,*

$$\widetilde{\mathcal{D}}^{\sigma_\ell,\underline{\tau},\sigma_r}_{i} = \sum_{j=1}^{r} \tau_i(x_j)\partial_{x_j} - \widehat{\kappa}_i - \sum_{k=1}^{i-1} r^+_{ki} - \sum_{k=i+1}^{n} r^-_{ik} \tag{6.4}$$

*with*

$$\widehat{\kappa}_i = \kappa_i - \frac{1}{4}\sum_{\lambda\in\Sigma}\Big(\frac{1+\xi_{2\lambda}}{1-\xi_{2\lambda}}\Big)\,\mathrm{mtp}(\lambda)\tau_i(t_\lambda).$$

REMARK 6.5. It follows from Corollary 6.4(4) that $\widetilde{\mathcal{D}}^{\sigma_\ell,\underline{\tau},\sigma_r}_{i} \in \mathbb{D}_M((V_\ell \otimes \underline{U} \otimes V_r^*)^{\mathfrak{m}})^W$ is obtained from $\mathcal{D}^{\sigma_\ell,\underline{\tau},\sigma_r}_{i}$ by replacing the core $\kappa^{\mathrm{core}}$ (5.1) of $\kappa$ by

$$\widehat{\kappa}^{\mathrm{core}} := \kappa^{\mathrm{core}} - \frac{1}{4}\sum_{\lambda\in\Sigma}\Big(\frac{1+\xi_{2\lambda}}{1-\xi_{2\lambda}}\Big)\,\mathrm{mtp}(\lambda)t_\lambda. \tag{6.5}$$

REMARK 6.6. If the conjecture in Remark 4.20 is valid, then the differential operators $\widetilde{H}^{\sigma_\ell,\underline{\tau},\sigma_r}$, $\widetilde{\mathcal{D}}^{\sigma_\ell,\underline{\tau},\sigma_r}_{i}$ $(1 \leq i \leq n)$ already pairwise commute in $\mathbb{D}_M((V_\ell \otimes \underline{U} \otimes V_r^*)^{\mathfrak{m}})^W$. This is indeed the case when $G$ is real split, see [**20**, §6].

We now prove Theorem 6.3 in a couple of steps.

LEMMA 6.7. *For $k \in \mathbb{Z}_{\geq 0}$,*

$$\Delta^{(k)}(\Omega) = \sum_{j=1}^{k+1}\Omega_j + 2\sum_{1\leq j<j'\leq k+1}\sum_{s=1}^{\dim(\mathfrak{h})}(z_s)_j(z_s)_{j'} + 2\sum_{1\leq j<j'\leq k+1}\sum_{\alpha\in R}(e_{-\alpha})_j(e_\alpha)_{j'}$$

*in $U(\mathfrak{g})^{\otimes(k+1)}$, with the subindices indicating in which tensor component of $U(\mathfrak{g})^{\otimes(k+1)}$ the elements are placed.*

PROOF. This follows by induction to $k$ using the expression (5.2) for $\Omega$. □

COROLLARY 6.8. *For $0 \leq i \leq n$ we have*

$$D_{\Omega,i}^{\sigma_\ell,\underline{\tau},\sigma_r} = L_{\widetilde{\Pi}(\Omega)}^{\sigma_{\ell;n},\sigma_r} - 2\sum_{j=i+1}^{n}\sum_{s=1}^{\dim(\mathfrak{h})}\tau_j(z_s)L_{\widetilde{\Pi}(z_s)}^{\sigma_{\ell;n},\sigma_r} - 2\sum_{j=i+1}^{n}\sum_{\alpha\in R}\tau_j(e_{-\alpha})L_{\widetilde{\Pi}(e_\alpha)}^{\sigma_{\ell;n},\sigma_r}$$
$$+ \sum_{j=i+1}^{n}\tau_j(\Omega) + 2\sum_{i+1\leq j<j'\leq n}(\tau_j\otimes\tau_{j'})(\varpi),$$

*viewed as differential operators in $\mathbb{D}\big((V_\ell\otimes\underline{U}\otimes V_r^*)^{\mathfrak{m}}, V_\ell\otimes\underline{U}\otimes V_r^*\big)$.*

PROOF. For $0 \leq i < n$ this follows directly from Definition 4.10, Lemma 6.7, the identity $S(\Omega) = \Omega$ and (5.1). For $i = n$ it follows from Definition 4.10 and (4.10). □

Write for $\lambda \in \Sigma$,

$$\varpi_\lambda^r := 2(\mathrm{id}\otimes\mathrm{pr}_{\mathfrak{k}})(\varpi_\lambda) = \sum_{\alpha\in R_\lambda} e_{-\alpha}\otimes y_\alpha \in (\mathfrak{g}\otimes\mathfrak{k})^M.$$

Furthermore, for $X = \sum_t a_t\otimes b_t \in \mathfrak{g}\otimes\mathfrak{k}$ we write $(\tau_i\otimes\sigma_{\ell;n})(X) := \sum_t \tau_i(a_t)\sigma_{\ell;n}(b_t)$, viewed as endomorphism of $V_\ell\otimes\underline{U}\otimes V_r^*$ (cf. Remark 2.3).

LEMMA 6.9. *We have for $1 \leq i \leq n$,*

(6.6)
$$\frac{1}{2}(D_{\Omega,i}^{\sigma_\ell,\underline{\tau},\sigma_r} - D_{\Omega,i-1}^{\sigma_\ell,\underline{\tau},\sigma_r}) = \sum_{j=1}^{r}\tau_i(x_j)\partial_{x_j} - \frac{1}{2}\tau_i(\Omega) - \sum_{k=i+1}^{n}(\tau_i\otimes\tau_k)(\varpi)$$
$$+(\tau_i\otimes\sigma_{\ell;n})(\varpi_{\mathfrak{m}}) + \sum_{\lambda\in\Sigma}\Big(\frac{(\tau_i\otimes\sigma_{\ell;n})(\varpi_\lambda^r) + \xi_\lambda(\tau_i\otimes\sigma_r^*)(\varpi_\lambda^r)}{1-\xi_{2\lambda}}\Big),$$

*viewed as differential operators in $\mathbb{D}\big((V_\ell\otimes\underline{U}\otimes V_r^*)^{\mathfrak{m}}, V_\ell\otimes\underline{U}\otimes V_r^*\big)$.*

PROOF. All identities below are in $\mathbb{D}((V_\ell\otimes\underline{U}\otimes V_r^*)^{\mathfrak{m}}, V_\ell\otimes\underline{U}\otimes V_r^*)$, i.e., they hold true when acting on functions $f \in C^\infty(A_{\mathrm{reg}}; (V_\ell\otimes\underline{U}\otimes V_r^*)^{\mathfrak{m}})$. By Corollary 6.8,

$$\frac{1}{2}(D_{\Omega,i}^{\sigma_\ell,\underline{\tau},\sigma_r} - D_{\Omega,i-1}^{\sigma_\ell,\underline{\tau},\sigma_r}) = \sum_{s=1}^{\dim(\mathfrak{h})}\tau_i(z_s)L_{\widetilde{\Pi}(z_s)}^{\sigma_{\ell;n},\sigma_r} + \sum_{\alpha\in R}\tau_i(e_{-\alpha})L_{\widetilde{\Pi}(e_\alpha)}^{\sigma_{\ell;n},\sigma_r}$$
$$(6.7)\qquad - \frac{1}{2}\tau_i(\Omega) - \sum_{k=i+1}^{n}(\tau_i\otimes\tau_k)(\varpi).$$

Use that $L_{\widetilde{\Pi}(x_j)}^{\sigma_{\ell;n},\sigma_r} = \partial_{x_j}$ for $1 \leq j \leq r$, $L_{\widetilde{\Pi}(z_s)}^{\sigma_{\ell;n},\sigma_r} = \sigma_{\ell;n}(z_s)$ for $r+1 \leq s \leq \dim(\mathfrak{h})$, $L_{\widetilde{\Pi}(e_\alpha)}^{\sigma_{\ell;n},\sigma_r} = \sigma_{\ell;n}(e_\alpha)$ for $\alpha \in R_0$, and

$$L_{\widetilde{\Pi}(e_\alpha)}^{\sigma_{\ell;n},\sigma_r} = \frac{\sigma_{\ell;n}(y_\alpha) + \xi_\lambda\sigma_r^*(y_\alpha)}{1-\xi_{2\lambda}}$$

for $\alpha \in R_\lambda$ with $\lambda \in \Sigma$ (the last formula is a consequence of (4.6)). Substitute these equations in (6.7) and group the terms which act through $\tau_i$, through $\tau_i$ and $\tau_k$, through $\tau_i$ and $\sigma_{\ell;n}$, and through $\tau_i$ and $\sigma_r^*$. This directly leads to the desired result. □

**Proof of Theorem 6.3:** We write the degree zero terms of the differential operator on the right-hand side of (6.6) in terms of local factors. By the fact that $\varpi_{\mathfrak{m}} \in S^2(\mathfrak{m})$, $\Omega_{\mathfrak{m}} = \mu(\varpi_{\mathfrak{m}})$ and the definition of $\sigma_{\ell;n}$, we have

$$(\tau_i \otimes \sigma_{\ell;n})(\varpi_{\mathfrak{m}}) = (\sigma_\ell \otimes \tau_i)(\varpi_{\mathfrak{m}}) + \sum_{k=1}^{i-1} (\tau_k \otimes \tau_i)(\varpi_{\mathfrak{m}}) + \tau_i(\Omega_{\mathfrak{m}}) + \sum_{k=i+1}^{n} (\tau_i \otimes \tau_k)(\varpi_{\mathfrak{m}})$$

in $\mathrm{End}(V_\ell \otimes \underline{U} \otimes V_r^*)$. Similarly, we have for $\alpha \in R \setminus R_0$,

$$\tau_i(e_{-\alpha})\sigma_{\ell;n}(y_\alpha) = \sigma_\ell(y_\alpha)\tau_i(e_{-\alpha}) + \sum_{k=1}^{i-1} \tau_k(y_\alpha)\tau_i(e_{-\alpha}) + \tau_i(e_{-\alpha}y_\alpha) + \sum_{k=i+1}^{n} \tau_i(e_{-\alpha})\tau_k(y_\alpha).$$

Substituting these identities into (6.6) and comparing the resulting expression with the explicit formula for $\mathcal{D}_i^{\sigma_\ell,\underline{\mathcal{I}},\sigma_r}$ from Definition 6.2, formula (6.2) will follow from the identities

$$\begin{aligned}
r_{ki}^{+} &= -(\tau_k \otimes \tau_i)(\varpi_{\mathfrak{m}}) - \sum_{\lambda\in\Sigma} \frac{1}{1-\xi_{2\lambda}} \sum_{\alpha\in R_\lambda} \tau_k(y_\alpha)\tau_i(e_{-\alpha}), \qquad 1 \le k < i,\\
r_{ik}^{-} &= (\tau_i \otimes \tau_k)(\varpi') - \sum_{\lambda\in\Sigma} \frac{1}{1-\xi_{2\lambda}} \sum_{\alpha\in R_\lambda} \tau_i(e_{-\alpha})\tau_k(y_\alpha), \qquad i < k \le n,\\
\tau_i(\kappa^{\mathrm{core}}) &= -\frac{1}{2}\tau_i(\Omega_{\mathfrak{m}}) + \frac{1}{2}\tau_i(\Omega') - \sum_{\lambda\in\Sigma} \frac{1}{1-\xi_{2\lambda}} \sum_{\alpha\in R_\lambda} \tau_i(e_{-\alpha}y_\alpha)
\end{aligned}$$

as endomorphisms of $V_\ell \otimes \underline{U} \otimes V_r^*$. Only the second equality requires proof; it follows from the fact that

$$\begin{aligned}
\varpi' - \sum_{\lambda\in\Sigma} \frac{1}{1-\xi_{2\lambda}} \sum_{\alpha\in R_\lambda} e_{-\alpha}\otimes y_\alpha &= \sum_{j=1}^{r} x_j \otimes x_j + \sum_{\lambda\in\Sigma}\sum_{\alpha\in R_\lambda} e_{-\alpha} \otimes \Big(\frac{(1-\xi_{2\lambda})e_\alpha - y_\alpha}{1-\xi_{2\lambda}}\Big)\\
&= \sum_{j=1}^{r} x_j \otimes x_j - \sum_{\lambda\in\Sigma}\sum_{\alpha\in R_\lambda} e_{-\alpha} \otimes \Big(\frac{\xi_{2\lambda}e_\alpha + c_\alpha e_{\theta\alpha}}{1-\xi_{2\lambda}}\Big)\\
&= \sum_{j=1}^{r} x_j \otimes x_j + \sum_{\lambda\in\Sigma} \frac{1}{1-\xi_{-2\lambda}} \sum_{\alpha\in R_\lambda} (e_{-\alpha} - c_\alpha^{-1}e_{-\theta\alpha})\otimes e_\alpha\\
&= r^-,
\end{aligned}$$

where we have used Lemma 3.2 for the third and fourth equality. It is clear that both sides of (6.2) lie in $\mathbb{D}_M((V_\ell \otimes \underline{U} \otimes V_r^*)^{\mathfrak{m}})^W$. This completes the proof of the theorem. □

The local factors $r^\pm$ and $\kappa^{\mathrm{core}}$ occurring in $\mathcal{D}_i^{\sigma_\ell,\underline{\mathcal{I}},\sigma_r} \in \mathbb{D}_M((V_\ell \otimes \underline{U} \otimes V_r^*)^{\mathfrak{m}})^W$ $(1 \le i \le n)$ decompose as follows.

Proposition 6.10. *We have*

$$\begin{aligned}
r^\pm &= \pm r + (\mathrm{id} \otimes \theta)r_{21},\\
\kappa^{\mathrm{core}} &= \mu((1\otimes\theta)r_{21}) + \frac{1}{4}\sum_{\lambda\in\Sigma}\Big(\frac{1+\xi_{2\lambda}}{1-\xi_{2\lambda}}\Big)z_\lambda + \frac{1}{4}\sum_{\lambda\in\Sigma}\Big(\frac{1+\xi_{2\lambda}}{1-\xi_{2\lambda}}\Big)\mathrm{mtp}(\lambda)t_\lambda
\end{aligned}$$

*with* $r \in (\mathcal{R} \otimes (\mathfrak{g}\otimes\mathfrak{g})^M)^W$ *given by* (5.10). *Furthermore,* $(\mathrm{id}\otimes\theta)r_{21} = (\theta\otimes\mathrm{id})r$.

PROOF. The alternative expressions for $r^{\pm}$ follow by a straightforward computation (the derivation is similar to the last computation in the proof of Theorem 6.3). Proposition 5.7 implies that

$$
\begin{aligned}
\kappa^{\mathrm{core}} :=& \frac{1}{4}\sum_{\lambda\in\Sigma}\Big(\frac{1+\xi_{2\lambda}}{1-\xi_{2\lambda}}\Big)z_\lambda - \frac{1}{2}\Omega_{\mathfrak{m}} \\
&+\frac{1}{2}\sum_{j=1}^{r}x_j^2 + \frac{1}{4}\sum_{\lambda\in\Sigma}\Big(\frac{1+\xi_{2\lambda}}{1-\xi_{2\lambda}}\Big)\,\mathrm{mtp}(\lambda)t_\lambda - \sum_{\lambda\in\Sigma}\frac{\mu((\mathrm{id}\otimes\theta)(\varpi_\lambda))}{1-\xi_{2\lambda}},
\end{aligned} \tag{6.8}
$$

from which the alternative expression for $\kappa^{\mathrm{core}}$ easily follows. The symmetry $(\mathrm{id}\otimes\theta)r_{21} = (\theta\otimes\mathrm{id})r$ follows from the fact that $(\theta\otimes\mathrm{id})(\varpi_\lambda)\in S^2(\mathfrak{g})$ for $\lambda\in\Sigma$, cf. Remark 5.4. □

Combined with (6.5) we conclude that

$$
\widehat{\kappa}^{\mathrm{core}} = \mu((1\otimes\theta)r_{21}) + \frac{1}{4}\sum_{\lambda\in\Sigma}\Big(\frac{1+\xi_{2\lambda}}{1-\xi_{2\lambda}}\Big)z_\lambda. \tag{6.9}
$$

REMARK 6.11. For real split $G$ an explanation of the folded structure of $r^{\pm}$ and $\kappa^{\mathrm{core}}$, as described by Proposition 6.10, is given in [**20**, §6] using quantum field theoretic arguments and the theory of $n$-point spherical functions.

**6.2. Coupled classical dynamical Yang-Baxter-reflection equations.** As a consequence of the commutativity of the $M$-restricted asymptotic boundary KZB operators (cf. Corollary 6.4) we obtain coupled classical dynamical Yang-Baxter-reflection equations for its local factors $r^{\pm}$ and $\widehat{\kappa}$. To formulate these equations, we define the classical dynamical reflection term $\mathcal{A}\in\mathcal{R}\otimes(U(\mathfrak{k})\otimes U(\mathfrak{g})^{\otimes 2}\otimes U(\mathfrak{k}))^M$ by

$$
\mathcal{A} := [\widehat{\kappa}_1 + r_{12}^-, \widehat{\kappa}_2 + r_{12}^+] - \sum_{j=1}^{r}\big((x_j)_1\partial_{x_j}(\widehat{\kappa}_2 + r_{12}^+) - (x_j)_2\partial_{x_j}(\widehat{\kappa}_1 + r_{12}^-)\big), \tag{6.10}
$$

with the sublabels numbering the two $U(\mathfrak{g})$ tensor legs within $U(\mathfrak{k})\otimes U(\mathfrak{g})^{\otimes 2}\otimes U(\mathfrak{k})$, and we define mixed classical dynamical Yang-Baxter terms $\mathrm{MYB}(s)\in\mathcal{R}\otimes(U(\mathfrak{g})^{\otimes 3})^M$ $(1\le s\le 3)$ by

(6.11)

$$
\mathrm{MYB}(1) := [r_{12}^+, r_{13}^+] + [r_{12}^+, r_{23}^-] - [r_{13}^+, r_{23}^-] - \sum_{j=1}^{r}\big((x_j)_2\partial_{x_j}(r_{13}^+) - (x_j)_3\partial_{x_j}(r_{12}^+)\big),
$$

$$
\mathrm{MYB}(2) := [r_{12}^-, r_{13}^+] + [r_{12}^-, r_{23}^+] + [r_{13}^-, r_{23}^+] - \sum_{j=1}^{r}\big((x_j)_1\partial_{x_j}(r_{23}^+) - (x_j)_3\partial_{x_j}(r_{12}^-)\big),
$$

$$
\mathrm{MYB}(3) := -[r_{12}^+, r_{13}^-] + [r_{12}^-, r_{23}^-] + [r_{13}^-, r_{23}^-] - \sum_{j=1}^{r}\big((x_j)_1\partial_{x_j}(r_{23}^-) - (x_j)_2\partial_{x_j}(r_{13}^-)\big).
$$

In the following we will write

$$
\begin{aligned}
\mathcal{A}_{ij} &:= (\sigma_\ell\otimes\tau_i\otimes\tau_j\otimes\sigma_r^*)(\mathcal{A}), && 1\le i<j\le n,\\
\mathrm{MYB}(s)_{ijk} &:= (\tau_i\otimes\tau_j\otimes\tau_k)(\mathrm{MYB}(s)), && 1\le i<j<k\le n,
\end{aligned}
$$

which we view as element in $\mathcal{R}\otimes \mathrm{End}_M((V_\ell\otimes \underline{U}\otimes V_r^*)^{\mathfrak{m}})$ (the number $n$ of $G$-representations in the tensor product representation $\underline{U}$ will always be clear from the context).

THEOREM 6.12. *For $n\geq 2$ and $1\leq i<j\leq n$,*

(6.12)
$$\Big(\mathcal{A}_{ij}+\sum_{k=1}^{i-1}\mathrm{MYB}(1)_{kij}+\sum_{k=i+1}^{j-1}\mathrm{MYB}(2)_{ikj}+\sum_{k=j+1}^{n}\mathrm{MYB}(3)_{ijk}\Big)_{(V_\ell\otimes\underline{U}\otimes V_r^*)^M}=0.$$

*The same holds true when $\widehat{\kappa}$ in $\mathcal{A}$ is replaced by $\kappa$.*

PROOF. By a lengthy computation one shows for $1\leq i<j\leq n$,

(6.13)
$$\begin{aligned}[\widetilde{\mathcal{D}}_i^{\sigma_\ell,\underline{\tau},\sigma_r},\widetilde{\mathcal{D}}_j^{\sigma_\ell,\underline{\tau},\sigma_r}]&=\sum_{s=1}^{r}\big([\tau_j(x_s),r_{ij}^-]-[\tau_i(x_s),r_{ij}^+]\big)\partial_{x_s}\\&\quad+\mathcal{A}_{ij}+\sum_{k=1}^{i-1}\mathrm{MYB}(1)_{kij}+\sum_{k=i+1}^{j-1}\mathrm{MYB}(2)_{ikj}+\sum_{k=j+1}^{n}\mathrm{MYB}(3)_{ijk}\end{aligned}$$

in $\mathbb{D}_M((V_\ell\otimes\underline{U}\otimes V_r^*)^{\mathfrak{m}})$. Using the explicit expression for $r^\pm$ one verifies that

$$[1\otimes h,r^-]=[h\otimes 1,r^+]\qquad \forall\, h\in\mathfrak{a},$$

hence the first-order term in the right-hand side of (6.13) vanishes and we are left with

(6.14)
$$[\widetilde{\mathcal{D}}_i^{\sigma_\ell,\underline{\tau},\sigma_r},\widetilde{\mathcal{D}}_j^{\sigma_\ell,\underline{\tau},\sigma_r}]=\mathcal{A}_{ij}+\sum_{k=1}^{i-1}\mathrm{MYB}(1)_{kij}+\sum_{k=i+1}^{j-1}\mathrm{MYB}(2)_{ikj}+\sum_{k=j+1}^{n}\mathrm{MYB}(3)_{ijk}$$

in $\mathbb{D}_M((V_\ell\otimes\underline{U}\otimes V_r^*)^{\mathfrak{m}})$. The result now follows since the left-hand side vanishes when acting on $C^\infty(A_{\mathrm{reg}};(V_\ell\otimes\underline{U}\otimes V_r^*)^M)$ by Corollary 6.4.

Finally, replacing the role of $\widetilde{\mathcal{D}}_i^{\sigma_\ell,\underline{\tau},\sigma_r}$ by $\mathcal{D}_i^{\sigma_\ell,\underline{\tau},\sigma_r}$ gives the result when $\widehat{\kappa}$ in $\mathcal{A}$ is replaced by $\kappa$. □

For $n=2$ we obtain the following special case of Theorem 6.12.

COROLLARY 6.13. *The classical dynamical reflection term $\mathcal{A}$ (see (6.10)) acts as zero on $(V_\ell\otimes U_1\otimes U_2\otimes V_r^*)^M$,*

$$\mathcal{A}|_{(V_\ell\otimes U_1\otimes U_2\otimes V_r^*)^M}=0. \tag{6.15}$$

The equation (6.15) is called the classical dynamical reflection equation for $\widehat{\kappa}$ relative to the pair $(r^+,r^-)$.

For $n=3$ we have

COROLLARY 6.14. *The classical dynamical reflection term $\mathcal{A}$ (see (6.10)) and the mixed classical dynamical Yang-Baxter terms $\mathrm{MYB}(s)\in\mathcal{R}\otimes(U(\mathfrak{g})^{\otimes 3})^M$ (see (6.11)) satisfy for $1\leq i<j\leq 3$,*

$$\big(\mathcal{A}_{ij}+\mathrm{MYB}(k)\big)|_{(V_\ell\otimes U_1\otimes U_2\otimes U_3\otimes V_r^*)^M}=0 \tag{6.16}$$

*where $\{k\}=\{1,2,3\}\setminus\{i,j\}$ and $\mathrm{MYB}(k)$ acts in the natural way on the tensor components $U_1\otimes U_2\otimes U_3$ within $(V_\ell\otimes U_1\otimes U_2\otimes U_3\otimes V_r^*)^M$.*

The three equations (6.16) are called the coupled classical dynamical Yang-Baxter-reflection equations for the triple $(r^+, r^-, \widehat{\kappa})$.

The classical dynamical reflection type equation (6.15) does not guarantee that $\mathcal{A}_{ij}$ in (6.16) vanishes when acting on $(V_\ell \otimes U_1 \otimes U_2 \otimes U_3 \otimes V_r^*)^M$. Hence in general the reflection term in the coupled classical dynamical Yang-Baxter-reflection equations (6.16) cannot be decoupled from the Yang-Baxter type terms.

DEFINITION 6.15. We call the equations (6.12) for $n > 3$ the *higher-order coupled classical dynamical Yang-Baxter-reflection equations* for the triple $(r^+, r^-, \widehat{\kappa})$.

REMARK 6.16. We expect that more generally,

$$\mathcal{A}_{ij} + \sum_{k=1}^{i-1} \mathrm{MYB}(1)_{kij} + \sum_{k=i+1}^{j-1} \mathrm{MYB}(2)_{ikj} + \sum_{k=j+1}^{n} \mathrm{MYB}(3)_{ijk} \tag{6.17}$$

vanishes when acting on $(V_\ell \otimes \underline{U} \otimes V_r^*)^{\mathfrak{m}}$ for all $n \geq 2$. This is the case when $G$ is real split, see [**20**, Thm. 6.26] and Remark 4.20. In the real split case we have $\mathfrak{m} = 0$ and hence we conclude that in this case (6.17) vanishes identically on $V_\ell \otimes \underline{U} \otimes V_r^*$ for all $n \geq 2$. These equations decouple, and are equivalent to the equations

$$\begin{aligned} \mathcal{A}_{12} &= 0 \quad \text{in } \mathcal{R} \otimes \mathrm{End}(V_\ell \otimes U_1 \otimes U_2 \otimes V_r^*), \\ \mathrm{MYB}(k) &= 0 \quad \text{in } \mathcal{R} \otimes \mathrm{End}(V_\ell \otimes U_1 \otimes U_2 \otimes U_3 \otimes V_r^*) \quad \text{for} \quad k = 1, 2, 3, \end{aligned} \tag{6.18}$$

which are the classical dynamical reflection equation for $\widehat{\kappa}$ relative to $(r^+, r^-)$ and the mixed classical dynamical Yang-Baxter equation for $(r^+, r^-)$ from [**20**, Thm. 6.31].

## 7. Example: $\mathrm{SU}(p, r)$.

As an example, we work out some of the results in more detail for the real simple Lie group $\mathrm{SU}(p, r)$.

Write $e_{ij}^{(m)} \in \mathrm{GL}(m; \mathbb{C})$ for the matrix unit with a one at entry $(i, j)$ and zeros everywhere else. Write $I_m \in \mathrm{GL}(m; \mathbb{C})$ for the unit matrix, and $I_m^- := \sum_{i=1}^m e_{i,m+1-i}^{(m)} \in \mathrm{GL}(m; \mathbb{C})$ for the antidiagonal $m \times m$-matrix with ones on the antidiagonal. Fix $1 \leq r \leq p$ and write

$$C := \begin{pmatrix} \frac{1}{\sqrt{2}} I_r & 0 & -\frac{1}{\sqrt{2}} I_r^- \\ 0 & I_{p-r} & 0 \\ \frac{1}{\sqrt{2}} I_r^- & 0 & \frac{1}{\sqrt{2}} I_r \end{pmatrix} \in \mathrm{GL}(p + r; \mathbb{C})$$

with 0 the zero matrix of the appropriate size, and consider $G := \{C^{-1} g C \mid g \in \mathrm{SU}(p, r)\}$. Then

$$\mathfrak{g}_\mathbb{R} = \{X \in \mathfrak{sl}(p + r; \mathbb{C}) \mid X^* J_{p,r} + J_{p,r} X = 0\}$$

with $X^* = \overline{X}^T$ the adjoint of $X$ and

$$J_{p,r} := \begin{pmatrix} 0 & 0 & -I_r^- \\ 0 & I_{p-r} & 0 \\ -I_r^- & 0 & 0 \end{pmatrix}.$$

The formula $\theta_\mathbb{R}(X) := J_{p,r} X J_{p,r}$ $(X \in \mathfrak{g}_\mathbb{R})$ defines a Cartan involution of $\mathfrak{g}_\mathbb{R}$, and the diagonal matrices in $\mathfrak{g}_\mathbb{R}$ form a $\theta_\mathbb{R}$-stable maximally nonabelian Cartan subalgebra $\mathfrak{h}_{\mathbb{R}}$ of $\mathfrak{g}_{\mathbb{R}}$ with associated $\mathfrak{p}_{\mathbb{R}}$-component

$$\mathfrak{a}_{\mathbb{R}} = \bigoplus_{j=1}^{r} \mathbb{R}(e_{jj}^{(p+r)} - e_{j'j'}^{(p+r)}) \quad \text{with} \quad j' := p+r+1-j.$$

The root system $R \subset \mathfrak{h}^*$ of $\mathfrak{g} = \mathfrak{sl}(p+r;\mathbb{C})$ is $R = \{\epsilon_i - \epsilon_j \mid 1 \leq i \neq j \leq p+r\}$ with $\epsilon_i(h) := \lambda_i$ for $h = \sum_{j=1}^{p+r} \lambda_j e_{jj}^{(p+r)} \in \mathfrak{h}$. Take $R^+ = \{\epsilon_i - \epsilon_j \mid 1 \leq i < j \leq p+r\}$ as the set of positive roots. We will use the shorthand notation $\alpha_{ij} := \epsilon_i - \epsilon_j$ for the roots. Write $\alpha_i := \alpha_{i,i+1}$ $(1 \leq i \leq p+q-1)$ for the associated basis elements of $R$, then the involution ${}^t\theta \in \mathrm{Aut}(R)$ is determined by

$$\theta(\alpha_i) = \begin{cases} -\alpha_{i'} & (1 \leq i \leq r), \\ \alpha_i & (r+1 \leq i < p), \\ -\alpha_{i'} & (p \leq i < p+r-1), \end{cases}$$

which in turn determines the Satake diagram of $G$.

Write $\mathfrak{a}_{\mathbb{R}}^* = \bigoplus_{j=1}^{r} \mathbb{R} f_j$ with $f_j \in \mathfrak{a}_{\mathbb{R}}^*$ defined by $f_j(e_{ii}^{(p+r)} - e_{i'i'}^{(p+r)}) := \delta_{i,j}$. The restricted root system $\Sigma$ is a root system of type $\mathrm{BC}_r$ with the positive restricted roots given by

$$\Sigma^+ = \{f_i \pm f_j\}_{1 \leq i < j \leq r} \cup \{f_j, 2f_j\}_{1 \leq j \leq r}.$$

In fact, for $1 \leq i < j \leq r$ and $1 \leq k \leq r$ we have

$$\begin{aligned} R_0 &= \{\alpha_{\ell m}\}_{r+1 \leq \ell \neq m \leq p}, \\ R_{f_i - f_j} &= \{\alpha_{ij}, \alpha_{j'i'}\}, \qquad & R_{f_i + f_j} &= \{\alpha_{ij'}, \alpha_{ji'}\}, \\ R_{f_k} &= \{\alpha_{k\ell}, \alpha_{\ell k'}\}_{r+1 \leq \ell \leq p}, \qquad & R_{2f_k} &= \{\alpha_{kk'}\}, \end{aligned}$$

(for $r = 1$ there are no restricted roots of the form $f_i \pm f_j$ $(i \neq j)$, and hence the middle line drops out). Write $\xi_i := \xi_{f_i}$ for the multiplicative character of $A$ associated to $f_i \in \mathfrak{a}_{\mathbb{R}}^*$. The dependence on $A$ of the local factors $r^{\pm}, \kappa$ of the asymptotic boundary KZB operators $\widetilde{\mathcal{D}}_{i;M}^{\sigma_\ell,\mathcal{I},\sigma_r}$ $(1 \leq i \leq n)$, as well as of the potential $V^{\sigma_\ell,\mathcal{I},\sigma_r}$ of the Schrödinger operator $\widetilde{H}_M^{\sigma_\ell,\mathcal{I},\sigma_r}$, are then described in terms of the multiplicative characters $\xi_i\xi_j^{\pm 1}$ $(1 \leq i < j \leq r)$ and $\xi_k$ $(1 \leq k \leq r)$, which are the characters associated to the indivisible positive restricted roots of $\Sigma$. The equivariance of the operators is with respect to the hyperoctahedral group $W \simeq S_r \ltimes (\pm 1)^r$.

Recall that $\mathfrak{k} = \mathfrak{m} \oplus \mathfrak{q}$ with $\mathfrak{q} = \bigoplus_{\alpha \in R^+ \setminus R_0^+} \mathbb{C} y_\alpha$. To further describe the local factors of $\widetilde{\mathcal{D}}_i^{\sigma_\ell,\mathcal{I},\sigma_r}$ and the potential $V^{\sigma_\ell,\mathcal{I},\sigma_r}$ of $\widetilde{H}^{\sigma_\ell,\mathcal{I},\sigma_r}$ one needs the explicit description of $\mathfrak{m}$ and of the standard basis elements $y_\alpha$ $(\alpha \in R^+ \setminus R_0^+)$ of $\mathfrak{q}$. We have

$$\mathfrak{m} = \mathfrak{t} \oplus \bigoplus_{r+1 \leq i \neq j \leq p} \mathbb{C} e_{ij}^{(p+r)}$$

with $\mathfrak{t}$ the $(+1)$-eigenspace of $\theta|_{\mathfrak{h}}$, which is explicitly described by

$$\mathfrak{t} = \Big\{ \sum_{i=1}^{r} \lambda_i (e_{ii}^{(p+r)} + e_{i'i'}^{(p+r)}) + \sum_{j=r+1}^{p} \mu_j e_{jj}^{(p+r)} \mid 2\sum_{i=1}^{r} \lambda_i + \sum_{j=r+1}^{p} \mu_j = 0 \Big\}.$$

The elements $y_\alpha$ ($\alpha \in R^+ \setminus R_0^+$) are explicitly given by

$$
\begin{aligned}
y_{\alpha_{ij}} &= e_{ij}^{(p+r)} + e_{i'j'}^{(p+r)}, & y_{\alpha_{j'i'}} &= e_{ji}^{(p+r)} + e_{j'i'}^{(p+r)},\\
y_{\alpha_{ij'}} &= e_{ij'}^{(p+r)} + e_{i'j}^{(p+r)}, & y_{\alpha_{ji'}} &= e_{ji'}^{(p+r)} + e_{j'i}^{(p+r)},\\
y_{\alpha_{k\ell}} &= e_{k\ell}^{(p+r)} - e_{k'\ell}^{(p+r)}, & y_{\alpha_{\ell k'}} &= e_{\ell k'}^{(p+r)} - e_{\ell k}^{(p+r)},\\
y_{\alpha_{kk'}} &= e_{kk'}^{(p+r)} + e_{k'k}^{(p+r)}
\end{aligned}
$$

for $1 \leq i < j \leq r$, $1 \leq k \leq r$ and $r+1 \leq \ell \leq p$ (the 1st–4th line describes $y_\alpha$ for the roots $\alpha \in R^+_{f_i - f_j}$, $\alpha \in R^+_{f_i+f_j}$, $\alpha \in R^+_{f_k}$ and $\alpha \in R^+_{2f_k}$ respectively).

Finally, since $\mathfrak{s}(\mathfrak{gl}(p) \times \mathfrak{gl}(r)) = \mathrm{Ad}(C)\mathfrak{k}$, the basis elements of $\mathfrak{k}$ can be expressed in terms of the standard basis elements of $\mathfrak{s}(\mathfrak{gl}(p) \times \mathfrak{gl}(r))$, and vice versa. For instance, the center of $\mathfrak{k}$ is spanned by

$$
\mathrm{Ad}(C^{-1})\begin{pmatrix} \frac{1}{p}I_p & 0 \\ 0 & -\frac{1}{r}I_r \end{pmatrix} = \begin{pmatrix} \big(\frac{1}{2p} - \frac{1}{2r}\big)I_r & 0 & -\big(\frac{1}{2p} + \frac{1}{2r}\big)I_r^- \\ 0 & \frac{1}{p}I_{p-r} & 0 \\ -\big(\frac{1}{2p} + \frac{1}{2r}\big)I_r^- & 0 & \big(\frac{1}{2p} - \frac{1}{2r}\big)I_r \end{pmatrix},
$$

which can be expressed as

$$
\left( \Big(\frac{1}{2p} - \frac{1}{2r}\Big)\sum_{i=1}^{r}(e_{ii}^{(p+r)} + e_{i'i'}^{(p+r)}) + \frac{1}{p}\sum_{j=r+1}^{p} e_{jj}^{(p+r)} \right) - \Big(\frac{1}{2p} + \frac{1}{2r}\Big)\sum_{i=1}^{r} y_{\alpha_{ii'}},
$$

where the first term lies in $\mathfrak{t}$ and the second term lies in $\mathfrak{q}$.

## Acknowledgment

The first author is grateful to ETH-ITS Zurich for the hospitality during 2019-2020 academic year. The authors thank Hadewijch de Clercq and two referees for comments and for pointing out a few typos.

## References

[1] Denis Bernard, *On the Wess-Zumino-Witten models on the torus*, Nuclear Phys. B **303** (1988), no. 1, 77–93, DOI 10.1016/0550-3213(88)90217-9. MR952765

[2] J. Cardy, *Boundary Conformal Field Theory*, Encyclopedia of Math. Phys. (Elsevier, 2006).

[3] William Casselman and Dragan Miličić, *Asymptotic behavior of matrix coefficients of admissible representations*, Duke Math. J. **49** (1982), no. 4, 869–930, DOI 10.1215/S0012-7094-82-04943-2. MR683007

[4] Jacques Dixmier, *Enveloping algebras*, Graduate Studies in Mathematics, vol. 11, American Mathematical Society, Providence, RI, 1996. Revised reprint of the 1977 translation, DOI 10.1090/gsm/011. MR1393197

[5] Pavel Etingof and Frédéric Latour, *The dynamical Yang-Baxter equation, representation theory, and quantum integrable systems*, Oxford Lecture Series in Mathematics and its Applications, vol. 29, Oxford University Press, Oxford, 2005. MR2142557

[6] Pavel Etingof and Olivier Schiffmann, *Twisted traces of intertwiners for Kac-Moody algebras and classical dynamical R-matrices corresponding to generalized Belavin-Drinfeld triples*, Math. Res. Lett. **6** (1999), no. 5-6, 593–612, DOI 10.4310/MRL.1999.v6.n6.a1. MR1739218

[7] P. Etingof, O. Schiffmann, and A. Varchenko, *Traces of intertwiners for quantum groups and difference equations*, Lett. Math. Phys. **62** (2002), no. 2, 143–158, DOI 10.1023/A:1021619920915. MR1952123

[8] Pavel Etingof and Alexander Varchenko, *Traces of intertwiners for quantum groups and difference equations. I*, Duke Math. J. **104** (2000), no. 3, 391–432, DOI 10.1215/S0012-7094-00-10433-4. MR1781477

[9] Giovanni Felder, *Conformal field theory and integrable systems associated to elliptic curves*, Proceedings of the International Congress of Mathematicians, Vol. 1, 2 (Zürich, 1994), Birkhäuser, Basel, 1995, pp. 1247–1255. MR1404026

[10] Giovanni Felder and Christian Wieczerkowski, *Conformal blocks on elliptic curves and the Knizhnik-Zamolodchikov-Bernard equations*, Comm. Math. Phys. **176** (1996), no. 1, 133–161. MR1372821

[11] Harish-Chandra, *Spherical functions on a semisimple Lie group. I*, Amer. J. Math. **80** (1958), 241–310, DOI 10.2307/2372786. MR94407

[12] G. J. Heckman and E. M. Opdam, *Root systems and hypergeometric functions. I*, Compositio Math. **64** (1987), no. 3, 329–352. MR918416

[13] Gerrit Heckman and Henrik Schlichtkrull, *Harmonic analysis and special functions on symmetric spaces*, Perspectives in Mathematics, vol. 16, Academic Press, Inc., San Diego, CA, 1994. MR1313912

[14] James E. Humphreys, *Introduction to Lie algebras and representation theory*, Graduate Texts in Mathematics, vol. 9, Springer-Verlag, New York-Berlin, 1978. Second printing, revised. MR499562

[15] Anthony W. Knapp, *Lie groups beyond an introduction*, Progress in Mathematics, vol. 140, Birkhäuser Boston, Inc., Boston, MA, 1996, DOI 10.1007/978-1-4757-2453-0. MR1399083

[16] Willard Miller Jr., Sarah Post, and Pavel Winternitz, *Classical and quantum superintegrability with applications*, J. Phys. A **46** (2013), no. 42, 423001, 97, DOI 10.1088/1751-8113/46/42/423001. MR3119484

[17] M. A. Ol′šaneckiĭ and A. M. Perelomov, *Quantum systems that are connected with root systems, and radial parts of Laplace operators* (Russian), Funktsional. Anal. i Prilozhen. **12** (1978), no. 2, 57–65, 96. MR498899

[18] Nicolai Reshetikhin, *Degenerate integrability of quantum spin Calogero-Moser systems*, Lett. Math. Phys. **107** (2017), no. 1, 187–200, DOI 10.1007/s11005-016-0897-8. MR3598882

[19] Olivier Schiffmann, *On classification of dynamical r-matrices*, Math. Res. Lett. **5** (1998), no. 1-2, 13–30, DOI 10.4310/MRL.1998.v5.n1.a2. MR1618367

[20] J.V. Stokman, N. Reshetikhin, *N-point spherical functions and asymptotic boundary KZB equations*, Invent. Math. (2022), `https://doi.org/10.1007/s00222-022-01102-3`.

[21] Jasper V. Stokman, *Folded and contracted solutions of coupled classical dynamical Yang-Baxter and reflection equations*, Indag. Math. (N.S.) **32** (2021), no. 6, 1372–1411, DOI 10.1016/j.indag.2021.07.003. MR4334172

[22] Garth Warner, *Harmonic analysis on semi-simple Lie groups. I*, Die Grundlehren der mathematischen Wissenschaften, Band 188, Springer-Verlag, New York-Heidelberg, 1972. MR0498999

[23] Garth Warner, *Harmonic analysis on semi-simple Lie groups. II*, Die Grundlehren der mathematischen Wissenschaften, Band 189, Springer-Verlag, New York-Heidelberg, 1972. MR0499000

[24] J.-B. Zuber, *CFT, BCFT, ADE and all that*, Quantum symmetries in theoretical physics and mathematics (Bariloche, 2000), Contemp. Math., vol. 294, Amer. Math. Soc., Providence, RI, 2002, pp. 233–266, DOI 10.1090/conm/294/04974. MR1907190



Yau Center for Mathematical Sciences, Tsinghua University, Bejing, China & ITMO University, Kronverskii Ave. 49, Saint Petersburg, 197101, Russia & KdV Institute for Mathematics, University of Amsterdam, Science Park 105-107, 1098 XG Amsterdam, The Netherlands

*Email address*: `reshetik@math.berkeley.edu`

KdV Institute for Mathematics, University of Amsterdam, Science Park 105-107, 1098 XG Amsterdam, The Netherlands.

*Email address*: `j.v.stokman@uva.nl`

Contemporary Mathematics
Volume **780**, 2022
https://doi.org/10.1090/conm/780/15692

# Elementary symmetric polynomials and martingales for Heckman-Opdam processes

Margit Rösler and Michael Voit

Abstract. We consider the generators $L_k$ of Heckman-Opdam diffusion processes in the compact and non-compact case in $N$ dimensions for root systems of type $A$ and $B$, with a multiplicity function of the form $k = \kappa k_0$ with some fixed value $k_0$ and a varying constant $\kappa \in [0, \infty[$. Using elementary symmetric functions, we present polynomials which are simultaneous eigenfunctions of the $L_k$ for all $\kappa \in ]0, \infty[$. This leads to martingales associated with the Heckman-Opdam diffusions $(X_{t,1}, \ldots, X_{t,N})_{t\geq 0}$. As our results extend to the freezing case $\kappa = \infty$ with a deterministic limit after some renormalization, we find formulas for the expectations $\mathbb{E}(\prod_{j=1}^{N}(y - X_{t,j}))$, $y \in \mathbb{C}$.

## 1. Introduction

In the theory of classical random matrix ensembles there exist several formulas regarding determinants as follows: Let $X$ be a random variable with values in some space of $N \times N$-matrices over $\mathbb{F} = \mathbb{R}, \mathbb{C}, \mathbb{H}$. Then the expectations $\mathbb{E}(\det(X - yI_N))$ are classical orthogonal polynomials of degree $N$ in $y \in \mathbb{C}$. Such results for the Hermite, Laguerre, and Jacobi ensembles can for example be found in [**DG**, **FG**, **A**], where the expectations above can be expressed via classical Hermite, Laguerre and Jacobi polynomials of degree $N$. These results were extended to the spectra of $\beta$-ensembles associated with $N$-dimensional time-homogeneous diffusion processes in [**KVW**, **V**] by using some martingales constructed from these diffusions via elementary symmetric polynomials. The generators of the diffusions were Dunkl-Bessel Laplacians and (symmetric) Heckman-Opdam Laplacians in the compact $BC$ case.

In the present paper we study corresponding results for Heckman-Opdam Laplacians in the non-compact $BC$ case as well Heckman-Opdam Laplacians of type $A$ in the compact and noncompact setting. Together with [**KVW**, **V**], the present paper covers the most important examples related with Calogero-Moser-Sutherland particle models. The basic ideas here are similar to those in [**KVW**, **V**]; however, while the approach in [**KVW**, **V**] is mainly based on Ito calculus, we now focus on an algebraic point of view.

The idea is as follows. Let $(X_t)_{t\geq 0}$ be a time-homogeneous diffusion on a suitable closed set $C \subset \mathbb{R}^N$ (such as a Weyl chamber or a fundamental alcove), where the paths are reflected at the boundary $\partial C$. Then the generator $L$ of the

2020 *Mathematics Subject Classification.* 60J60, 33C67, 82C23, 60B20.

associated transition semigroup is an elliptic partial differential operator whose domain is contained in the space of functions on $\mathbb{R}^N$ which admit corresponding symmetries on $\partial C$. We are now interested in functions $f : [0,\infty[\times\mathbb{R}^N \to \mathbb{C}$ for which $(f(t,X_t))_{t\geq 0}$ is a martingale (w.r.t. the canonical filtration), which essentially means that $f$ is $L$-space-time-harmonic, i.e., $(\frac{\partial}{\partial t} + L)f = 0$; see Section III.10 of [**RW**]. Examples of such harmonic functions can be given in terms of eigenfunctions of $L$. For a general background in stochastic analysis we also recommend [**P**].

In the framework of Heckman-Opdam theory (see [**HS**], [**HO**]), we fix some crystallographic root system $R$ (with associated Weyl group $W$) in $\mathbb{R}^N$ and choose $C$ as an associated Weyl chamber or fundamental alcove. We consider a non-negative multiplicity function $k \geq 0$ on $R$ of the form $k = \kappa \cdot k_0$ with some fixed multiplicity $k_0$ and a constant $\kappa > 0$. We now study the $W$-invariant Heckman-Opdam Laplace operators $L_\kappa := L_{\kappa k_0}$ as generators of diffusions (see [**Sch1, Sch2, RR1**]), where the parameter $\kappa$ is varying. We also study the renormalized generators $\widetilde{L}_\kappa := \frac{1}{\kappa}L_\kappa$ of the renormalized diffusions $(\widetilde{X}_t := X_{t/\kappa})_{t\geq 0}$. In suitable coordinates, these renormalized generators then have the form

$$\widetilde{L}_\kappa f = \frac{1}{\kappa}\Delta f + Hf$$

with some second-order differential operator $\Delta$ (often a classical Laplacian) and some first-order drift operator $H$, where both operators are independent of $\kappa$. This also works for $\kappa = \infty$ where $(\widetilde{X}_t)_{t\geq 0}$ is the deterministic solution of some ODE associated with $H$. We now use elementary symmetric polynomials to construct simultanous eigenfunctions $f$ of $\widetilde{L}_\kappa$ for all $\kappa \in ]0,\infty]$. In the next step we use martingales associated with these functions $f$ together with information in the deterministic case $\kappa = \infty$ in order to derive formulas for

$$\mathbb{E}\Big(\prod_{j=1}^{N}(\widetilde{X}_{t,i} - y)\Big) \qquad (y \in \mathbb{C}). \tag{1.1}$$

For some values of $\kappa$, the operators $L_\kappa$ admit an interpretation as Laplace-Beltrami operators on symmetric spaces. In these cases, (1.1) leads to determinantal formulas for Brownian motions on the symmetric spaces which are closely related, for instance, to some identities in [**R**]. Considering $t \to \infty$ in the compact cases, this also leads to determinantal formulas for the uniform distributions on compact symmetric spaces.

The paper is organized as follows: In Section 2 we first recapitulate some well-known facts on Heckman-Opdam hypergeometric functions and polynomials, and the associated Laplacians in the compact and non-compact setting. We there also prove that in the non-compact crystallographic case, the Heckman-Opdam processes admit arbitrary exponential moments for arbitrary deterministic starting conditions. In the main part of the paper, we then concentrate on specific root systems. Section 3 is devoted to the compact case of type $A$, which is related to Calogero-Moser-Sutherland particle models on the torus. Here, for instance, our results lead to determinantal formulas for Brownian motions and the uniform distribution on the unitary groups $U(N)$ and $SU(N)$. In Section 4 we study the non-compact case of type $A$. Section 5 then contains the non-compact case of type $BC$.

## 2. Heckman-Opdam theory

Here we briefly collect some facts from Heckman-Opdam theory; see [**HS, HO**] for a general background.

Let $(\mathfrak{a}, \langle\,.\,,\,.\,\rangle)$ be a Euclidean space of dimension $N$ with norm $|x| = \sqrt{\langle x, x\rangle}$. We identify $\mathfrak{a}$ with its dual space via the given scalar product. Let $R$ be a crystallographic, possibly not reduced root system in $\mathfrak{a}$ with associated finite reflection group $W$. Thus in particular, $R$ spans $\mathfrak{a}$. We fix a positive subsystem $R_+ \subset R$ and a $W$-invariant multiplicity $k : R \to [0, \infty[$. The Cherednik operators associated with $R_+$ and $k$ are defined as

$$D_\xi(k)f(x) = \partial_\xi f(x) + \sum_{\alpha\in R_+} \frac{k_\alpha\langle\alpha,\xi\rangle}{1 - e^{-\langle\alpha,x\rangle}}(f(x) - f(\sigma_\alpha(x)) - \langle\rho(k), \xi\rangle f(x) \tag{2.1}$$

for $\xi \in \mathfrak{a}$, with the (weighted) Weyl vector $\rho(k) := \frac{1}{2}\sum_{\alpha\in R_+} k_\alpha\alpha$. The operators $D_\xi(k)$, $\xi \in \mathfrak{a}$ commute, and there is a $W$-invariant tubular neighbourhood $U$ of $\mathfrak{a}$ in $\mathfrak{a}_\mathbb{C} = \mathfrak{a} + i\mathfrak{a}$ and a unique analytic function $(\lambda, z) \mapsto G(\lambda, k; z)$ on $\mathfrak{a}_\mathbb{C} \times U$, the so called Opdam-Cherednik kernel, which satisfies

$$G(\lambda, k; 0) = 1 \quad \text{and} \quad D_\xi(k)G(\lambda, k;\,.\,) = \langle\lambda, \xi\rangle\, G(\lambda, k;\,.\,) \quad \text{for all } \xi \in \mathfrak{a}. \tag{2.2}$$

The hypergeometric function associated with $R$ is defined by

$$F(\lambda, k; z) = \frac{1}{|W|}\sum_{w\in W} G(\lambda, k; w^{-1}z). \tag{2.3}$$

It is $W$-invariant in $\lambda$ and $z$. To introduce the associated Heckman-Opdam polynomials, we need the weight lattice and the set of dominant weights,

$$P = \{\lambda \in \mathfrak{a} : \langle\lambda, \alpha^\vee\rangle \in \mathbb{Z}\ \forall\alpha \in R\}, \quad P_+ = \{\lambda \in P : \langle\lambda, \alpha^\vee\rangle \geq 0\ \forall\alpha \in R_+\} \supset R_+,$$

where $\alpha^\vee = \frac{2\alpha}{\langle\alpha,\alpha\rangle}$. $P_+$ carries the usual dominance order. Let

$$\mathcal{T} := \mathrm{span}_\mathbb{C}\{e^{i\lambda},\ \lambda \in P\}$$

be the space of trigonometric polynomials associated with $R$. The orbit sums

$$M_\lambda = \sum_{\mu\in W\lambda} e^{i\mu}, \quad \lambda \in P_+$$

form a basis of the subspace $\mathcal{T}^W$ of $W$-invariant polynomials in $\mathcal{T}$. For $Q^\vee := \mathrm{span}_\mathbb{Z}\{\alpha^\vee,\ \alpha \in R\}$, consider the compact torus $T = \mathfrak{a}/2\pi Q^\vee$ with the weight function

$$\delta_k(x) := \prod_{\alpha\in R_+}\left|\sin\Big(\frac{\langle\alpha, x\rangle}{2}\Big)\right|^{2k_\alpha}. \tag{2.4}$$

The Heckman-Opdam polynomials associated with $R_+$ and $k$ are given by

$$P_\lambda(k; z) = M_\lambda(z) + \sum_{\nu<\lambda} c_{\lambda\nu}(k)M_\nu(z) \quad (\lambda \in P_+\,, z \in \mathfrak{a}_\mathbb{C})$$

where the coefficients $c_{\lambda\nu}(k) \in \mathbb{R}$ are uniquely determined by the condition that $P_\lambda(k;\,.\,)$ is orthogonal to $M_\nu$ in $L^2(T, \delta_k)$ for all $\nu \in P_+$ with $\nu < \lambda$. It is known that $\{P_\lambda(k,\,.\,), \lambda \in P_+\}$ is an orthonormal basis of $L^2(T, \delta_k)^W$ of all $W$-invariant functions from $L^2(T, \delta_k)$. According to [**HS**], the normalized polynomials

$$R_\lambda(k, z) := P_\lambda(k; z)/P_\lambda(k; 0)$$

can be expressed in terms of the hypergeometic function as

$$R_\lambda(k,z) = F(\lambda+\rho(k),k;iz). \tag{2.5}$$

Note that our notion slightly differs from [**HS, HO**], where the polynomials $P_\lambda$ are defined as exponential polynomials on the torus $i\mathfrak{a}/2\pi iQ^\vee$. We next introduce the Heckman-Opdam Laplacian

$$\Delta_k := \sum_{j=1}^N D_{\xi_j}(k)^2 \; - \; |\rho(k)|^2$$

with some orthonormal basis $\xi_1,\dots,\xi_N$ of $\mathfrak{a}$. The operator $\Delta_k$ is independent of the choice of this basis. Denote by $L_k$ the restriction of $\Delta_k$ to $W$-invariant functions. According to [**Sch2**],

$$L_k f(x) = \Delta f(x) + \sum_{\alpha\in R_+} k_\alpha \coth\Big(\frac{\langle\alpha,x\rangle}{2}\Big)\cdot\partial_\alpha f(x). \tag{2.6}$$

We notice that by construction, for all $\lambda\in\mathfrak{a}_\mathbb{C}$, the hypergeometric functions

$$F_\lambda := F(\lambda,k;\,.\,)$$

are eigenfunctions of $L_k$ with eigenvalues

$$\sum_{j=1}^N \langle\lambda,\xi_j\rangle^2 - |\rho(k)|^2.$$

The operator $L_k$ is independent of the choice of $R_+$ and generates a Feller diffusion on the closed Weyl chamber $\overline{\mathfrak{a}_+}\subset\mathfrak{a}$ associated with $R_+$ (called a radial Heckman-Opdam process), where the paths are reflected at the boundaries, see [**Sch1, Sch2**]. The transition probabilities of this process, with starting point $y\in\overline{\mathfrak{a}_+}$, are given by

$$p_t^W(x,y)d\mu(x)$$

with the $W$-invariant heat kernel

$$p_t^W(x,y) = \int_{i\mathfrak{a}} e^{-\frac{1}{2}(|\lambda|^2+|\rho|^2)}F_\lambda(x)F_\lambda(-y)d\nu'(\lambda),$$

where $\rho=\rho(k)$, $\nu'$ is the symmetric Plancherel measure as in [**Sch2**], and

$$d\mu(x) = d\mu_k(x) = \prod_{\alpha\in R_+} \big|2\sinh\langle\frac{\alpha}{2},x\rangle\big|^{2k_\alpha}\,dx.$$

In the compact case we take the factor $i$ in (2.5) into account as in [**RR1**] and consider the operator

$$\widehat{L}_k f(x) := \Delta f(x) + \sum_{\alpha\in R_+} k_\alpha \cot\Big(\frac{\langle\alpha,x\rangle}{2}\Big)\cdot\partial_\alpha f(x). \tag{2.7}$$

$\widehat{L}_k$ generates a Feller diffusion on a compact fundamental alcove of the affine Weyl group $W_{aff} = W\ltimes 2\pi Q^\vee$ in $\mathfrak{a}$, again with reflecting boundaries. The trigonometric polynomials $P_\lambda$ ($\lambda\in P_+$) are eigenfunctions of $\widehat{L}_k$ with eigenvalues

$$-\langle\lambda,\lambda+2\rho(k)\rangle\le 0.$$

They were used in [**RR1**] to construct the transition densities of the diffusions with generators $\widehat{L}_k$. For root systems of type $BC$, the $P_\lambda$ are multivariate Jacobi polynomials, and the associated Feller diffusions are multivariate Jacobi processes which were studied in [**Dem**].

Before turning to details for root systems of type $A$ and $BC$, we conclude this section with an integrability result which assures the existence of exponential moments for the radial Heckman-Opdam processes in the non-compact, crystallographic case. This will be needed in Sections 4 and 5.

LEMMA 2.1. *For each* $y \in \mathfrak{a}$ *and* $\beta \in \mathfrak{a}$,

$$\int_{\mathfrak{a}} e^{\langle \beta, x\rangle}\, p_t^W(x,y)d\mu(x) < \infty. \tag{2.8}$$

PROOF. For $y=0$ this is obvious from Theorem 5.2 of [**Sch2**]. For general $y$, we employ the $L^p$-theory for the hypergeometric transform developed in [**NPP**]. For suitable functions on $\mathfrak{a}$ and on $i\mathfrak{a}$ respectively, the hypergeometric transform and the inverse hypergeometric transform are given by

$$\mathcal{H}f(\lambda) = \int_{\mathfrak{a}} f(x)F_\lambda(-x)d\mu(x), \quad \mathcal{I}(g)(x) = \int_{i\mathfrak{a}} g(\lambda)F_\lambda(x)d\nu'(\lambda).$$

We denote by $\mathbb{C}[\mathfrak{a}_\mathbb{C}]$ the space of polynomial functions on $\mathfrak{a}_\mathbb{C}$ and by $\partial(q)$ the constant coefficient differential operator associated with $q \in \mathbb{C}[\mathfrak{a}_\mathbb{C}]$. Moreover, for $x \in \mathfrak{a}$ we denote by $C(x) := co(W.x)$ the convex hull of the $W$-orbit of $x$ in $\mathfrak{a}$. For an exponent $0 < p \le 2$, the $W$-invariant $L^p$-Schwartz space is given by

$$\mathcal{C}^p(\mathfrak{a})^W = \Big\{ f \in C^\infty(\mathfrak{a})^W : \sup_{x\in\mathfrak{a}} \frac{(1+|x|)^n}{F_0(x)^{2/p}}\, |\partial(q)f(x)| < \infty \;\forall n \in \mathbb{N}_0,\, q \in \mathbb{C}[\mathfrak{a}_\mathbb{C}]\Big\}.$$

Moreover, we consider the $W$-invariant Schwartz space $\mathcal{S}(\mathfrak{a}_{\epsilon_p})^W$ with $\epsilon_p = \frac{2}{p}-1$, which consists of all $W$-invariant continuous functions on the closed tube $\mathfrak{a}_{\epsilon_p} = C(\epsilon_p\rho) + i\mathfrak{a} \subset \mathfrak{a}_\mathbb{C}$ which are holomorphic in its interior and satisfy

$$\sup_{\lambda\in\mathfrak{a}_{\epsilon_p}} (1+|\lambda|)^n \big|\partial(q)g(\lambda)\big| < \infty \quad \text{for all } n \in \mathbb{N}_0,\, q \in \mathbb{C}[\mathfrak{a}_\mathbb{C}]. \tag{2.9}$$

By Theorem 5.6 of [**NPP**], the hypergeometric transform $\mathcal{H}$ is a topological isomorphism from $\mathcal{C}^p(\mathfrak{a})^W$ onto $\mathcal{S}(\mathfrak{a}_{\epsilon_p})^W$ with inverse $\mathcal{I}$.

We claim that for fixed $y \in \mathfrak{a}$, the function

$$g(\lambda) := e^{\frac{1}{2}\langle \lambda,\lambda\rangle}F_\lambda(-y)$$

belongs to $\mathcal{S}(\mathfrak{a}_{\epsilon_p})$ for each $p \in ]0,2]$; here $\langle\,.\,,\,.\,\rangle$ denotes the bilinear extension of the given scalar product to $\mathfrak{a}_\mathbb{C}$. As soon as this is proven, it will follow that $p_t^W(\,.\,,y) = e^{-\frac{1}{2}|\rho|^2}\,\mathcal{I}(g)$ belongs to $\mathcal{C}^p(\mathfrak{a})^W$ for each $p\in]0,2]$. In order to check that $g \in \mathcal{S}(\mathfrak{a}_{\epsilon_p})$, we only have to verify the growth condition (2.9). Let $q \in \mathbb{C}[\mathfrak{a}_\mathbb{C}]$. Then in view of Theorem 3.4 (and Remark 3.2) of [**Sch2**],

$$|\partial_\lambda(q)F_\lambda(-y)| \le C(1+|y|)^{\deg q}F_0(-y)\cdot e^{\max_{w\in W}\operatorname{Re}\langle w\lambda,-y\rangle} \quad (\lambda\in\mathfrak{a}_\mathbb{C}).$$

This shows that $\partial_\lambda(q)F_\lambda(-y)$ is bounded as a function of $\lambda$ on $\mathfrak{a}_{\epsilon_p}$. Moreover, $\partial(q)e^{\frac{1}{2}\langle\lambda,\lambda\rangle} = \widetilde{q}(\lambda)e^{\frac{1}{2}\langle\lambda,\lambda\rangle}$ with some polynomial $\widetilde{q}$, and

$$\big|e^{\frac{1}{2}\langle\lambda,\lambda\rangle}\big| \asymp e^{-\frac{1}{2}|\lambda|^2} \quad \text{on } \mathfrak{a}_{\epsilon_p}.$$

Therefore $\partial(q)g(\lambda)$ decays exponentially as $|\lambda|\to\infty$ within $\mathfrak{a}_{\epsilon_p}$. It follows that $g\in\mathcal{S}(\mathfrak{a}_{\epsilon_p})$ and thus $p_t^W(\,.\,,y)\in\mathcal{C}^p(\mathfrak{a})^W$. In particular, for each $p\in]0,2]$ there exists a constant $C_p > 0$ such that

$$p_t^W(x,y) \le C_p F_0(x)^{2/p} \quad \text{for all } x \in \mathfrak{a}.$$

From [**Sch2**] we know that in the closed chamber $\overline{\mathfrak{a}_+}$, $F_0(x) \asymp q_0(x)e^{-\langle \rho, x\rangle}$ with a certain positive polynomial $q_0$. Hence there exists a nonnegative polynoimal $q_p$ (depending on $p$), such that

$$p_t^W(x,y) \le q_p(x)e^{-\frac{2}{p}\langle \rho, x\rangle} \quad \text{for all } x \in \overline{\mathfrak{a}_+}.$$

Now fix $\beta \in \mathfrak{a}$ and note that $\rho$ is contained in the open chamber $\mathfrak{a}_+$. Choosing $p > 0$ small enough, we therefore obtain

$$\int_{\overline{\mathfrak{a}_+}} e^{\langle \beta, x\rangle} p_t^W(x,y)d\mu(x) \le \int_{\overline{\mathfrak{a}_+}} q_p(x)e^{\langle \beta, x\rangle - \frac{2}{p}\langle \rho, x\rangle + 2\langle \rho, x\rangle} dx < \infty.$$

This yields the assertion. □

## 3. The compact case of type $A_{N-1}$

In this section we study Heckman-Opdam processes of type $A_{N-1}$ in the compact setting. The generators of these processes are the Hamiltonians of interacting particle models of Calogero-Sutherland type with $N$ particles on the torus $\mathbb{T} := \{z \in \mathbb{C} : |z| = 1\}$; see [**LV**] for the background. These processes are diffusions on some fundamental domain of $W = S_N$ in $\mathbb{T}^N$. It will however be convenient to consider also associated diffusions on $\mathbb{R}^N$ with $2\pi$-periodicity such that the diffusions on $\mathbb{T}^N$ appear as images under $x \mapsto e^{ix}$. To introduce the processes on $\mathbb{R}^N$, we consider the root system $R = A_{N-1} = \{\pm(e_i - e_j) : 1 \le i < j \le N\}$ in $\mathbb{R}^N$ with positive subsystem $R_+ = \{e_i - e_j : i < j\}$ and fix a multiplicity parameter $k \in ]0, \infty[$. Let

$$\omega := (1, \ldots, 1)^T \in \mathbb{R}^N.$$

Then $Q^\vee = \mathbb{Z}^N \cap (\mathbb{R}\omega)^\perp$, and a fundamental domain for the action of $W_{aff} = W \ltimes 2\pi Q^\vee$ in $\mathbb{R}^N$ is given by

$$\begin{aligned} C_N &= \{x \in \mathbb{R}^N : 0 \le \langle \alpha, x\rangle \le 2\pi \ \forall \alpha \in R_+\} \\ &= \{x \in \mathbb{R}^N : x_1 \le x_2 \le \ldots \le x_N \le x_1 + 2\pi\}. \end{aligned}$$

We consider the $W$-invariant Heckman-Opdam Laplacian

$$\widehat{L}_k f(x) = \Delta f(x) + k \sum_{j=1}^N \sum_{l \ne j} \cot\Big(\frac{x_j - x_l}{2}\Big) \frac{\partial}{\partial x_j} f(x) \tag{3.1}$$

with reflecting boundaries, i.e. with domain

$$D(\widehat{L}_k) = \{f|_{C^N} : f \in C^2(\mathbb{R}^N) \text{ invariant under } W_{aff}\}.$$

$\widehat{L}_k$ is the generator of a Feller semigroup of transition operators on $C_N$, c.f. [**RR1**]. Associated Feller diffusions $(X_{t,k})_{t\ge 0}$ with continuous paths (which are reflected at the boundary of $C_N$) are called Heckman-Opdam processes of type $A_{N-1}$ on $C_N$. We also consider the renormalized generators

$$\widetilde{L}_k := \frac{1}{k}\widehat{L}_k = \frac{1}{k}\Delta + \sum_{j=1}^N \sum_{l \ne j} \cot\Big(\frac{x_j - x_l}{2}\Big) \frac{\partial}{\partial x_j}$$

which degenerate for $k \to \infty$ into

$$\widetilde{L}_\infty = \sum_{j=1}^{N} \sum_{l \neq j} \cot\Big(\frac{x_j - x_l}{2}\Big) \frac{\partial}{\partial x_j}.$$

For $k \in ]0, \infty[$, the process $\widetilde{X}_k := (\widetilde{X}_{t,k})_{t \geq 0}$ with $\widetilde{X}_{t,k} := X_{t/k,k}$ is a Feller diffusion associated with $\widetilde{L}_k$. It can be also described as solution of the SDE

$$d\widetilde{X}_{t,k,j} = \frac{\sqrt{2}}{\sqrt{k}} dB_{t,j} + \sum_{l \neq j} \cot\Big(\frac{\widetilde{X}_{t,k,j} - \widetilde{X}_{t,k,l}}{2}\Big) dt \qquad (j = 1, \ldots, N) \tag{3.2}$$

with some $N$-dimensional Brownian motion $(B_{t,1}, \ldots, B_{t,N})_{t \geq 0}$. Moreover, for deterministic starting conditions, $\widetilde{L}_\infty$ is the generator of a deterministic process whose paths are the solution of some initial value problem for the ODE

$$\frac{dx_j}{dt}(t) = \sum_{l \neq j} \cot\Big(\frac{x_j(t) - x_l(t)}{2}\Big) \qquad (j = 1, \ldots, N). \tag{3.3}$$

As for Dunkl processes in [**AV**], one can show that for initial data in the interior of $C_N$, the solution $(\widetilde{X}_{t,\infty})_{t \geq 0}$ of this ODE exists for all $t \geq 0$ in the interior of $C_N$. We also point out that in the Dunkl setting, the ODE analogous to (3.2) has unique solutions for all $t \geq 0$ even for starting points at the boundary of the chamber, see [**VW**]. We expect that such a result is also true in the present setting.

We need the following stationary solutions of (3.3):

LEMMA 3.1. *For each* $x_1 \in \mathbb{R}$,

$$\Big(x_1, x_1 + \frac{1}{N}2\pi, x_1 + \frac{2}{N}2\pi, \ldots, x_1 + \frac{N-1}{N}2\pi\Big) \in C_N$$

*is a stationary solution of* (3.3).

PROOF. Assume first that $N$ is odd. As the cotangent is odd and $\pi$-periodic, we have for $j = 1, \ldots, N$ that

$$\sum_{l \neq j} \cot\Big(\frac{(j-l)\pi}{N}\Big) = \sum_{l=1}^{N-1} \cot\Big(\frac{l\pi}{N}\Big) = \sum_{l=1}^{(N-1)/2} \Big(\cot\Big(\frac{l\pi}{N}\Big) + \cot\Big(\frac{(N-l)\pi}{N}\Big)\Big) = 0.$$

If $N$ is even, then our computation leads to the additional term $\cot((N\pi/2)/N) = 0$ in the last sum and thus to the same result. This yields the claim. □

The generator $\widetilde{L}_k$ and the associated diffusion $\widetilde{X}_k$ on $\mathbb{R}^N$ can be decomposed into two independent parts, namely the center of gravity and the process of the distances of neighboring particles. This reflects the fact that the usual representation of the symmetric group $S_N$ on $\mathbb{R}^N$ decomposes into two irreducible components. More precisely, consider the center-of-gravity process $\widetilde{X}_k^{cg} := (\widetilde{X}_{t,k}^{cg})_{t \geq 0}$ with

$$\widetilde{X}_{t,k}^{cg} := \frac{1}{N}(\widetilde{X}_{t,k,1} + \ldots + \widetilde{X}_{t,k,N}) \cdot \omega$$

which is the orthogonal projection of $\widetilde{X}_{t,k}$ onto $\mathbb{R}\omega$. Then the diffusion

$$\widetilde{X}_k^{diff} := \widetilde{X}_k - \widetilde{X}_k^{cg}$$

lives on the orthogonal complement $(\mathbb{R}\omega)^\perp \subset \mathbb{R}^N$.

LEMMA 3.2. *Let $k \in ]0,\infty[$. If $\widetilde{X}_k$ starts in some deterministic point, then the processes $\widetilde{X}_k^{diff}$ and $\widetilde{X}_k^{cg}$ are stochastically independent.*

PROOF. By the SDE (3.2) we have

$$d\widetilde{X}_{t,k,j}^{cg} = \frac{\sqrt{2}}{N\sqrt{k}} d\Big(\sum_{l=1}^{N} B_{t,l}\Big) =: \frac{\sqrt{2}}{\sqrt{Nk}}\, dB_t \tag{3.4}$$

with some one-dimensional Brownian motion $(B_t)_{t\geq 0}$ while $\widetilde{X}_k^{diff}$ satisfies

$$d\widetilde{X}_{t,k,j}^{diff} = \frac{\sqrt{2}}{\sqrt{k}} d\Big(B_{t,j} - \frac{1}{N}\sum_{l=1}^{N} B_{t,l}\Big) + F_j(\widetilde{X}_{t,k}^{diff})\, dt \quad (j=1,\ldots,N)$$

where the processes $(B_{t,j} - \frac{1}{N}\sum_{l=1}^{N} B_{t,l})_{t\geq 0}$ are stochastically independent of $(B_t)_{t\geq 0}$, and the $F_j$ are concrete continuous functions. This implies the claim by the very definition of solutions of SDEs. □

We next transfer all data from $\mathbb{R}^N$ to the torus $\mathbb{T}^N$ via $z_j := e^{ix_j}$ for $j = 1,\ldots,N$, or for short, $z := e^{ix} \in \mathbb{T}^N$. A short computation shows that in $z$-coordinates, the operator $L_k$ is given by the diffusion operator

$$\begin{aligned} H_k &= -\sum_{j=1}^{N}\Big(z_j\frac{\partial}{\partial z_j}\Big)^2 - k\sum_{j=1}^{N}\sum_{l\neq j}\frac{z_j+z_l}{z_j-z_l}\cdot z_j\frac{\partial}{\partial z_j} \\ &= -\sum_{j=1}^{N} z_j^2\frac{\partial^2}{\partial z_j^2} - (1-k(N-1))\sum_{j=1}^{N} z_j\frac{\partial}{\partial z_j} - 2k\sum_{j=1}^{N}\sum_{l\neq j}\frac{z_j^2}{z_j-z_l}\frac{\partial}{\partial z_j}, \end{aligned} \tag{3.5}$$

acting on permutation invariant functions from $C^2(\mathbb{T}^N)$. The operator $H_k$ appears in a prominent way in the particle models of Calogero-Sutherland type on $\mathbb{T}$; see Section 2 of [LV]. It is obtained from the Calogero-Sutherland Hamiltonian by conjugation with the ground state. The operator $H_k$ is the generator of the Feller diffusions $Z_k := (Z_{t,k} := e^{iX_{t,k}})_{t\geq 0}$ on the alcove

$$\mathbb{A}_N := \{e^{ix} : x \in C_N\} \subset \mathbb{T}^N.$$

Further, the operators $\widetilde{H}_k := \frac{1}{k}H_k$ generate the diffusions $\widetilde{Z}_k := (\widetilde{Z}_{t,k} := e^{i\widetilde{X}_{t,k}})_{t\geq 0}$ for $k \in ]0,\infty[$. Clearly, this also works for $k=\infty$ where $\widetilde{Z}_\infty$ is deterministic for deterministic initial data.

REMARK 3.3. Besides the generators in (3.5), also the operators

$$D_k := \sum_{j=1}^{N} z_j^2\frac{\partial^2}{\partial z_j^2} + 2k\sum_{j=1}^{N}\sum_{l\neq j}\frac{z_j^2}{z_j-z_l}\frac{\partial}{\partial z_j} \tag{3.6}$$

appear in the literature; see e.g. [**F**, **St**, **OO**]. Here $-H_k = D_k + E_k$ with the Euler operator

$$E_k := (1-k(N-1))\sum_{j=1}^{N} z_j\frac{\partial}{\partial z_j} \tag{3.7}$$

which commutes with $D_k$. The $-D_k$ are generators of diffusions with additional drift on $\mathbb{T}$ which rotates the complete system at some fixed speed. Clearly, the subsequent results can be easily translated to $-D_k$. From a stochastic point of view, the diffusions associated with the operators $H_k$ seem to be the most natural ones.

We recall that the (symmetric) eigenfunctions of $H_k$ are Jack polynomials. To become precise, we introduce the following notations. We write

$$\Lambda_N^+ = \{\lambda \in \mathbb{Z}_+^N : \lambda_1 \geq \cdots \geq \lambda_N\}$$

for the set of partitions of length at most $N$. Denote by $C_\lambda^\alpha$, $\lambda \in \Lambda_N^+$, the Jack polynomials of index $\alpha > 0$ in $N$ variables with the normalization

$$(z_1 + \cdots + z_N)^m = \sum_{|\lambda|=m} C_\lambda^\alpha(z) \quad (m \in \mathbb{Z}_+); \tag{3.8}$$

see [**St**, **BF**]. The $C_\lambda^\alpha$ are symmetric and homogeneous of degree $|\lambda| := \lambda_1 + \cdots + \lambda_N$. Moreover, by [**St**, **BF**], the $C_\lambda^\alpha$ with index $\alpha = 1/k$ are eigenfunctions of $D_k$ with eigenvalues

$$d_\lambda(k) = \sum_{j=1}^N \lambda_j(\lambda_j - 1 + 2k(N-j)).$$

In addition, as $C_\lambda^\alpha$ is homogeneous of degree $|\lambda|$,

$$E_k C_\lambda^\alpha = (1 - k(N-1))|\lambda|\, C_\lambda^\alpha.$$

In summary we obtain:

LEMMA 3.4. *For $\lambda \in \Lambda_n^+$, $k \in ]0,\infty[$ and $\alpha = 1/k$, $C_\lambda^\alpha$ is an eigenfunction of $\widetilde{H}_k = \frac{1}{k} H_k$ with eigenvalue*

$$-\sum_{j=1}^N \lambda_j\Big(\frac{\lambda_j}{k} + N + 1 - 2j\Big) \leq 0.$$

It is well-known (c.f. [**HO**]) that the polynomials $C_\lambda^{1/k}$, $\lambda \in \Lambda_n^+$ form a complete orthogonal system of $L^2(\mathbb{T}^N, \mu_k)^W$ with the probability measure

$$d\mu_k(z) = \phi_k(z)dz, \quad \phi_k(z) := c_k \cdot \prod_{j,l:l\neq j} |z_j - z_l|^k \cdot \mathbf{1}_{\mathbb{A}_N}(z), \tag{3.9}$$

where $c_k > 0$ is a normalization constant and $dz$ denotes the Haar measure on $\mathbb{T}^N$. Notice that these measures appear also in the context of circular $\beta$-ensembles in random matrix theory.

We also notice that the elementary symmetric polynomials $e_l$ $(l = 0, \ldots, N)$ in $N$ variables, which are determined by

$$\prod_{j=1}^N (y - x_j) = \sum_{l=0}^N (-1)^l e_l(x)\, y^{N-l} \quad (y \in \mathbb{C}),$$

are Jack polynomials for all $k > 0$ up to normalization. More precisely, by (3.8),

$$e_l(x) = \frac{1}{l!} C_\lambda^\alpha(x) \text{ with } \lambda = 1^l = (\underbrace{1,\ldots,1}_{l \text{ times}}, 0, \ldots, 0) \quad (l = 0, \ldots, N). \tag{3.10}$$

Thus in view of Lemma 3.4, the polynomials $e_l$ are eigenfunctions of $\widetilde{H}_k$ with eigenvalues $-l\big(\frac{1}{k} + N - l\big)$. Moreover, by a continuity argument or by direct computation, the $e_l$ are also eigenfunctions of

$$\widetilde{H}_\infty = (N-1)\sum_{j=1}^N z_j \frac{\partial}{\partial z_j} - 2\sum_{j=1}^N \sum_{l\neq j} \frac{z_j^2}{z_j - z_l}\frac{\partial}{\partial z_j}$$

with eigenvalues $-l(N-l)$.

We now use these properties to construct martingales from our processes $\widetilde{Z}_k$ for $k \in ]0,\infty[$. We recall that by Dynkin's formula (see e.g. Section III.10 of [**RW**]), the following holds for the generator $L$ of a Feller semigroup and an arbitrary associated Feller process $(X_t)_{t\geq 0}$: if $f$ is a bounded eigenfunction of $L$ with eigenvalue $r \in \mathbb{R}$, then the process $(e^{-rt}f(X_t))_{t\geq 0}$ is a martingale w.r.t. the canonical filtration. We thus have:

Corollary 3.5. *For $k \in ]0,\infty[$ and $z \in \mathbb{A}_N$ consider the diffusion $(\widetilde{Z}_{t,k})_{t\geq 0}$ on $\mathbb{A}_N$ with start in $z$. Then, for $l = 0, 1, \ldots, N$, the process $(e^{l(1/k+N-l)t}e_l(\widetilde{Z}_{t,k}))_{t\geq 0}$ is a martingale. In particular, for $t \geq 0$,*

$$\mathbb{E}(e_l(\widetilde{Z}_{t,k})) = e^{-l(1/k+N-l)t}\, e_l(z).$$

This statement also holds in the deterministic case $k = \infty$, where we also obtain additional information for $t \to \infty$:

Corollary 3.6. *For each starting point $z$ in the interior of $\mathbb{A}_N$, the deterministic process $(\widetilde{Z}_{t,\infty})_{t\geq 0}$ satisfies*

$$e_l(\widetilde{Z}_{t,\infty}) = e^{-l(N-l)t}e_l(z) \qquad (l = 0, 1, \ldots, N). \tag{3.11}$$

*Moreover, the limit $Z := \lim_{t\to\infty} \widetilde{Z}_{t,\infty} \in \mathbb{T}^N$ exists and is given by*

$$Z = (Z_1, Z_1 e^{2\pi i/N}, \ldots, Z_1 e^{2\pi i(N-1)/N}) \tag{3.12}$$

*where $Z_1 \in \mathbb{T}$ is as follows: If $z = (e^{ix_1}, \ldots, e^{ix_N})$ with $(x_1, \ldots, x_N) \in C_N$, then*

$$Z_1 = e^{ix_0} \quad \textit{with} \quad x_0 = \frac{x_1 + \ldots + x_N - \pi(N-1)}{N}. \tag{3.13}$$

Proof. Eq. (3.11) is obvious. As each $\zeta \in \mathbb{A}_N$ is uniquely determined by the elementary symmetric functions $e_l(\zeta)$, (3.11) and a continuity argument imply that $Z = (Z_1, \ldots, Z_N) := \lim_{t\to\infty} \widetilde{Z}_{t,\infty} \in \mathbb{A}_N$ exists, and that

$$\prod_{j=1}^N (y - Z_j) = y^N - e_N(Z) = y^N - z_1 z_2 \cdots z_N\,.$$

This shows that $Z$ has the form as stated in (3.12) for some $Z_1 \in \mathbb{T}$ with $Z_1^N = z_1 \cdots z_n$. To identify $Z_1$, we write the initial condition as $z = (e^{ix_1}, \ldots, e^{ix_N})$ with $(x_1, \ldots, x_N) \in C_N$. In the $x$-coordinates, our process $(\widetilde{X}_{t,\infty} = (x_{t,1}, \ldots, x_{t,N}))_{t\geq 0}$ satisfies the ODE (3.3). This ODE yields that $x_{t,1} + \ldots + x_{t,N}$ is independent of $t \in [0,\infty]$. Therefore, the form of $Z = (e^{ix_{\infty,1}}, \ldots, e^{ix_{\infty,N}})$ in (3.12) yields

$$x_1 + \ldots + x_N = x_{\infty,1} + \ldots + x_{\infty,N} = \sum_{j=0}^{N-1}(x_{\infty,1} + j\cdot 2\pi/N) = Nx_{\infty,1} + (N-1)\pi.$$

This implies (3.13). □

In the next step we use the decomposition of the processes $\widetilde{X}_k = \widetilde{X}_k^{diff} + \widetilde{X}_k^{cg}$. Eq. (3.4) and the expectations of the geometric Brownian motion imply

$$(3.14)\quad \mathbb{E}(e^{-il\widetilde{X}^{cg}_{t,k,j}}) = \mathbb{E}\left(e^{-\frac{il\sqrt{2}}{\sqrt{Nk}}B_t - il\widetilde{X}^{cg}_{0,k,j}}\right) = e^{-\frac{tl^2}{Nk} - il\widetilde{X}^{cg}_{0,k,j}} \quad (j,l = 1,\ldots,N).$$

This yields:

COROLLARY 3.7. *Let $k \in ]0,\infty[$ and $x \in C_N$ with $x_1 + \ldots + x_N = 0$. Let $z = e^{ix} \in \mathbb{A}_N$, and consider the diffusion $(\widetilde{Z}_{t,k})_{t\geq 0}$ on $\mathbb{A}_N$ with start in $z$. Then, for $t \geq 0$ and $l = 1,\ldots,N$,*

$$\mathbb{E}(e_l(e^{i\widetilde{X}^{diff}_{t,k}})) = e^{-l(N-l+(N+l)/(Nk))t} e_l(z).$$

PROOF. By our initial conditions and Eq. (3.4) we have $\widetilde{X}^{cg}_{t,k,j} = \widetilde{X}^{cg}_{t,k,1}$ for all $t,k$ and $j = 1,\ldots,N$. Hence, the stochastic independence of $\widetilde{X}_k^{diff}, \widetilde{X}_k^{cg}$, Corollary 3.5, (3.14), and the initial condition imply

$$\begin{aligned}\mathbb{E}(e_l(e^{i\widetilde{X}^{diff}_{t,k}})) &= \mathbb{E}(e_l(e^{i\widetilde{X}_{t,k} - i\widetilde{X}^{cg}_{t,k}})) = \mathbb{E}(e_l(\widetilde{Z}_{t,k}) \cdot e^{-il\widetilde{X}^{cg}_{t,k,1}})\\ &= \mathbb{E}(e_l(\widetilde{Z}_{t,k})) \cdot \mathbb{E}(e^{-il\widetilde{X}^{cg}_{t,k,1}}) = e^{-l(1/k+N-l)t}\, e_l(z) \cdot e^{-\frac{tl^2}{Nk}}\end{aligned}$$

as claimed. □

EXAMPLE 3.8. For the starting configuration $x = (0, 2\pi/N, \ldots, (N-1)2\pi/N)$ and $z = e^x$, we have $e_1(z) = \ldots = e_{N-1}(z) = 0$ and $e_N(z) = (-1)^{N-1}$. Hence, by Corollary 3.7,

$$\mathbb{E}\big(e_l(e^{i\widetilde{X}^{diff}_{t,k}})\big) = 0 \quad (l = 1,\ldots,N-1), \quad \text{and} \quad \mathbb{E}\big(e_N(e^{i\widetilde{X}^{diff}_{t,k}})\big) = (-1)^{N-1} e^{-2Nt/k}$$

for all $t \geq 0$. As $e_0 = 1$, we conclude that in this case for all $y \in \mathbb{C}$ and $t \geq 0$,

$$(3.15)\qquad \mathbb{E}\Big(\prod_{j=1}^{N}(y - e^{i\widetilde{X}^{diff}_{t,k,j}})\Big) = \mathbb{E}\Big(\sum_{l=0}^{N} y^{N-l}(-1)^l e_l(e^{i\widetilde{X}^{diff}_{t,k}})\Big) = y^N - e^{-2Nt/k}.$$

We point out that this result differs from the case $k = \infty$ where $x$ is a stationary solution of the associated ODE by Lemma 3.1, and thus $\mathbb{E}\big(\prod_{j=1}^N (y - e^{i\widetilde{X}^{diff}_{t,\infty,j}})\big) = y^N - 1$ is independent from $t$.

We next study the limit $t \to \infty$ for $k \in ]0,\infty[$ similar to the limit results for $k = \infty$ in Corollary 3.6. We need the following well-known result, which follows easily for instance from the explicit formulas for the densities of our diffusions in [**RR1**]:

LEMMA 3.9. *Let $k \in ]0,\infty[$. Then for each starting point in $\mathbb{A}_N$, the process $(\widetilde{Z}_{t,k})_{t\geq 0}$ converges in distribution for $t \to \infty$ to the probability measure $\mu_k$ on $\mathbb{A}_N$ from (3.9).*

This observation and Corollary 3.5 for $t \to \infty$ imply:

COROLLARY 3.10. *Let $Z = (Z_1,\ldots,Z_N)$ be an $\mathbb{A}_N$-valued random variable with the distribution $\mu_k$ of a circular $\beta$-ensemble. Then for each $y \in \mathbb{C}$,*

$$\mathbb{E}\Big(\prod_{j=1}^{N}(y - Z_j)\Big) = y^N.$$

PROOF.

$$\mathbb{E}\Big(\prod_{j=1}^{N}(y-Z_j)\Big)=\mathbb{E}\Big(\sum_{l=0}^{N}y^{N-l}(-1)^l e_l(Z)\Big)=y^N.$$

□

A corresponding result can be also stated under the condition that $Z$ only takes values in the alcove

$$\mathbb{A}_N^1 := \{z\in\mathbb{A}_N : z_1\cdots z_N=1\}.$$

For this consider the process $\widetilde{X}_k^{diff}$ in the decomposition $\widetilde{X}_k=\widetilde{X}_k^{diff}+\widetilde{X}_k^{cg}$ above. Then $e^{i\widetilde{X}_k^{diff}}$ is a diffusion on $\mathbb{A}_N^1$ which converges for $t\to\infty$ in distribution to the conditional probability measure $\mu_k^1\in M^1(\mathbb{A}_N^1)$ of $\mu_k$ under the condition $\mathbb{A}_N^1$. (This is, up to normalization, the measure with the density (2.4)). Eq. (3.15) now leads to:

COROLLARY 3.11. *Let $Z$ be an $\mathbb{A}_N^1$-valued random variable with distribution $\mu_k^1$. Then for each $y\in\mathbb{C}$,*

$$\mathbb{E}\Big(\prod_{j=1}^{N}(y-Z_j)\Big)=y^N.$$

REMARK 3.12. It can be shown and is well known that the measures $\mu_k$ tend for $k\to\infty$ weakly to the probability measure $\mu_\infty$ which appears as image of the uniform distribution on $\mathbb{T}$ under the mapping

$$\mathbb{T}\to\mathbb{A}_N,\quad z\to(z,z\cdot e^{2\pi i/N},\ldots,z\cdot e^{2\pi i(N-1)/N}).$$

Corollary 3.10 and continuity show that a random variable $Z=(Z_1,\ldots,Z_N)$ on $\mathbb{A}_N$ with distribution $\mu_\infty$ also satisfy $\mathbb{E}\big(\prod_{j=1}^N(y-Z_j)\big)=y^N$. A corresponding result holds also in the situation of Corollary 3.11.

Please notice that this differs from the situation in the end of Example 3.8 for $k=\infty$ where we have a purely deterministic situation and also a slightly different result.

Corollaries 3.5, 3.7, 3.10, and 3.11 have applications to Brownian motions and uniform probabilities on compact symmetric spaces of type $A$.

For a first example, consider the space $C(U(N))$ of all conjugacy classes of $U(N)$, which can be identified with $\mathbb{A}_N$ up to the cyclic group $\mathbb{Z}_N$, i.e., $C(U(N))\sim\mathbb{A}_N/\mathbb{Z}_N$. In fact, the conjugacy classes are characterized via the ordered spectra of matrices from $U(N)$ where, say, the eigenvalue with the smallest nonnegative argument has the first position. On the other hand, elements in $\mathbb{A}_N$ describe configurations of ordered points on $\mathbb{T}$ where the position of the first entry is arbitrary. It is also well known that the pushforward of the uniform distribution (i.e., the normalized Haar measure) of $U(N)$ under the natural projection $U(N)\mapsto C(U(N))\sim\mathbb{A}_N/\mathbb{Z}_N$ agrees with the pushforward of $\mu_k$ with $k=1$ under the canonical mapping $\mathbb{A}_N\mapsto\mathbb{A}_N/\mathbb{Z}_N$. Corollary 3.10 thus leads to the following.

COROLLARY 3.13. *Let $Z$ be a uniformly distributed $U(N)$-valued random variable. Then for each $y\in\mathbb{C}$, $\mathbb{E}\big(\det(yI_N-Z)\big)=y^N$.*

The same procedure also works for $SU(N)$, where the space $C(SU(N))$ of conjugacy classes corresponds to $\mathbb{A}_N^1/\mathbb{Z}_N$, and the pushforward of the uniform distribution corresponds to $\mu_k^1\in M^1(\mathbb{A}_N^1)$ with $k=1$. Hence, by Corollary 3.11:

COROLLARY 3.14. *Let $Z$ be an $SU(N)$-valued random variable which is uniformly distributed. Then for $y \in \mathbb{C}$, $\mathbb{E}(\det(yI_N - Z)) = y^N$.*

Corollaries 3.13 and 3.14 are special cases of well-known general formulas for integrals of polynomials on unitary groups w.r.t. uniform distributions in [**CS**]. On the other hand, our approach leads to generalizations of these formulas for Brownian motions on $U(N)$ and $SU(N)$.

For $k = 1/2$ or $k = 2$, our results above are related to the compact symmetric spaces $U(n)/O(n)$ and $U(2n)/Sp(n)$ associated with the root system $A_{N-1}$.

## 4. The non-compact case of type $A_{N-1}$

In this section we start with the $W$-invariant Heckman-Opdam Laplacians

$$L_k f(x) = \Delta f(x) + k \sum_{j=1}^{N} \sum_{l \neq j} \coth\Big(\frac{x_j - x_l}{2}\Big) \frac{\partial}{\partial x_j} f(x) \tag{4.1}$$

for $k \in [0, \infty[$ on the Weyl chamber

$$C_N^A := \{x \in \mathbb{R}^N : x_1 \geq x_2 \geq \ldots \geq x_N\}$$

of type $A_{N-1}$. As in Section 3, $L_k$ is the generator of a Feller diffusion $(X_{t,k})_{t \geq 0}$ on $C_N^A$ with reflecting boundaries. Again we study the renormalized generators $\widetilde{L}_k := \frac{1}{k} L_k$ which degenerate for $k \to \infty$ into

$$\widetilde{L}_\infty = \sum_{j=1}^{N} \sum_{l \neq j} \coth\Big(\frac{x_j - x_l}{2}\Big) \frac{\partial}{\partial x_j}.$$

The process $\widetilde{X}_k := (\widetilde{X}_{t,k} := X_{t/k,k})_{t \geq 0}$ for $k \in ]0, \infty[$ then solves the SDE

$$d\widetilde{X}_{t,k,j} = \frac{\sqrt{2}}{\sqrt{k}} dB_{t,j} + \sum_{l \neq j} \coth\Big(\frac{\widetilde{X}_{t,k,j} - \widetilde{X}_{t,k,l}}{2}\Big) dt \qquad (j = 1, \ldots, N) \tag{4.2}$$

which degenerates for $k = \infty$ into the ODE

$$\frac{dx_j}{dt}(t) = \sum_{l \neq j} \coth\Big(\frac{x_j(t) - x_l(t)}{2}\Big) \qquad (j = 1, \ldots, N). \tag{4.3}$$

Again, for initial data in the interior of the chamber $C_N^A$, the solution $(\widetilde{X}_{t,\infty})_{t \geq 0}$ of these differential equations exists for all $t \geq 0$ in the interior of $C_N^A$.

As in the preceding section, we next decompose the diffusions $\widetilde{X}_k$ into the center-of-gravity process $\widetilde{X}_k^{cg} := (\widetilde{X}_{t,k}^{cg})_{t \geq 0}$ with

$$\widetilde{X}_{t,k}^{cg} := \frac{1}{N} (\widetilde{X}_{t,k,1} + \ldots + \widetilde{X}_{t,k,N}) \cdot \omega$$

and the process $\widetilde{X}_k^{diff} := \widetilde{X}_k - \widetilde{X}_k^{cg}$ on the Weyl chamber

$$C_{N,0}^A := \{x \in C_N^A : x_1 + \ldots + x_N = 0\}$$

which describes the distances of the particles.

As in the proof of Lemma 3.2, the processes $\widetilde{X}_k^{diff}$ and $\widetilde{X}_k^{cg}$ are stochastically independent.

In the next step we observe that the processes $\widetilde{X}$, $\widetilde{X}_k^{diff}$, $\widetilde{X}_k^{cg}$ admit arbitrary exponential moments for arbitrary deterministic starting points. In fact, for $\widetilde{X}_k^{diff}$

this follows from Lemma 2.1. Moreover, $\widetilde{X}_k^{cg}$ is a classical one-dimensional Brownian motion up to scaling and has therefore arbitrary exponential moments. Finally, the independence of $\widetilde{X}_k^{diff}$ and $\widetilde{X}_k^{cg}$ ensures that $\widetilde{X}_k = \widetilde{X}_k^{diff} + \widetilde{X}_k^{cg}$ has this property as well.

With the existence of exponential moments in mind, we now follow Section 3 and observe that the trigonometric elementary symmetric polynomials

$$\widetilde{e}_l(x) := e_l(e^x) \quad (l = 0, \dots, N,\ x \in C_N^A)$$

are eigenfunctions of $\widetilde{L}_k$ for all $k$ with eigenvalues $l\big(\frac{1}{k} + N - l\big) \geq 0$. This leads to martingales for the diffusions $(\widetilde{X}_{t,k})_{t\geq 0}$ on $C_N^A$, similar to Corollaries 3.5 and 3.6 as follows:

COROLLARY 4.1. *For $k \in ]0, \infty[$ and $x \in C_N^A$ consider the diffusion $(\widetilde{X}_{t,k})_{t\geq 0}$ on $C_N^A$ with start in $x$. Then, for $l = 0, 1, \dots, N$, the process $\big(e^{-l(1/k+N-l)t}\, \widetilde{e}_l(\widetilde{X}_{t,k})\big)_{t\geq 0}$ is a martingale. In particular, for $t \geq 0$,*

$$\mathbb{E}\big(\widetilde{e}_l(\widetilde{X}_{t,k})\big) = e^{l(1/k+N-l)t}\, \widetilde{e}_l(x).$$

*This result also holds for $k = \infty$. More precisely, the solution $(\widetilde{X}_{t,\infty})_{t\geq 0}$ of the ODE* (4.3) *with start $x$ in the interior of $C_N^A$ satisfies*

$$\widetilde{e}_l(\widetilde{X}_{t,\infty}) = e^{l(N-l)t}\, \widetilde{e}_l(x) \qquad (l = 0, 1, \dots, N). \tag{4.4}$$

Using the independence of the processes $\widetilde{X}_k^{diff}$ and $\widetilde{X}_k^{cg}$, we also obtain the following analog of Corollary 3.7.

COROLLARY 4.2. *For $k \in ]0, \infty[$ and $x \in C_{N,0}^A$ consider the diffusion $\widetilde{X}_k^{diff}$ on the chamber $C_{N,0}^A$ with start in $x$. Then, for $t \geq 0$ and $l = 1, \dots, N$,*

$$\mathbb{E}\big(\widetilde{e}_l(\widetilde{X}_{t,k}^{diff})\big) = e^{l(N-l+(N+l)/(Nk))t}\, \widetilde{e}_l(x).$$

*This holds also for $k = \infty$ where the expectations can be omitted.*

Corollary 4.2 can be restated as:

COROLLARY 4.3. *For $k \in ]0, \infty[$ and $x \in C_{N,0}^A$ consider the diffusion $\widetilde{X}_k^{diff}$ on $C_{N,0}^A$ with start in $x$. Then, for $t \geq 0$ and $y \in \mathbb{C}$,*

$$\mathbb{E}\Big(\prod_{j=1}^N (y - e^{\widetilde{X}_{k,t,j}^{diff}})\Big) = P_{t,N,k,x}(y)$$

*with the polynomial*

$$P_{t,N,k,x}(y) := \sum_{l=0}^N y^{N-l}(-1)^l e^{l(N-l+(N+l)/(Nk))t}\, \widetilde{e}_l(x).$$

For general starting points $x \in C_{N,0}^A$, these polynomials do not seem to have particularly nice properties. For the starting configuration $x = 0 \in C_{N,0}^A$, which is of particular interest here, we get

$$P_{t,N,k,0}(y) = \sum_{l=0}^N \binom{N}{l} (-1)^l y^{N-l} e^{l(N-l+(N+l)/(Nk))t}.$$

We do not have much information about these polynomials. This is a contrast to the Bessel processes of type $A$ where in a corresponding formula classical Hermite polynomials appear; see [**KVW**].

EXAMPLE 4.4. (1) Let $N = 2$. We try to solve the ODE (4.3) with the singular starting point $x = 0 \in C_2^A$. In fact, (4.3) yields that $x_1(t) = -x_2(t)$ and that $y(t) := x_1(t) - x_2(t)$ satisfies $\frac{dy}{dt}(t) = 2\coth(y(t)/2)$. On the other hand, (4.4) for $N = 2$, $l = 1$ suggests that

$$e^{x_1(t)} + e^{-x_1(t)} = 2e^t$$

and thus $x_1(t) = \operatorname{arcosh}(e^t)$. It is now easily checked that in fact,

$$t \mapsto (\operatorname{arcosh}(e^t), -\operatorname{arcosh}(e^t))$$

is continuous on $[0, \infty)$ and solves (4.3) for $t > 0$. Moreover, for $k = \infty$ we have $x^{diff}(t) = x(t)$, and

$$P_{t,2,\infty,0}(y) = y^2 - 2e^t\, y + 1.$$

These polynomials have the zeros $e^{\pm x_1(t)}$ as claimed.

(2) Let $N = 3$. We again try to solve the ODE (4.3) with $x = 0 \in \partial C_3^A$. Here symmetry arguments imply that the solution of the ODE (4.3) must have the form $x(t) = (x_1(t), 0, -x_1(t))$. On the other hand, formula (4.4) with $N = 2$, $l = 1, 2$ suggests that

$$e^{x_1(t)} + e^{-x_1(t)} + 1 = 3e^{2t}$$

and thus $x_1(t) = \operatorname{arcosh}\big(3(e^{2t} - 1)/2\big)$. This indeed gives a solution as in example (1). Moreover, the polynomial

$$P_{t,3,\infty,0}(y) = y^3 - 3e^{2t}y^2 + 3e^{2t}y - 1$$

has the zeros 1 and $e^{\pm x_1(t)}$ as claimed.

Unfortunately, we have no closed formulas for $x(t)$ for general dimension $N$ even for the starting point $x = 0 \in C_N^A$.

We finally mention that Corollaries 4.2 and 4.3 for $k = 1/2, 1, 2$ have applications to Brownian motions on the noncompact symmetric spaces of type A, i.e. on $GL(N, \mathbb{F})/U(N, \mathbb{F})$ for $\mathbb{F} = \mathbb{R}, \mathbb{C}, \mathbb{H}$ similar to the results in the end of Section 3.

## 5. The non-compact case of type $BC_N$

We here start with the nonreduced root system

$$R = BC_N = \{\pm e_i, \pm 2e_i, \pm(e_i \pm e_j);\ 1 \le i < j \le N\} \subset \mathbb{R}^N$$

for $N \ge 2$ with the multiplicity $k = (k_1, k_2, k_3)$, where $k_1, k_2 \ge 0$, $k_3 > 0$ are the values on the roots $e_i, 2e_i, e_i \pm e_j$. As indicated in the introduction, we now reparametrize the multiplicity $k$. This will be not natural at a first glance, but it will turn out to be useful in the end. We here mainly follow the notations in [**Dem, V**] and define

$$\kappa = k_3, \quad q = N - 1 + \frac{1 + 2k_1 + 2k_2}{2k_3}, \quad p = N - 1 + \frac{1 + 2k_2}{2k_3}. \tag{5.1}$$

Then $\kappa > 0$, $q \ge p \ge N - 1 + 1/2\kappa$, and

$$k = (k_1, k_2, k_3) = \kappa \cdot \Big(q - p, k_{0,2}, 1\Big) \quad \text{with} \quad k_{0,2} := p - (N-1) - \frac{1}{2\kappa}. \tag{5.2}$$

We now regard $p, q$ as fixed parameters and $\kappa > 0$ as a varying parameter, and denote the multiplicity $k$ in (5.2) by $k_\kappa$. The associated $W$-invariant Heckman-Opdam Laplacian (2.6) is

$$
\begin{aligned}
L_\kappa = \Delta + \sum_{i=1}^N \kappa\Big((q-p)\coth(\frac{x_i}{2}) + 2k_{0,2}\coth(x_i) \\
+ \sum_{j:j\neq i}\big(\coth(\frac{x_i - x_j}{2}) + \coth(\frac{x_i + x_j}{2})\big)\Big)\partial_i
\end{aligned} \tag{5.3}
$$

on the Weyl chamber

$$C_N^B := \{x \in \mathbb{R}^N : x_1 \geq x_2 \geq \ldots \geq x_N \geq 0\}$$

of type $B_N$, associated with $R_+ = \{e_i, 2e_i, e_i - e_j : 1 \leq i < j \leq N\}$. As in the preceding sections, the operators $L_\kappa$ are the generators of diffusions $(X_{t,\kappa})_{t\geq 0}$ on $C_N^B$ where the paths are reflected at $\partial C_N^B$. The renormalized operators

$$\widetilde{L}_\kappa := \frac{1}{\kappa}L_\kappa$$

then are the generators of the diffusions $(\widetilde{X}_{t,\kappa} := X_{t/\kappa,\kappa})_{t\geq 0}$ which may be regarded as solutions of the SDE

$$
\begin{aligned}
d\widetilde{X}_{t,\kappa,j} = & \frac{\sqrt{2}}{\sqrt{\kappa}}dB_{t,j} + \sum_{l\neq j}\Big(\coth\Big(\frac{\widetilde{X}_{t,\kappa,j} - \widetilde{X}_{t,\kappa,l}}{2}\Big) + \coth\Big(\frac{\widetilde{X}_{t,\kappa,j} + \widetilde{X}_{t,\kappa,l}}{2}\Big)\Big)dt \\
& + \Big((q-p)\coth(\frac{\widetilde{X}_{t,\kappa,j}}{2}) + 2k_{0,2}\coth(\widetilde{X}_{t,\kappa,j})\Big)dt \qquad (j = 1, \ldots, N).
\end{aligned} \tag{5.4}
$$

For $\kappa \to \infty$, the generator degenerates into

$$
\begin{aligned}
\widetilde{L}_\infty = \sum_{i=1}^N \Big((q-p)\coth(\frac{x_i}{2}) + 2(p - (N-1))\coth(x_i) \\
+ \sum_{j:j\neq i}\Big(\coth(\frac{x_i - x_j}{2}) + \coth(\frac{x_i + x_j}{2})\Big)\Big)\partial_i,
\end{aligned} \tag{5.5}
$$

and (5.4) becomes the ODE

$$
\begin{aligned}
\frac{dx_j}{dt}(t) = & \sum_{l\neq j}\Big(\coth\Big(\frac{x_j(t) - x_l(t)}{2}\Big) + \coth\Big(\frac{x_j(t) + x_l(t)}{2}\Big)\Big) \\
& + (q-p)\coth(\frac{x_j(t)}{2}) + 2\big(p - (N-1)\big)\coth x_j(t) \qquad (j = 1, \ldots, N).
\end{aligned} \tag{5.6}
$$

Again, for initial data in the interior of the chamber $C_N^B$, the solution $(\widetilde{X}_{t,\infty})_{t\geq 0}$ of these differential equations exists for $t \geq 0$ in the interior of $C_N^B$.

We now turn to eigenfunctions of the operators $\widetilde{L}_\kappa$ and consider the Heckman-Opdam hypergeometric functions $F_{BC}(\lambda, k_\kappa; x)$ associated with $BC_N$ in the variable $x \in \mathbb{R}^N$, with spectral parameter $\lambda \in \mathbb{C}^N$ and multiplicity $k_\kappa$. By Section 2, for $\kappa \in ]0, \infty[$, the functions $x \mapsto F_{BC}(\lambda, k_\kappa; x)$ are eigenfunctions of the renormalized

Laplacian $\widetilde{L}_\kappa$ with the eigenvalues

$$r_{\lambda,\kappa} := \frac{1}{\kappa}\Big(\sum_{j=1}^N \lambda_j^2 - |\rho(\kappa)|^2\Big),$$

where

$$\rho(\kappa) = \frac{1}{2}\sum_{\alpha\in R_+} k_\kappa(\alpha)\alpha \quad \text{with} \quad \rho(\kappa)_j = \frac{\kappa}{2}\big((q-p) - 2k_{0,2} + 2(N-j)\big). \tag{5.7}$$

We further consider the associated (normalized) Heckman-Opdam polynomials $R_\lambda(k,\,.\,) = R_\lambda^{BC}(k,\,.\,)$, as introduced in (2.5), which are indexed by $P_+ = \Lambda_N^+$ (the set of partitions of length at most $N$). These are multivariate generalizations of the classical Jacobi polynomials which are well-studied in the literature, see for instance [**BO**, **L**, **RR2**]. The polynomials $\widetilde{R}_\lambda$ defined by

$$\widetilde{R}_\lambda(\cos x) := R_\lambda(k; x)$$

form an orthogonal basis of $L^2(\mathbb{A}_N, w_k)$ on

$$\mathbb{A}_N := \{x \in \mathbb{R}^N \mid -1 \le y_1 \le \dots \le y_N \le 1\}$$

with the weight function

$$w_k(y) := \prod_{i=1}^N (1-y_i)^{k_1+k_2-1/2}(1+y_i)^{k_2-1/2} \cdot \prod_{i<j} |y_i - y_j|^{2k_3}. \tag{5.8}$$

Here we are mainly interested in the fact that for $k = k_\kappa$ and $\lambda \in \Lambda_N^+$, the exponential polynomials

$$H_\lambda(x) := \widetilde{R}_\lambda(\cosh x) = F_{BC}(\lambda + \rho(k), k; x)$$

are eigenfunctions of $\widetilde{L}_\kappa$ with the eigenvalues

$$\begin{aligned} r_\lambda &= \frac{1}{\kappa}\langle \lambda, \lambda + 2\rho(k)\rangle \\ &= \frac{1}{\kappa}\sum_{j=1}^N \lambda_j(\lambda_j + k_1 + 2k_2 + 2k_3(N-j)) \\ &= \sum_{j=1}^N \lambda_j\Big(\frac{\lambda_j - 1}{\kappa} + p + q + 2 - 2j\Big). \end{aligned} \tag{5.9}$$

We now consider the partitions $\lambda(n) := 1^n \in \Lambda_+^N$ for $n = 0, \dots, N$. It is known (see Section 5 of [**V**] and in particular Lemma 5.1 there) that the Jacobi polynomials $\widetilde{R}_{\lambda(n)}$ are of the form

$$\widetilde{R}_{\lambda(n)} = \sum_{l=0}^n c_{n,l}(p,q)\cdot e_l \quad \text{with} \quad c_{n,n}(p,q) \neq 0, \tag{5.10}$$

where the $e_l$ are again the elementary symmetric polynomials in $N$ variables and the coefficients $c_{n,l}(p,q) \in \mathbb{R}$ depend on $p, q$ only and not on $\kappa$. This observation will be crucial in the following and is the reason for our parametrization of $k$ by $p, q, \kappa$ above. For more details on the $c_{n,l}(p,q)$ we refer to [**V**]. In summary, the functions

$H_{\lambda(n)}$ with $n = 0, \dots, N$ are independent of $\kappa$ and simultanous eigenfunctions of the operators $\widetilde{L}_\kappa$ for all $\kappa \in ]0, \infty[$ with the eigenvalues

$$r_n = n(p+q-n+1).$$

Clearly, this observation also holds for $\kappa = \infty$. This implies:

LEMMA 5.1. *For each starting point $x \in C_N^B$ of the processes $(\widetilde{X}_{t,\kappa})_{t\geq 0}$ with $\kappa \in ]0, \infty]$, the processes*

$$\left(e^{-r_n t} \cdot H_{\lambda(n)}(\widetilde{X}_{t,\kappa})\right)_{t\geq 0} \tag{5.11}$$

*are martingales for $n = 0, \dots, N$, where the numbers $r_n$ and the functions $H_{\lambda(n)}$ do not depend on $\kappa$.*

*In particular, for $x$ in the interior of $C_N^B$, the solution $(\widetilde{X}_{t,\infty})_{t\geq 0}$ of the ODE (5.6) with $\widetilde{X}_{0,\infty} = x$ satisfies*

$$H_{\lambda(n)}(\widetilde{X}_{t,\infty}) = e^{r_n t} H_{\lambda(n)}(x) \quad \textit{for} \quad t \geq 0,\ n = 0, \dots, N.$$

PROOF. This follows from our preceding considerations and the fact that the random variables $H_{\lambda(n)}(\widetilde{X}_{t,\kappa})$ are integrable for $\kappa < \infty$ and $t > 0$ by Lemma 2.1. □

We may invert (5.10) and write the elementary symmetric polynomials $e_l$ as linear combinations of the $\widetilde{R}_{\lambda(n)}$ for $l, n = 0, \dots, N$ with coefficients independent of $\kappa$. Lemma 5.1 thus implies:

COROLLARY 5.2. *Fix some deterministic starting point $x \in C_N^B$ as well as the parameters $p, q$. Consider the associated diffusions $(\widetilde{X}_{t,\kappa})_{t\geq 0}$ for $\kappa \in ]0, \infty[$. Then there are coefficients $a_{n,l} \in \mathbb{R}$ for $0 \leq l \leq n \leq N$ such that*

$$\mathbb{E}\big(e_n(\cosh(\widetilde{X}_{t,\kappa}))\big) = \sum_{l=0}^{n} a_{n,l}\, e^{r_l t}$$

*with $r_0 = 0$ where the coefficients $a_{n,l}$ and the exponents $r_l$ depend on $p, q$ and $x$ only and not on $\kappa$. The same holds for $\kappa = \infty$ and starting points $x$ in the interior of $C_N^B$.*

EXAMPLE 5.3. For $x = 0 \in C_N^B$, we have $H_{\lambda(n)}(0) = 1$ and thus

$$\mathbb{E}\big(H_{\lambda(n)}(\widetilde{X}_{t,\kappa})\big) = e^{r_n t} \qquad (n = 0, \dots, N,\ t \geq 0,\ \kappa \in ]0, \infty[).$$

We finally turn to an application concerning a determinantal formula. Again we fix some starting point $x \in C_N^B$ as well as $p, q$ and consider the associated diffusions $(\widetilde{X}_{t,\kappa})_{t\geq 0}$ for $\kappa \in ]0, \infty[$. Then by Corollary 5.2, for all $t \geq 0$ and $y \in \mathbb{C}$,

$$\begin{aligned}
\mathbb{E}\Big(\prod_{j=1}^{N}(y - \cosh(\widetilde{X}_{t,\kappa,j}))\Big) &= \sum_{n=0}^{N}(-1)^n\, \mathbb{E}\big(e_n(\cosh(\widetilde{X}_{t,\kappa}))\big) \cdot y^{N-n} \\
&= \sum_{n=0}^{N}(-1)^n\, e_n\big(\cosh(\widetilde{X}_{t,\infty})\big) \cdot y^{N-n} \\
&= \prod_{j=1}^{N}\big(y - \cosh(\widetilde{X}_{t,\infty,j})\big) =: D_{t,x}(y).
\end{aligned} \tag{5.12}$$

It is an interesting task to find particularly nice starting points $x \in C_N^B$ for which $D_{t,x}(y)$ can be determined explicitly.

For instance, in the setting of multivariate Bessel processes of types $A_{N-1}$ or $B_N$ and start in the origin, $D_{t,0}(y)$ is a classical one-dimensional Hermite or Laguerre polynomial of degree $N$ in $y$, which is scaled by a factor $\sqrt{t}$; for the details see [**KVW**]. Moreover, in the setting of Heckman-Opdam Jacobi processes of type $BC$ on the compact alcove $\mathbb{A}_N$, and with the same paramatrization of the multiplicity $k$ by $p, q, \kappa$ as here, there is a (unique) stationary solution $x_0$ of the ODE in the interior of $\mathbb{A}_N$, whose coordinates are the ordered zeroes of some classical one-dimensional Jacobi polynomial of degree $N$ in $y$. The indices of this Jacobi polynomial are determined by $p, q$. This means that for this particular starting point $x_0$, the function $D_{t,x_0}(y)$ is just this specific Jacobi polynomial and is independent of $t \geq 0$ (due to stationarity). We refer to [**V**] for further details.

We expect that in our non-compact $BC$ setting, $D_{t,x}(y)$ should be of particular interest when the associated ODE (5.6) starts in $x = 0 \in \partial C_N^B$, where we expect that the corresponding initial value problem is uniquely solvable, as in the Dunkl setting in [**VW**]. It seems that the explicit solution of this initial value problem is more involved than in the cases considered in [**KVW**, **V**].

## References


[A] K. Aomoto, *Jacobi polynomials associated with Selberg integrals*, SIAM J. Math. Anal. **18** (1987), no. 2, 545–549, DOI 10.1137/0518042. MR876291

[AV] Sergio Andraus and Michael Voit, *Limit theorems for multivariate Bessel processes in the freezing regime*, Stochastic Process. Appl. **129** (2019), no. 11, 4771–4790, DOI 10.1016/j.spa.2018.12.011. MR4013880

[BF] T. H. Baker and P. J. Forrester, *The Calogero-Sutherland model and generalized classical polynomials*, Comm. Math. Phys. **188** (1997), no. 1, 175–216, DOI 10.1007/s002200050161. MR1471336

[BO] R. J. Beerends and E. M. Opdam, *Certain hypergeometric series related to the root system BC*, Trans. Amer. Math. Soc. **339** (1993), no. 2, 581–609, DOI 10.2307/2154288. MR1123450

[CS] Benoît Collins and Piotr Śniady, *Integration with respect to the Haar measure on unitary, orthogonal and symplectic group*, Comm. Math. Phys. **264** (2006), no. 3, 773–795, DOI 10.1007/s00220-006-1554-3. MR2217291

[Dem] N. Demni, *β-Jacobi processes*, Adv. Pure Appl. Math. **1** (2010), no. 3, 325–344, DOI 10.1515/APAM.2010.019. MR2719370

[DG] Persi Diaconis and Alex Gamburd, *Random matrices, magic squares and matching polynomials*, Electron. J. Combin. **11** (2004/06), no. 2, Research Paper 2, 26. MR2120097

[F] P. J. Forrester, *Log-gases and random matrices*, London Mathematical Society Monographs Series, vol. 34, Princeton University Press, Princeton, NJ, 2010, DOI 10.1515/9781400835416. MR2641363

[FG] Peter J. Forrester and Alex Gamburd, *Counting formulas associated with some random matrix averages*, J. Combin. Theory Ser. A **113** (2006), no. 6, 934–951, DOI 10.1016/j.jcta.2005.09.001. MR2244124

[HO] G. Heckman, E. Opdam, Jacobi polynomials and hypergeometric functions associated with root systems. In: Encyclopedia of Special Functions, Part II: Multivariable Special Functions, eds. T.H. Koornwinder, J.V. Stokman, Cambridge University Press, Cambridge, 2021.

[HS] Gerrit Heckman and Henrik Schlichtkrull, *Harmonic analysis and special functions on symmetric spaces*, Perspectives in Mathematics, vol. 16, Academic Press, Inc., San Diego, CA, 1994. MR1313912

[HW] David G. Hobson and Wendelin Werner, *Non-colliding Brownian motions on the circle*, Bull. London Math. Soc. **28** (1996), no. 6, 643–650, DOI 10.1112/blms/28.6.643. MR1405497

[KN] Rowan Killip and Irina Nenciu, *Matrix models for circular ensembles*, Int. Math. Res. Not. **50** (2004), 2665–2701, DOI 10.1155/S1073792804141597. MR2127367

[KVW] M. Kornyik, M. Voit, and J. Woerner, *Some martingales associated with multivariate Bessel processes*, Acta Math. Hungar. **163** (2021), no. 1, 194–212, DOI 10.1007/s10474-020-01096-5. MR4217966

[L] Michel Lassalle, *Polynômes de Jacobi généralisés* (French, with English summary), C. R. Acad. Sci. Paris Sér. I Math. **312** (1991), no. 6, 425–428. MR1096625

[LV] Luc Lapointe and Luc Vinet, *Exact operator solution of the Calogero-Sutherland model*, Comm. Math. Phys. **178** (1996), no. 2, 425–452. MR1389912

[NPP] E. K. Narayanan, A. Pasquale, and S. Pusti, *Asymptotics of Harish-Chandra expansions, bounded hypergeometric functions associated with root systems, and applications*, Adv. Math. **252** (2014), 227–259, DOI 10.1016/j.aim.2013.10.027. MR3144230

[OO] Andrei Okounkov and Grigori Olshanski, *Asymptotics of Jack polynomials as the number of variables goes to infinity*, Internat. Math. Res. Notices **13** (1998), 641–682, DOI 10.1155/S1073792898000403. MR1636541

[P] Philip E. Protter, *Stochastic integration and differential equations: A new approach*, 2nd ed., Applications of Mathematics (New York), vol. 21, Springer-Verlag, Berlin, 2004. Stochastic Modelling and Applied Probability. MR2020294

[R] E. M. Rains, *Combinatorial properties of Brownian motion on the compact classical groups*, J. Theoret. Probab. **10** (1997), no. 3, 659–679, DOI 10.1023/A:1022601711176. MR1468398

[RR1] Heiko Remling and Margit Rösler, *The heat semigroup in the compact Heckman-Opdam setting and the Segal-Bargmann transform*, Int. Math. Res. Not. IMRN **18** (2011), 4200–4225, DOI 10.1093/imrn/rnq239. MR2836019

[RR2] Heiko Remling and Margit Rösler, *Convolution algebras for Heckman-Opdam polynomials derived from compact Grassmannians*, J. Approx. Theory **197** (2015), 30–48, DOI 10.1016/j.jat.2014.07.005. MR3351538

[RW] L. C. G. Rogers and David Williams, *Diffusions, Markov processes, and martingales. Vol. 1*, Cambridge Mathematical Library, Cambridge University Press, Cambridge, 2000. Foundations; Reprint of the second (1994) edition, DOI 10.1017/CBO9781107590120. MR1796539

[Sch1] Bruno Schapira, *The Heckman-Opdam Markov processes*, Probab. Theory Related Fields **138** (2007), no. 3-4, 495–519, DOI 10.1007/s00440-006-0034-1. MR2299717

[Sch2] Bruno Schapira, *Contributions to the hypergeometric function theory of Heckman and Opdam: sharp estimates, Schwartz space, heat kernel*, Geom. Funct. Anal. **18** (2008), no. 1, 222–250, DOI 10.1007/s00039-008-0658-7. MR2399102

[St] Richard P. Stanley, *Some combinatorial properties of Jack symmetric functions*, Adv. Math. **77** (1989), no. 1, 76–115, DOI 10.1016/0001-8708(89)90015-7. MR1014073

[V] Michael Voit, *Some martingales associated with multivariate Jacobi processes and Aomoto's Selberg integral*, Indag. Math. (N.S.) **31** (2020), no. 3, 398–410, DOI 10.1016/j.indag.2020.02.005. MR4092353

[VW] Michael Voit and Jeannette H. C. Woerner, *The differential equations associated with Calogero-Moser-Sutherland particle models in the freezing regime*, Hokkaido Math. J. **51** (2022), no. 1, 153–174, DOI 10.14492/hokmj/2020-307. MR4406927

Institut für Mathematik, Universität Paderborn, Warburger Str. 100, D-33102 Paderborn, Germany

*Email address*: roesler@math.upb.de

Fakultät Mathematik, Technische Universität Dortmund, Vogelpothsweg 87, D-44221 Dortmund, Germany

*Email address*: michael.voit@math.tu-dortmund.de

Contemporary Mathematics
Volume **780**, 2022
https://doi.org/10.1090/conm/780/15693

# Conformal hypergeometry and integrability

Volker Schomerus

ABSTRACT. Conformal field theories play a central role in modern theoretical physics with many applications to the understanding of phase transitions, gauge theories and even the quantum physics of gravity, through Maldacena's celebrated holographic duality. The key analytic tool in the field is the so-called conformal partial wave expansion, i.e. a Fourier-like decomposition of physical quantities into a basis of partial waves for the conformal group $SO(1, d + 1)$. While the general theory of these partial waves remains largely unexplored, the very simplest specimens are known to be (close relatives of) multivariate $BC_2$-symmetric Heckman-Opdam hypergeometric functions. Further advances in the theory of conformal partial waves are expected to exploit a deep relation with (super-)integrability and in particular with certain limits of Gaudin integrable models.

## 1. Introduction

During the last century physicists gradually learned to appreciate the importance of conformal symmetry for our understanding of nature. By definition, conformal symmetry is made up of angle preserving transformations. The most commonly observed (conformal) symmetries are rotations and translations. These are present in all homogeneous and isotropic systems. Beyond that, dilations (scale transformations) are the most prominent examples of conformal transformations. Scale invariance is an important phenomenon in physics that refers to situations in which the features of a system do not change upon zooming in or out. The first known occurrence in nature is the critical point for the liquid-vapour transition in the phase diagram of water. Many more of these points have been discovered in the meantime. Some of them can be realized in spin systems such as the famous Ising model.

Particle physicists started to appreciate conformal field theories (CFTs) through the influential work of Wilson on the renormalization group [**74**]. By construction, scale invariance occurs at all fixed points of renormalization group transformations. This gives CFTs a very special status within the space of all theories, [1] and makes them a good starting point to explore, in particular, non-perturbative phenomena in quantum field theory. More recently, CFTs became the main actors on another stage, Juan Maldacena's celebrated AdS/CFT duality [**47**]. Its most general formulation proposes an equivalence between non-gravitational quantum systems in $d$

[1]Throughout this introduction I will not distinguish clearly between scale and conformal symmetry. I will briefly come back to the issue in section 2.

dimensions and gravitational systems in one dimension higher. In this context, a $d$-dimensional CFT is like a holographic screen for a gravitational theory in Anti-deSitter (AdS) space. As we make progress in dealing with CFTs, the AdS/CFT duality can be applied to provide novel views on the deep mysteries of a quantum theory of gravity, such as the phenomenon of Hawking radiation and black hole information.

The very brief history of scale/conformal invariance in physics that I presented in the first two paragraphs may serve one main purpose: it explains why a significant part of modern theoretical physics research is devoted to CFTs and why the interest ranges over so many different communities, from condensed matter theory and modern material science all the way to particle physics and quantum gravity.

Invariance under conformal transformations turns out to be a surprisingly powerful symmetry principle that strongly constrains physical observables such as the correlations between different measurements. These constraints may be exposed with the help of the most important analytical tool of CFT, the so-called *conformal partial wave* expansions. One should think of these as an extension of Fourier analysis, but for systems with conformal rather than mere translational symmetry. Just as plane waves in Fourier analysis are uniquely characterized by their behaviour under translations, conformal partial waves are determined by conformal symmetry alone, i.e. they belong to the realm of group and representation theory.

When conformal field theories were considered for the first time about 50 years ago, the relevant group theory was very far from being developed, which halted initial progress. This changed only after physicists started to focus on $d = 2$ dimensions where conformal symmetry becomes infinite dimensional. For that case a very beautiful theory was developed with many deep relations to mathematics and integrable systems. But this is not our topic here. Instead, what I shall describe below is relevant for CFTs in dimension $d > 2$. At the end of the last century, and in particular with the discovery of the AdS/CFT duality, the case for studying CFTs in $d > 2$ had become so strong that physicists made another effort to develop at least some of its group theoretic foundations. The influential work of Dolan and Osborn [**16, 17**] on conformal partial waves for scalar 4-point functions and some subsequent work by other authors, see in particular [**35, 44, 45, 57**], eventually provided a basis for some new breakthrough developments in higher dimensional CFT, e.g. for the 3-dimensional critical Ising model.

At the time, physicists did not know that essentially all the required results on conformal partial waves were already contained in the seminal papers of Heckman and Opdam, see [**30, 31, 54**]. Somewhat surprisingly, before it was uncovered in [**37**], the profound relation went unnoticed between the study of integrable quantum mechanics of Calogero-Moser-Sutherland type on the one hand and conformal partial waves on the other. The main intent of this text is to help reduce the apparent and unfortunate divide between the physics communities that work on conformal field theories on the one hand and mathematicians with expertise in modern group theory, harmonic analysis, hypergeometry and integrability on the other. The study of CFTs continues to challenge profound advances of its group theoretic ingredients. So it is not too late for the relevant communities to join forces.

After these introductory remarks, let me describe the plan of this short note. The next section is meant as a smooth introduction to some basic notions of critical systems and CFT. In the example of the critical Ising model I will outline how statistical systems can become critical and possibly conformal. In addition, I shall introduce the notions of (scalar and spinning) fields, correlation functions, scaling weights, conformal partial waves etc. and I will sketch the main idea of the so-called conformal bootstrap program. Section 2 remains very descriptive, without precise definitions. Nevertheless, I hope that it can build a first intuition into the basic concepts of the area, in a spirit that is relatively close to the physicists' approach. In section 3, I then rephrase the basic setup in a more mathematical language. There, the notions of fields, correlation functions and conformal partial waves are embedded into group theory and harmonic analysis. Once this is achieved, we can walk along well trodden paths into the world of hypergeometry and integrable systems. We will see the first examples of the relation between conformal partial waves and hypergeometry at the end of section 3. As I will argue there, conformal partial waves for scalar 4-point functions can be identified with the wave functions of hyperbolic Calogero-Moser-Sutherland models for the root system $BC_2$, i.e. these conformal partial waves are (close relatives of) $BC_2$ Heckman-Opdam hypergeometric functions. I conclude section 3 with a lightning overview of extensions which are more or less well understood, or at least have been actively researched already.

The final section is then devoted to a new important direction about which very little seems to be known for now: multipoint correlation functions. In a CFT, correlations of up to $N = 3$ local measurements are essentially determined by conformal symmetry. Hence, the first non-trivial correlations that can be analysed with well studied conformal partial waves appears when $N = 4$. The territory we shall begin to explore in the final section lies just one step beyond. For $N > 4$, conformal partial waves were largely unexplored until very recently. Section 4 embeds the profound mathematical challenges in gaining control of these objects into the context of $N$-site Gaudin integrable models for the conformal group [**6**]. As I will discuss in detail, the study of the relevant conformal partial waves requires us to consider certain limits of the Gaudin integrable models which extend the so-called caterpillar or bending flow limits previously considered in [**13, 14, 67**]. In the final subsection I highlight an intriguing new appearance of the lemniscatic elliptic Calogero-Moser-Sutherland model of [**20**] in the context of $N$-point scalar conformal blocks or, equivalently, in limits of $N$-site Gaudin integrable models.

While sections 2 and 3 of this note survey developments from several decades and do not contain any new results, section 4 is based on some very recent research in collaboration with Ilija Buric, Sylvain Lacroix, Jeremy Mann and Lorenzo Quintavalle [**6, 7**]. The content of section 4 goes significantly further than what I had reported at the meeting in December. When I gave my talk I had the vague hope that it could trigger fruitful interaction, but I was surprised to see how quickly this materialized. In fact, some of the new material that is now included in the final subsection was initiated through the interactions with participants of the workshop, and in particular with Pavel Etingof.

## 2. Conformal field theory and partial waves

**2.1. The critical Ising model.** As we have discussed in the introduction, conformal field theories are used to describe the critical behavior of many statistical systems. One of the most well known is the Ising model. Let us use this simple model to illustrate some of the key concepts. The Ising model can be defined on a $d$-dimensional hypercubic lattice with lattice sites given by $x_n = n^\mu \epsilon_\mu \in \mathbb{R}^d$. Here the summation over $\mu = 1, \dots, d$ is understood and $\epsilon_\mu = \epsilon e_\mu$ coincide with the unit vectors $e_\mu$ of $\mathbb{R}^d$ up to a common scale factor $\epsilon$. We refer to $\epsilon$ as the lattice spacing. The coefficients $n^\mu$ are integers.

To each lattice site we now assign a spin $\sigma_n \in \{\pm 1\}$. Any such choice of spin assignments on all lattice sites is referred to as a spin configuration. The object $\sigma$ is (a discrete version) of what physicists call a *field*. In order to define our theory we must specify a measure on the on the space of configurations. For the Ising model, this measure is

$$P_\zeta(\{\sigma\}) = \mathcal{N} e^{\zeta \sum_{\langle nm \rangle} \sigma_n \sigma_m} . \tag{2.1}$$

Here the sum extends over nearest neighbors $\langle nm \rangle$, i.e. over all pairs of lattice sites that are a distance $|x_n - x_m| = \epsilon$ apart.

The nature of a typical configuration selected by the measure (2.1) depends very much on the value of the parameter $\zeta$. If we set $\zeta = 0$, for example, neighboring spins do not to interact, i.e. the value the measure assumes is the same on all spin configurations. Hence, if we measure the spin at site $n$, this won't tell us anything about the value of the spin at another site $m$, in other words there is no correlation between the value $\sigma_n$ of the spin at site $n$ and the value of $\sigma_m$ at another sites $m \neq n$ of the lattice. In the opposite case with $\zeta = \infty$, configurations in which spins assume different values are infinitely suppressed, i.e. the measure only allows for the two constant configurations $\sigma_n = 1$ or $\sigma_n = -1$ for all sites $n$. In this case, the correlations between spins are as strong as they can be. We only need to measure the spin $\sigma_n$ at one site $n$ to know its value at all the other sites $m \neq n$. Once we go away from this extreme case, spin flips are possible but still costly. In other words, while the measure still assumes its maximum value on the two constant configurations, it becomes less strongly peaked as we decrease $\zeta$. Consequently, spin configurations with large 'islands' of aligned spins will start to play a role. As we reduce the value of $\zeta$, smaller islands contribute. It turns out that, for some very special value $\zeta = \zeta_c$, islands of all sizes appear with equal probability. At this very special point, the system appears to be self-similar on scales that are large compared to the lattice spacing $\epsilon$ (and small compared to the overall size $L$ of the lattice if we consider a finite lattice). It is this 'critical' value $\zeta = \zeta_c$ of the parameter $\zeta$ that we want to focus on.

In order to probe the Ising model, physicists measure correlations between spins. The simplest interesting example is the correlation between two spins $\sigma_n$ and $\sigma_m$ at two different lattice sites $n, m$. The quantity is defined as

$$\langle \sigma_n \sigma_m \rangle_\zeta = \sum_{\{\sigma\}} P_\zeta(\{\sigma\})\, \sigma_n \sigma_m . \tag{2.2}$$

To make this sum well defined, one can imagine that it is carried out on a finite lattice with $L^d$ lattice sites, even though criticality requires $L$ being sent to infinity at the end. As $\zeta$ approaches $\zeta_c$ the system becomes self-similar in the sense described above and hence this correlation function must take the form

$$\langle \sigma_n \sigma_m \rangle_{\zeta_c} \sim \frac{1}{|x_n - x_m|^{2\Delta_\sigma}} . \tag{2.3}$$

Here, $|x_n - x_m|$ denotes the Euclidean distance between the lattice points $x_n$ and $x_m$ in $\mathbb{R}^d$. The $\sim$ means the function on the right hand side approximates the correlation function on intermediate scales or, equivalently, in the limit of an infinite lattice size $L$ and vanishing lattice spacing $\epsilon$. The parameter $\Delta_\sigma$ that describes how the correlation function falls off as we increase the distance between the two lattice sites is referred to as the *scaling weight* of $\sigma$. It is a very important observable of the critical system.

Determining the value of $\Delta_\sigma$ is one of the central challenges of CFT. In $d = 2$ dimensions, it is possible to give an exact value. As was observed in [**3**], the scaling weight of the spin field in the critical 2-dimensional Ising model can be calculated using representation theory of the Virasoro algebra with central charge $c = 1/2$. This approach famously gives

$$\Delta_\sigma^{[2\mathrm{D}]} = 1/8 \; . \tag{2.4}$$

For the 3-dimensional critical Ising model, the exact value of the scaling weight $\Delta_\sigma$ is not known, but there exist good numerical simulations (NSs) of the correlation (2.2) near $\zeta = \zeta_c$ that give the approximate result [**29**]

$$\Delta_\sigma^{[3\mathrm{D}],\mathrm{NS}} \sim 0.51813(5) \; . \tag{2.5}$$

Let us point out right away that a more precise value of this exponent is known by now. However, this precision determination is based not on numerical simulation of the Ising model, but rather on the conformal bootstrap program that we shall outline below.

Before we go there, let us briefly mention that the scaling weight $\Delta_\sigma$ of the spin field $\sigma$ is not the only field in the Ising model. Obviously, it is possible to construct many similar quantities. To give just one additional example, let us introduce the object

$$\varphi_n^\mu = \sigma(x_n)\nabla^\mu\sigma(x_n) = \sigma(x_n)\sigma(x_n + \epsilon_\mu) - 1 \; . \tag{2.6}$$

Note that the definition of $\varphi^\mu$ depends on the choice of a direction $\mu$ in the $d$-dimensional space. Physicists refer to $\varphi^\mu$ as a vector field as opposed to the spin field $\sigma$ which is scalar, i.e. does not single out any direction. One can go further and consider fields that are sensitive to the value of the Ising spin at more than two sites. The relevant sites may fall into a local cluster or be located along lines, surfaces, etc. One calls the associated objects local, line, surface operators, respectively. These operators can be inserted into correlation functions of the form (2.2), just as we did for the spin field $\sigma$. It turns out that the 2-point correlations of local operators also fall off with some characteristic exponent $\Delta_\varphi$ whose value depends on the field $\varphi$ that is inserted.

**2.2. Conformal field theory.** Critical systems are included in the wider class of homogeneous and isotropic systems which are invariant under translations $x^\mu \mapsto x^\mu + a^\mu$ and rotations $x^\mu \mapsto O^\mu_\nu x^\nu$ of points $x = (x^\mu) \in \mathbb{R}^d$. As we have described above, the characteristic feature of critical system is self-similarity, i.e. at the critical point the symmetry is enhanced to include scale transformations $x^\mu \mapsto \lambda x^\mu$ with $\lambda \in \mathbb{R}^+$. It turns out, however, that the symmetry of many critical systems, including the critical Ising model, is further enhanced to the $d$-dimensional conformal group $SO(1, d+1)$. In addition to translations, rotations and scale transformations, the conformal group possesses $d$ generators of so-called special conformal transformations. This enhancement has dramatic consequences as we shall discuss now.

To begin exploring these consequences let us point out that in a conformal field theory, correlation functions of three fields are fixed by symmetry, up to some overall constant. This follows from the well known fact that conformal transformation can be used to map any set of three points in $\mathbb{R}^d$ to the values $x_1 = 0, x_2 = \vec{e}_1, x_3 = \infty$. Hence, a 3-point function of scalar fields $\phi_i$ of scaling weight $\Delta_i$ must take the form

$$(2.7) \quad \langle \phi_1(x_1)\phi_2(x_2)\phi_3(x_3)\rangle = \frac{\gamma_{123}}{|x_{12}|^{\Delta_{12,3}}|x_{23}|^{\Delta_{23,1}}|x_{13}|^{\Delta_{13,2}}} = \gamma_{123}$$

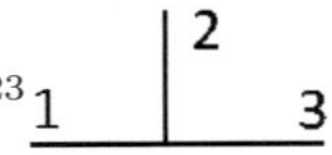


where $x_{ij} = x_i - x_j$ and $\Delta_{12,3} = \Delta_1 + \Delta_2 - \Delta_3$ etc. In more mathematical terms, the $x_i$-dependent function on the right hand side may be thought of as Clebsch-Gordon coefficients for the conformal group and the prefactors $\gamma_{123}$ provide a multiplication table for scalar fields of the theory. This is the interpretation we have tried to stress in the symbolic representation of the 3-point function on the right hand side.

What is so special about conformal field theories is indeed that their 'multiplication table' is really just a set of constants. To avoid confusion we need to stress that the 'multiplication' does not close on scalar fields, i.e. 3-point functions of two scalars and one vector field, for example, can be non-vanishing. But this does not alter the basic conclusion that conformal symmetry guarantees that there exists a 'multiplication table' of constants. It was already observed in the early days of CFT that the multiplication table of conformal field theories must obey stringent consistency conditions that are reminiscent of associativity constraints on multiplication tables in algebra. This observation is the main content of Polyakov's conformal bootstrap program [**61**]. The concrete implementation of this program, however, is rather difficult and so far analytic solutions were only found for 2-dimensional conformal field theories. Nevertheless, the modern conformal bootstrap has led to profound new insights on conformal field theories in $d \geq 3$, see below.

**2.3. Conformal partial waves and the bootstrap.** The most basic associativity constraints arise from the study of 4-point correlation functions. These are no longer determined by conformal symmetry alone simply because one can form two independent conformally invariant cross ratios from four points in $\mathbb{R}^d$. We shall parametrize these through two variables $u_1$ and $u_2$. In order to analyse 4-point functions it is useful to expand them in terms of a 'basis' of conformal partial waves

$$(2.8) \qquad \langle \prod_{i=1}^{4} \phi_i(x_i)\rangle \sum_{\Delta,l} \gamma_{12;\Delta,l}\gamma_{34;\Delta,l} G_{\Delta,l}(u_1,u_2)$$

Here the sum extends over all primary (= non-derivative) fields of scaling weight $\Delta$ and spin $l$, i.e. over fields that possess $l$ symmetric vector indices $\mu_1, \ldots, \mu_l$. Note that the coefficients $\gamma$ are part of the multiplication table, though they also include multiplication of scalar fields into fields that carry spin, unlike the coefficients we introduced in eq. (2.7). Generalizing the graphical notation we introduced in eq. (2.7) we can write such 3-point functions as

$$(2.9) \qquad \langle \phi_1(x_1)\phi_2(x_2)\Phi_3(x_3)\rangle = \gamma_{12;\Delta,l}$$

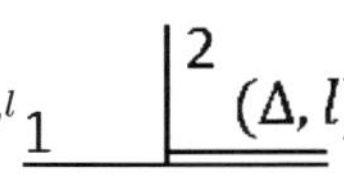


The Clebsch-Gordan coefficients for two scalar and one spinning field are certainly known explicitly. But since we won't need such detail, we shall content ourselves with the graphical notation. The conformal partial waves that appear in the expansion (2.8) of the scalar 4-point function can be considered as 4-point Clebsch-Gordan maps of the form

(2.10) $G_{\Delta,l} =$ 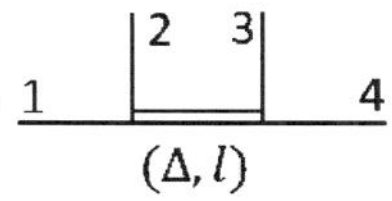 .



The graphical notation in this equation emphases the construction of the partial wave by contraction of two (3-point) Clebsch-Gordan coefficients. One can actually translate this into a precise integral formula for $G$,

$$G_{\Delta,l} \sim \int d^d x_0 x_{10}^{l+a-\Delta} x_{20}^{l-a-\Delta} x_{30}^{l-b+\Delta-d} x_{40}^{l_b+\Delta-d} (|\mu||\tilde{\mu}\,)^l\, Y_l^d \left(\frac{\mu\cdot\tilde{\mu}}{|\mu||\tilde{\mu}|}\right) \tag{2.11}$$

where

$$\mu = \frac{x_{10}}{x_{10}^2} - \frac{x_{20}}{x_{20}^2} \quad , \quad \tilde{\mu} = \frac{x_{40}}{x_{40}^2} - \frac{x_{30}}{x_{30}^2}, \tag{2.12}$$

the parameters $a, b$ are determined by the weights $\Delta_i$ as $2a = \Delta_2 - \Delta_1$, $2b = \Delta_3 - \Delta_4$ and $Y_l^d$ denotes zonal spherical functions. We have also used the notation $x^t = |x|^t$ whenever elements $x \in \mathbb{R}^d$ are raised to some power $t$. The integral over $x_0 \in \mathbb{R}^d$ takes care of the sum over intermediate states when we construct a 4-point invariant from two 3-point invariants. Since the relevant representations of the conformal group are infinite dimensional (see next section), the summation is most easily expressed as an integral over some insertion point $x_0$ of the intermediate field of weight $\Delta$ and spin $l$.

Before we discuss which special functions the integrals (2.11) integrate to we want to sketch one important application of conformal partial wave expansions that should suffice to illustrate their importance. The attentive reader might have noticed already that the expansion (2.8) breaks the symmetry of the correlation functions with respect to the permutation of the four scalar fields. The expansion we have displayed pairs the fields $\phi_1$ and $\phi_2$ together. This arbitrary choice is referred to as the choice of a *channel*. To match common convention, we shall use the term operator product expansion (OPE) channel from now on, though it would be more appropriate to call it a conformal partial wave (CPW) channel here. In any case, for $N = 4$ scalar fields there are to more channels, in addition to the one we have displayed, for which $\phi_1$ is paired with either $\phi_3$ or $\phi_4$. Note that the conformal partial waves $G$ do depend on the choice of the OPE channel. But after summing the partial waves with the coefficients $\gamma$, the symmetry between the four fields should be restored. This provides an important consistency condition on the coefficients $\gamma$ that is known as crossing symmetry condition. The conformal bootstrap [**61**] is designed to exploit these constraints in order to compute observables of the theory, such as the values of the scaling weights $\Delta$ or the coefficients $\gamma$. The program has had incredible success in $d = 2$ dimensions after the seminal paper of Belavin, Polyakov and Zamolodchikov [**3**]. During the last decade, the conformal bootstrap has seen an impressive revival in higher dimensions with new numerical as well as analytical incarnations, see [**18, 19, 60, 63, 64**] as well as [**1, 2, 25, 40–42**] for early contributions and [**58, 59, 70**] for reviews, though geared to physicists. This has produced many stunning new insights including record precision computations of critical exponents in the critical Ising model [**46**] through the numerical conformal bootstrap (nCB),

$$\Delta_\sigma^{[3D],\mathrm{nCB}} \sim 0.5181489(10) \ . \tag{2.13}$$

Note that the bootstrap analysis succeeded to increase the precision to which we know the scaling weight $\Delta_\sigma$ of the spin field of the Ising model in $d = 3$ dimensions by two orders of magnitude, as compared to all the clever numerical simulations that had been developed over a long time. This celebrated result has triggered a tremendous amount of new research that uses the bootstrap program to compute observables in (super)conformal field theories in higher dimensions.

The determination of $\Delta_\sigma$ for the $d = 3$ critical Ising model required very good knowledge of the conformal partial waves which need to be evaluated in a power series expansion around one special point $u_1^*, u_2^*$ in the space of cross ratios to very high orders. When conformal partial wave expansions were first introduced in [**23**, **24**] the partial waves remained elusive objects. The necessary mathematical results that made efficient evaluation of the partial waves possible were pretty much out of reach back then. Physicists wondered for more than three decades what kind of special functions conformal partial are in dimension $d > 2$.[2] They were interested to understand how partial waves depend on the cross ratios $u_i$, the parameters $\Delta, l$ of the 'intermediate field' and on the weights $\Delta_i$ of the external fields. When higher dimensional conformal field theories became a central research topic at the end of the last century, Dolan and Osborn found a way to address these questions. Their idea was to characterize the partial waves as eigenfunctions of Casimir differential operators [**16**]. By analysing these operators they were able to uncover many interesting properties and in particular to express conformal partial waves in even dimensions through sums of products of Gauss' hypergeometric functions [**16**, **17**]. After a number of relevant extensions [**35**, **44**, **45**, **57**], this line of research provided enough control over the conformal partial waves to implement a numerical version of Polyakov's bootstrap program.

The central observation of [**37**] implied later that all the results physicists had obtained on conformal partial waves for scalar 4-point functions must be contained in the seminal series of papers by Heckman and Opdam [**32**, **33**, **53**, **55**], see also [**30**, **56**]. From the vantage point of [**38**] it became clear that the conformal partial waves (2.11) are sums of Harish-Chandra functions associated with the root system $BC_2$ and very close relatives of the corresponding Heckman-Opdam hypergeometric functions. At this point we will leave the more physics oriented approach of the first section to explain the relation between conformal partial waves and hypergeometry in a more mathematical language that should be easier to follow than the original work on the subject.

## 3. Conformal partial waves and hypergeometry

**3.1. Group theoretic reformulation.** In order to understand the natural home of conformal partial waves in hypergeometry, we now want to start translating the physics ingredients we described above into a more mathematical language, mostly in the context of group theory. Recall that the symmetry group of a CFT is $G = SO(1, d+1)$ (or some covering thereof). The fields we worked with in the previous sections were inserted at points $x$ in the $d$-dimensional space. Since the action of conformal transformations is transitive, it is clear that we can think of space as a quotient of the conformal group by the stabilizer of a single point. The latter is the parabolic subgroup $P$ that is generated by dilations, rotations and

[2] In $d = 2$ dimensions it is easy to perform the integrals and to express them as a sum of products of Gauss' hypergeometric functions.

special conformal transformations. Dilations form a subgroup $SO(1,1) \subset G$ while rotations are elements of $SO(d) \subset G$. These act on $\mathbb{R}^d$ in the usual way and hence they act on the $d$-dimensional abelian subgroup $N \subset G$ that is generated by special conformal transformations. Altogether, the stabilizer $P$ of a single point takes the form $P = K \ltimes N$ where $K = SO(1,1) \times SO(d)$. Note that $K$ is not compact. From now on we shall think of a point $x$ as an element $x \in G/P$ in the quotient space of the conformal group $G$ and the parabolic subgroup $P$.

The next ingredient we need to transfer to mathematics is the notion of a field. Configurations of a scalar field $\phi$ of weight $\Delta$ are elements in the space $\Gamma_\Delta$ of sections of a line bundle over the quotient space $G/P$. More precisely, we define

$$\Gamma_\Delta = \Gamma_\Delta(G/P) = \{f : G \to \mathbb{C} \,|\, f(gp) = \chi_\Delta(p^{-1})f(g)\} \ . \tag{3.1}$$

The object $\chi_\Delta$ is a character of the parabolic subgroup $P \subset G$ that is trivial on the subgroups $N$ and $R = SO(d)$. On the generator $D$ of scale transformations, on the other hand, the character $\chi_\Delta$ assumes the value $\chi_\Delta(D) = \Delta$. Note that $D$ is an element of the conformal algebra $\mathfrak{g} = \mathfrak{so}(1, d+1)$ and we did not distinguish in notation between a character of the conformal group and its Lie algebra. Let us also stress that the linear spaces $\Gamma_\Delta$ of sections come equipped with the left regular action of the conformal group. For generic values of $\Delta$ the representation of $G$ on $\Gamma_\Delta$ is irreducible and infinite dimensional.

After this preparation we can now state that an $N$-point correlation function may be considered as a conformally invariant element in the tensor product of $N$ spaces $\Gamma_1, \ldots, \Gamma_N$,

$$\langle \varphi_1(x_1) \cdots \varphi_N(x_N) \rangle \in (\Gamma_1 \otimes \cdots \otimes \Gamma_N)^G \tag{3.2}$$

where the superscript $G$ is used to denote the subspace of $G$-invariant elements. To be more precise, the correlation function itself is not invariant, but there exists a map from the correlation function to the associated invariant. The latter is mediated by multiplication with a certain function $\Omega(x_i; \Delta_i)$ of the insertion points $x_i$. While $\Omega$ is not unique, there exists a canonical choice [**8**].

**3.2. Conformal partial waves as spherical functions.** For the moment let us focus on the case with $N = 4$ scalar field insertions. The space of 4-point conformal invariants can be evaluated in two steps. First we want to evaluate the tensor product of the tensor factors $\Gamma_1 = \Gamma_{\Delta_1}$ and $\Gamma_2 = \Gamma_{\Delta_2}$. The result we are going to spell out in a moment involves the following linear space

$$\Gamma_\Delta(G/K) \cong \{f : G \to \mathbb{C} \,|\, f(gk) = \chi_\Delta(k^{-1})f(g)\} \ . \tag{3.3}$$

The definition is similar to the construction in eq. (3.1), except that the right covariance condition only involves elements of the subgroup $K \subset P \subset G$ of the conformal group. In an abuse of notation we have denoted the character of this subgroup $K$ by the same symbol $\chi_\Delta$ as in the case of the parabolic subgroup $P$ above. Note that the two sets of characters are related by restriction since we assumed the character of $P$ to be trivial on the abelian subgroup $N \subset P$. With this notation we can now state [**15**]

$$\Gamma_i(G/P) \otimes \Gamma_j(G/P) \cong \Gamma_{\Delta_{ij}}(G/K) \tag{3.4}$$

where $\Delta_{ij} = \Delta_i - \Delta_j$. We can put this formula to use in the following analysis of conformal invariants in the 4-fold tensor product,

$$\begin{aligned}(\Gamma_{\Delta_1} \otimes \Gamma_{\Delta_2} \otimes \Gamma_{\Delta_3} \otimes \Gamma_{\Delta_4})^G &\cong (\Gamma_{\Delta_{12}}(G/K) \otimes \Gamma_{\Delta_{34}}(G/K))^G \\ &\cong \Gamma_{\Delta_{12},\Delta_{43}}(K\backslash G/K) .\end{aligned}$$

The final result is expressed in terms of the spaces

$$\Gamma_{\Delta,\Delta'}(K\backslash G/K) = \{f : G \to \mathbb{C} \,|\, f(k_l g k_r) = \chi_{\Delta_{21}}(k_l)\chi_{\Delta_{34}}(k_r^{-1})f(g)\} . \tag{3.5}$$

These consists of sections of a line bundle over the double coset space $G /\!\!/ K$. It is not difficult to see that the left action of $K$ in $G/K$ is transitive with stabilizer subgroup $B = SO(d-2)$. It follows that the double coset space has dimension

$$\dim(G /\!\!/ K) = \dim(G) - 2\dim(K) + \dim(B) = 2 .$$

The double quotient $G /\!\!/ K$ coincides with the space of conformally invariant cross ratios that can be build from $N = 4$ points in $\mathbb{R}^d$ for $d \geq 2$.

**3.3. Hypergeometry of conformal partial waves.** It is obvious that the Laplace operator on the conformal group $G$ descends to the space $\Gamma(K\backslash G/K)$. The associated radial Laplacian has been constructed and studied extensively in the mathematical literature [**28**, **34**]. In particular, it is well known [**51**, **52**] to be intimately related the trigonometric/hyperbolic Calogero-Moser-Sutherland Hamiltonian [**11**, **49**, **72**]

$$\begin{aligned} H_{\mathrm{CS}} = &-\sum_{i=1}^{2} \frac{\partial^2}{\partial u_i^2} + \frac{k_m(k_m-1)}{2}\left[\sinh^{-2}\left(\frac{u_1+u_2}{2}\right) + \sinh^{-2}\left(\frac{u_1-u_2}{2}\right)\right] \\ &+ \sum_{i=1}^{2}\left[k_l(k_l-1)\sinh^{-2}(u_i) + \frac{k_s(k_s+2k_l-1)}{4}\sinh^{-2}\left(\frac{u_i}{2}\right)\right] \end{aligned} \tag{3.6}$$

associated to the roots system $BC_2$ with multiplicity parameters given by

$$k_s = \Delta_{43} , \quad k_m = \frac{1}{2}(\Delta_{21} - \Delta_{43} + 1) , \quad k_l = \frac{1}{2}(d-2) \tag{3.7}$$

for the short, middle and long roots, respectively. These Hamiltonians have also been discussed in the talks of Margit Rösler and Eric Opdam at the meeting. The line of reasoning we have outlined here now allows to draw an important conclusion: In the case of scalar 4-point functions, the sought after conformal partial waves are eigenfunction of a Calogero-Moser-Sutherland Hamiltonian and hence they are closely related with Heckman-Opdam hypergeometric functions for the $BC_2$ root system. To be quite precise, the boundary conditions one has to impose in the case of conformal partial waves differ from those imposed in the work of Heckman and Opdam. The correct conditions are described in [**38**]. It turns out that conformal partial waves are a linear combination of four of the eight Harish-Chandra functions in the problem.

In spite of this minor difference, the relation with Heckman-Opdam hypergeometry can be used to deduce many important properties in conformal partial waves, see [**38**]. The position of poles and their residues can be obtained easily from those of the corresponding Harish-Chandra functions, see e.g. [**30**, **56**]. To obtain series expansions we realized the partial waves as the $q \mapsto 1^-$ limit of virtual Koornwinder polynomials [**62**] and employed the binomial expansions of Koornwinder polynomials from [**50**] to perform the degeneration with the help of [**43, 71**], see appendix A of [**38**] for details. Interesting shift equations for the various parameters of partial waves, finally, arise from the bispectral duality [**65**] between the trigonometric Calogero-Moser-Sutherland model with the rational Ruijsenaars-Schneider model for the same root system [**66**]. In this way, all properties of conformal partial waves physicists needed to prepare for applications in the modern conformal bootstrap can be recovered with relatively little effort from results mathematicians had obtained at the end of the last century.

The relation between conformal partial waves and Calogero-Moser-Sutherland models was first observed in [**37**] based on a direct analysis of the Casimir operators in the work of Dolan and Osborn. The direct group theoretical explanation I have outlined in the previous section has been described in [**68, 69**], with inspiration from earlier work in particular by Feher [**21**].

This group theoretic approach to conformal partial waves admits many relevant extensions. Among them is the extension to correlation functions of fields with spin, i.e. fields that transform non-trivially under the action of the rotation group $SO(d) \subset G$. In the general setup we described, this extension can be implemented by replacing the characters $\chi_\Delta$ of the parabolic subgroup $P$ or its subgroup $K = SO(1,1) \times SO(d)$ in eqs. (3.1) and (3.3), respectively, by some finite dimensional representations. The radial Laplacians on the associated vector bundles over the double coset $G \mathbin{/\!\!/} K$ have been worked out very explicitly [**68, 69**] for a few representations of the rotation group $SO(d)$ in $d = 3, 4$. A general theory of the Hamiltonians and their eigenfunctions remains to be developed, even though for applications in physics sufficient control has been achieved, see [**12, 39**] and further references therein.

Another very important extension is required to study superconformal field theories, i.e. quantum field theories in which the conformal Lie algebra $\mathfrak{g} = \mathfrak{so}(1, d+1)$ gets extended to a superconformal one. This is particularly important since most known conformal quantum field theories are actually supersymmetric, especially in dimension $d > 3$. For superconformal algebras of type I a general theory is obtained in [**9, 10**]. It reduces the theory of superconformal partial waves to that of ordinary conformal partial waves for fields with spin, see previous paragraph. The CFT literature contains few results in superconformal partial waves, except for very exceptional choices of field insertions, see e.g. [**9**] for references to the physics literature.

There is one more extension we want to briefly mention before we conclude this section. So far we have mostly discussed correlation functions of fields $\varphi(x)$ that are inserted at a point $x$ in the $d$-dimensional Euclidean space $x \in \mathbb{R}^d$ or its conformal compactification. But physics applications require to also consider operators that are localized along higher dimensional geometries such as lines, surfaces etc. One such example are the famous Wilson line operators in conformal gauge theories. Such non-local operators (fields) can be treated within the same framework, except that the stabilizer group $P$ of a point $x$ gets replaces by the stabilizer group $P_p = SO(1, p+1) \times SO(d-p) \ltimes \mathbb{R}^d$ of some $p$-dimensional submanifold. In this context, 2-point correlation functions of non-local fields can be treated pretty much in the same way as the 4-point function of local fields.[3] The theory of conformal partial waves for 2-point correlation functions of non-local fields in CFT has been developed systematically in [**36**]. Not surprisingly, it can be reduced to the study of the Laplace operator on double quotients of the form $K_p\backslash G/K_{p'}$ with $K_p = SO(1,p+1)\times SO(d-p)$. Therefore it is related to Heckman-Opdam hypergeometric functions for root systems $BC_n$ with the rank $n$ given by

$$n = min\,(p+2, p'+2, d-p, d-p') \ . \tag{3.8}$$

Note that the rank is $n = 2$ in case at least one of the non-local fields is inserted along a surface of dimension $p = 0$ and $d \geq 2$. Higher ranks appear e.g. for correlation functions of two line operators (i.e. $p = p' = 1$) in 4-dimensional conformal field theories etc. The theory of the associated conformal partial waves is relatively well developed, but this time only due to its relations with $BC_n$ Heckman-Opdam theory [**36**].

## 4. Integrability of multipoint conformal partial waves

Now that have described the link between conformal partial waves and hypergeometry at the example of scalar 4-point functions and have sketched a few extensions at the end of the previous section, I want to dedicate the final section to one challenging new direction that is only beginning develop. In the study of concrete conformal field theories physicists often compute correlation functions of more than $N = 4$ fields. And even though the conformal bootstrap in its original formulation is focused on 4-point functions, one could very well imagine an alternative program that is based multipoint partial waves. Note that associativity of an algebra can be studied either within the set of all products that one can form from four elements of some linear basis or, alternatively, by studying associativity for higher products of a set of generating elements in the algebra. Some first attempt to carry the conformal bootstrap into this direction was recently made in [**4**], though in a somewhat restricted setting.

From these short comments we can conclude that it is of significant interest to develop a theory of multipoint conformal partial waves for $N > 4$ insertion points. The following exposition of the subject in the context of integrability is based on several recent publications and work in progress [**5**–**7**]. There have also been other activities in physics to study multipoint partial waves, in particular directly through their integral representations. The most recent list of references to this research can be found in [**7**].

To set the stage, let us state the well-known fact that the number of independent conformal invariants one can build from $N$ points on $\mathbb{R}^d$ is given by

$$n_{cr}(N,d) = \begin{cases} \frac{1}{2}N(N-3) & N \leq d+2 \\ Nd - \frac{1}{2}(d+2)(d+1) & N > d+2 \end{cases} \tag{4.1}$$

Following the approach we have described above, a complete characterization of the associated conformal partial waves requires the same number of independent commuting differential operators. As in the case of the 4-point function, the set of differential operators depends on the choice of an OPE channel, see below for a precise definition.

[3] A non-local field that is inserted along a submanifold of dimension $p = 0$ may be considered as a 'compound' of two local fields.

**4.1. Gaudin integrable models.** The main idea for how to construct a complete set of commuting differential operators for each OPE channel of an N-point function exploits an interesting relation with Gaudin integrable systems. In order to make precise statements we need a bit of background on Gaudin models [**26**, **27**]. Let us begin with a central object, the so-called Lax matrix,

$$\mathcal{L}(w) = \sum_{i=1}^{N} \frac{\mathcal{T}_\alpha^{(i)} T^\alpha}{w - w_i} = \mathcal{L}_\alpha(w) T^\alpha \ . \tag{4.2}$$

Here, $w_i$ are a set of complex numbers, $T_\alpha$ denotes a basis of generators of the conformal Lie algebra in $d$ dimensions and $T^\alpha$ its dual basis with respect to an invariant bilinear form. The object $\mathcal{T}_\alpha^{(i)}$ denotes the set of right invariant vector fields on the space $\Gamma_{G/P}^{\Delta_i}$.[4]

Given some conformally invariant symmetric tensor $\kappa_p$ of degree $p$ one can construct a family $\mathcal{H}_p(w)$ of commuting operators as [**22**, **48**, **73**]

$$\mathcal{H}_p(w) = \kappa_p^{\alpha_1 \cdots \alpha_p} \mathcal{L}_{\alpha_1}(w) \cdots \mathcal{L}_{\alpha_p}(w) + \dots \ , \tag{4.3}$$

where the dots represent correction terms expressible as lower degree combinations of the Lax matrix components $\mathcal{L}_\alpha(w)$ and their derivatives with respect to $w$. For $p = 2$ such correction terms are absent. The correction terms are necessary to ensure that the families commute,

$$[\, \mathcal{H}_p(w) \, , \, \mathcal{H}_q(w') \,] = 0 \ , \tag{4.4}$$

for all $p, q$ and all $w, w' \in \mathbb{C}$. It is a well-known fact that these families commute with the diagonal action of the conformal algebra, i.e.

$$[\, \mathcal{T}_\alpha \, , \, \mathcal{H}_p(w) \,] = 0 \quad \text{where } \mathcal{T}_\alpha = \sum_{i=1}^{N} \mathcal{T}_\alpha^{(i)} \ . \tag{4.5}$$

Hence the commuting families $\mathcal{H}_p(w)$ of operators descend to differential operators on functions of the $n_{\text{cr}}(N,d)$ conformally invariant cross ratios.

**4.2. OPE limits of the Gaudin model.** At this point, the commuting families $\mathcal{H}_p(w)$ still contain $N$ continuous parameters $w_i, i = 1, \dots, N$. Without loss of generality we can set three of these complex numbers to some specific values, e.g. $w_1 = 0, w_{N-1} = 1, w_N = \infty$ so that we remain with $N - 3$ complex parameters our Gaudin Hamiltonians depend on. The set of differential equations which characterize conformal partial waves, on the other hand, should only depend on the discrete choice of an OPE channel. As we shall see now, any OPE channel determines a certain limit of the parameters $w_i$. In other words, the parameters $w_i$ are entirely fixed by the channel.

In order to make a precise statement, we need a bit of preparation. Let us pick some OPE channel $\mathcal{C}$ along with an (arbitrary) external edge in the diagram. The latter serves as a reference point and we shall assign it the label $N$. As this edge is external, it is attached to a unique vertex which we will denote by $\rho_*$. Such a choice of reference vertex defines a so-called rooted tree representation of the channel, see Figure 1 for a concrete example with $N = 8$. We draw the OPE diagram on a plane, with the vertex $\rho_*$ situated at the top and with each vertex having two downward edges attached. The representation on a plane forces us to

[4]Our discussion is tailored to the study of N-point functions of scalar fields. To accomodate tensor fields one needs to extend to more general flag manifolds.

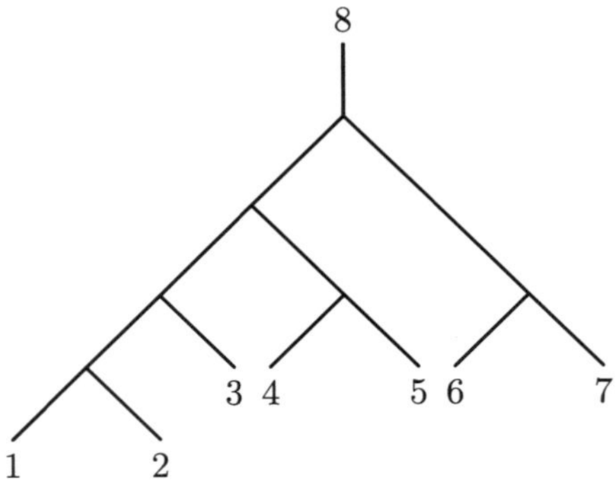


FIGURE 1. Plane rooted tree representation of an OPE diagram with 8 external fields.

make a choice of which edges are pointing towards the left and which edges are pointing towards the right: this choice is arbitrary, and gives rise to what is called a plane (or ordered) representation of the underlying rooted tree.

Obviously, each vertex $\rho \in V$ from the set $V$ of $N-2$ vertices of the OPE diagram defines a partition $\underline{N} = I_{\rho,1} \uplus I_{\rho,2} \uplus I_{\rho,3}$, with the sets $I_{\rho,j}$ formed by the labels of the external fields attached to the three branches of the vertex. We shall fix the labelling of the three branches using the plane rooted tree representation of the channel: choose the branch $I_{\rho,1}$ to be the one pointing to the bottom left while the branch $I_{\rho,2}$ is pointing the bottom right. By construction, the last branch $I_{\rho,3}$ then always points to the top and contains the reference point $N$. Let us also note that the vertices $\rho \neq \rho^*$ are associated with a unique internal edge, namely the one that points upward. Thereby the plane rooted tree representation allows us to identify the set $E$ of internal edges with $E \equiv V \setminus \{\rho^*\}$.

So far, vertices were just enumerated by an element $\rho \in V$. Once we have drawn the diagram in a plane, we can assign a sequence $s^\rho = (s_1^\rho, s_2^\rho, \dots, s_{\delta_\rho}^\rho)$ of elements $s_a^\rho \in \{1,2\}$ to each vertex $\rho$. This sequence $s^\rho$ encodes the path from $\rho_*$ to $\rho$. It tells us whether we have to move to the left (for $s_a^\rho = 1$) or right (for $s_a^\rho = 2$) every time we reach a new vertex until we arrive at $\rho$ after $\delta_\rho$ steps. We shall also refer to the length $\delta_\rho$ of the sequence as the depth of the vertex and to $s^\rho$ as the *binary sequence* of $\rho$. Note that the top vertex $\rho_*$ has depth $\delta_{\rho_*} = 0$.

In order to construct the limit of the Gaudin model associated with the OPE channel $\mathcal{C}$, we need now construct a polynomial $g_\rho(\varpi)$ for each vertex $\rho$. If $s^\rho$ is the binary sequence associated with the vertex $\rho$, the polynomial $g_\rho$ is defined as

$$g_\rho(\varpi) = \sum_{a=1}^{\delta_\rho} \varpi^{a-1} \delta_{s_a^\rho,2} \ . \tag{4.6}$$

Obviously the top vertex $\rho_*$ is assigned to $g_{\rho_*}(\varpi) = 0$. The vertices of depth $\delta_\rho = 1$ are associated with $g_{\rho_1}(\varpi) = 0$ or $g_{\rho_2}(\varpi) = 1$, depending on whether they are reached from $\rho_*$ by going down to the left ($\rho_1$) or to the right ($\rho_2$) etc.

Similarly, we can assign polynomials $f_i(\varpi)$ to each external edge $1 \leq i < N$ at the bottom of the plane rooted tree. Once again, we can encode the path from $\rho_*$ down to the edge $i$ by a binary sequence $s^i = (s_1^i, s_2^i, \dots, s_{\delta_i}^i)$. The length $\delta_i$ of the sequence $s^i$ is also referred to as the depth of the edge $i$. Now we introduce

$$f_i(\varpi) = \sum_{a=1}^{\delta_i} \varpi^{a-1} \delta_{s^i_a,2} + \varpi^{\delta_i} \delta_{s^i_{\delta_i},1} \ . \tag{4.7}$$

and set

$$f_N(\varpi) = \varpi^{-1} \tag{4.8}$$

for the external edge of the reference field at the top of the plane rooted tree. We have thus set up all the necessary notation that is needed to construct the relevant scaling limits of the $N$-site Gaudin model that remove the dependence on the continuous parameters $w_i$

The main idea in defining the channel dependent limit is to set $w_i = f_i(\varpi)$ and to send $\varpi$ to zero. When we perform this limit of the objects $\mathcal{H}^{(p)}(w; w_i)$ of the $N$-site Gaudin model we must adjust the spectral parameter $w$ and multiply with some appropriate powers of $\varpi$ in order to obtain sufficiently many independent commuting operator. More precisely, for given a vertex $\rho$ we introduce

$$\mathcal{H}^{(p)}_\rho(w) = \lim_{\varpi \to 0} \varpi^{pn_\rho} \mathcal{H}^{(p)}\left(\varpi^{n_\rho} w + g_\rho(\varpi), w_i = f_i(\varpi)\right) \ . \tag{4.9}$$

It can be shown that the set of these objects with $\rho \in E$ indeed contain $n_{\mathrm{cr}}(N,d)$ independent operators on the space of conformal invariants. These include operators of the form

$$\mathcal{C}^{(p)}_\rho = \lim_{w \to \infty} w^p \mathcal{H}^{(p)}_\rho(w) \quad for \quad \rho \neq \rho^* \ . \tag{4.10}$$

These measure the weight and spin of the field which is exchanged along the internal edge $\rho \in E$ above the vertex $\rho \neq \rho^*$. We refer to the operators $\mathcal{C}^{(p)}_\rho$ as Casimir operators of the edge $\rho$. For a given edge $\rho \neq \rho^*$ there can be relations between operators with different values of $p$. Taking these into account, the internal edge $\rho \neq \rho^*$ provides us with

$$n_{\mathrm{Cas}}(\rho, d) := \min\left(|I_{\rho,3}|, N - |I_{\rho,3}|, \mathrm{rank}_d\right) =: \mathfrak{d}_\rho \tag{4.11}$$

independent operators. Here $|I|$ denotes the order of the subset $I \subset \underline{N}$ and $\mathrm{rank}_d$ is the rank of the $d$-dimensional conformal algebra. It is easy to see that for $N > 4$, the integers $n_{\mathrm{Cas}}$ do not add up to $n_{\mathrm{cr}}$. This means that the Casimir operators do not suffice to characterize multipoint conformal partial waves. The missing operators can all be constructed from the objects (4.9). It turns out that given a vertex $\rho$ and an index $p$ one can prepare

$$n^{(p)}_{\mathrm{vert}}(\rho, d) = \max\left[\left((p-2) - \sum_{e \in \overline{E}_\rho}^{3} \Theta_0(p - 2\mathfrak{d}_e)(p - 2\mathfrak{d}_e - 1)\right), 0\right] , \tag{4.12}$$

additional independent *vertex differential operators.* Here $\Theta_0$ is the Heaviside step function with $\Theta_0(0) = 0$ and $\overline{E}_\rho$ consists of all external and internal edges that have one endpoint at the vertex $\rho$. For internal edges $e = \rho'$, the quantity $\mathfrak{d}_e$ was introduced as a shorthand in eq. (4.11) and for external edges we set $\mathfrak{d}_e = 1$. The formula applies for all $p$ as long as $d$ is odd. It requires some adjustment in even $d$, see [**7**] for details and a derivation.

Given this counting of independent commuting differential operators it is now possible to check that

$$n_{\rm cr}(N,d) = \sum_{\rho\neq\rho^*} n_{\rm Cas}(\rho,d) + \sum_{\rho}\sum_{p} n^{(p)}_{\rm vert}(\rho,d)\ , \tag{4.13}$$

i.e. that the set of independent differential operators suffices to completely characterize the multipoint conformal partial waves.

The OPE limits we described here are generalizations of the so-called caterpillar or bending flow limits that have been considered in the mathematical literature, see e.g. [**13, 14, 67**]. The caterpillar limit corresponds to the special case of an OPE limit for which the OPE diagram has the shape of a comb, see Figure 2. Following our conventions in [**6**], we enumerate the vertices $\rho = [r]$ by an integer $r = 1, \dots, N-2$ that is related to its depth $\delta_\rho = \delta_{[r]}$ as $\delta_{[r]} = N-2-r$.

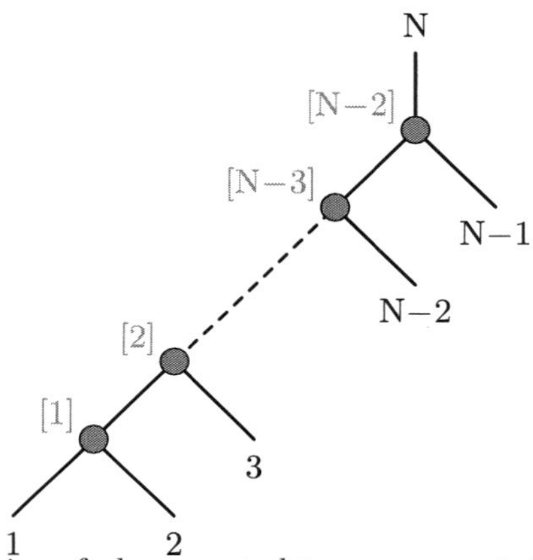


FIGURE 2. Choice of plane rooted tree representation of the comb channel OPE diagram with $N$ points.

One can compute the limit of the Gaudin model associated with this tree using the construction outlined in the previous subsection. For the polynomials $f_i$ that determine the parameters $w_i$ of the Gaudin model, one finds from eq. (4.7) that

$$w_i = f_i(\varpi) = \varpi^{N-1-i}, \qquad \forall i \in \{1, \cdots, N\}\ . \tag{4.14}$$

while all the vertex polynomials vanish, i.e. $g_{[r]} = 0$ for $r = 1, \dots, N-2$. Our general prescription (4.9) thus becomes

$$\mathcal{H}^{(p)}_{[r]}(w) = \lim_{\varpi\to 0} \varpi^{p(N-2-r)}\mathcal{H}^{(p)}\left(\varpi^{N-2-r}w; w_i = \varpi^{N-1-i}\right)\ . \tag{4.15}$$

In order to illustrate our results and also prepare for the discussion in the next subsection we want to take a closer look at the case of $N=6$ points for $d>4$. According to equation (4.11), the three internal edges $[r], r = 1,2,3$, in the comb channel diagram contribute

$$n_{\rm Cas}([1],d) = 2 = n_{\rm Cas}([3],d) \quad , \quad n_{\rm Cas}([2],d) = 3 \tag{4.16}$$

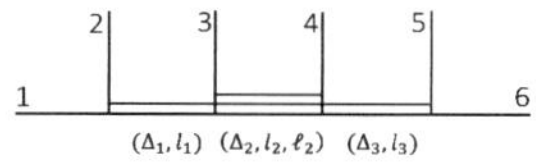


FIGURE 3. The two nontrivial vertices in the 6-point function are characterized by six quantum numbers each.

and hence a total number of 7 Casimir differential operators. Our result (4.12), on the other hand shows that the four vertices $[r], r = 1, \ldots, 4$, contribute

$$n^{(p)}_{\text{vert}}([1], d) = 0 = n^{(p)}_{\text{vert}}([4], d) \quad , \quad n^{(p)}_{\text{vert}}([2], d) = 1 = n^{(p)}_{\text{vert}}([3], d) \tag{4.17}$$

and hence a total of two vertex differential operators which are both of order four. When combined, the number of differential operators matches the number of $n_{\text{cr}} = 9$ cross ratios.

**4.3. Relation with lemniscatic elliptic CMS model.** The two inner vertices of the $N = 6$ point partial wave are among the simplest vertices that carry degrees of freedom, i.e. for which $n_{\text{vert}} \neq 0$. As one can infer from our general formula (4.12), it is in fact the unique vertex in $d > 4$ that gives rise to a single parameter. Let us focus on the vertex $\rho = [2]$ of the $N = 6$ point function, see our discussion at the end of the previous subsection, which connects the internal edges [1] and [2] with the external line $i = 3$. According to our list (4.16), the edge [2] admits two independent Casimir operators whose eigenvalues we parametrize through $\Delta_1$ and $l_1$. The internal edge [3], on the other side, admits three independent Casimir operators. Their eigenvalues can be parametrized by three parameters $\Delta_2, l_2, \ell_2$. The external scalar edge comes with one parameter, namely the scaling weight $\Delta_3$ of the external scalar field.[5] Let us note that for $d = 4$, the vertex which we consider here is the special case of a single variable vertex with one additional parameter $\ell_1 \neq 0$. On the other hand, if we specialize the non-trivial vertex of the $N = 6$ point function to $\ell_2 = 0$ it can be considered down to $d = 3$.

Our goal now is to identify the fourth order vertex differential operator for the single variable vertices. The easiest way to describe the findings of [**7**] involves a new algebra that is generated by operators $A, A^\dagger$ and $N$ which obey the following relations

$$[N, A^\dagger] = A^\dagger \quad , \quad [N, A] = -A\,, \tag{4.18}$$

$$AA^\dagger = R_\alpha(N+1) \quad , \quad A^\dagger A = R_\alpha(N) \tag{4.19}$$

where

$$R_\alpha(N) = \frac{N(N+2\alpha-1)}{(N+\alpha-1)(N+\alpha)}(N+\nu_1+2\alpha)(N-\nu_1-1)(N+\nu_2+2\alpha)(N-\nu_2-1)\,.$$

The algebra depends parametrically on the three spin parameters as $\alpha = \ell_2 + (d-3)/2$ and $\nu_k = l_k - \ell_2$, but is independent of the conformal weights $\Delta_i$. The family of algebras comes equipped with an involutive anti-automorphism $*$ defined by $N^* = N$ and $A^* = A^\dagger$, provided that the parameters obey

$$\bar{\alpha} = \alpha\,, \qquad \bar{\nu}_1 = -(2\alpha + 1 + \nu_1)\,, \qquad \bar{\nu}_2 = -(2\alpha + 1 + \nu_2)\,. \tag{4.20}$$

[5]This is different from the weight on the third internal edge which we also denoted by $\Delta_3$ in the previous subsection.

The vertex operator $H$ can now be written as a rational combination of the three generators as,

$$H^{(d;\Delta_i;l_i;\ell_i)} = B^\dagger B - \Gamma(N+\alpha)^2 + \frac{\alpha(\alpha-1)\,K}{(N+\alpha)^2-1} + E^{(d;\Delta_i;l_k;\ell_2)}. \tag{4.21}$$

Here, we defined the operators

$$B^\dagger := \frac{A-A^\dagger}{2\mathrm{i}} - \mathrm{i}(2\gamma_1\gamma_2 + \mathrm{i}\gamma_3)(N+\alpha)\,, \tag{4.22}$$

$$B := \frac{A-A^\dagger}{2\mathrm{i}} + \mathrm{i}(2\gamma_1\gamma_2 - \mathrm{i}\gamma_3)(N+\alpha)\,, \tag{4.23}$$

and the two parameters

$$\Gamma := \frac{1}{4}(1+4\gamma_1^2)(1+4\gamma_2^2)\,, \tag{4.24}$$

$$K := (\nu_1+\alpha)(\nu_2+\alpha)(\nu_1+\alpha+1)(\nu_2+\alpha+1)\,. \tag{4.25}$$

An expression for the constant term $E$ in our formula for the vertex operator (4.21) can be found in subsection 5.2 of [**5**]. The three parameters $\gamma_i$ that appear in the last few formulas are related to the conformal weights $\Delta_i$ of the three fields at the vertex as

$$\Delta_i = \frac{d}{2} + \mathrm{i}\gamma_i\ .$$

The operator $H$ is self-adjoint provided that the parameters $\gamma_i$ are real.

As one can easily check, the algebra generated by $A, A^\dagger$ and $N$ admits a representation in a basis $C_n^{(\alpha)}$ of the form

$$NC_n^{(\alpha)} := nC_n^{(\alpha)}, \tag{4.26}$$

$$AC_n^{(\alpha)} := (n+\nu_1+2\alpha)(n+\nu_2+2\alpha)\frac{n+2\alpha-1}{n+\alpha}C_{n-1}^{(\alpha)}\,, \tag{4.27}$$

$$A^\dagger C_n^{(\alpha)} := (n-\nu_1)(n-\nu_2)\frac{n+1}{n+\alpha}C_{n+1}^{(\alpha)}, \tag{4.28}$$

where eq. (4.27) applies to all integers $n > 0$, and $AC_0^{(\alpha)} = 0$ when $n = 0$, i.e. the state $C_0^{(\alpha)}$ is annihilated by the lowering operator $A$. Similarly, the action of the raising operator $A^\dagger$ vanishes if $n = \nu_1$ or $n = \nu_2$. Consequently, one can restrict the action of $A, A^\dagger$ and $N$ to the finite dimensional subspace that is spanned by $C_n$ for $n = 0, \ldots, \min(\nu_1, \nu_2)$. When the basis is realized as Gegenbauer polynomials $C^{(\alpha)}(s) = C^{(\alpha)}(1 - 2\mathcal{X})$, the operators $A, A^\dagger$ describe the action of second order differential operators in the variable $\mathcal{X}$. We then showed that the resulting fourth order differential operator $H = H_{\mathcal{X}}$ can be mapped to the Hamiltonian $L$ of a crystallographic elliptic Calogero-Moser-Sutherland model that was discovered originally by Etingof, Felder, Ma and Veselov [**20**]. The latter acts on functions in a variable $z$ on the lemniscatic elliptic curve. We found the variables $\mathcal{X}$ to be related to $z$ as

$$\mathcal{X} = \frac{\wp(1/2)^2}{\wp(1/2)^2 - \wp(z)^2}\ , \tag{4.29}$$

where $\wp$ is the Weierstrass elliptic function with lemniscatic periodicity conditions $\wp(z) = \wp(z+1) = \wp(z+i)$. Given this change of variables, our vertex Hamiltonian $H = H_{\mathcal{X}}$ and the elliptic CMS Hamiltonian $L = L_z$ of [**20**] are related as

$$\Theta^{-1} H_{\mathcal{X}} \Theta \sim L_z \tag{4.30}$$

where

$$\Theta \sim \mathcal{X}^{(l_1+l_2-2\ell_2+\Delta_3+(1-d)/2)/4}(1-\mathcal{X})^{(l_1+l_2-2\ell_2-\Delta_3+(1+d)/2)/4} . \tag{4.31}$$

The lemniscatic Calogero-Moser-Sutherland Hamiltonian $L$ involves seven parameters (multiplicities) whose values are determined by the three conformal weights $\Delta_i$, the spin parameters $l_1, l_2, \ell_2$ and the dimension $d$. Precise formulas can be found in [**7**]. A similar relation for a 3-point vertex with one scalar and two generic spinning legs of the $d = 4$ dimensional conformal group also holds. In that case, the dimension $d$ is fixed and the seven multiplicities of the lemniscatic model $L$ are obtained from the three conformal weights and four spin variables $l_1, \ell_1, l_2, \ell_2$, see [**7**] for precise formulas.

## 5. Concluding comments

This brings us to the end of this short note on higher dimensional CFT and its profound relation to multivariate hypergeometry. In the past, the interaction between these two fields has not lived up to its true potential. I hope that this text can make a modest contribution to developing a culture of mutual interest and understanding. It would clearly be very beneficial to join forces on overcoming some of the challenges that arise within CFT.

In my outline of concrete open problems I have focused on the study of multipoint partial waves, mostly because it is particularly timely and has seen some recent advances that are partly due to the very fruitful interactions I had during and after the workshop. Let me stress, however, that there are many other interesting problems that could profit from fresh mathematical input. One of these concerns the study of hypergeometric functions associated with (conformal) supergroups. I have only scratched this topic very briefly at the end of section 3. Given that most of the CFTs we know about are supersymmetric, progress in the theory superconformal partial waves is guaranteed to have many applications. While physicists have made some advances over the last years, very much remains to be done.

## Acknowledgment

I wish to thank my co-authors Ilija Buric, Misha Isachenkov, Sylvain Lacroix, Pedro Liendo, Yanick Linke, Jeremy Mann, Lorenzo Quintavalle, Evgeny Sobko for the inspiring collaboration on the material presented in this note. At the same time I am very grateful to the organizers of the workshop for giving me the opportunity to present our work and to Jasper for providing the occasion. I have profited very much from many insightful questions, comments and discussions with participants during and after the meeting.

## References

[1] Luis F. Alday, Agnese Bissi, and Tomasz Lukowski, *Large spin systematics in CFT*, JHEP **11** (2015), 101.

[2] Luis F. Alday and Alexander Zhiboedov, *Conformal bootstrap with slightly broken higher spin symmetry*, J. High Energy Phys. **6** (2016), 091, front matter+33, DOI 10.1007/JHEP06(2016)091. MR3538252

[3] A. A. Belavin, A. M. Polyakov, and A. B. Zamolodchikov, *Infinite conformal symmetry in two-dimensional quantum field theory*, Nuclear Phys. B **241** (1984), no. 2, 333–380, DOI 10.1016/0550-3213(84)90052-X. MR757857

[4] Carlos Bercini, Vasco Gonçalves, and Pedro Vieira, *Light-cone bootstrap of higher point functions and Wilson loop duality*, Phys. Rev. Lett. **126** (2021), no. 12, Paper No. 121603, 7, DOI 10.1103/physrevlett.126.121603. MR4238675

[5] Ilija Burić, Sylvain Lacroix, Jeremy A. Mann, Lorenzo Quintavalle, and Volker Schomerus, *Gaudin models and multipoint cnformal blocks: general theory*, J. High Energy Phys. **10** (2021), Paper No. 139, 46, DOI 10.1007/jhep10(2021)139. MR4339787

[6] Ilija Burić, Sylvain Lacroix, Jeremy A. Mann, Lorenzo Quintavalle, and Volker Schomerus, *From Gaudin integrable models to d-dimensional multipoint conformal blocks*, Phys. Rev. Lett. **126** (2021), no. 2, Paper No. 021602, 7, DOI 10.1103/physrevlett.126.021602. MR4207799

[7] Ilija Buríc, Sylvain Lacroix, Jeremy A. Mann, Lorenzo Quintavalle, and Volker Schomerus, *Gaudin Models and Multipoint Conformal Blocks II: Comb channel vertices in 3D and 4D*, (2021).

[8] Ilija Burić, Volker Schomerus, and Mikhail Isachenkov, *Conformal group theory of tensor structures*, J. High Energy Phys. **10** (2020), 004, 38, DOI 10.1007/jhep10(2020)004. MR4204101

[9] Ilija Burić, Volker Schomerus, and Evgeny Sobko, *Superconformal blocks: general theory*, J. High Energy Phys. **1** (2020), 159, 40, DOI 10.1007/jhep01(2020)159. MR4088160

[10] Ilija Burić, Volker Schomerus, and Evgeny Sobko, *The superconformal Xing equation*, J. High Energy Phys. **10** (2020), 147, 43, DOI 10.1007/jhep10(2020)147. MR4203958

[11] F. Calogero, *Solution of the one-dimensional $N$-body problems with quadratic and/or inversely quadratic pair potentials*, J. Mathematical Phys. **12** (1971), 419–436, DOI 10.1063/1.1665604. MR280103

[12] Alejandro Castedo Echeverri, Emtinan Elkhidir, Denis Karateev, and Marco Serone, *Seed conformal blocks in 4D CFT*, J. High Energy Phys. **2** (2016), 183, front matter+34, DOI 10.1007/jhep02(2016)183. MR3490242

[13] Alexander Chervov, Gregorio Falqui, and Leonid Rybnikov, *Limits of Gaudin algebras, quantization of bending flows, Jucys-Murphy elements and Gelfand-Tsetlin bases*, Lett. Math. Phys. **91** (2010), no. 2, 129–150, DOI 10.1007/s11005-010-0371-y. MR2586869

[14] Alexander Chervov, Gregorio Falqui, and Leonid Rybnikov, *Limits of Gaudin systems: classical and quantum cases*, SIGMA Symmetry Integrability Geom. Methods Appl. **5** (2009), Paper 029, 17, DOI 10.3842/SIGMA.2009.029. MR2506183

[15] V. K. Dobrev, G. Mack, V. B. Petkova, S. G. Petrova, and I. T. Todorov, *Harmonic Analysis on the n-Dimensional Lorentz Group and Its Application to Conformal Quantum Field Theory*, Lect. Notes Phys. **63** (1977), 1–280.

[16] F. A. Dolan and H. Osborn, *Conformal partial waves and the operator product expansion*, Nuclear Phys. B **678** (2004), no. 1-2, 491–507, DOI 10.1016/j.nuclphysb.2003.11.016. MR2022998

[17] F. A. Dolan and H. Osborn, *Conformal Partial Waves: Further Mathematical Results*, (2011).

[18] Sheer El-Showk, Miguel F. Paulos, David Poland, Slava Rychkov, David Simmons-Duffin, et al., *Solving the 3D Ising Model with the Conformal Bootstrap*, Phys.Rev. **D86** (2012), 025022.

[19] Sheer El-Showk, Miguel F. Paulos, David Poland, Slava Rychkov, David Simmons-Duffin, and Alessandro Vichi, *Solving the 3d Ising model with the conformal bootstrap II. c-minimization and precise critical exponents*, J. Stat. Phys. **157** (2014), no. 4-5, 869–914, DOI 10.1007/s10955-014-1042-7. MR3269694

[20] Pavel Etingof, Giovanni Felder, Xiaoguang Ma, and Alexander Veselov, *On elliptic Calogero-Moser systems for complex crystallographic reflection groups*, J. Algebra **329** (2011), 107–129, DOI 10.1016/j.jalgebra.2010.04.011. MR2769318

[21] L. Fehér and B. G. Pusztai, *Derivations of the trigonometric $BC_n$ Sutherland model by quantum Hamiltonian reduction*, Rev. Math. Phys. **22** (2010), no. 6, 699–732, DOI 10.1142/S0129055X10004065. MR2665762

[22] Boris Feigin, Edward Frenkel, and Nikolai Reshetikhin, *Gaudin model, Bethe ansatz and critical level*, Comm. Math. Phys. **166** (1994), no. 1, 27–62. MR1309540

[23] S. Ferrara, A. F. Grillo, G. Parisi, and R. Gatto, *The shadow operator formalism for conformal algebra. Vacuum expectation values and operator products*, Lett. Nuovo Cimento (2) **4** (1972), 115–120. MR356772

[24] S. Ferrara, A. F. Grillo, G. Parisi, and Raoul Gatto, *Covariant expansion of the conformal four-point function*, Nucl. Phys. **B49** (1972), 77–98, [Erratum: Nucl. Phys.B53,643(1973)].

[25] A. Liam Fitzpatrick, Jared Kaplan, David Poland, and David Simmons-Duffin, *The analytic bootstrap and AdS superhorizon locality*, J. High Energy Phys. **12** (2013), 004, front matter+34, DOI 10.1007/JHEP12(2013)004. MR3138193

[26] M. Gaudin, *Diagonalisation d'une classe d'Hamiltoniens de spin* (French, with English summary), J. Physique **37** (1976), no. 10, 1089–1098, DOI 10.1051/jphys:0197600370100108700. MR421442

[27] Michel Gaudin, *La fonction d'onde de Bethe* (French), Collection du Commissariat à l'Énergie Atomique: Série Scientifique. [Collection of the Atomic Energy Commission: Science Series], Masson, Paris, 1983. MR693905

[28] Harish-Chandra, *Collected papers*, Springer, 1984.

[29] Martin Hasenbusch, *Finite size scaling study of lattice models in the three-dimensional Ising universality class*, Phys. Rev. B **82** (2010), 174433.

[30] Gerrit Heckman and Henrik Schlichtkrull, *Harmonic analysis and special functions on symmetric spaces*, Perspectives in Mathematics, vol. 16, Academic Press, Inc., San Diego, CA, 1994. MR1313912

[31] G. J. Heckman, *Root systems and hypergeometric functions. I-II*, Compositio Mathematica **64** (1987), no. 3, 329–373 and 353–373.

[32] G. J. Heckman, *Root systems and hypergeometric functions. II*, Compositio Math. **64** (1987), no. 3, 353–373. MR918417

[33] G. J. Heckman and E. M. Opdam, *Root systems and hypergeometric functions. I*, Compositio Math. **64** (1987), no. 3, 329–352. MR918416

[34] Sigurdur Helgason, *Differential geometry, Lie groups, and symmetric spaces*, Pure and Applied Mathematics, vol. 80, Academic Press, Inc. [Harcourt Brace Jovanovich, Publishers], New York-London, 1978. MR514561

[35] Matthijs Hogervorst and Slava Rychkov, *Radial Coordinates for Conformal Blocks*, Phys. Rev. **D87** (2013), 106004.

[36] Mikhail Isachenkov, Pedro Liendo, Yannick Linke, and Volker Schomerus, *Calogero-Sutherland approach to defect blocks*, J. High Energy Phys. **10** (2018), 204, front matter+43, DOI 10.1007/jhep10(2018)204. MR3879680

[37] Mikhail Isachenkov and Volker Schomerus, *Superintegrability of d-dimensional conformal blocks*, Phys. Rev. Lett. **117** (2016), no. 7, 071602, 5, DOI 10.1103/PhysRevLett.117.071602. MR3630927

[38] Mikhail Isachenkov and Volker Schomerus, *Integrability of conformal blocks. Part I. Calogero-Sutherland scattering theory*, J. High Energy Phys. **7** (2018), 180, front matter + 65, DOI 10.1007/jhep07(2018)180. MR3858543

[39] Denis Karateev, Petr Kravchuk, and David Simmons-Duffin, *Weight shifting operators and conformal blocks*, J. High Energy Phys. **2** (2018), 081, front matter+80, DOI 10.1007/jhep02(2018)081. MR3789589

[40] Apratim Kaviraj, Kallol Sen, and Aninda Sinha, *Analytic bootstrap at large spin*, J. High Energy Phys. **11** (2015), 083, front matter+32, DOI 10.1007/JHEP11(2015)083. MR3455012

[41] Apratim Kaviraj, Kallol Sen, and Aninda Sinha, *Universal anomalous dimensions at large spin and large twist*, JHEP **07** (2015), 026.

[42] Zohar Komargodski and Alexander Zhiboedov, *Convexity and Liberation at Large Spin*, JHEP **1311** (2013), 140.

[43] Tom H. Koornwinder, *Okounkov's BC-type interpolation MacDonald polynomials and their $q = 1$ limit*, Sém. Lothar. Combin. **72** (2014/15), Art. B72a, 27. MR3383151

[44] Filip Kos, David Poland, and David Simmons-Duffin, *Bootstrapping Mixed Correlators in the 3D Ising Model*, JHEP **11** (2014), 109.

[45] Filip Kos, David Poland, and David Simmons-Duffin, *Bootstrapping the $O(N)$ vector models*, JHEP **06** (2014), 091.

[46] Filip Kos, David Poland, David Simmons-Duffin, and Alessandro Vichi, *Precision islands in the Ising and $O(N)$ models*, J. High Energy Phys. **8** (2016), 036, front matter+15, DOI 10.1007/JHEP08(2016)036. MR3555601

[47] Juan Maldacena, *The large-N limit of superconformal field theories and supergravity*, Internat. J. Theoret. Phys. **38** (1999), no. 4, 1113–1133, DOI 10.1023/A:1026654312961. Quantum gravity in the southern cone (Bariloche, 1998). MR1705508

[48] A. I. Molev, *Feigin-Frenkel center in types $B$, $C$ and $D$*, Invent. Math. **191** (2013), no. 1, 1–34, DOI 10.1007/s00222-012-0390-7. MR3004777

[49] J. Moser, *Three integrable Hamiltonian systems connected with isospectral deformations*, Advances in Math. **16** (1975), 197–220, DOI 10.1016/0001-8708(75)90151-6. MR375869

[50] A. Okounkov, *BC-type interpolation Macdonald polynomials and binomial formula for Koornwinder polynomials*, Transform. Groups **3** (1998), no. 2, 181–207, DOI 10.1007/BF01236432. MR1628453

[51] M. A. Olshanetsky and A. M. Perelomov, *Classical integrable finite-dimensional systems related to Lie algebras*, Phys. Rep. **71** (1981), no. 5, 313–400, DOI 10.1016/0370-1573(81)90023-5. MR615898

[52] M. A. Olshanetsky and A. M. Perelomov, *Quantum integrable systems related to Lie algebras*, Phys. Rep. **94** (1983), no. 6, 313–404, DOI 10.1016/0370-1573(83)90018-2. MR708017

[53] E. M. Opdam, *Root systems and hypergeometric functions iii*, Compositio Mathematica **67** (1988), no. 1, 21–49 (eng).

[54] Opdam, E. M., *Root systems and hypergeometric functions III-IV*, Compositio Mathematica **67** (1988), no. 1-2, 21–49 and 191–209.

[55] E. M. Opdam, *Root systems and hypergeometric functions. IV*, Compositio Math. **67** (1988), no. 2, 191–209. MR951750

[56] Eric M. Opdam, *Lecture notes on Dunkl operators for real and complex reflection groups*, MSJ Memoirs, vol. 8, Mathematical Society of Japan, Tokyo, 2000. With a preface by Toshio Oshima. MR1805058

[57] João Penedones, Emilio Trevisani, and Masahito Yamazaki, *Recursion relations for conformal blocks*, J. High Energy Phys. **9** (2016), 070, front matter+49, DOI 10.1007/JHEP09(2016)070. MR3557973

[58] David Poland, Slava Rychkov, and Alessandro Vichi, *The conformal bootstrap: theory, numerical techniques, and applications*, Rev. Modern Phys. **91** (2019), no. 1, 015002, 74, DOI 10.1103/RevModPhys.91.015002. MR3942977

[59] David Poland and David Simmons-Duffin, *The conformal bootstrap*, Nature Phys. **12** (2016), no. 6, 535–539.

[60] David Poland, David Simmons-Duffin, and Alessandro Vichi, *Carving Out the Space of 4D CFTs*, JHEP **05** (2012), 110.

[61] A. M. Polyakov, *Non-Hamiltonian approach to conformal quantum field theory* (Russian, with English summary), Ž. Èksper. Teoret. Fiz. **66** (1974), 23–42; English transl., Soviet Physics JETP **39** (1974), 10–18. MR0395598

[62] Eric M. Rains, *$\mathrm{BC}_n$-symmetric polynomials*, Transform. Groups **10** (2005), no. 1, 63–132, DOI 10.1007/s00031-005-1003-y. MR2127341

[63] Riccardo Rattazzi, Slava Rychkov, and Alessandro Vichi, *Bounds in 4D conformal field theories with global symmetry*, J. Phys. A **44** (2011), no. 3, 035402, 24, DOI 10.1088/1751-8113/44/3/035402. MR2749080

[64] Riccardo Rattazzi, Vyacheslav S. Rychkov, Erik Tonni, and Alessandro Vichi, *Bounding scalar operator dimensions in 4D CFT*, J. High Energy Phys. **12** (2008), 031, 49, DOI 10.1088/1126-6708/2008/12/031. MR2469909

[65] S. N. M. Ruijsenaars, *Action-angle maps and scattering theory for some finite-dimensional integrable systems. I. The pure soliton case*, Comm. Math. Phys. **115** (1988), no. 1, 127–165. MR929148

[66] S. N. M. Ruijsenaars and H. Schneider, *A new class of integrable systems and its relation to solitons*, Ann. Physics **170** (1986), no. 2, 370–405, DOI 10.1016/0003-4916(86)90097-7. MR851627

[67] Leonid Rybnikov, *Cactus group and monodromy of Bethe vectors*, Int. Math. Res. Not. IMRN **1** (2018), 202–235, DOI 10.1093/imrn/rnw259. MR3801430

[68] Volker Schomerus and Evgeny Sobko, *From spinning conformal blocks to matrix Calogero-Sutherland models*, J. High Energy Phys. **4** (2018), 052, front matter+28, DOI 10.1007/jhep04(2018)052. MR3801209

[69] Volker Schomerus, Evgeny Sobko, and Mikhail Isachenkov, *Harmony of spinning conformal blocks*, J. High Energy Phys. **3** (2017), 085, front matter+22, DOI 10.1007/JHEP03(2017)085. MR3650732

[70] David Simmons-Duffin, *The Conformal Bootstrap*, Theoretical Advanced Study Institute in Elementary Particle Physics: New Frontiers in Fields and Strings, 2 2016.

[71] Jasper V. Stokman and Tom H. Koornwinder, *Limit Transitions for BC Type Multivariable Orthogonal Polynomials*, 2005.
[72] Bill Sutherland, *Exact results for a quantum many body problem in one dimension. 2.*, Phys. Rev. **A5** (1972), 1372–1376.
[73] D. Talalaev, *Quantization of the Gaudin system*, (2004).
[74] Kenneth G. Wilson, *Renormalization group and critical phenomena. 1. Renormalization group and the Kadanoff scaling picture*, Phys. Rev. B **4** (1971), 3174–3183.

Deutsches Elektronen Synchroton DESY, Notkestrasse 85, 22607 Hamburg, Germany and Universität Hamburg, Luruper Chausee 149, 22761 Hamburg, Germany
*Email address*: volker.schomerus@desy.de

Contemporary Mathematics
Volume **780**, 2022
https://doi.org/10.1090/conm/780/15694

# Determinant of $\mathbb{F}_p$-hypergeometric solutions under ample reduction

Alexander Varchenko

ABSTRACT. We consider the KZ differential equations over $\mathbb{C}$ in the case, when the hypergeometric solutions are one-dimensional integrals. We also consider the same differential equations over a finite field $\mathbb{F}_p$. We study the polynomial solutions of these differential equations over $\mathbb{F}_p$, constructed in a previous work joint with V. Schechtman and called the $\mathbb{F}_p$-hypergeometric solutions.

The dimension of the space of $\mathbb{F}_p$-hypergeometric solutions depends on the prime number $p$. We say that the KZ equations have ample reduction for a prime $p$, if the dimension of the space of $\mathbb{F}_p$-hypergeometric solutions is maximal possible, that is, equal to the dimension of the space of solutions of the corresponding KZ equations over $\mathbb{C}$. Under the assumption of ample reduction, we prove a determinant formula for the matrix of coordinates of basis $\mathbb{F}_p$-hypergeometric solutions. The formula is analogous to the corresponding formula for the determinant of the matrix of coordinates of basis complex hypergeometric solutions, in which binomials $(z_i - z_j)^{M_i+M_j}$ are replaced with $(z_i - z_j)^{M_i+M_j-p}$ and the Euler gamma function $\Gamma(x)$ is replaced with a suitable $\mathbb{F}_p$-analog $\Gamma_{\mathbb{F}_p}(x)$ defined on $\mathbb{F}_p$.

CONTENTS



2020 *Mathematics Subject Classification.* Primary 13A35; Secondary 33C60, 32G20.

*Key words and phrases.* KZ equations, reduction to characteristic $p$, $\mathbb{F}_p$-hypergeometric solutions.

The author was supported in part by NSF grant DMS-1954266.

## 1. Introduction

The KZ equations were introduced in [**KZ**] as the differential equations satisfied by conformal blocks on sphere in the Wess-Zumino-Witten model of conformal field theory. The hypergeometric solutions of the KZ equations were constructed more than 30 years ago, see [**SV1**, **SV2**]. The KZ equations and the hypergeometric solutions are related to many subjects in algebra, representation theory, theory of integrable systems, enumerative geometry, to name a few.

The polynomial solutions of the KZ equations over the finite field $\mathbb{F}_p$ with a prime number $p$ of elements were constructed recently in [**SV3**], see also [**V4**, **V5**, **V6**, **V7**]. We call these solutions the *$\mathbb{F}_p$-hypergeometric solutions*. The general problem is to understand relations between the hypergeometric solutions of the KZ equations over $\mathbb{C}$ and the $\mathbb{F}_p$-hypergeometric solutions and observe how the remarkable properties of hypergeometric solutions are reflected in the properties of the $\mathbb{F}_p$-hypergeometric solutions. This program is in the first stages, where we consider essential examples and study the corresponding $\mathbb{F}_p$-hypergeometric solutions by direct methods.

In this paper we consider the KZ differential equations in the case, when the hypergeometric solutions over $\mathbb{C}$ are one-dimensional integrals.

The dimension of the space of $\mathbb{F}_p$-hypergeometric solutions depends on the prime number $p$. We say that the KZ equations have ample reduction for a prime $p$, if the dimension of the space of $\mathbb{F}_p$-hypergeometric solutions is maximal possible, that is, equal to the dimension of the space of solutions of the corresponding KZ equations over $\mathbb{C}$. Under the assumption of ample reduction, we prove a determinant formula for the matrix of coordinates of basis $\mathbb{F}_p$-hypergeometric solutions. The formula is analogous to the corresponding formula for the determinant of the matrix of coordinates of basis complex hypergeometric solutions, see [**V1**], in which binomials $(z_i - z_j)^{M_i+M_j}$ are replaced with $(z_i - z_j)^{M_i+M_j-p}$ and the Euler gamma function $\Gamma(x)$ is replaced with a suitable $\mathbb{F}_p$-analog $\Gamma_{\mathbb{F}_p}(x)$ defined on $\mathbb{F}_p$.

In Section 2 we describe our KZ equations and their reduction modulo $p$. We define the hypergeometric solutions over $\mathbb{C}$ and $\mathbb{F}_p$-hypergeometric solutions. The ample reduction is defined in Section 2.5.

As mentioned earlier, the $\mathbb{F}_p$-hypergeometric solutions are polynomials. In Section 3 we give a formula for their coefficients.

In Section 4, we consider the particular case of our KZ equations, whose space of solutions over $\mathbb{C}$ is one-dimensional, with the basis solution given by the Euler beta integral. We describe the corresponding $\mathbb{F}_p$-hypergeometric solution, which we call the *$\mathbb{F}_p$-beta integral*.

In Section 5 we consider an arbitrary polynomial solution (not necessarily $\mathbb{F}_p$-hypergeometric) of our KZ equations over $\mathbb{F}_p$ and describe its leading term with respect to the lexicographical ordering of monomials, see Theorem 5.3. It turns out that the notion of leading term and the formula for the leading term in Theorem 5.3 are useful in studying polynomial solutions of the KZ equations over $\mathbb{F}_p$. The notion of leading term replaces, to some extend, the notion of initial condition, when the differential equations are over $\mathbb{F}_p$.

The module of $\mathbb{F}_p$-hypergeometric solutions has a natural basis. In Section 6 we describe the leading terms of the basis $\mathbb{F}_p$-hypergeometric solutions. Section 7 contains our main result, Theorem 7.2, describing the determinant of coordinates of the basis $\mathbb{F}_p$-hypergeometric solutions, under assumption of ample reduction.

In Section 2.6 we give an example of KZ equations and a prime $p$, such that the space of complex solutions is one-dimensional, the space of polynomial solutions over $\mathbb{F}_p$ is one-dimensional, and the KZ equations have no $\mathbb{F}_p$-hypergeometric solutions. In Section 8 we show that if the reduction of our KZ equations is ample for a prime $p$, then all polynomial solutions are $\mathbb{F}_p$-hypergeometric.

## 2. KZ equations

**2.1. Description of equations.** In this paper the numbers $p$, $q$ *are prime numbers, $n$ is a positive integer,* $p > n \geqslant 2$, $p > q$. We fix *a vector* $(m_1, \dots, m_n) \in \mathbb{Z}^n_{>0}$, *such that* $m_i < q$ *for all* $i = 1, \dots, n$, and study the system of equations for a column vector $I(z) = (I_1(z), \dots, I_n(z))$:

$$\frac{\partial I}{\partial z_i} = \frac{1}{q}\sum_{j\neq i}\frac{\Omega_{ij}}{z_i - z_j}I, \quad i = 1, \dots, n, \qquad m_1 I_1(z) + \dots + m_n I_n(z) = 0, \tag{2.1}$$

where $z = (z_1, \dots, z_n)$, the $n \times n$-matrices $\Omega_{ij}$ have the form:

$$\Omega_{ij} = \begin{pmatrix} & \overset{i}{\vdots} & & \overset{j}{\vdots} & \\ i\cdots & -m_j & \cdots & m_j & \cdots \\ & \vdots & & \vdots & \\ j\cdots & m_i & \cdots & -m_i & \cdots \\ & \vdots & & \vdots & \end{pmatrix}, \tag{2.2}$$

and all other entries are zero. This joint system of *differential and algebraic equations* is called the *system of KZ equations* in this paper.

REMARK. System of equations (2.1) is the system of standard KZ differential equations with parameter $q$, associated with the Lie algebra $\mathfrak{sl}_2$ and the subspace of singular vectors of weight $\sum_{i=1}^n m_i - 2$ of the tensor product $V_{m_1} \otimes \dots \otimes V_{m_n}$, where $V_{m_i}$ is the irreducible $m_i + 1$ dimensional $\mathfrak{sl}_2$-module, up to a gauge transformation, see this example in [**V3**, Section 1.1].

We consider system (2.1) over the field $\mathbb{C}$ and over the field $\mathbb{F}_p$ with $p$ elements.

**2.2. Solutions over $\mathbb{C}$.** Consider the *master function*

$$\Phi(t, z_1, \dots, z_n) = \prod_{a=1}^{n}(t - z_a)^{-m_a/q} \tag{2.3}$$

and the column $n$-vector of hypergeometric integrals

$$I^{(\gamma)}(z) = (I_1(z), \dots, I_n(z)), \tag{2.4}$$

where

$$I_j = \int \Phi(t, z_1, \dots, z_n)\,\frac{dt}{t - z_j}, \qquad j = 1, \dots, n. \tag{2.5}$$

The integrals $I_j$, $j = 1, \dots, n$, are over an element $\gamma$ of the first homology group of the algebraic curve with affine equation

$$y^q = (t - z_1)^{m_1} \dots (t - z_n)^{m_n}.$$

Starting from such $\gamma$, chosen for given values $\{z_1, \dots, z_n\}$, the vector $I^{(\gamma)}(z)$ can be analytically continued as a multivalued holomorphic function of $z$ to the complement in $\mathbb{C}^n$ of the union of the diagonal hyperplanes $z_i = z_j$, $i \neq j$.

The complex vector space of such integral solutions is the $n-1$-dimensional vector space of all solutions of system (2.1). See these statements in the example in [**V3**, Section 1.1], also in [**SliV1**], see also the determinant formula (7.1).

**2.3. $\mathbb{F}_p$-Integrals.** Let $P(x_1, \dots, x_k)$ be a polynomial with coefficients in an $\mathbb{F}_p$-module,

$$P(x_1, \dots, x_k) = \sum_{d_1, \dots, d_k} c_{d_1,\dots,d_k}\, x_1^{d_1} \dots x_k^{d_k}.$$

Let $(l_1, \dots, l_k) \in \mathbb{Z}^k_{>0}$. The coefficient $c_{l_1 p-1,\dots,l_k p-1}$ is called the *$\mathbb{F}_p$-integral over cycle* $[l_1, \dots, l_k]_p$ and denoted by

$$\int_{[l_1,\dots,l_k]_p} P(x_1, \dots, x_k)\, dx_1 \dots dx_k .$$

We have an analog of Stokes' Theorem:

$$\int_{[l_1,\dots,l_k]_p} \frac{\partial P}{\partial x_i}(x_1, \dots, x_k)\, dx_1 \dots dx_k \,=\, 0$$

for any $[l_1, \dots, l_k]_p$.

**2.4. Solutions over $\mathbb{F}_p$.** Polynomial solutions of system (2.1), considered over the field $\mathbb{F}_p$, were constructed in [**SV3**].

For $i = 1, \dots, n$, choose the least positive integers $M_i$ such that

$$M_i \equiv -\frac{m_i}{q} \qquad (\mathrm{mod}\ p)\,. \tag{2.6}$$

Let

$$\Phi_p(t, z, M) := \prod_{i=1}^{n} (t - z_i)^{M_i}, \tag{2.7}$$

$$P(t, z) := \Big( \frac{\Phi_p(t, z)}{t - z_1}, \dots, \frac{\Phi_p(t, z)}{t - z_n} \Big) \,=\, \sum_i P^i(z)\, t^i\,, \tag{2.8}$$

where $P(t, z)$ is considered as a column $n$-vector of polynomials in $t, z_1, \dots, z_n$ and $P^i(z)$ as column $n$-vectors of polynomials in $z_1, \dots, z_n$ with coefficients in $\mathbb{F}_p$.

For a positive integer $l$, denote

$$I^{[l]}(z) \,:=\, \int_{[l]_p} \Big( \frac{\Phi_p(t, z)}{t - z_1}, \dots, \frac{\Phi_p(t, z)}{t - z_n} \Big)\, dt. \tag{2.9}$$

THEOREM 2.1 ([**SV3**, Theorem 1.2]). *For any positive integer $l$, the vector of polynomials $I^{[l]}(z)$ is a solution of KZ system* (2.1).

The solutions $I^{[l]}(z)$ given by this construction are called the *$\mathbb{F}_p$-hypergeometric solutions* of equations (2.1).

REMARK. The polynomial $\Phi_p(t, z)$ is an $\mathbb{F}_p$-analog of the master function $\Phi(t, z)$. The polynomial $P(t, z)$ is an analog of the integrand of integral (2.4). The transformation $P(t, z) \to I^{[l]}(z)$ is an analog of the integral and the index $[l]_p$ is an analog of the integration cycle.

Denote $\mathbb{F}_p[z^p] := \mathbb{F}_p[z_1^p, \dots, z_n^p]$. The set of all polynomial solutions of system (2.1) with coefficients in $\mathbb{F}_p$ is a module over the ring $\mathbb{F}_p[z^p]$ since equations (2.1) are linear and $\frac{\partial z_i^p}{\partial z_j} = 0$ in $\mathbb{F}_p[z]$ for all $i, j$.

The $\mathbb{F}_p[z^p]$-module

$$\mathcal{M} = \Big\{ \sum_l c_l(z) I^{[l]}(z) \mid c_l(z) \in \mathbb{F}_p[z^p] \Big\}, \tag{2.10}$$

spanned by $\mathbb{F}_p$-hypergeometric solutions, is called the *module of $\mathbb{F}_p$-hypergeometric solutions.*

The range for the index $l$ is defined by the inequalities $0 < lp-1 \leqslant \sum_{i=1}^n M_i - 1$. Hence $l = 1, \dots, r$, where

$$r := \Big[ \sum_{i=1}^n M_i/p \Big], \tag{2.11}$$

the integer part of the number $\sum_{i=1}^n M_i/p$.

THEOREM 2.2. *The $\mathbb{F}_p$-hypergeometric solutions $I^{[l]}(z)$, $l = 1, \dots, r$, are linearly independent over $\mathbb{F}_p[z^p]$.*

PROOF. The proof coincides with the proof of [**V5**, Theorem 3.1], see also the proof of [**SliV1**, Theorem 3.2]. Other proofs see in [**V6**, Section 4.1] and in Section 6.3. □

LEMMA 2.3. *We have $r < n$.* □

**2.5. Ample reduction.** We say that system (2.1) has *ample reduction* for a prime $p$ if

$$\Big[ \sum_{i=1}^n M_i/p \Big] = n - 1\,, \tag{2.12}$$

that is, the rank $r = \Big[ \sum_{i=1}^n M_i/p \Big]$ of the module $\mathcal{M}$ of $\mathbb{F}_p$-hypergeometric solutions takes the possible maximum value $n-1$.

EXAMPLE 2.4. On the one hand, if $q > n$, $p = lq + q - 1$ for some $l \in \mathbb{Z}_{>0}$ and $m_i = 1$, $i = 1, \dots, n$. Then $M_i = ((q-1)p - 1)/q = p - l - 1$ and

$$\Big[ \sum_{i=1}^n \frac{M_i}{p} \Big] = \Big[ n \frac{(q-1)p - 1}{pq} \Big] = \Big[ n - \frac{n}{q} - \frac{n}{pq} \Big] = n - 1.$$

Hence under these assumptions system (2.1) has ample reduction.

On the one hand, if $q > n$, $p = lq+1$ for some $l \in \mathbb{Z}_{>0}$ and $m_i = 1$, $i = 1, \dots, n$. Then $M_i = (p-1)/q = l$ and

$$\Big[ \sum_{i=1}^n \frac{M_i}{p} \Big] = \Big[ n \frac{p-1}{pq} \Big] = 0.$$

Hence under these assumptions system (2.1) does not have ample reduction.

In this example and in general the ampleness depends on the residue of $p$ modulo $q$.

LEMMA 2.5. *If system (2.1) has ample reduction for a prime p, then for any $l = 1, \dots, n-1$ and any subset $I \subset \{1, \dots, n\}$ with $|I| = l$, we have*

$$(l-1)p < \sum_{i \in I} M_i < lp. \tag{2.13}$$

PROOF. The second inequality holds since $0 < M_i < p$ for any $i$. Assume that the first inequality is not true and $\sum_{i\in I} M_i \leqslant (l-1)p$ for some $l$ and $I$. Then

$$\sum_{i=1}^n M_i \leqslant (l-1)p + \sum_{i\in \bar I} M_i < (l-1)p + (n-l)p = (n-1)p,$$

where $\bar I$ is the complement of $I$. That contradicts to the ampleness of the reduction. □

**2.6. Example.** Let $n=2$, $p=3$, $q=2$, $m_1=m_2=1$. Then $M_1=M_2=1$. The KZ equations take the form

$$\frac{\partial I}{\partial z_1} = \frac{\Omega}{z_1 - z_2} I, \qquad \frac{\partial I}{\partial z_2} = \frac{\Omega}{z_2 - z_1} I, \qquad \Omega = \begin{pmatrix} 1 & -1 \\ -1 & 1 \end{pmatrix}.$$

The polynomial $(z_1 - z_2)^2 \begin{pmatrix} 1 \\ -1 \end{pmatrix}$ is a solution. At the same time $r = [(M_1+M_2)/p] = 0$ and there are no $\mathbb{F}_p$-hypergeometric solutions.

## 3. Coefficients of polynomials

**3.1. Coefficients of $\mathbb{F}_p$-hypergeometric solutions.** For $l = 1, \dots, r$, the coordinates of the column vector $I^{[l]}(z) = (I_1^{[l]}(z), \dots, I_1^{[l]}(z))$ are homogeneous polynomials in $z_1, \dots, z_n$ of degree

$$\delta_l := \sum_{j=1}^n M_j - lp. \tag{3.1}$$

Let

$$I^{[l]}(z) = \sum_{d_1,\dots,d_n} I^{[l]}_{d_1,\dots,d_n} z_1^{d_1} \dots z_n^{d_n}, \qquad I^{[l]}_{d_1,\dots,d_n} \in \mathbb{F}_p^n.$$

LEMMA 3.1. *The coefficient $I^{[l]}_{d_1,\dots,d_n} \in \mathbb{F}_p^n$ is nonzero if and only if*

$$\sum_{i=1}^n d_i = \delta_l, \quad \text{and} \quad d_i \leqslant M_i \quad \textit{for} \quad i = 1, \dots, n, \tag{3.2}$$

*moreover,*

$$I^{[l]}_{d_1,\dots,d_n} = (-1)^{\delta_l} \prod_{j=1}^n \binom{M_j}{d_j} \Big(1 - \frac{d_1}{M_1}, \dots, 1 - \frac{d_n}{M_n}\Big). \tag{3.3}$$

*If $(I^{[l]}_{d_1,\dots,d_n;1}, \dots, I^{[l]}_{d_1,\dots,d_n;n})$ are coordinates of $I^{[l]}_{d_1,\dots,d_n}$, then*

$$\sum_{i=1}^n m_i I^{[l]}_{d_1,\dots,d_n;i} = \sum_{i=1}^n M_i I^{[l]}_{d_1,\dots,d_n;i} = 0. \tag{3.4}$$

PROOF. The first statements follow from formulas (2.7) and (2.8). Formula (3.4) follows from formulas (3.3), (2.6). □

**3.2. Coefficients and singular vectors.** Consider the Lie algebra $\mathfrak{sl}_2$ over the field $\mathbb{F}_p$ with standard generators $e, f, h$ and relations $[e, f] = h$, $[h, e] = 2e$, $[h, f] = -2f$.

For $m \in \mathbb{Z}_{\geqslant 0}$, $m < p$, let $V_m$ be the irreducible $\mathfrak{sl}_2$-module over $\mathbb{F}_p$ with highest weight $m$, basis $f^j v_m$, $j = 0, \dots, m$, and standard $\mathfrak{sl}_2$-action.

Consider the $\mathfrak{sl}_2$-module $\otimes_{j=1}^n V_{m_j}$. For $i = 1, \dots, n$, let

$$f^{(i)}v := v_{m_1} \otimes \dots \otimes v_{m_{i-1}} \otimes f v_{m_i} \otimes v_{m_{i+1}} \otimes \dots \otimes v_{m_n} \in \otimes_{j=1}^n V_{m_j} . \tag{3.5}$$

Then

$$hf^{(i)}v = \Big(\sum_{j=1}^n m_j - 2\Big) f^{(i)}v , \qquad ef^{(i)}v = m_i \, v_{m_1} \otimes \dots \otimes v_{m_i} \otimes \dots \otimes v_{m_n} .$$

Denote by $V[-2]$ the $n$-dimensional subspace of $\otimes_{j=1}^n V_{m_j}$ generated by $f^{(i)}v$, $i = 1, \dots, n$. Denote

$$\mathrm{Sing} V[-2] = \Big\{ \sum_{i=1}^n c_i f^{(i)} v \ \Big| \ \sum_{i=1}^n c_i m_i = 0 \Big\} \subset V[-2] .$$

The $n-1$-dimensional subspace $\mathrm{Sing}V[-2] \subset V[-2]$ is the kernel of the restriction to $V[-2]$ of the operator $e : \otimes_{j=1}^n V_{m_j} \to \otimes_{j=1}^n V_{m_j}$ . Denote

$$\mathrm{Sing}V[-2][z] \ := \ \mathrm{Sing}V[-2] \otimes_{\mathbb{F}_p} \mathbb{F}_p[z] .$$

Define an isomorphism of vector spaces

$$\iota : \mathbb{F}_p^n \to V[-2], \qquad (c_1, \dots, c_n) \mapsto \sum_{i=1}^n c_i f^{(i)} v . \tag{3.6}$$

Then an $\mathbb{F}_p$-hypergeometric solution $I^{[l]}(z)$ is identified with the polynomial

$$\iota I^{[l]}(z) := \sum_{d_1, \dots, d_n} \iota I^{[l]}_{d_1, \dots, d_n} \, z_1^{d_1} \dots z_n^{d_n} \in \mathrm{Sing}V[-2][z] .$$

**3.3. Operators $\Omega_{ij}^{\mathfrak{sl}_2}$.** The isomorphism $\iota$ identifies a linear operator $\Omega_{ij} : \mathbb{F}_p^n \to \mathbb{F}_p^n$, appearing in system (2.1), with a linear operator on $V[-2]$, which we denote by $\iota\Omega_{ij}$. Namely, the linear operator $\iota\Omega_{ij}$ is the restriction to $V[-2]$ of the Casimir operator on $\otimes_{j=1}^n V_{m_j}$ defined by the formula

$$\Omega_{ij}^{\mathfrak{sl}_2} \ := \ \frac{1}{2} h^{(i)} h^{(j)} + e^{(i)} f^{(j)} + f^{(i)} e^{(j)} \ - \ \frac{m_i m_j}{2} \, \mathrm{Id} ,$$

where for $x \in \mathfrak{sl}_2$ we define the operator $x^{(i)}$ on $\otimes_{j=1}^n V_{m_j}$ by

$$x^{(i)} := 1 \otimes \dots \otimes 1 \otimes x \otimes 1 \otimes \dots \otimes 1$$

with $x$ at the $i$th position. Notice that each $\Omega_{ij}^{\mathfrak{sl}_2}$ preserves $\mathrm{Sing}V[-2]$.

## 4. $\mathbb{F}_p$-Beta integral and KZ equations for $n = 2$

**4.1. Solutions over $\mathbb{C}$.** Consider the system of KZ equations (2.1) over $\mathbb{C}$ for $n = 2$. Then the master function is

$$\Phi(t, z_1, z_2) = (t - z_1)^{-m_1/q} (t - z_2)^{-m_2/q} , \tag{4.1}$$

and the one-dimensional space of solutions is generated by the 2-column vector

$$I(z_1,z_2)=\int_{z_1}^{z_2}\Phi(t,z)\Big(\frac{f^{(1)}v}{t-z_1}+\frac{f^{(2)}v}{t-z_2}\Big)dt. \tag{4.2}$$

To determine this integral we assume that $z_1, z_2$ are real, $z_1 < z_2$, and fix a univalued branch on $[z_1, z_2]$ of each of the factors $(t-z_1)^{-m_1/q}$, $(t-z_2)^{-m_2/q}$. Then

(4.3)

$$I(z_1,z_2)$$
$$=(z_2-z_1)^{-m_1/q}(z_1-z_2)^{-m_2/q}\frac{\Gamma(-m_1/q+1)\Gamma(-m_1/q+1)}{\Gamma(-m_1/q-m_2/q+1)}\Big(\frac{f^{(1)}v}{-m_1/q}-\frac{f^{(2)}v}{-m_2/q}\Big),$$

where $(z_k-z_l)^{-m_l/q}$ is the value of the chosen branch of the function $(t-z_l)^{-m_l/q}$ at $t=z_k$.

REMARK. Calculation of each coordinate of the vector $I(z_1,z_2)$ is reduced to the beta integral,

$$\int_0^1 t^{\alpha-1}(1-t)^{\beta-1}dt=\frac{\Gamma(\alpha)\Gamma(\beta)}{\Gamma(\alpha+\beta)}\ , \tag{4.4}$$

after change of variables. Formula (4.3) shows how the beta integral appears in hypergeometric solutions of KZ equations.

If $(-m_1/q, -m_2/q)=(M_1,M_2)$, where $M_1, M_2$ are positive integers, then

$$I(z_1,z_2)=(z_2-z_1)^{M_1}(z_1-z_2)^{M_2}\frac{\Gamma(M_1+1)\,\Gamma(M_2+1)}{\Gamma(M_1+M_2+1)}\Big(\frac{f^{(1)}v}{M_1}-\frac{f^{(2)}v}{M_2}\Big). \tag{4.5}$$

**4.2. Factorial and gamma functions.** Recall that $p$ is an odd prime number. The $p$-adic factorial function is defined on positive integers by

$$(x!)_p=\prod_{1\leqslant j\leqslant x,\ (j,p)=1} j\ .$$

The Morita $p$-adic gamma function is the unique continuous function of a $p$-adic integer $x$ (with values in $\mathbb{Z}_p$) such that

$$\Gamma_p(x)=(-1)^x\prod_{1\leqslant j<x,\ (j,p)=1} j\,,$$

for positive integers $x$. Thus $\Gamma_p(x+1)=(-1)^x(x!)_p$ for positive integers $x$.

Define the function

$$\Gamma_{\mathbb{F}_p}:\mathbb{Z}_{\geqslant 0}\to\mathbb{F}_p$$

by setting $\Gamma_{\mathbb{F}_p}(0)=1$, $\Gamma_{\mathbb{F}_p}(1)=-1$ and mapping an integer $x>1$ to the image of the integer $\Gamma_p(x)$ in $\mathbb{F}_p$.

LEMMA 4.1. *We have* $\Gamma_{\mathbb{F}_p}(x+p)=\Gamma_{\mathbb{F}_p}(x)$ *for all* $x$.

PROOF. The lemma follows from Wilson's theorem, $(p-1)!\equiv -1 \pmod p$. □

We extend the function $\Gamma_{\mathbb{F}_p}$ to the set $\mathbb{Z}$ by periodicity, $\Gamma_{\mathbb{F}_p}(x+p)=\Gamma_{\mathbb{F}_p}(x)$. Then we get

$$\Gamma_{\mathbb{F}_p}(x)\,\Gamma_{\mathbb{F}_p}(1-x)\ =\ (-1)^x$$

also by Wilson's theorem.

LEMMA 4.2. *Let $A, B$ be positive integers such that $A < p$, $B < p$, $p \leqslant A + B$. Then we have an identity in $\mathbb{F}_p$,*

$$
\begin{aligned}
B\binom{B-1}{A+B-p} &= B\binom{B-1}{p-A-1} \\
&= (-1)^{A+1}\,\frac{A!\,B!}{(A+B-p)!} = (-1)^{A}\,\frac{\Gamma_{\mathbb{F}_p}(A+1)\Gamma_{\mathbb{F}_p}(B+1)}{\Gamma_{\mathbb{F}_p}(A+B-p+1)}\,.
\end{aligned}
\tag{4.6}
$$

□

PROOF. We have

$$
\begin{aligned}
B\binom{B-1}{p-A-1} &= \frac{B(B-1)\cdots(A+B-p+1)}{1\cdots(p-A-1)} \\
&= \frac{B\cdots(A+B-p+1)(A+B-p)!A!}{(-1)^{p-A-1}(p-1)(p-2)\cdots(A+1)\,A!(A+B-p)!} \\
&= (-1)^{A+1}\,\frac{A!\,B!}{(A+B-p)!} = (-1)^{A}\,\frac{\Gamma_{\mathbb{F}_p}(A+1)\,\Gamma_{\mathbb{F}_p}(B+1)}{\Gamma_{\mathbb{F}_p}(A+B-p+1)}\,.
\end{aligned}
$$

□

**4.3. $\mathbb{F}_p$-hypergeometric solutions.** Consider the system of KZ equations (2.1) over $\mathbb{F}_p$ for $n = 2$. Assume that system (2.1) has ample reduction for a prime $p$. Then the integers $M_1$, $M_2$, introduced in (2.6), satisfy the inequalities

$$
0 < M_1,\ M_2,\ M_1 + M_2 - p + 1 < p\,. \tag{4.7}
$$

In this case the module $\mathcal{M}$ of $\mathbb{F}_p$-hypergeometric solutions is of rank one and generated by $I^{[1]}(z_1, z_2)$. The solution $I^{[1]}(z_1, z_2)$ is the coefficient of $t^{p-1}$ in the Taylor expansion of the polynomial 2-vector

$$
P(t, z_1, z_2) = (t - z_1)^{M_1}(t - z_2)^{M_2}\Big(\frac{f^{(1)}v}{t - z_1} + \frac{f^{(2)}v}{t - z_2}\Big), \tag{4.8}
$$

cf. (4.2)

THEOREM 4.3. *We have*

$$
\begin{aligned}
I^{[1]}(z_1, z_2) &= (-1)^{M_2}(z_2 - z_1)^{M_1+M_2-p}\,\frac{\Gamma_{\mathbb{F}_p}(M_1{+}1)\,\Gamma_{\mathbb{F}_p}(M_2{+}1)}{\Gamma_{\mathbb{F}_p}(M_1 + M_2 - p + 1)}\Big(\frac{f^{(1)}v}{M_1} - \frac{f^{(2)}v}{M_2}\Big) \\
&= (-1)^{M_1}(z_1 - z_2)^{M_1+M_2-p}\,\frac{\Gamma_{\mathbb{F}_p}(M_1{+}1)\,\Gamma_{\mathbb{F}_p}(M_2{+}1)}{\Gamma_{\mathbb{F}_p}(M_1 + M_2 - p + 1)}\Big(\frac{f^{(2)}v}{M_2} - \frac{f^{(1)}v}{M_1}\Big),
\end{aligned}
\tag{4.9}
$$

*cf.* (4.5).

PROOF. Make the transformation

$$
P(t, z_1, z_2) \ \mapsto\ P(t + z_1, z_1, z_2) = t^{M_1}(t + z_1 - z_2)^{M_2}\Big(\frac{f^{(1)}v}{t} + \frac{f^{(2)}v}{t + z_1 - z_2}\Big).
$$

This change of variables does not change the coefficient of $t^{p-1}$ in the Taylor expansion by Lucas theorem, see [**Lu**] and the proof of [**V5**, Lemma 5.2]. Hence

$$
I^{[1]}(z_1, z_2) \ = \ (z_1 - z_2)^{M_1+M_2-p}\Big(\binom{M_2}{p - M_1} f^{(1)}v + \binom{M_2 - 1}{p - M_1 - 1} f^{(2)}v\Big).
$$

Then
(4.10)

$$\binom{M_2}{p-M_1} f^{(1)}v + \binom{M_2-1}{p-M_1-1} f^{(2)}v = M_2 \binom{M_2-1}{p-M_1-1} \Big(\frac{f^{(1)}v}{p-M_1} + \frac{f^{(2)}v}{M_2}\Big).$$

Now the theorem follows from Lemma 4.2. □

REMARK. For positive integers $a, b$ satisfying the inequalities $a < p$, $b < p$, $p-1 \leqslant a+b$, we have the $\mathbb{F}_p$-*beta integral formula*

$$\int_{[1]_p} t^a(1-t)^b dt \;=\; -\,\frac{a!\,b!}{(a+b-p+1)!}\,, \tag{4.11}$$

which follows from Lemma 4.2.

Formula (4.4) for the beta integral is the one-dimensional case of the $n$-dimensional Selberg integral formula, see [**Se**]. In [**RV1, RV2**] we develop $\mathbb{F}_p$-analogs of the $n$-dimensional Selberg integral formulas.

## 5. Leading term of a polynomial solution

**5.1. Lexicographical ordering.** For a permutation $\sigma = (\sigma_1, \dots, \sigma_n) \in S_n$ denote by $>_\sigma$ the *lexicographical ordering* of monomials

$$z_1^{d_1} \dots z_n^{d_n}\,, \qquad d_1, \dots, d_n \in \mathbb{Z}_{\geqslant 0}\,,$$

relative to the ordering $(\sigma_1, \dots, \sigma_n)$ of the integers $(1, \dots, n)$. So $z_{\sigma_1} >_\sigma z_{\sigma_2} >_\sigma \dots >_\sigma z_{\sigma_{n-1}} >_\sigma z_{\sigma_n}$ and so on.

For a nonzero polynomial

$$f(z) = \sum_{d_1,\dots,d_n} a_{d_1,\dots,d_n} z_1^{d_1} \dots z_n^{d_n}$$

let $f_\sigma(z)$ be the summand $a_{d_1,\dots,d_n} z_1^{d_1} \dots z_n^{d_n}$ corresponding to the largest monomial entering $f(z)$ with a nonzero coefficient. We call $f_\sigma(z)$ the $\sigma$-*leading term* of $f(z)$, the corresponding $a_{d_1,\dots,d_n}$ the $\sigma$-*leading coefficient*, the corresponding $z_1^{d_1} \dots z_n^{d_n}$ the $\sigma$-*leading monomial.*

In particular, consider elements $f(z) = (f_1(z), \dots, f_n(z)) \in (\mathbb{F}_p[z])^n$ as polynomials in $z$ with coefficients in $\mathbb{F}_p^n$. Then for any $\sigma \in S_n$ a nonzero element $f(z)$ has $\sigma$-leading term $f_\sigma(z) = a_{d_1,\dots,d_n} z_1^{d_1} \dots z_n^{d_n}$ and $\sigma$-leading coefficient $a_{d_1,\dots,d_n} \in \mathbb{F}_p^n$.

**5.2. Leading term of a polynomial solution.** Consider the lexicographical ordering $>_{\rm id}$ of monomials corresponding to the identity permutation id $\in S_n$. Hence $z_1 >_{\rm id} z_2 >_{\rm id} \dots >_{\rm id} z_n$ and so on.

LEMMA 5.1. *Let $I(z) = (I_1(z), \dots, I_n(z))$ be a polynomial solution of system (2.1) over $\mathbb{F}_p$ (not necessarily an $\mathbb{F}_p$-hypergeometric solution). Let $Cz_1^{d_1} \dots z_n^{d_n}$, $C = (C_1, \dots, C_n) \in \mathbb{F}_p^n$, be the* id*-leading term of $I(z)$. Then*
(5.1)

$$\sum_{j=1}^n m_j C_j = 0, \qquad \sum_{l=j+1}^n \Omega_{jl}\, C \;=\; q d_j\, C\,, \quad j = 1, \dots, n-1\,, \quad d_n \equiv 0 \;(\mathrm{mod}\; p)\,,$$

*where $C$ is considered as a column vector.*

PROOF. Rewrite the KZ equations as

$$\Big(\prod_{k\neq j}(z_j-z_k)\Big)\frac{\partial I}{\partial z_j}-\frac{1}{q}\sum_{l\neq j}\Big(\prod_{k\notin\{l,j\}}(z_j-z_k)\Big)\Omega_{jl}I=0, \tag{5.2}$$

$j=1,\dots,n$. Now the lemma follows from calculating the leading term of the left-hand side in (5.2) and equating it to zero. □

For $j\neq l$ introduce the $n\times n$-matrices

$$\Omega^M_{jl}=\begin{pmatrix} & \overset{j}{\vdots} & & \overset{l}{\vdots} & \\ j\cdots & M_l & \cdots & -M_l & \cdots \\ & \vdots & & \vdots & \\ l\cdots & -M_j & \cdots & M_j & \cdots \\ & \vdots & & \vdots & \end{pmatrix} \tag{5.3}$$

with all other entries equal to zero. For $j=1,\dots,n-1$, denote $\Omega^M_j=\sum_{l=j+1}^n\Omega^M_{jl}$.

COROLLARY 5.2. *Let $I(z)$ be a polynomial solution of system (2.1) over $\mathbb{F}_p$. Then the* id*-leading term $Cz_1^{d_1}\dots z_n^{d_n}$ of $I(z)$ satisfies the system of equations:*

$$\sum_{j=1}^n M_jC_j=0,\quad \Omega^M_jC=d_j\,C,\quad j=1,\dots,n-1,\quad d_n\equiv 0 \ (\mathrm{mod}\ p). \tag{5.4}$$

□

THEOREM 5.3. *Let a pair $C=(C_1,\dots,C_n)$, $(d_1,\dots,d_n)\in\mathbb{F}_p^n$ be a solution of system*

$$\sum_{j=1}^n M_jC_j=0,\quad \Omega^M_jC=d_j\,C,\quad j=1,\dots,n-1,\quad d_n=0. \tag{5.5}$$

*Let the index $i$ be such that*

$$C_i\neq 0\quad\text{and}\quad C_j=0,\quad j=1,\dots,i-1. \tag{5.6}$$

*Then*

$$\begin{aligned} d_j&=M_j\,, & j&=1,\dots,i-1\,,\\ d_i&=\sum_{j=i}^n M_j\,, & d_i&\neq M_i\,,\\ d_j&=0\,, & j&=i+1,\dots,n\,, \end{aligned} \tag{5.7}$$

$$\sum_{l=i+1}^n M_l\neq 0,\qquad C_j=-\frac{M_i}{\sum_{l=i+1}^n M_l}C_i\,,\qquad j=i+1,\dots,n. \tag{5.8}$$

*Conversely, if a pair $C=(C_1,\dots,C_n)$, $(d_1,\dots,d_n)\in\mathbb{F}_p^n$ has properties (5.6), (5.7), (5.8), then it is a solution of system (5.5).*

PROOF. For any $j=1,\dots,n-1$, we have

(5.9)

$$\begin{aligned}&\Omega^M_j(C_1,\dots,C_n)=(0,\dots,0,\\ &\qquad M_{j+1}(C_j-C_{j+1})+\dots+M_n(C_j-C_n),M_j(C_{j+1}-C_j),\dots,M_j(C_n-C_j)),\end{aligned}$$

where in the right-hand side 0 is repeated $j-1$-times,

Assume (5.5) and (5.6). First we check that $d_j = M_j$, $j = 1, \dots, i-1$. Indeed,

$$\Omega_j^M C = \Big(0, \dots, 0, -\sum_{l=j+1}^{n} M_l C_l,\ M_j C_{j+1}, \dots, M_j C_n\Big). \tag{5.10}$$

$$= \Big(0, \dots, 0, \quad 0 \quad, M_j C_{j+1}, \dots, M_j C_n\Big) = M_j\, C. \tag{5.11}$$

Here we used that $\sum_{l=j+1}^n M_l C_l = \sum_{l=1}^n M_l C_l = 0$. Hence if $\Omega_j^M C = d_j C$, the $d_j = M_j$.

We also have

(5.12)

$$\Omega_i^M C = (0, \dots, 0, \sum_{l=i+1}^{n} M_l (C_i - C_l),\ M_i(C_{i+1} - C_i), \dots, M_j(C_n - C_i))$$

$$= (0, \dots, 0, \quad C_i \sum_{l=i}^{n} M_l \quad, M_i(C_{i+1} - C_i), \dots, M_j(C_n - C_i)) = d_i C\,.$$

Hence

$$d_i = \sum_{l=i}^{n} M_l\,, \qquad M_i(C_j - C_i) = C_j \sum_{l=i}^{n} M_l\,, \qquad j = i+1, \dots, n\,. \tag{5.13}$$

The second equality in (5.13) implies $-M_i C_i = C_j \sum_{l=i+1}^n M_l$. Hence $\sum_{l=i+1}^n M_l \neq 0$. This inequality and the first equality in (5.13) imply $d_i \neq M_i$. Also the second equality in (5.13) implies that $C_{i+1} = \dots = C_n$ and, therefore, $\Omega_j^M C = 0$ for $j > i$. Hence $d_j = 0$ for $j = i+1, \dots, n-1$. Thus we deduced (5.7) and (5.8).

The proof that (5.6), (5.7), (5.8) imply (5.5) is straightforward. □

COROLLARY 5.4. *Let a pair $C = (C_1, \dots, C_n)$, $(d_1, \dots, d_n) \in \mathbb{F}_p^n$ be a solution of system (5.4) such that $C \neq 0$. Then $\sum_{j=1}^n d_j = \sum_{j=1}^n M_j$ .* □

COROLLARY 5.5. *Let $I(z) = (I_1(z), \dots, I_n(z))$ be a homogeneous polynomial solution of system (2.1) over $\mathbb{F}_p$ of degree $d$, then*

$$d \equiv \sum_{j=1}^{n} M_j \pmod{p}, \tag{5.14}$$

*cf. Example* 2.6. □

COROLLARY 5.6. *Let a pair $C = (C_1, \dots, C_n)$, $(d_1, \dots, d_n) \in \mathbb{F}_p^n$ be a solution of system (5.4) such that $C \neq 0$. Then $(d_1, \dots, d_n)$ is uniquely determined by $C$. If $c \in \mathbb{F}_p^\times$, then the pair $cC, (d_1, \dots, d_n)$ is also a solution of system (5.4). The number of equivalence classes of solutions $C, (d_1, \dots, d_n)$ of system (5.4), modulo the equivalence relation $C \mapsto cC$, equals $n-1$ decreased by the number of indices $i$, $1 \leqslant i \leqslant n-2$, such that $\sum_{j=i+1} M_j = 0$.* □

**5.3. Example.** Let $n = 2$, $p = 3$, $q = 2$, $m_1 = m_2 = 1$, $M_1 = M_2 = 1$. Then system (2.1) has a (non-$\mathbb{F}_p$-hypergeometric) polynomial solution $I(z) = (z_1 - z_2)^2 \begin{pmatrix} 1 \\ -1 \end{pmatrix}$, see Example 2.6. The leading id-term of $I(z)$ equals $z_1^2 \begin{pmatrix} 1 \\ -1 \end{pmatrix}$ in agreement with Theorem 5.3.

In this example the module of all polynomial solutions is one-dimensional and generated by $I(z)$. This follows from the fact that the id-leading term of any polynomial solution $J(z)$ has the form $cz_1^{a_1}z_2^{a_2}\cdot z_1^2\begin{pmatrix}1\\-1\end{pmatrix}$, where $c\in\mathbb{F}_p^\times$, $a_1,a_2\in p\mathbb{Z}$, by Theorem 5.3.

## 6. Leading term of an $\mathbb{F}_p$-hypergeometric solution

**6.1. Leading term of $I^{[l]}(z)$.** Using the isomorphism $\iota$, defined in (3.6), we consider $\mathbb{F}_p$-hypergeometric solutions as polynomials in $z$ with coefficients in $\mathrm{Sing}V[-2]$.

Recall that for $l=1,\dots,r$, the $\mathbb{F}_p$-hypergeometric solution $I^{[l]}(z)$ is a homogeneous polynomial in $z$ of degree $\delta_l=\sum_{j=1}^n M_j-lp$.

We describe the leading term $I^{[l]}_{\mathrm{id}}(z)$ of $I^{[l]}(z)$ with respect to the lexicographical ordering $>_{\mathrm{id}}$.

THEOREM 6.1. *Given $l=1,\dots,r$, denote by $i=i(l)$ the unique positive integer such that*

$$0\leqslant\sum_{j=i}^n M_j-lp<M_i\,. \tag{6.1}$$

*Then*

$$I^{[l]}_{\mathrm{id}}(z)=(-1)^{\sum_{j=1}^i M_j}\,\frac{\Gamma_{\mathbb{F}_p}(M_i+1)\,\Gamma_{\mathbb{F}_p}\Big(\sum_{j=i+1}^n M_j-(l-1)p+1\Big)}{\Gamma_{\mathbb{F}_p}\Big(\sum_{j=i}^n M_j-lp+1\Big)} \tag{6.2}$$
$$\times\left(\frac{\sum_{j=i+1}^n f^{(j)}v}{\sum_{j=i+1}^n M_j}-\frac{f^{(i)}v}{M_i}\right)z_1^{M_1}z_2^{M_2}\dots z_{i-1}^{M_{i-1}}z_i^{\sum_{j=i}^n M_j-lp},$$

*where $f^{(k)}v$ are introduced in* (3.5).

PROOF. The theorem follows from formula (3.3), Corollary 5.2, Theorem 5.3 and Lemma 4.2, where Lemma 4.2 is applied with $A=M_i$ and $B=\sum_{j=i+1}^n M_j-(l-1)p$. □

REMARK. Let $\sigma=(\sigma_1,\dots,\sigma_n)\in S_n$. The leading term $I^{[l]}_\sigma(z)$ of $I^{[l]}(z)$ with respect to the lexicographical ordering $>_\sigma$ is obtained from the right-hand side of formula (6.2) by simultaneous reordering of variables $z_1,\dots,z_n$ and parameters $M_1,\dots,M_n$.

**6.2. Indices $i(l)$.** In Theorem 6.1 we define the numbers $i(l)$, $l=1,\dots,r$. The leading term $I^{[l]}_{\mathrm{id}}(z)$ of the $\mathbb{F}_p$-hypergeometric solution $I^{[l]}(z)$ is a monomial in $z_1,\dots,z_{i(l)}$.

LEMMA 6.2. *We have*

$$1\leqslant i(r)<i(r-1)<\dots<i(1)<n\,. \tag{6.3}$$

□

COROLLARY 6.3. *If system* (2.1) *has ample reduction for a prime $p$, then*

$$i(l)=n-l,\qquad l=1,\dots,n-1. \tag{6.4}$$

□

**6.3. Corollary of Theorem 6.1.** The $\mathbb{F}_p$-hypergeometric solutions $I^{[l]}(z)$, $l=1,\dots,r$, are linearly independent over the field $\mathbb{F}_p(z)$ of rational functions in $z$ with coefficients in $\mathbb{F}_p$.

This follows from the fact that the leading coefficients of $I^{[l]}(z)$, $l=1,\dots,r$, are linear independent over the field $\mathbb{F}_p$.

**6.4. Example.** Let $q=3$, $p=13$, $(m_1,\dots,m_6)=(2,2,2,1,1,1)$. Then $(M_1,\dots,M_6)=(8,8,8,4,4,4)$, $r=\big[\sum_j M_j/p\big]=2$. We have two $\mathbb{F}_p$-hypergeometric solutions $I^{[l]}(z)$, $l=1,2$, of degrees 23 and 10, respectively.

Consider the identity element $\mathrm{id}=(1,2,3,4,5,6)$ of $S_6$ and the elements $s_{3,4}=(1,2,4,3,5,6)$, $\sigma=(6,5,4,3,2,1)\in S_6$. Then the leading terms are

$$I^{[1]}_{\mathrm{id}}(z)=-\binom{8}{7}\Big((1-\frac{7}{8})f^{(3)}v+f^{(4)}v+f^{(5)}v+f^{(6)}v\Big)z_1^8z_2^8z_3^7\,,$$
$$I^{[1]}_{s_{3,4}}(z)=-\binom{4}{3}\Big((1-\frac{3}{4})f^{(3)}v+f^{(5)}v+f^{(6)}v\Big)z_1^8z_2^8z_4^4z_3^3\,,$$
$$I^{[1]}_{\sigma}(z)=-\binom{8}{3}\Big((1-\frac{3}{8})f^{(2)}v+f^{(1)}v\Big)z_6^4z_5^4z_4^4z_3^8z_2^3\,,$$

$$I^{[2]}_{\mathrm{id}}(z)=\binom{8}{2}\Big((1-\frac{2}{8})f^{(2)}v+f^{(3)}v+f^{(4)}v+f^{(5)}v+f^{(6)}v\Big)z_1^8z_2^2\,,$$
$$I^{[2]}_{s_{3,4}}(z)=\binom{8}{2}\Big((1-\frac{2}{8})f^{(2)}v+f^{(3)}v+f^{(4)}v+f^{(5)}v+f^{(6)}v\Big)z_1^8z_2^2\,,$$
$$I^{[2]}_{\sigma}(z)=\binom{4}{2}\Big((1-\frac{2}{4})f^{(4)}v+f^{(3)}v+f^{(2)}v+f^{(1)}v\Big)z_6^4z_5^4z_4^2\,.$$

## 7. Determinant of $\mathbb{F}_p$-hypergeometric solutions

**7.1. Determinant over $\mathbb{C}$.** Consider the system of KZ equations (2.1) over $\mathbb{C}$. Then the space of solutions is $n-1$-dimensional.

Recall the master function $\Phi(t,z)$ introduced in (2.4). Consider the hypergeometric integrals

$$I^{(j)}(z)=\int_{z_j}^{z_{j+1}}\Phi(t,z)\sum_{j=1}^{n}\frac{f^{(j)}v}{t-z_j}\,dt,\qquad j=1,\dots,n-1.$$

To determine the integrals we assume that $z_1,\dots,z_n$ are real, $z_1<\dots<z_n$, and for every $l=1,\dots,n$, we fix a univalued branch of the function $(t-z_l)^{-m_l/q}$ on the rays $\{t\in\mathbb{R}\mid t<z_l\}$ and $\{t\in\mathbb{R}\mid t>z_l\}$.

The integrals $I^{(j)}$, $j=1,\dots,n-1$, form a basis of solutions of system (2.1) due to Theorem 7.1.

THEOREM 7.1 (**[V1, V2, V3]**). *We have*

$$(7.1)\quad \det_{j,l=1}^{n-1}\Big(\frac{-m_{l+1}}{q}\int_{z_j}^{z_{j+1}}\Phi(t,z)\frac{dt}{t-z_{l+1}}\Big)
=\frac{\Gamma(-m_1/q+1)\cdots\Gamma(-m_n/q+1)}{\Gamma(-m_1/q-\dots-m_n/q+1)}\prod_{1\leqslant j,\,l\leqslant n,\ j\neq l}(z_j-z_l)^{-m_l/q}\,.$$

Below we prove an analog of this theorem for $\mathbb{F}_p$-hypergeometric solutions.

**7.2. Determinant over $\mathbb{F}_p$.** Introduce a basis of $\mathrm{Sing}V[-2]$,

$$w_j = \frac{f^{(j)}v}{M_j} - \frac{f^{(j+1)}v}{M_{j+1}}, \qquad j = 1, \dots, n-1 . \tag{7.2}$$

Expand each of the $\mathbb{F}_p$-hypergeometric solutions $I^{[l]}(z)$ with respect to this basis,

$$I^{[l]}(z) = \sum_{j=1}^{n-1} c_j^l(z) w_j, \tag{7.3}$$

where $c_j^l(z)$ are scalar homogeneous polynomials of degree $\sum_{j=1}^n M_j - lp$.

THEOREM 7.2. *Assume that system* (2.1) *has ample reduction for a prime* $p$. *Then we have the square matrix* $c(z) = (c_j^l(z))_{l,j=1,\dots,n-1}$ *with nonzero determinant, given by the following formula*

$$\det c(z) = \frac{\Gamma_{\mathbb{F}_p}(M_1+1)\cdots\Gamma_{\mathbb{F}_p}(M_n+1)}{\Gamma_{\mathbb{F}_p}(M_1+\dots+M_n-(n-1)p+1)} \prod_{1\leqslant i<j\leqslant n} (-1)^{M_j}(z_j-z_i)^{M_i+M_j-p}. \tag{7.4}$$

Notice that

$$\frac{\Gamma_{\mathbb{F}_p}(M_1+1)\cdots\Gamma_{\mathbb{F}_p}(M_n+1)}{\Gamma_{\mathbb{F}_p}(M_1+\dots+M_n-(n-1)p+1)} = (-1)^{n-1}\frac{M_1!\cdots M_n!}{(M_1+\dots+M_n-(n-1)p)!} . \tag{7.5}$$

The theorem is proved in Sections 7.3, 7.4.

**7.3. Preliminary remarks.**

LEMMA 7.3. *The function* $\det c(z)$ *is a nonzero homogeneous polynomial of degree*

$$(n-1)\sum_{j=1}^n M_j - \frac{n(n-1)}{2}\,p \tag{7.6}$$

*with* id*-leading term*

$$\det{}_{\mathrm{id}} c(z) = \mathrm{const}\ z_1^{(n-1)(M_1-p)+\sum_{j=2}^n M_j} z_2^{(n-2)(M_2-p)+\sum_{j=3}^n M_j} \dots z_{n-1}^{M_{n-1}-p+M_n}, \tag{7.7}$$

*where*

$$\mathrm{const} = (-1)^{n(n-1)/2+\sum_{j=1}^{n-1}(n-j)M_j}\ \frac{\Gamma_{\mathbb{F}_p}(M_1+1)\cdots\Gamma_{\mathbb{F}_p}(M_n+1)}{\Gamma_{\mathbb{F}_p}(M_1+\dots+M_n-(n-1)p+1)}, \tag{7.8}$$

PROOF. If $\det c(z)$ is nonzero, then it is of degree

$$\deg(\det c(z)) = \sum_{l=1}^{n-1}\Big(\sum_{j=1}^n M_j - lp\Big),$$

which gives (7.6). By Theorem 6.1 we have

$$(7.9)\quad I^{[l]}_{\mathrm{id}}(z) = (-1)^{\sum_{j=1}^{n-l} M_j} \frac{\Gamma_{\mathbb{F}_p}(M_{n-l}+1)\,\Gamma_{\mathbb{F}_p}\Big(\sum_{j=n-l+1}^{n} M_j - (l-1)p+1\Big)}{\Gamma_{\mathbb{F}_p}\Big(\sum_{j=n-l}^{n} M_j - lp+1\Big)}$$

$$\times \left(\frac{\sum_{j=n-l+1}^{n} f^{(j)}v}{\sum_{j=n-l+1}^{n} M_j} - \frac{f^{(n-l)}v}{M_{n-l}}\right) z_1^{M_1} z_2^{M_2} \dots z_{n-l-1}^{M_{n-l-1}} z_{n-l}^{\sum_{j=n-l}^{n} M_j - lp},$$

Expanding vectors $\left(\frac{\sum_{j=n-l+1}^{n} f^{(j)}v}{\sum_{j=n-l+1}^{n} M_j} - \frac{f^{(n-l)}v}{M_{n-l}}\right)$, $l=1,\dots,n-1$, with respect to the basis $w_1,\dots,w_{n-1}$, we obtain a square $(n-1)\times(n-1)$-matrix $C(\mathrm{id}) = (C(\mathrm{id})^l_j)$ of coefficients of the expansion. The matrix is triangular with respect to the main anti-diagonal. The anti-diagonal entries are

$$C(\mathrm{id})^l_{n-l} = (-1)^{\sum_{j=1}^{n-l} M_j+1} \frac{\Gamma_{\mathbb{F}_p}(M_{n-l}+1)\,\Gamma_{\mathbb{F}_p}\Big(\sum_{j=n-l+1}^{n} M_j - (l-1)p+1\Big)}{\Gamma_{\mathbb{F}_p}\Big(\sum_{j=n-l}^{n} M_j - lp+1\Big)}.$$

Hence

$$\det C(\mathrm{id}) = (-1)^{n(n-1)/2+\sum_{j=1}^{n-1}(n-j)M_j} \frac{\Gamma_{\mathbb{F}_p}(M_1+1)\cdots\Gamma_{\mathbb{F}_p}(M_n+1)}{\Gamma_{\mathbb{F}_p}(M_1+\dots+M_n-(n-1)p+1)}.$$

These formulas imply the lemma. □

### 7.4. Differential equations for $\det c(z)$.

LEMMA 7.4. *The polynomial* $\det c(z)$ *satisfies the system of scalar differential equations*

$$(7.10)\qquad \frac{\partial y}{\partial z_i} = \sum_{j\neq i} \frac{M_i+M_j}{z_i-z_j}\, y, \qquad i=1,\dots,n.$$

PROOF. The operators $\Omega^M_{ij}$ defined in (5.3) preserve $\mathrm{Sing}V[-2]$ and

$$(7.11)\qquad \mathrm{Tr}\,\big|_{\mathrm{Sing}[-2]}\Omega^M_{ij} = M_i + M_j.$$

The polynomial $\det c(z)$, as the determinant of solutions, satisfies the system of equations

$$(7.12)\qquad \frac{\partial \det c(z)}{\partial z_i} = \sum_{j\neq i} \frac{\mathrm{Tr}\,\big|_{\mathrm{Sing}[-2]}\Omega^M_{ij}}{z_i-z_j} \det c(z), \qquad i=1,\dots,n.$$

This proves the lemma. □

LEMMA 7.5. *The polynomial*

$$y^0(z) = \prod_{1\leqslant i<j\leqslant n} (z_i-z_j)^{M_i+M_j-p}$$

*is a solution of system* (7.12). □

LEMMA 7.6. *For any* $i\neq j$, *the polynomial* $\det c(z)$ *is divisible by* $(z_i-z_j)^{M_i+M_j-p}$.

PROOF. We will prove that $\det c(z)$ is divisible by $(z_{n-1}-z_n)^{M_{n-1}+M_n-p}$. The divisibility by $(z_i-z_j)^{M_i+M_j-p}$ for other $i\neq j$ is proved by reordering variables.

Introduce new variables $u_1,\dots,u_n$ by the equaions: $u_n=z_1+\dots+z_n$,

$$z_{j+1}-z_j=u_1u_2\dots u_j,\qquad j=1,\dots,n-1. \tag{7.13}$$

The variables $z_1,\dots,z_n$ are polynomials in $u_1,\dots,u_n$. Hence if $I(z)$ is a polynomial solution of system (2.1), then $I(z(u))$ is a polynomial solution of the transformed differential KZ equations, which take the form:

$$\frac{\partial I}{\partial u_i}=\Big(\frac{\sum_{j>i}\Omega^M_{ij}}{u_i}+\operatorname{Reg}_i(u)\Big)I,\quad i=1,\dots,n-1,\qquad \frac{\partial I}{\partial u_n}=0, \tag{7.14}$$

where $\operatorname{Reg}_i(u)$ is an operator-valued function of $u$ regular at the point $u_1=\dots=u_n=0$. See this statement in [**V7**, Proposition 2.2.3]. In particular, we have

$$\frac{\partial I}{\partial u_{n-1}}=\Big(\frac{\Omega^M_{n-1,n}}{u_{n-1}}+\operatorname{Reg}_{n-1}\Big)I. \tag{7.15}$$

Hence the polynomial $\det c(z(u))$ a solution of the system of differential equations;

$$\frac{\partial y}{\partial u_i}=\Big(\frac{\sum_{j>i}(M_i+M_j)}{u_i}+\operatorname{Tr}\operatorname{Reg}_i(u)\Big)y,\quad i=1,\dots,n-1,\qquad \frac{\partial y}{\partial u_n}=0, \tag{7.16}$$

cf. formulas (7.11), (7.12). In particular, we have

$$\frac{\partial I}{\partial u_{n-1}}=\Big(\frac{M_{n-1}+M_n}{u_{n-1}}+\operatorname{Reg}_{n-1}\Big)I. \tag{7.17}$$

Hence the polynomial $\det c(z(u))$ is divisible by a monomial $u^d_{n-1}$, where $d$ is a nonnegative solution of the congruence

$$d\equiv M_{n-1}+M_n \pmod p.$$

The number $M_{n-1}+M_n-p$ is the smallest nonnegative solution. Hence $\det c(z(u))$ is divisible by $u_{n-1}^{M_{n-1}+M_n-p}$, that is,

$$\det c(z(u))=F(u_1,\dots,u_n)\,u_{n-1}^{M_{n-1}+M_n-p},$$

where $F(u_1,\dots,u_n)$ is a polynomial. Thus

$$\det c(z) =F\Big(z_2-z_1,\frac{z_3-z_2}{z_2-z_1},\dots,\frac{z_n-z_{n-1}}{z_{n-1}-z_{n-2}},z_1+\dots+z_n\Big)\Big(\frac{z_n-z_{n-1}}{z_{n-1}-z_{n-2}}\Big)^{M_{n-1}+M_n-p}.$$

Hence $\det c(z)$ is divisible $(z_n-z_{n-1})^{M_{n-1}+M_n-p}$. The lemma is proved. □

PROOF OF THEOREM 7.2. The theorem follows from Lemmas 7.6 and 7.3. □

**7.5. Remark on the initial value problem.**

COROLLARY 7.7. *Assume that system (2.1) has ample reduction for a prime $p$. For $x=(x_1,\dots,x_n)\in\mathbb{F}_p^n$ with distinct coordinates and $w\in\operatorname{Sing}V[-2]$, there exist a unique vector $(c_1,\dots,c_{n-1})\in\mathbb{F}_p^{n-1}$ such that*

$$w=\sum_{l=1}^{n-1}c_lI^{[l]}(x)\,. \tag{7.18}$$

Denote $(\mathbb{F}_p^n)^o = \{x \in \mathbb{F}_p^n \mid x \text{ has distinct coordinates}\}$. We have an isomorphism of the two trivial bundles

$$\mathrm{Sing} V[-2] \times (\mathbb{F}_p^n)^o \to (\mathbb{F}_p^n)^o \quad \text{and} \quad \mathbb{F}_p^{n-1} \times (\mathbb{F}_p^n)^o \to (\mathbb{F}_p^n)^o,$$

which sends $(w, x)$ to $((c_1, \dots, c_{n-1}), x)$.

## 8. Properties of $\mathbb{F}_p$-hypergeometric solutions

In this section we add more properties of $\mathbb{F}_p$-hypergeometric solutions.

**8.1. Uniqueness property.** Given $l = 1, \dots, r$, the $\mathbb{F}_p$-hypergeometric solution $I^{[l]}(z)$ has degree $\sum_{j=1}^n M_j - lp$ and id-leading term

$$(8.1)\quad (-1)^{M_{i(l)}+1} \frac{\Gamma_{\mathbb{F}_p}(M_{i(l)}+1)\,\Gamma_{\mathbb{F}_p}\Big(\sum_{j=i(l)+1}^n M_j - (l-1)p + 1\Big)}{\Gamma_{\mathbb{F}_p}\Big(\sum_{j=i(l)}^n M_j - lp + 1\Big)}$$
$$\times \left( \frac{f^{(i(l))}v}{M_{i(l)}} - \frac{\sum_{j=i(l)+1}^n f^{(j)}v}{\sum_{j=i(l)+1}^n M_j} \right) z_1^{M_1} z_2^{M_2} \dots z_{i(l)-1}^{M_{i(l)-1}} z_{i(l)}^{\sum_{j=i(l)}^n M_j - lp},$$

where the number $i(l)$ is defined in Theorem 6.1.

THEOREM 8.1. *If $I(z)$ is a homogeneous polynomial solution of system* (2.1) *with* id*-leading term* (8.1), *then $I(z) = I^{[l]}(z)$.*

PROOF. By Theorem 5.3 the id-leading term of the difference $I(z) - I^{[l]}(z)$ has the form

$$(8.2)\quad (0, \dots, 0, C_k, \dots, C_n) z_1^{a_1} \dots z_{k-1}^{a_{k-1}} z_k^{a_k},$$

for some $k$, where $C_k \neq 0$ and

$$(8.3)\quad a_j \equiv M_j \pmod{p}, \quad j = 1, \dots, k-1, \qquad a_k \equiv \sum_{j=k}^n M_j \pmod{p}, \qquad a_k \not\equiv M_k,$$

$$(8.4)\quad \sum_{j=1}^k a_j = \sum_{j=1}^n M_j - lp,$$

$$(8.5)\quad z_1^{M_1} z_2^{M_2} \dots z_{i(l)-1}^{M_{i(l)-1}} z_{i(l)}^{\sum_{j=i(l)}^n M_j - lp} >_{\mathrm{id}} z_1^{a_1} \dots z_{k-1}^{a_{k-1}} z_k^{a_k}.$$

The inequality $k \geqslant i(l)$ is impossible due to (8.3), (6.1). The inequality $k < i(l)$ is impossible due to (8.3), (8.5). □

**8.2. $L$-admissible solutions and filtration on space of all polynomial solutions.** Let $L = (L_1, \dots, L_n) \in \mathbb{Z}_{\geqslant 0}^n$. Let $I(z)$ be a polynomial in $z$ with coefficients in $\mathbb{F}_p^n$. We say that $I(z)$ is *$L$-admissible* if

$$(8.6)\quad \frac{\partial^{L_j+1} I}{\partial z_i^{L_j+1}}(z) = 0, \qquad j = 1, \dots, n.$$

Denote by $\mathcal{M}_L$ the $\mathbb{F}_p[z^p]$-module of all $L$-admissible polynomial solutions of (2.1).

For example, $\mathcal{M}_{(0,\dots,0)}$ consists of polynomial solutions $I(z) = (I_1(z), \dots, I_n(z))$ lying in $(\mathbb{F}_p[z^p])^n$, in other words, it consists of all $I(z)$, such that

$$\sum_{j\neq i} \frac{\Omega_{ij}}{z_i - z_j} I(z) = 0\,, \quad i = 1,\dots,n, \qquad m_1 I_1(z) + \dots + m_n I_n(z) = 0.$$

In particular, if system (2.1) has constant solutions then they lie in $\mathcal{M}_{(0,\dots,0)}$.

The modules $\mathcal{M}_L$ form a filtration on the module of all polynomial solutions of system (2.1). Namely, if $L = (L_1,\dots,L_n)$, $L' = (L'_1,\dots,L'_n)$ and $L_j \leqslant L'_j$ for all $j$, then $\mathcal{M}_L \subset \mathcal{M}_{L'}$.

THEOREM 8.2. *Let $L = (M_1+1,\dots,M_n+1)$. Then $\mathcal{M}_L$ coincides with the module $\mathcal{M}$ of $\mathbb{F}_p$-hypergeometric solutions of system* (2.1).

PROOF. Any $\mathbb{F}_p$-hypergeometric solution $I^{[l]}(z)$ lies in $\mathcal{M}_L$ by construction. We show that any element $I(z) \in \mathcal{M}_L$ is a linear combination of the $\mathbb{F}_p$-hypergeometric solutions with coefficients in $\mathbb{F}_p[z^p]$. Since system (2.1) is homogeneous, it is enough to assume that $I(z)$ is a homogeneous polynomial.

Assume that $I(z)$ is a homogeneous polynomial. Let

$$(0,\dots,0,C_i,\dots,C_n) z_1^{d_1} \dots z_n^{d_n}$$

be the id-leading term of $I(z)$, where $1 \leqslant i < n$, $C_i \neq 0$. Divide each $d_j$ by $p$ with remainder,

$$d_j = q_j p + r_j\,, \qquad 0 \leqslant r_j < p, \qquad j = 1,\dots,n\,.$$

Then $r_j \leqslant M_j$, $j = 1,\dots,n$, since $I(z) \in \mathcal{M}_L$, and $(r_1,\dots,r_n)$ has the form $(M_1,\dots,M_{i-1},r_i,0,\dots,0)$, $r_i \neq M_i$, by Theorem 5.3. We have

$$r_i = \sum_{j=i}^{n} M_j - lp$$

for some positive integer $l$, by Corollary 5.4.

Consider the $\mathbb{F}_p$-hypergeometric solution $I^{[l]}(z)$. By Theorem 5.3 the id-leading term of $I^{[l]}(z)$ is

$$(0,\dots,0,C_i^l,\dots,C_n^l)\, z_1^{M_1} \dots z_{i-1}^{M_{i-1}} z_i^{r_i}\,,$$

where

$$(0,\dots,0,C_i,\dots,C_n) = c\,(0,\dots,0,C_i^l,\dots,C_n^l)\,. \tag{8.7}$$

for some $c \in \mathbb{F}_p^\times$. Both $I(z)$ and $cz_1^{q_1p}\dots z_n^{q_np} I^{[l]}(z)$ belong to $\mathcal{M}_L$ and have the same leading term. Hence the leading monomial of the difference $I(z) - cz_1^{q_1p}\dots z_n^{q_np} I^{[l]}(z)$ is lexicographically smaller than the leading monomial $z_1^{d_1}\dots z_n^{d_n}$ of $I(z)$. Notice that the difference is also a homogeneous polynomial.

Iterating this procedure, which decreases the leading monomial, we present $I(z)$ as a linear combination of the $\mathbb{F}_p$-hypergeometric solutions with coefficients in $\mathbb{F}_p[z^p]$. □

### 8.3. Ample reduction.

THEOREM 8.3. *Assume that system* (2.1) *has ample reduction for a prime $p$. Then any polynomial solution $I(z)$ of system* (2.1) *belongs to the module of $\mathbb{F}_p$-hypergeometric solutions.*

PROOF. System (2.1) is homogeneous. Hence it is enough to prove the theorem assuming that $I(z)$ is a homogeneous polynomial solution.

Let $I_{\rm id}(z) = Cz_1^{d_1}\dots z_n^{d_n}$ be the id-leading term of $I(z)$. By Corollary 5.2 the id-leading term has the form described in Theorem 5.3. The ampleness of the reduction implies there exists an $\mathbb{F}_p$-hypergeometric solution $I^{[l]}(z)$, whose id-leading $I^{[l]}_{\rm id}(z)$ has the property

$$I_{\rm id}(z) = cz_1^{a_1}\dots z_n^{a_n} I^{[l]}_{\rm id}(z)\,,$$

where $c\in\mathbb{F}_p$ and $a_1,\dots,a_n\in p\mathbb{Z}_{\geqslant 0}$. The difference $I(z) - cz_1^{a_1}\dots z_n^{a_n}I^{[l]}(z)$ is a homogeneous polynomial solution of system (2.1) with id-leading term $>_{\rm id}$-smaller than the id-leading term of $I(z)$. Iteration of this procedure implies the theorem. □

## Acknowledgment

The author thanks Ivan Charednik and Alexey Slinkin for useful discussions.

## References


[KZ] V. Knizhnik and A. Zamolodchikov, *Current algebra and the Wess-Zumino model in two dimensions*, Nucl. Phys. **B247** (1984), 83–103

[Lu] E. Lucas, *Theorie des Fonctions Numeriques Simplement Periodiques*, American Journal of Mathematics. 1 (2) (1878) 184–196, doi:10.2307/2369308, JSTOR 2369308, MR 1505161

[RV1] R. Rimányi, A. Varchenko, *The $\mathbb{F}_p$-Selberg integral*, arXiv:2011.14248, 1–19, 2020.

[RV2] R. Rimányi, A. Varchenko, *The $\mathbb{F}_p$-Selberg integral of type $A_n$*, Lett. Math. Phys., **111**, 71 (2021), https://doi.org/10.1007/s11005-021-01417-x

[Se] A. Selberg, *Bemerkninger om et multipelt integral*, Norsk Mat. Tidsskr., **26** (1944), 71–78

[SV1] V. Schechtman and A. Varchenko, Integral Representations of N-Point Conformal Correlators in the WZW Model, Bonn, Max-Planck Institute, 1989, 1–22

[SV2] V. Schechtman, A. Varchenko, *Arrangements of Hyperplanes and Lie Algebra Homology*, Invent. Math. **106** (1991), 139–194

[SV3] V. Schechtman, A. Varchenko, *Solutions of KZ differential equations modulo p*, The Ramanujan Journal, **48** (3), 2019, 655–683, https://doi.org/10.1007/s11139-018-0068-x, arXiv:1707.02615

[SliV1] A. Slinkin, A. Varchenko, *Hypergeometric Integrals Modulo p and Hasse–Witt matrices*, Arnold Math J., **7** (2021), 267–311, https://doi.org/10.1007/s40598-020-00168-2

[V1] A. Varchenko, *Beta-Function of Euler, Vandermonde Determinant, Legendre Equation and Critical Values of Linear Functions of Configuration of Hyperplanes*, I. Izv. Akademii Nauk USSR, Seriya Mat., 53:6 (1989), 1206–1235; II, Izv. Akademii Nauk USSR, Seriya Mat. 54:1 (1990), 146–158

[V2] A. Varchenko, *Determinant Formula for Selberg Integrals*, Funct. Analysis and its Applications, 25 (1991), no. 4, 88–89

[V3] Alexander Varchenko, *Special functions, KZ type equations, and representation theory*, CBMS Regional Conference Series in Mathematics, vol. 98, Published for the Conference Board of the Mathematical Sciences, Washington, DC; by the American Mathematical Society, Providence, RI, 2003, DOI 10.1090/cbms/098. MR2016311

[V4] A. Varchenko, *Solutions modulo p of Gauss-Manin differential equations for multidimensional hypergeometric integrals and associated Bethe ansatz*, Mathematics 2017, 5(4), 52; doi:10.3390/math5040052, 1–18

[V5] A. Varchenko, *Hyperelliptic integrals modulo p and Cartier-Manin matrices*, Pure and Applied Math Quarterly, **16** (2020), n. 3, 315–336

[V6] A. Varchenko, *Remarks on the Gaudin model modulo p*, Journal of Singularities, **18** (2018), 486–499, arXiv:1708.06264

[V7] A. Varchenko, *An invariant subbundle of the KZ connection mod p and reducibility of $\widehat{\mathfrak{sl}_2}$ Verma modules mod p*, Mathematical Notes, **109**, no. 3 (2021), 386–397, DOI: 10.1134/S0001434621030068

Department of Mathematics, University of North Carolina at Chapel Hill, Chapel Hill, North Carolina 27599-3250; and Faculty of Mathematics and Mechanics, Lomonosov Moscow State University, Leninskiye Gory 1, 119991 Moscow GSP-1, Russia
*Email address*: anv@email.unc.edu

Contemporary Mathematics
Volume **780**, 2022
https://doi.org/10.1090/conm/780/15695

# Notes on solutions of KZ equations modulo $p^s$ and $p$-adic limit $s \to \infty$

Alexander Varchenko,
with an appendix by Steven Sperber and Alexander Varchenko

Abstract. We consider the differential KZ equations over $\mathbb{C}$ in the case, when the hypergeometric solutions are one-dimensional hyperelliptic integrals of genus $g$. In this case the space of solutions of the differential KZ equations is a $2g$-dimensional complex vector space.

We also consider the same differential equations modulo $p^s$, where $p$ is an odd prime number and $s$ is a positive integer, and over the field $\mathbb{Q}_p$ of $p$-adic numbers.

We describe a construction of polynomial solutions of the differential KZ equations modulo $p^s$. These polynomial solutions have integer coefficients and are $p^s$-analogs of the hyperelliptic integrals. We call them the $p^s$-hypergeometric solutions. We consider the space $\mathcal{M}_{p^s}$ of all $p^s$-hypergeometric solutions, which is a module over the ring of polynomial quasi-constants modulo $p^s$. We study basic properties of $\mathcal{M}_{p^s}$, in particular its natural filtration, and the dependence of $\mathcal{M}_{p^s}$ on $s$.

We show that the $p$-adic limit of $\mathcal{M}_{p^s}$ as $s \to \infty$ gives us a $g$-dimensional vector space of solutions of the differential KZ equations over the field $\mathbb{Q}_p$. The solutions over $\mathbb{Q}_p$ are power series at a certain asymptotic zone of the KZ equations.

In the appendix written jointly with Steven Sperber we consider all asymptotic zones of the KZ equations in the special case $g = 1$ of elliptic integrals. It turns out that in this case the $p$-adic limit of $\mathcal{M}_{p^s}$ as $s \to \infty$ gives us a one-dimensional space of solutions over $\mathbb{Q}_p$ at every asymptotic zone. We apply Dwork's theory of the classical hypergeometric function over $\mathbb{Q}_p$ and show that our germs of solutions over $\mathbb{Q}_p$ defined at different asymptotic zones analytically continue into a single global invariant line subbundle of the associated KZ connection. Notice that the corresponding KZ connection over $\mathbb{C}$ does not have proper nontrivial invariant subbundles, and therefore our invariant line subbundle is a new feature of the KZ equations over $\mathbb{Q}_p$.

Also in the appendix we follow Dwork and describe the Frobenius transformations of solutions of the KZ equations for $g = 1$. Using these Frobenius transformations we recover the unit roots of the zeta functions of the elliptic curves defined by the affine equations $y^2 = \beta\, x(x-1)(x-\alpha)$ over the finite field $\mathbb{F}_p$. Here $\alpha, \beta \in \mathbb{F}_p^\times, \alpha \neq 1$. Notice that the same elliptic curves considered over $\mathbb{C}$ are used to construct the complex holomorphic solutions of the KZ equations for $g = 1$.

2020 *Mathematics Subject Classification.* Primary 13A35; Secondary 11G25, 14G10, 33C60, 32G20.

*Key words and phrases.* KZ equations, reduction modulo $p^s$, $p^s$-hypergeometric solutions, $p$-adic limit, Frobenius transformations, unit roots.

Supported in part by NSF grant DMS-1954266.

## Contents



## 1. Introduction

**1.1.** The KZ equations were introduced in [**KZ**] as the differential equations satisfied by conformal blocks on sphere in the Wess-Zumino-Witten model of conformal field theory. The solutions of the KZ equations in the form of multidimensional hypergeometric integrals were constructed more than 30 years ago, see [**SV1**]. The KZ equations and the hypergeometric solutions are related to many subjects in algebra, representation theory, theory of integrable systems, enumerative geometry.

The polynomial solutions of the KZ equations over the finite field $\mathbb{F}_p$ of a prime number $p$ of elements were constructed relatively recently in [**SV2**], see also [**V4**]-[**V8**], [**RV1**, **RV2**]. These solutions were called the $\mathbb{F}_p$-hypergeometric solutions. The general problem is to understand relations between the hypergeometric solutions of the KZ equations over $\mathbb{C}$ and the $\mathbb{F}_p$-hypergeometric solutions and observe how the remarkable properties of hypergeometric solutions are reflected in the properties of the $\mathbb{F}_p$-hypergeometric solutions. For example, the $\mathbb{F}_p$-hypergeometric solutions inherit some determinant properties of the hypergeometric solutions and some Selberg integral properties, see [**V8**, **RV1**, **RV2**].

This program is in the first stages, where we consider essential examples and study the corresponding $\mathbb{F}_p$-hypergeometric solutions by direct methods.

In this paper we consider the differential KZ equations over $\mathbb{C}$ in the case, when the hypergeometric solutions are one-dimensional hyperelliptic integrals of genus $g$. In this case the space of solutions of the differential KZ equations is a $2g$-dimensional complex vector space. We also consider the same differential equations modulo $p^s$, where $p$ is an odd prime number and $s$ is a positive integer, and over the field $\mathbb{Q}_p$ of $p$-adic numbers.

We give a construction of polynomial solutions of the differential KZ equations modulo $p^s$ for positive integers $s$. We call such solutions the *$p^s$-hypergeometric solutions.* This construction is a straightforward modification of the construction in [**SV2**] of polynomial solutions modulo $p$.

In this paper we consider the space $\mathcal{M}_{p^s}$ of all $p^s$-hypergeometric solutions, which is a module over the ring of polynomial quasi-constants modulo $p^s$. We study basic properties of $\mathcal{M}_{p^s}$, in particular its natural filtration, and dependence of $\mathcal{M}_{p^s}$ on $s$.

We show that the $p$-adic limit of $\mathcal{M}_{p^s}$ as $s \to \infty$ gives us a $g$-dimensional vector space of solutions of the differential KZ equations over the field $\mathbb{Q}_p$. The solutions over $\mathbb{Q}_p$ are power series at a certain asymptotic zone of the KZ equations. This is the main result of the paper, see Lemma 9.5 and Theorem 10.7.

**1.2.** In the appendix written jointly with Steven Sperber we consider all six asymptotic zones of the KZ equations in the special case $g = 1$ of elliptic integrals. It turns out that in this case the $p$-adic limit of $\mathcal{M}_{p^s}$ as $s \to \infty$ gives us a one-dimensional space of solutions over $\mathbb{Q}_p$ at every asymptotic zone. We apply Dwork's theory of the classical hypergeometric function over $\mathbb{Q}_p$ and show that our germs of solutions over $\mathbb{Q}_p$ defined at different asymptotic zones analytically continue into a single global invariant line subbundle of the associated KZ connection. Notice that the corresponding KZ connection over $\mathbb{C}$ does not have proper nontrivial invariant subbundles, and therefore our invariant line subbundle is a new feature of the KZ equations over $\mathbb{Q}_p$.

Following Dwork we show that our line subbundle is spanned at any point of the base by the germs of all solutions of the KZ equations bounded in their discs of convergence. This statement gives a definition of the line subbundle independent of asymptotic zones and analytic continuation.

Also in the appendix we follow Dwork and describe the Frobenius transformations of solutions of the KZ equations for $g = 1$. Using these Frobenius transformations we recover the unit roots of the zeta functions of the elliptic curves defined by the affine equations $y^2 = \beta\, x(x-1)(x-\alpha)$ over the finite field $\mathbb{F}_p$. Here $\alpha, \beta \in \mathbb{F}_p^\times, \alpha \neq 1$. Notice that the same elliptic curves considered over $\mathbb{C}$ are used to construct the complex holomorphic solutions of the KZ equations for $g = 1$.

In the end of Section A.10 we argue that the KZ equations for $g = 1$ contain more arithmetic information than the associated hypergeometric differential equation (1.2) for the hypergeometric function $I(z)$ in (1.1), studied in [**Dw**].

**1.3.** Our $p$-adic limit of $\mathcal{M}_{p^s}$ as $s \to \infty$ is similar to the $p$-adic limit in the following classical example, see [**Ig, Ma, Cl, BV1**]. Consider the elliptic integral

$$I(z) = \frac{1}{\pi}\int_1^\infty \frac{dx}{\sqrt{x(x-1)(x-z)}} = \sum_{k=0}^{\infty} \binom{-1/2}{k}^2 z^k. \tag{1.1}$$

It satisfies the hypergeometric differential equation

$$z(1-z)I'' + (1-2z)I' - (1/4)I = 0. \tag{1.2}$$

The coefficients of the power series $I(z)$ are $p$-adic integers and the power series $I(z)$ converges $p$-adically for $|z|_p < 1$, where $|z|_p$ is the $p$-adic norm of $z \in \mathbb{Q}_p$. One may show that for any positive integer $s$ the polynomial

$$I_{(p^s-1)/2}(z) = \sum_{k=0}^{(p^s-1)/2} \binom{(p^s-1)/2}{k}^2 z^k \tag{1.3}$$

is a solution of the differential equation (1.2) modulo $p^s$. Thus we get a sequence $(I_{(p^s-1)/2}(z))_{s=1}^\infty$ of polynomials with integer coefficients, each of which is a solution of the differential equation (1.2) modulo $p^s$, and the $p$-adic limit of the sequence, as $s$ tends to $\infty$, is the $p$-adic power series solution $I(z)$ of the differential equation (1.2).

The $p^s$-hypergeometric solutions of our differential KZ equations are analogs of the polynomials $I_{(p^s-1)/2}(z)$ with an analogous $p$-adic limit. The difference is that the construction of the $p^s$-hypergeometric solutions does not indicate the analogous $p$-adic limiting solutions $I(z)$, and the analogous limiting $p$-adic power series solutions $I(z)$ can be discovered only after rewriting the $p^s$-hypergeometric solutions in a suitable asymptotic zone of the differential KZ equations.

In the simplest example of our differential KZ equations, the $p$-adic solution is the 3-vector

$$I(u_1,u_2)=u_1^{-3/2}\sum_{k=0}^{\infty}\binom{-1/2}{k+1}\binom{-3/2}{k}\Big(\frac{k+1}{-1/2-k},1,\frac{-1/2}{-1/2-k}\Big)u_2^k\,, \tag{1.4}$$

while the sequence $(I_{(p^s-3)/2}(u_1,u_2))_{s=1}^{\infty}$ of the $p^s$-hypergeometric solutions modulo $p^s$ of the same equations is given by the formula

$$\begin{aligned} I_{(p^s-3)/2}(u_1,u_2) &= u_1^{(p^s-3)/2}\sum_{k=0}^{(p^s-3)/2}\binom{(p^s-1)/2}{k+1}\binom{(p^s-3)/2}{k} \\ &\times \Big(\frac{k+1}{(p^s-1)/2-k},1,\frac{(p^s-1)/2}{(p^s-1)/2-k}\Big)u_2^k\,, \end{aligned} \tag{1.5}$$

see Section 9.7.

The sum $\sum_{k=0}^{(p^s-3)/2}$ in (1.5) is the truncation of the sum $\sum_{k=0}^{\infty}$ in (1.4), similar to what happens in (1.1) and (1.3). A new feature appears when we compare the prefactor $u_1^{-3/2}$ and the sequence of prefactors $\big(u_1^{(p^s-3)/2}\big)_{s=1}^{\infty}$. As $s\to\infty$ the sequence of prefactors $\big(u_1^{(p^s-3)/2}\big)_{s=1}^{\infty}$ tends $p$-adically to the prefactor $u^{-3/2}$ multiplied by a Teichmuller constant on a suitable domain in $\mathbb{Z}_p$, where $\mathbb{Z}_p$ is the ring of $p$-adic integers, see Section 10.3 and Theorem 10.5.

**1.4.** The paper is organized as follows. In Section 2 we define our system of KZ equations. In Section 3 we describe its complex solutions as hyperelliptic integrals. In Section 4 we describe the $p^s$-hypergeometric solutions of our KZ equations modulo $p^s$ and define the filtered module $\mathcal{M}_{p^s}$ of all $p^s$-hypergeometric solutions. In Section 5 we prove the independence of the module $\mathcal{M}_{p^s}$ from some arithmetic data involved in its definition. In Section 6 we discuss the properties of the operator $\mathcal{M}_{p^s}\to\mathcal{M}_{p^s}$ of multiplication by $p$. In Section 7 we calculate the coefficients of the Taylor expansion of the $p^s$-hypergeometric solutions. In Section 8 we relate the operator $\mathcal{M}_{p^s}\to\mathcal{M}_{p^s}$ of multiplication by $p$ and the Cartier-Manin matrix associated with the hyperelliptic curve defined by the affine equation $y^2=(x-z_1)\cdots(x-z_n)$. In Section 9 we consider one of the asymptotic zones of our KZ equations. Using the coordinates in that asymptotic zone we describe the $p$-adic limit of the $p^s$-hypergeometric solutions in Section 10. In Appendix A we apply Dwork's theory in [**Dw**] to the case $g=1$. In Section A.12 we discuss open problems related to the case of an arbitrary $g$.

## 2. KZ equations

Let $\mathfrak{g}$ be a simple Lie algebra with an invariant scalar product. The *Casimir element* is

$$\Omega = \sum_i h_i \otimes h_i \quad \in \ \mathfrak{g} \otimes \mathfrak{g},$$

where $(h_i) \subset \mathfrak{g}$ is an orthonormal basis. Let $V = \otimes_{i=1}^n V_i$ be a tensor product of $\mathfrak{g}$-modules, $\kappa \in \mathbb{C}^\times$ a nonzero number. The *differential KZ equations* is the system of differential equations on a $V$-valued function $I(z_1, \dots, z_n)$,

$$\frac{\partial I}{\partial z_i} \ = \ \frac{1}{\kappa} \sum_{j \neq i} \frac{\Omega_{i,j}}{z_i - z_j} I, \qquad i = 1, \dots, n,$$

where $\Omega_{i,j} : V \to V$ is the Casimir operator acting in the $i$th and $j$th tensor factors, see [**KZ, EFK**].

This system is a system of Fuchsian first order linear differential equations. The equations are defined on the complement in $\mathbb{C}^n$ to the union of all diagonal hyperplanes.

The object of our discussion is the following particular case.

Let *$p$ be an odd prime number, $n = 2g + 1$ an odd positive integer, $p > n \geqslant 2$.* We study the system of equations for a column vector $I(z) = (I_1(z), \dots, I_n(z))$ :

$$\text{(2.1)} \qquad \frac{\partial I}{\partial z_i} \ = \ \frac{1}{2} \sum_{j \neq i} \frac{\Omega_{ij}}{z_i - z_j} I, \quad i = 1, \dots, n, \qquad I_1(z) + \dots + I_n(z) = 0,$$

where $z = (z_1, \dots, z_n)$, the $n \times n$-matrices $\Omega_{ij}$ have the form:

$$\text{(2.2)} \qquad \Omega_{ij} \ = \ \begin{pmatrix} & \overset{i}{\vdots} & & \overset{j}{\vdots} & \\ i \cdots & -1 & \cdots & 1 & \cdots \\ & \vdots & & \vdots & \\ j \cdots & 1 & \cdots & -1 & \cdots \\ & \vdots & & \vdots & \end{pmatrix},$$

and all other entries are zero. This joint system of *differential and algebraic equations* will be called the *system of KZ equations* in this paper.

System (2.1) is the system of the differential KZ equations with parameter $\kappa = 2$ associated with the Lie algebra $\mathfrak{sl}_2$ and the subspace of singular vectors of weight $2g - 1$ of the tensor power $(\mathbb{C}^2)^{\otimes(2g+1)}$ of two-dimensional irreducible $\mathfrak{sl}_2$-modules, up to a gauge transformation, see this example in [**V3**, Section 1.1].

We consider system (2.1) over the field $\mathbb{C}$. We also consider the same system of equations modulo $p^s$ and over the field $\mathbb{Q}_p$ of $p$-adic numbers.

## 3. Complex solutions

Consider the *master function*

$$\text{(3.1)} \qquad \Phi(x, z) = \prod_{a=1}^{n} (x - z_a)^{-1/2}$$

and the column $n$-vector of hyperelliptic integrals

$$\text{(3.2)} \qquad I^{(\gamma)}(z) = (I_1(z), \dots, I_n(z)), \qquad I_j = \int \frac{\Phi(x, z)}{x - z_j}\, dx\,.$$

The integrals $I_j$, are over an element $\gamma$ of the first homology group of the algebraic curve with affine equation

$$y^2 = (x - z_1) \dots (x - z_n) .$$

Starting from such $\gamma$, chosen for given values $\{z_1, \dots, z_n\}$, the vector $I^{(\gamma)}(z)$ can be analytically continued as a multivalued holomorphic function of $z$ to the complement in $\mathbb{C}^n$ of the union of the diagonal hyperplanes $z_i = z_j$, $i \neq j$.

THEOREM 3.1. *The vector* $I^{(\gamma)}(z)$ *is a solution of system* (2.1).

Theorem 3.1 is a classical statement. Much more general algebraic and differential equations satisfied by analogous multidimensional hypergeometric integrals were considered in [**SV1**]. Theorem 3.1 is discussed as an example in [**V3**, Section 1.1].

PROOF. The theorem follows from Stokes' theorem and the two identities:

$$-\frac{1}{2}\Big(\frac{\Phi(x,z)}{x-z_1} + \dots + \frac{\Phi(x,z)}{x-z_n}\Big) = \frac{\partial \Phi}{\partial x}(x,z)\,, \tag{3.3}$$

$$\Big(\frac{\partial}{\partial z_i} - \frac{1}{2}\sum_{j\neq i}\frac{\Omega_{i,j}}{z_i - z_j}\Big)\Big(\frac{\Phi(x,z)}{x-z_1}, \dots, \frac{\Phi(x,z)}{x-z_n}\Big) = \frac{\partial \Psi^i}{\partial x}(x,z), \tag{3.4}$$

where $\Psi^i(x,z)$ is the column $n$-vector $(0,\dots,0,-\frac{\Phi(x,z)}{x-z_i},0,\dots,0)$ with the nonzero element at the $i$-th place. □

THEOREM 3.2 ([**V1**, Formula (1.3)]). *All solutions of system* (2.1) *have this form. Namely, the complex vector space of solutions of the form* (3.2) *is* $n-1$-*dimensional.*

This theorem follows from the determinant formula for multidimensional hypergeometric integrals in [**V1**], in particular, from [**V1**, Formula (1.3)].

## 4. Solutions modulo $p^s$

**4.1. Leading terms.** For a ring $R$ denote $R[z] = R[z_1, \dots, z_n]$. For a positive integer $t$ denote $R[z^{p^t}] = R[z_1^{p^t}, \dots, z_n^{p^t}]$.

Consider the lexicographical ordering of monomials $z_1^{d_1} \dots z_n^{d_n}$, so we have $z_1 > \dots > z_n$ and so on. For a nonzero polynomial $f(z) = \sum_{d_1,\dots,d_n} a_{d_1,\dots,d_n} z_1^{d_1} \dots z_n^{d_n}$ let $f_{\mathfrak{l}}(z)$ be the nonzero summand $a_{d_1,\dots,d_n} z_1^{d_1} \dots z_n^{d_n}$ with the largest monomial $z_1^{d_1} \dots z_n^{d_n}$. We call $f_{\mathfrak{l}}(z)$ the *leading term* of $f(z)$, the coefficient $a_{d_1,\dots,d_n}$ – the *leading coefficient*, the monomial $z_1^{d_1} \dots z_n^{d_n}$ – the *leading monomial*.

Let $s$ be a positive integer. An element $a \in \mathbb{Z}/p^s\mathbb{Z}$ has a unique presentation $a = a_0 + a_1 p + \dots + a_{s-1}p^{s-1}$, where $a_i \in \{0, \dots, p-1\}$. An element $a$ is invertible if and only if $a_0 \neq 0$.

Denote $\mathbb{F}_p = \mathbb{Z}/p\mathbb{Z}$.

Let $\pi_s$ denote the homomorphisms $\mathbb{Z} \to \mathbb{Z}/p^s\mathbb{Z}$, $\mathbb{Z}[z] \to (\mathbb{Z}/p^s\mathbb{Z})[z]$, $\mathbb{Z}[z]^n \to (\mathbb{Z}/p^s\mathbb{Z})[z]^n$ and for $t < s$ let $\pi_{s,t}$ denote the homomorphisms $\mathbb{Z}/p^s\mathbb{Z} \to \mathbb{Z}/p^t\mathbb{Z}$, $(\mathbb{Z}/p^s\mathbb{Z})[z] \to (\mathbb{Z}/p^t\mathbb{Z})[z]$, $(\mathbb{Z}/p^s\mathbb{Z})[z]^n \to (\mathbb{Z}/p^t\mathbb{Z})[z]^n$.

**4.2. Quasi-constants.** We say that a polynomial $f(z) \in \mathbb{Z}[z]$ is a *quasi-constant modulo* $p^s$ if $\frac{\partial f}{\partial z_i} \in p^s\mathbb{Z}[z]$ for $i = 1, \dots, n$. The quasi-constants modulo $p^s$ form a subring of $\mathbb{Z}[z]$ denoted by $\mathbb{Z}[z]_{p^s}$. For example $(z_1 + z_2)^{p^s} \in \mathbb{Z}[z]_{p^s}$.

LEMMA 4.1. *As a $\mathbb{Z}$-module, the ring $\mathbb{Z}[z]_{p^s}$ is spanned by the monomials $p^{s-t} z_1^{d_1} \dots z_n^{d_n}$, where $t$ is the maximal integer such that $t \leqslant s$ and $p^t$ divides every $d_1, \dots, d_n$.*

For example, $z_1^{lp^s}$ and $p^{s-1} z_1^p z_2^{p^2}$ are such monomials.

PROOF. Let $f(z) = \sum_d c_d z_1^{d_1} \dots z_n^{d_n} \in \mathbb{Z}[z]_{p^s}$. We show that each summand $c_d z_1^{d_1} \dots z_n^{d_n}$ is a multiple of a monomial of Lemma 4.1. Indeed, let $c_{d^0} z_1^{d_1^0} \dots z_n^{d_n^0}$ be the leading term of $f(z)$. Then all first partial derivatives of it must lie in $p^s\mathbb{Z}[z]$ Hence $c_{d^0} \in p^{s-t}\mathbb{Z}$, where $t$ is the maximal integer such that $t \leqslant s$ and $p^t$ divides every $d_1^0, \dots, d_n^0$. Subtracting the leading term from $f(z)$ and repeating the reasoning we prove the lemma. □

LEMMA 4.2. *Let $f(z)$ be a quasi-constant modulo $p^s$ and $t \in \mathbb{Z}_{\geqslant 0}$. Then $p^t f(z)$ is a quasi-constant modulo $p^r$ for any $1 \leqslant r \leqslant s + t$.* □

The rings of quasi-constants form a decreasing filtration, $\mathbb{Z}[z]_p \supset \mathbb{Z}[z]_{p^2} \supset \dots$.

**4.3. Solutions of system (2.1) modulo $p^s$.** We say that a column $n$-vector $I(z) \in \mathbb{Z}[z]^n$ of polynomials with integer coefficients is a *solution of system* (2.1) *modulo* $p^s$, if $\pi_s I(z) \in (\mathbb{Z}/p^s\mathbb{Z})[z]^n$ satisfies system (2.1).

LEMMA 4.3. *Let $I(z)$ be a solution of system* (2.1) *modulo $p^s$.*

(i) *Let $t \in \mathbb{Z}_{\geqslant 0}$. Then $p^t I(z)$ is a solution of system* (2.1) *modulo $p^r$ for any $1 \leqslant r \leqslant s + t$.*
(ii) *Let $f(z)$ be a quasi-constant modulo $p^s$. Then $f(z)I(z)$ is a solution of system* (2.1) *modulo $p^s$.*
(iii) *Let $1 \leqslant t < s$ and $I(z) \in p^t\mathbb{Z}[z]^n$. Let $f(z)$ be a quasi-constant modulo $p^{s-t}$. Then $f(z)I(z)$ is a solution of system* (2.1) *modulo $p^s$.*

□

**4.4. $p^s$-Hypergeometric solutions.** Let $M$ be the least positive integers such that

$$M \equiv -\frac{1}{2} \pmod{p^s}. \tag{4.1}$$

We have

$$M = \frac{p^s - 1}{2} = \frac{p-1}{2}\Big(1 + p + \dots + p^{s-1}\Big).$$

Introduce the *master polynomial*

$$\Phi_{p^s}(x, z) = \prod_{i=1}^{n} (x - z_i)^M \in \mathbb{Z}[x, z]. \tag{4.2}$$

Let

$$P_{p^s}(x, z) = \Big(\frac{\Phi_{p^s}(x, z)}{x - z_1}, \dots, \frac{\Phi_{p^s}(x, z)}{x - z_n}\Big) = \sum_i P_{p^s}^i(z)\, x^i, \tag{4.3}$$

where $P_{p^s}(x,z)$ is a column $n$-vector of polynomials in $x, z_1, \dots, z_n$ and $P^i_{p^s}(z)$ are $n$-vectors of polynomials in $z_1, \dots, z_n$ with coefficients in $\mathbb{Z}$. For a positive integer $l$, denote

$$I^{[lp^s-1]}_{p^s}(z) \,=\, P^{lp^s-1}_{p^s}(z)\,.$$

THEOREM 4.4. *For any positive integer $l$, the vector of polynomials $I^{[lp^s-1]}_{p^s}(z) \in \mathbb{Z}[z]^n$ is a solution of system (2.1) modulo $p^s$.*

PROOF. We have the following modifications of identities (3.3), (3.4):

$$M\Big(\frac{\Phi_{p^s}(x,z)}{x-z_1} + \dots + \frac{\Phi_{p^s}(x,z)}{x-z_n}\Big) \,=\, \frac{\partial \Phi_{p^s}}{\partial x}(x,z)\,, \tag{4.4}$$

$$\Big(\frac{\partial}{\partial z_i} + M\sum_{j\neq i}\frac{\Omega_{i,j}}{z_i-z_j}\Big)\Big(\frac{\Phi_{p^s}(x,z)}{x-z_1}, \dots, \frac{\Phi_{p^s}(x,z)}{x-z_n}\Big) \,=\, \frac{\partial \Psi^i_{p^s}}{\partial x}(x,z), \tag{4.5}$$

where $\Psi^i_{p^s}(x,z)$ is the column $n$-vector $(0,\dots,0,-\frac{\Phi_{p^s}(x,z)}{x-z_i},0,\dots,0)$ with the nonzero element at the $i$-th place. The theorem follows from identities (4.4), (4.5). □

REMARK. In [**SV2**] it was explained on how to construct polynomial solutions modulo $p$ of an arbitrary system differential KZ equations, associated with any Kac-Moody algebra and any tensor product of highest weight representations. The same construction gives polynomial solutions modulo $p^s$. The details will be provided elsewhere.

The range for the index $l$ is defined by the inequalities $0 < lp^s - 1 \leqslant n\frac{p^s-1}{2}$. Hence $l = 1, \dots, g$. The solutions $I^{[lp^s-1]}_{p^s}(z)$, $l = 1, \dots, g$, given by this construction, will be called the *$p^s$-hypergeometric solutions in $\mathbb{Z}[z]^n$*. For $t = 1, \dots, s-1$ and $l = 1, \dots, g$, the vector $p^{s-t} I^{[lp^t-1]}_{p^t}(z)$ is a solution of system (2.1) modulo $p^s$, see Lemma 4.3. Such solutions also will be called *$p^s$-hypergeometric solutions in $\mathbb{Z}[z]^n$*.

**4.5. Modules.** Consider the increasing filtration

$$0 = \mathcal{M}^0_{p^s} \subset \mathcal{M}^1_{p^s} \subset \dots \subset \mathcal{M}^{s-1}_{p^s} \subset \mathcal{M}^s_{p^s} = \mathcal{M}_{p^s}\,, \tag{4.6}$$

where

$$\mathcal{M}^t_{p^s} \,=\, \Big\{\pi_s\Big(\sum_{r=1}^{t}\sum_{l=1}^{g} c_{r,l}(z)\, p^{s-r} I^{[lp^r-1]}_{p^r}(z)\Big) \mid c_{r,l}(z) \in \mathbb{Z}[z]_{p^r}\Big\}, \tag{4.7}$$

$t = 1, \dots, s$. We have $\mathcal{M}_{p^s} \subset (\mathbb{Z}/p^s\mathbb{Z})[z]^n$. Every element of $\mathcal{M}_{p^s}$ is a polynomial solution of system (2.1) with coefficients in $\mathbb{Z}/p^s\mathbb{Z}$, see Lemma 4.3. The set $\mathcal{M}_{p^s}$ is a module over the ring $\mathbb{Z}[z]_{p^s}$ of quasi-constants modulo $p^s$, where $f(z) \in \mathbb{Z}[z]_{p^s}$ acts by multiplication by $\pi_s f(z)$. Each $\mathcal{M}^t_{p^s}$ is an $\mathbb{Z}[z]_{p^s}$-submodule of $\mathcal{M}_{p^s}$.

Each $\mathcal{M}^t_{p^s}$ is also a module over the larger ring $\mathbb{Z}[z]_{p^t}$ of quasi-constants modulo $p^t$, where $f(z) \in \mathbb{Z}[z]_{p^t}$ acts by multiplication by $\pi_s f(z)$.

The elements of $\mathcal{M}_{p^s}$ will be called the *$p^s$-hypergeometric solutions* in $(\mathbb{Z}/p^s\mathbb{Z})[z]^n$.

## 5. Independence of modules from the choice of $M$

**5.1. More general construction of solutions.** For $i = 1, \dots, n$, let $M_i$ be a positive integer such that

$$M_i \equiv -\frac{1}{2} \pmod{p^s}. \tag{5.1}$$

Denote $\vec{M} = (M_1, \dots, M_n)$. Consider the *master polynomial*

$$\Phi(x, z, \vec{M}) = \prod_{i=1}^{n} (x - z_i)^{M_i} \in \mathbb{Z}[x, z], \tag{5.2}$$

and the Taylor expansion

$$P(x, z, \vec{M}) = \Big(\frac{\Phi(x, z, \vec{M})}{x - z_1}, \dots, \frac{\Phi(x, z, \vec{M})}{x - z_n}\Big) = \sum_i P^i(z, \vec{M})\, x^i,$$

where $P^i(z, \vec{M})$ are $n$-vectors of polynomials in $z_1, \dots, z_n$ with coefficients in $\mathbb{Z}$. For a positive integer $l$, denote

$$I^{[lp^s-1]}(z, \vec{M}) \;=\; P^{lp^s-1}(z, \vec{M}).$$

THEOREM 5.1. *For any positive integers $l, t$, $t \leqslant s$, the vector of polynomials $I^{[lp^t-1]}(z, \vec{M}) \in \mathbb{Z}[z]^n$ is a solution of system* (2.1) *modulo $p^t$.*

PROOF. The theorem follows from straightforward modifications of identities (4.4), (4.5). □

**5.2. More modules.** Consider the increasing filtration

$$0 = \mathcal{M}^0_{p^s}(\vec{M}) \subset \mathcal{M}^1_{p^s}(\vec{M}) \subset \dots \subset \mathcal{M}^{s-1}_{p^s}(\vec{M}) \subset \mathcal{M}^s_{p^s}(\vec{M}) = \mathcal{M}_{p^s}(\vec{M}), \tag{5.3}$$

where

$$\mathcal{M}^t_{p^s}(\vec{M}) \;=\; \Big\{\pi_s\Big(\sum_{r=1}^{t}\sum_{l\geqslant 1} c_{r,l}(z)\, p^{s-r} I^{[lp^r-1]}(z, \vec{M})\Big) \mid c_{r,l}(z) \in \mathbb{Z}[z]_{p^r}\Big\}, \tag{5.4}$$

$t = 1, \dots, s$. We have $\mathcal{M}_{p^s}(\vec{M}) \subset (\mathbb{Z}/p^s\mathbb{Z})[z]^n$. Every element of $\mathcal{M}_{p^s}(\vec{M})$ is a polynomial solution of system (2.1) with coefficients in $\mathbb{Z}/p^s\mathbb{Z}$, see Lemma 4.3. The set $\mathcal{M}_{p^s}(\vec{M})$ is a module over the ring $\mathbb{Z}[z]_{p^s}$ of quasi-constants modulo $p^s$, where $f(z) \in \mathbb{Z}[z]_{p^s}$ acts by multiplication by $\pi_s f(z)$. Each $\mathcal{M}^t_{p^s}(\vec{M})$ is an $\mathbb{Z}[z]_{p^s}$-submodule of $\mathcal{M}_{p^s}(\vec{M})$.

Each $\mathcal{M}^t_{p^s}(\vec{M})$ is also a module over the larger ring $\mathbb{Z}[z]_{p^t}$ of quasi-constants modulo $p^t$, where $f(z) \in \mathbb{Z}[z]_{p^t}$ acts by multiplication by $\pi_s f(z)$.

THEOREM 5.2. *Filtration* (5.3) *does not depend on the choice of $\vec{M} = (M_1, \dots, M_n)$, satisfying congruences* (5.1). *Moreover, filtration* (5.3) *coincides with filtration* (4.6).

For $s = 1$ the statement is [**SliV**, Theorem 3.1].

PROOF. First we show that $\mathcal{M}_{p^s}(\vec{M})$ and filtration (5.3) do not depend on the choice of $\vec{M}$. Let $\vec{M} = (M_1, \dots, M_n)$, $\vec{M}' = (M'_1, \dots, M'_n)$ be two vectors satisfying congruences (5.1). We say that $\vec{M}' \geqslant \vec{M}$ if $M'_i \geqslant M_i$ for all $i$. The vector $(\frac{p^s-1}{2}, \dots, \frac{p^s-1}{2})$ is the minimal vector with respect to this partial order. To show that $\mathcal{M}_{p^s}(\vec{M})$ and filtration (5.3) do not depend on the choice of $\vec{M}$ it is enough to show that the filtrations are the same for a vector $\vec{M}$ and for a vector $\vec{M}' = \vec{M} + (0,\dots,0,\frac{p^s-1}{2},0,\dots,0)$, where the nonzero element stays at the $j$-th position for some $j$. Then

$$P(x,z,\vec{M}') = P(x,z,\vec{M})\cdot(x-z_j)^{p^s} = P(x,z,\vec{M})\sum_{a=0}^{p^s}(-1)^{p^s-a}\binom{p^s}{a}x^a z_j^{p^s-a}.$$

Recall that $P(x,z,\vec{M}) = \sum_i P^i(z,\vec{M})\,x^i$, $P(x,z,\vec{M}') = \sum_i P^i(z,\vec{M}')\,x^i$, For any $r \leqslant s$ and $l$ we have

$$P^{lp^r-1}(z,\vec{M}') = \sum_{a=0}^{p^s}(-1)^{p^s-a}\binom{p^s}{a} z_j^{p^s-a}P^{lp^r-a-1}(z,\vec{M}). \tag{5.5}$$

We are interested in this formula, since $I^{[lp^r-1]}(z,\vec{M}') = P^{lp^r-1}(z,\vec{M}')$ is a solution of system (2.1) modulo $p^r$.

LEMMA 5.3. *Let $b,c \in \mathbb{Z}_{>0}$ be such that $bp^c \leqslant p^s$, $p \not| b$. Then $p^{s-c}$ is the maximal power of $p$ dividing $\binom{p^s}{bp^c}$.*

PROOF. For $a \leqslant p^s$ we have $a\binom{p^s}{a} = p^s\binom{p^s-1}{a-1}$, and $p \not| \binom{p^s-1}{a-1}$ by Lucas' theorem, [**Lu**]. □

LEMMA 5.4. *Let $a \in \{0,\dots,p^s\}$, $a = bp^c$, $p \not| b$. Consider the vector*

$$V = (-1)^{p^s-a}\binom{p^s}{a} z_j^{p^s-a}P^{lp^r-a-1}(z,\vec{M})$$

*appearing in* (5.5)*. If $lp^r \leqslant a$, then $V = 0$. For $lp^r > a$, we write $lp^r - a = vp^u$, where $u = \min(r,c)$. Then*

$$V \;=\; d(z)\,p^{r-u}I^{[vp^u-1]}(z,\vec{M}), \tag{5.6}$$

*where $d(z) = z_j^{p^s-a}(-1)^{p^s-a}\binom{p^s}{a}/p^{r-u}$ is a quasi-constant modulo $p^u$.*

PROOF. The lemma follows from Lemma 5.3. □

COROLLARY 5.5. *For any $r = 1,\dots,s$, we have $\mathcal{M}^r_{p^s}(\vec{M}') \subset \mathcal{M}^r_{p^s}(\vec{M})$.* □

LEMMA 5.6. *For any $r = 1,\dots,s$, we have $\mathcal{M}^r_{p^s}(\vec{M}') \supset \mathcal{M}^r_{p^s}(\vec{M})$.*

PROOF. Let $w$ be the greatest integer such that $wp^r \leqslant \deg_t P(t,z,\vec{M})$. Then $w' = w + p^{s-r}$ is the greatest integer such that $w'p^r \leqslant \deg_t P(t,z,\vec{M}')$. Comparing the coefficients in (5.5) and using Lemma 5.5, we observe that for any $l = 1,\dots,w$ we have

$$\begin{aligned} I^{[(l+p^{s-r})p^r-1]}(z,\vec{M}') &= I^{[lp^r-1]}(z,\vec{M}) \\ &+ \sum_{m=1}^{l-1} c_{r,m}(z)I^{[mp^r-1]}(z,\vec{M}) + \sum_{k=1}^{r-1}\sum_{m\geqslant 1} c_{k,m}(z)\,p^{r-k}I^{[mp^k-1]}(z,\vec{M}), \end{aligned} \tag{5.7}$$

where $c_{i,j}(z) \in \mathbb{Z}[z]_{p^i}$. This triangular system of equations with respect to $I^{[lp^r-1]}(z,\vec{M})$, $l = 1,\dots,w$, can be written as

(5.8)

$$I^{[lp^r-1]}(z,\vec{M}) = \sum_{m\geqslant 1} c'_{r,m}(z)I^{[mp^r-1]}(z,\vec{M}') + \sum_{k=1}^{r-1}\sum_{m\geqslant 1} c'_{k,m}(z)\,p^{r-k}I^{[mp^k-1]}(z,\vec{M}),$$

$l = 1, \ldots, w$, for suitable $c'_{i,j}(z) \in \mathbb{Z}[z]_{p^i}$. Applying the previous reasoning to the sum $\sum_{k=1}^{r-1} \sum_{m\geqslant 1} c_{k,m}(z)\, p^{r-k} I^{[mp^k-1]}(z, \vec{M})$, we obtain

(5.9)

$$I^{[lp^r-1]}(z, \vec{M}) = \sum_{m\geqslant 1} c'_{r,m}(z) I^{[mp^r-1]}(z, \vec{M}') + \sum_{k=1}^{r-1} \sum_{m\geqslant 1} c''_{k,m}(z)\, p^{r-k} I^{[mp^k-1]}(z, \vec{M}'),$$

$l = 1, \ldots, w$, for suitable $c''_{i,j}(z) \in \mathbb{Z}[z]_{p^i}$. This proves the lemma. □

COROLLARY 5.7. *The module $\mathcal{M}_{p^s}(\vec{M})$ and filtration* (5.3) *do not depend on the choice of $\vec{M} = (M_1, \ldots, M_n)$, satisfying congruences* (5.1). □

LEMMA 5.8. *Let $\vec{M}^{\min} = (\frac{p^s-1}{2}, \ldots, \frac{p^s-1}{2})$. Then $\mathcal{M}_{p^s}(\vec{M}^{\min}) = \mathcal{M}_{p^s}$.*

PROOF. The proof of the lemma is a straightforward modification of the proof of Corollary 5.5 and Lemma 5.6. □

Theorem 5.2 is proved. □

## 6. Filtrations and homomorphisms

**6.1. Reduction from modulo $p^s$ to modulo $p^{s-m}$.** If $I(z)$ is a polynomial solution of system (2.1) modulo $p^s$, then $I(z)$ is also a polynomial solution of system (2.1) modulo $p^{s-m}$ for any $1 \leqslant m < s$. This defines a map

(6.1)

$$\mathfrak{r}_{s,s-m} : \mathcal{M}_{p^s}(\vec{M}) \to \mathcal{M}_{p^{s-m}}(\vec{M}),$$

$$\pi_s\left(\sum_{r=1}^{s}\sum_{l\geqslant 1} c_{r,l}(z)\, p^{s-r} I^{[lp^r-1]}(z, \vec{M})\right) \mapsto \pi_{s-m}\left(\sum_{r=m+1}^{s}\sum_{l\geqslant 1} c_{r,l}(z)\, p^{s-r} I^{[lp^r-1]}(z, \vec{M})\right),$$

where $\vec{M}$ is a vector with coordinates satisfying congruences (5.1). See these sums in (5.4). In the last sum we have $p^{s-r} I^{[lp^r-1]}(z, \vec{M}) = p^{s-m-(r-m)} I^{[(lp^m)p^{r-m}-1]}(z, \vec{M})$ and a solution $I^{[lp^r-1]}(z, \vec{M})$ modulo $p^r$ also can be considered as a solution $I^{[(lp^m)p^{r-m}-1]}(z, \vec{M})$ modulo $p^{r-m}$.

For any $r = 1, \ldots, s$, the submodule $\mathcal{M}^r_{p^s}(\vec{M}) \subset \mathcal{M}_{p^s}(\vec{M})$ is mapped by $\mathfrak{r}_{s,s-m}$ to the submodule $\mathcal{M}^{r-m}_{p^{s-m}}(\vec{M})$. The induced map

$$\mathfrak{r}_{s,s-m} : \mathcal{M}^r_{p^s}(\vec{M}) \to \mathcal{M}^{r-m}_{p^t}(\vec{M}) \tag{6.2}$$

is a homomorphism of $\mathbb{Z}[z]_{p^r}$-modules. Thus the map (6.1) is a homomorphism of filtered modules decreasing the index of filtration by $m$.

By Theorem 5.2 we have $\mathcal{M}_{p^s}(\vec{M}) = \mathcal{M}_{p^s}$. Hence homomorphism (6.1) also can be considered as a homomorphism of filtered modules,

$$\mathfrak{r}_{s,s-m} : \mathcal{M}_{p^s} \to \mathcal{M}_{p^{s-m}}\,, \tag{6.3}$$

decreasing the index of filtration by $m$.

It is rather nontrivial to write a formula for this map in terms of the generators $I^{[lp^r-1]}(z)$ of these modules.

**6.2. Multiplication by $p^m$.** If $I(z)$ is a polynomial solution of system (2.1) modulo $p^s$, then for any positive integer $m$ the polynomial $p^m I(z)$ is a polynomial solution of system (2.1) modulo $p^{s+m}$. In particular, multiplication by $p^m$ defines a map

$$(6.4)\ \mathfrak{p}_{s,s+m} : \mathcal{M}^t_{p^s} \to \mathcal{M}^t_{p^{s+m}}\,,$$

$$\pi_s\Big(\sum_{r=1}^{t}\sum_{l=1}^{g} c_{r,l}(z)\, p^{s-r} I^{[lp^r-1]}_{p^r}(z)\Big) \mapsto \pi_{s+m}\Big(\sum_{r=1}^{t}\sum_{l=1}^{g} c_{r,l}(z)\, p^{s+m-r} I^{[lp^r-1]}_{p^r}(z)\Big)$$

for any $t = 1, \dots, s$. See these sums in (4.7). Clearly this map is an isomorphism of filtered $\mathbb{Z}[z]_{p^t}$-modules.

**6.3. The composition of homomorphisms.** For $m < s$ denote by $\mathfrak{c}_{s,m}$ the composition $\mathfrak{p}_{s-m,m}\mathfrak{r}_{s,s-m}$,

$$(6.5)\qquad \mathfrak{c}_{s,m} \ :\ \mathcal{M}_{p^s} \to \mathcal{M}_{p^s}\,, \qquad I(z) \mapsto p^m I(z)\,.$$

For any $t = 1, \dots, s$, this map induces a homomorphism $\mathcal{M}^t_{p^s} \to \mathcal{M}^{t-m}_{p^s}$ of $\mathbb{Z}[z]_{p^t}$-modules.

We have $\mathfrak{c}_{s,m} = (\mathfrak{c}_{s,1})^m$ for $m < s$ and $\mathfrak{c}_{s,m} = (\mathfrak{c}_{s,1})^m = 0$ for $m \geqslant s$.

As we know, the module $\mathcal{M}_{p^s}$ is generated by the elements $\pi_s(p^{s-r} I^{[lp^r-1]}_{p^r}(z))$, $r = 1, \dots, s$, $l = 1, \dots, g$. For $l = 1, \dots, g$, we have

(6.6)

$$\mathfrak{c}_{s,1} \ :\ \pi_s(I^{[lp^s-1]}(z)) \ \mapsto\ \pi_s(pI^{[lp^s-1]}(z)) = \pi_s\Big(\sum_{r=1}^{s-1}\sum_{k=1}^{g} c^{l,s}_{r,k}(z)\, p^{s-r} I^{[kp^r-1]}_{p^r}(z)\Big)$$

for suitable coefficients $c^{l,s}_{r,k}(z) \in \mathbb{Z}[z]_{p^r}$.

The set of the coefficients $(c^{l,s}_{r,k}(z))_{l,s,r,k}$ determines the homomorphisms $\mathfrak{c}_{s,m}$ for all $s, m$. In what follows we shall describe the coefficients $c^{l,s}_{s-1,k}(z)$ for all $l$, $s$, $k$, see Theorem 8.4.

**6.4. Graded modules and homomorphisms.** Denote

$$(6.7)\qquad \mathrm{gr}\mathcal{M}_{p^s} \ =\ \oplus_{t=1}^{s}\, \mathrm{gr}\mathcal{M}^t_{p^s}\,, \qquad \mathrm{gr}\mathcal{M}^t_{p^s} \ =\ \mathcal{M}^t_{p^s}\Big/\mathcal{M}^{t-1}_{p^s}\,.$$

LEMMA 6.1. *For any $t = 1, \dots, s$, the action of $\mathbb{Z}[z]_{p^t}$ on $\mathcal{M}^t_{p^s}$ makes $\mathrm{gr}\mathcal{M}^t_{p^s}$ an $\mathbb{F}_p[z^{p^t}]$-module. Multiplication by $p$ on $\mathcal{M}^t_{p^s}$ induces a homomorphism*

$$(6.8)\qquad \mathrm{gr}\mathfrak{c}_{s,1} \ :\ \mathrm{gr}\mathcal{M}^t_{p^s} \to \mathrm{gr}\mathcal{M}^{t-1}_{p^s}$$

*of $\mathbb{F}_p[z^{p^t}]$-modules.* □

## 7. Coefficients of solutions

**7.1. Homogeneous polynomials.** For $l = 1, \dots, g$, the solution $I^{[lp^s-1]}(z) = (I^{[lp^s-1]}_1, \dots, I^{[lp^s-1]}_n)$ is a homogeneous polynomial in $z$ of degree

$$(7.1)\qquad \delta_l = (2g+1)\,\frac{p^s-1}{2}\ -\ lp^s = (g-l)p^s + \frac{p^s-1}{2} - g\,.$$

Notice that $(-1)^{\delta_l} = (-1)^{s\frac{p-1}{2}+l}$.

**7.2. Formula for coefficients.** Recall that $M = \frac{p^s-1}{2}$. Projection of this integer to $\mathbb{Z}/p^s\mathbb{Z}$ is invertible. Let

$$I^{[lp^s-1]}(z) = \sum_{d_1,\dots,d_n} I^{[lp^s-1]}_{d_1,\dots,d_n} z_1^{d_1}\dots z_n^{d_n}, \qquad I^{[lp^s-1]}_{d_1,\dots,d_n} \in \mathbb{Z}^n.$$

LEMMA 7.1 ([**V8**, Lemma 3.1]). *We have*

$$I^{[lp^s-1]}_{d_1,\dots,d_n} = (-1)^{\delta_l}\prod_{i=1}^n \binom{M}{d_i}\Big(1-\frac{d_1}{M},\dots,1-\frac{d_n}{M}\Big). \tag{7.2}$$

*The sum of coordinates of this vector is divisible by $p^s$.* □

LEMMA 7.2 (cf. [**V8**, Theorem 6.1]). *For $l=1,\dots,g$, the leading term of the $p^s$-hypergeometric solution $I^{[lp^s-1]}(z)$ is*

$$I^{[lp^s-1]}_l(z) = (-1)^{\delta_l}\binom{M}{l}\Big(0,\dots,0,\frac{l}{M},1,\dots,1\Big) z_1^M\dots z_{2g-2l}^M z_{2g-2l+1}^{M-l}, \tag{7.3}$$

*where 0 is repeated $2g-2l$ times and 1 is repeated $2l$ times.*

PROOF. The lemma follows from Lemma 7.1. □

LEMMA 7.3. *The projections to $\mathbb{Z}/p^s\mathbb{Z}$ of the integers $\binom{M}{l}$, $\binom{M}{l}\frac{l}{M}$ are invertible.*

PROOF. The invertibility of $\binom{M}{l}$ follows from Lucas' theorem, [**Lu**]. □

## 8. Multiplication by $p$ and Cartier-Manin matrix

### 8.1. Linear independence.

LEMMA 8.1. *The projections of the $p^s$-hypergeometric solutions $I^{[lp^s-1]}(z) \in \mathbb{Z}[z]^n$, $l=1,\dots,g$, to $\mathbb{F}_p[z]^n$ are linearly independent over $\mathbb{F}_p[z]$, that is, if*

$$\sum_{l=1}^g c_l(z) I^{[lp^s-1]}(z) \in p\mathbb{Z}[z]^n \tag{8.1}$$

*for some $c_l(z)\in\mathbb{Z}[z]$, then all $c_l(z)\in p\mathbb{Z}[z]$.*

PROOF. Recall that the projection $\mathbb{Z}[z]^n \to \mathbb{F}_p[z]^n$ is denoted by $\pi_1$. By Lemma 7.3, the leading coefficient of $\pi_1(c_l(z)I^{[lp^s-1]}(z))$ equals the product of the leading coefficient of $\pi_1(c_l(z))$ and the leading coefficient of $\pi_1(I^{[lp^s-1]}(z))$, if $\pi_1(c_l(z))$ is nonzero. In that case the leading coefficient of $\pi_1(c_l(z)I^{[lp^s-1]}(z))$ is a nonzero multiple of the nonzero vector $\pi_1((0,\dots,0,\frac{l}{M},1,\dots,1))$.

If relation (8.1) holds and some of the coefficients $c_l(z)$ have nonzero projections $\pi_1(c_l(z))$, then for several values of such indices $l$ the sum of the corresponding leading coefficients has to be equal to zero, which is impossible due to the fact that the vectors $\pi_1((0,\dots,0,\frac{l}{M},1,\dots,1))$ are linear independent over $\mathbb{F}_p$. □

COROLLARY 8.2. *The projections of the $p^s$-hypergeometric solutions $I^{[lp^s-1]}(z)\in \mathbb{Z}[z]^n$, $l=1,\dots,g$, to $\mathbb{F}_p[z]^n$ are linearly independent over $\mathbb{F}_p[z^{p^s}]$.* □

Denote by

$$\mathrm{gr}_t : \mathcal{M}^t_{p^s} \to \mathrm{gr}\mathcal{M}^t_{p^s} \tag{8.2}$$

the natural projection. Then the elements $\mathrm{gr}_t(\pi_s(p^{s-t}I^{[lp^t-1]}(z)))$, $l=1,\dots,g$, generate the $\mathbb{F}_p[z^{p^t}]$-module $\mathrm{gr}\mathcal{M}^t_{p^s}$.

COROLLARY 8.3. *For* $t = 1, \dots, s$, *the* $\mathbb{F}_p[z^{p^t}]$*-module* $\mathrm{gr}\mathcal{M}^t_{p^s}$ *is a free module of rank* $g$ *with a basis* $\mathrm{gr}_t(\pi_s(p^{s-t}I^{[lp^t-1]}(z)))$, $l = 1, \dots, g$. □

Denote the basis vectors of the $\mathbb{F}_p[z^{p^t}]$-module $\mathrm{gr}\mathcal{M}^t_{p^s}$ by

$$v^l_{s,t} := \mathrm{gr}_t(\pi_s(p^{s-t}I^{[lp^t-1]}(z))), \qquad l = 1, \dots, g. \tag{8.3}$$

**8.2. Cartier-Manin matrices.** Let

$$f(x,z) = (x - z_1) \dots (x - z_n), \qquad n = 2g + 1.$$

Consider the hyperelliptic curve $X$ defined by the affine equation

$$y^2 = (x - z_1) \dots (x - z_n).$$

Consider the space $\Omega^1(X)$ of regular 1-forms on $X$ with basis $\frac{x^{i-1}dx}{y}$, $i = 1, \dots, g$. Define the Cartier map $\mathcal{C} : \Omega^1(X) \to \Omega^1(X)$ as follows. We have

$$\frac{x^{i-1}dx}{y} = \frac{x^{i-1}y^{p-1}dx}{y^{p-1}y} = \frac{x^{i-1}f(x)^{(p-1)/2}dx}{y^p}.$$

Let $x^{i-1}f(x,z)^{(p-1)/2} = \sum_j c^j_i(z)x^j$. Define

$$\mathcal{C} \; : \; \frac{x^{i-1}dx}{y} \mapsto \sum_{j=1}^{g} c_i^{jp-1}(z)\frac{x^{j-1}dx}{y},$$

see [**AH**]. The map $\mathcal{C}$ is identified with the $g \times g$-matrix $(\mathcal{C}^j_i(z))^g_{i,j=1}$,

$$\mathcal{C}^j_i(z) \; = \; c_i^{jp-1}(z).$$

**8.3. Matrix of** $\mathrm{gr}\mathfrak{c}_{s,1}$**.** Recall that multiplication of solutions by $p$ defines a map

$$\mathrm{gr}\mathfrak{c}_{s,1} \; : \; \mathrm{gr}\mathcal{M}^t_{p^s} \to \mathrm{gr}\mathcal{M}^{t-1}_{p^s}, \tag{8.4}$$

where $\mathrm{gr}\mathcal{M}^t_{p^s}$ is a free $\mathbb{F}_p[z^{p^t}]$-module with a basis $(v^l_{s,t})^g_{l=1}$ and $\mathrm{gr}\mathcal{M}^{t-1}_{p^s}$ is a free $\mathbb{F}_p[z^{p^{t-1}}]$-module with a basis $(v^l_{s,t-1})^g_{l=1}$, see (8.3). The map (8.4) is a homomorphism of $\mathbb{F}_p[z^{p^t}]$-modules.

THEOREM 8.4. *The matrix of* $\mathrm{gr}\mathfrak{c}_{s,1}$ *is the Cartier-Manin matrix* $\mathcal{C}(z^{p^{t-1}})$,

$$\mathrm{gr}\mathfrak{c}_{s,1} \; : \; v^l_{s,t} \; \mapsto \; \sum_{m=1}^{g} v^m_{s,t-1}\mathcal{C}^l_m(z^{p^{t-1}}), \qquad l = 1, \dots, g. \tag{8.5}$$

PROOF. The problem is to express modulo $p^{s-t+2}\mathbb{Z}[z]^n$ the element $p \cdot p^{s-t}I^{[lp^t-1]}_{p^t}(z)$ in terms of the elements $p^{s-t+1}I^{[mp^{t-1}-1]}_{p^{t-1}}(z)$, $m = 1, \dots, g$. In other words, we need to express $I^{[lp^t-1]}_{p^t}(z)$ in terms of $I^{[mp^{t-1}-1]}_{p^{t-1}}(z)$, $m = 1, \dots, g$, modulo $p\mathbb{Z}[z]^n$.

By definition, the vector $I^{[lp^t-1]}_{p^t}(z)$ is the coefficient of $x^{lp^t-1}$ in the Taylor expansion of the polynomial

$$P_{p^t}(x,z) = P_{p^{t-1}}(x,z)\prod_{i=1}^{n}(x - z_i)^{p^{t-1}(p-1)/2},$$

while the vector $I^{[mp^{t-1}-1]}_{p^{t-1}}(z)$ is the coefficient of $x^{mp^{t-1}-1}$ in the Taylor expansion of the polynomial $P_{p^{t-1}}(x,z)$, see notations in Section 4.4. We have

$$\prod_{i=1}^n (x-z_i)^{p^{t-1}(p-1)/2} \equiv \prod_{i=1}^n (x^{p^{t-1}} - z_i^{p^{t-1}})^{(p-1)/2} \qquad \text{mod } p\,.$$

Hence $I^{[lp^t-1]}_{p^t}(z) \equiv \sum_{m=1}^g I^{[mp^{t-1}-1]}_{p^{t-1}}(z)\,\mathcal{C}^l_m(z^{p^{t-1}})$ mod $p$. Theorem 8.4 is proved. □

COROLLARY 8.5. *The matrix of* $\mathrm{gr}\mathfrak{c}_{s,m}\ :\ \mathrm{gr}\mathcal{M}^t_{p^s} \to \mathrm{gr}\mathcal{M}^{t-m}_{p^s}$ *is the product of Cartier-Manin matrices* $\mathcal{C}(z^{p^{t-1}})\mathcal{C}(z^{p^{t-2}})\cdots\mathcal{C}(z^{p^{t-m}})$. *Moreover, this statement, applied to the map*

$$\mathrm{gr}\mathfrak{c}_{s,s-1}\ :\ \mathrm{gr}\mathcal{M}^s_{p^s} \to \mathrm{gr}\mathcal{M}^1_{p^s} = \mathcal{M}^1_{p^s}\,,$$

*can be reformulated as follows. For any* $l=1,\dots,g$, *the solution* $I^{lp^s-1}_{p^s}(z)$ *modulo* $p^s$ *of system* (2.1), *projected to* $\mathbb{F}_p[z]^n$, *equals the projection to* $\mathbb{F}_p[z]^n$ *of the solution*

$$\sum_{m_1,\dots,m_{s-1}=1}^g I^{[m_1p-1]}_p(z)\mathcal{C}^{m_2}_{m_1}(z^p)\cdots\mathcal{C}^{m_{s-1}}_{m_{s-2}}(z^{p^{s-2}})\mathcal{C}^l_{m_{s-1}}(z^{p^{s-1}})$$

*modulo* $p$ *of system* (2.1). □

See these sums in [**V5**, Section 8].

## 9. Change of variables

**9.1. Change of the variable $x$.** Change the variable $x$ and set $x = v + z_n$. Then

$$\tilde\Phi_{p^s}(v,z) := \Phi_{p^s}(v+z_n,z) = \Big(\prod_{i=1}^{n-1}(v-(z_i-z_n))\Big)^{(p^s-1)/2} v^{(p^s-1)/2}\,. \tag{9.1}$$

Let

(9.2)

$$\begin{aligned}\tilde P_{p^s}(v,z) &:= P_{p^s}(v+z_n,z)\\ &= \Big(\frac{\tilde\Phi_{p^s}(v,z)}{v-(z_1-z_2)},\dots,\frac{\tilde\Phi_{p^s}(v,z)}{v-(z_{n-1}-z_n)},\frac{\tilde\Phi_{p^s}(v,z)}{v}\Big) = \sum_i \tilde P^i_{p^s}(z)\,v^i\,,\end{aligned}$$

where $\tilde P^i_{p^s}(z)$ are $n$-vectors of polynomials in $z$ with integer coefficients. For a positive integer $l$, denote

$$\tilde I^{[lp^s-1]}_{p^s}(z) := \tilde P^{lp^s-1}_{p^s}(z)\,. \tag{9.3}$$

The polynomial $\tilde I^{[lp^s-1]}_{p^s}(z)$ is nonzero if $l=1,\dots,g$. Notice that every polynomial $\tilde I^{[lp^s-1]}_{p^s}(z)$ is a function of differences $z_i-z_n$, $i=1,\dots,n-1$.

Consider the increasing filtration

$$0 = \tilde{\mathcal{M}}^0_{p^s} \subset \tilde{\mathcal{M}}^1_{p^s} \subset \cdots \subset \tilde{\mathcal{M}}^{s-1}_{p^s} \subset \tilde{\mathcal{M}}^s_{p^s} = \tilde{\mathcal{M}}_{p^s}\,, \tag{9.4}$$

where

$$(9.5)\qquad \tilde{\mathcal{M}}_{p^s} = \Big\{\pi_s\Big(\sum_{r=1}^{s}\sum_{l=1}^{g} c_{r,l}(z)\, p^{s-r}\tilde{I}^{[lp^r-1]}_{p^r}(z)\Big) \mid c_{r,l}(z)\in\mathbb{Z}[z]_{p^r}\Big\},$$

$$(9.6)\qquad \tilde{\mathcal{M}}^t_{p^s} = \Big\{\pi_s\Big(\sum_{r=1}^{t}\sum_{l=1}^{g} c_{r,l}(z)\, p^{s-r}\tilde{I}^{[lp^r-1]}_{p^r}(z)\Big) \mid c_{r,l}(z)\in\mathbb{Z}[z]_{p^r}\Big\},$$

$t=1,\dots,s$.

THEOREM 9.1. *For any $l$, the vector of polynomials $\tilde{I}^{[lp^s-1]}_{p^s}(z)\in\mathbb{Z}[z]^n$ is a solution of system (2.1) modulo $p^s$. For any $t=1,\dots,s$ we have $\tilde{\mathcal{M}}^t_{p^s}=\mathcal{M}^t_{p^s}$.*

PROOF. The proof is the same as the proof of Theorem 5.2 and the proof of [**V5**, Lemma 5.2]. In the proof of Theorem 9.1 the following Lemma 9.2 is used instead of Lemma 5.3. □

LEMMA 9.2. *Let $r=0,\dots,s-1$ and $m\not|\, p$, then $\binom{mp^r+lp^s-1}{lp^s-1}$ is divisible by $p^{s-r}$.*

PROOF. We have $\binom{mp^r+lp^s-1}{lp^s-1}=\binom{mp^r+lp^s-1}{mp^r}=\frac{lp^s}{mp^r}\binom{mp^r+lp^s-1}{mp^r-1}$. □

**9.2. Change of variables $z$.** We introduce the new variables $u_1,\dots,u_n$ by the formulas:

(9.7)

$$u_1=z_1-z_n,\quad u_2=\frac{z_2-z_n}{z_1-z_n},\quad \dots\quad u_{n-1}=\frac{z_{n-1}-z_n}{z_{n-2}-z_n},\quad u_n=z_1+\dots+z_n,$$

or

$$z_i-z_n=u_1\cdots u_i,\qquad i=1,\dots,n-1,\qquad z_1+\dots+z_n=u_n\,.$$

For any $l,s$ we denote $u=(u_1,\dots,u_{n-1})$,

$$(9.8)\qquad \hat{I}^{[lp^s-1]}_{p^s}(u) := \tilde{I}^{[lp^s-1]}_{p^s}(z(u))\,.$$

Each $\hat{I}^{[lp^s-1]}_{p^s}(u)$ is an $n$-vector of polynomials in $u$ with coefficients in $\mathbb{Z}^n$.

Each $\hat{I}^{[lp^s-1]}_{p^s}(u)$ is a solution of system (2.1) modulo $p^s$, in which the change of variables $z=z(u)$ is performed.

**9.3. Change of variables in the KZ equations.** It is known that system (2.1) of the differential KZ equations has suitable asymptotic zones with appropriate local coordinates, in which the differential KZ equations have singularities only at the coordinate hyperplanes. See a definition of the asymptotic zones, for example, in [**V2**]. The coordinates $u$ defined in (9.7) are local coordinates in one of the asymptotic zones. In these coordinates, system (2.1) takes the form,

$$(9.9)\qquad \frac{\partial I}{\partial u_n}=0,\qquad \frac{\partial I}{\partial u_i}=\frac12\Big(\frac{\Omega_i}{u_i}+\operatorname{Reg}_i(u)\Big)I,\quad i=1,\dots,n-1,$$
$$I_1+\dots+I_n=0,$$

where $\Omega_i=\sum_{i\leqslant k<l\leqslant n}\Omega_{k,l}$ and $\operatorname{Reg}_i(u)$ is an $n\times n$-matrix depending on $u$ and regular at the origin $u=0$. The origin is a regular singular point of system (9.9) and one may expand solutions at the origin in suitable series in the variables $u$.

Any polynomial $\hat{I}^{[lp^s-1]}_{p^s}(u)$ is a solution of system (9.9) modulo $p^s$. We will expand the polynomial $\hat{I}^{[lp^s-1]}_{p^s}(u)$ at $u=0$ and show that this expansion has a

$p$-adic limit as $s \to \infty$. In that way we will construct a $g$-dimensional space of $p$-adic solutions of system (9.9), which is the same as the original system (2.1) of the differential KZ equations up to the change of variables, $z = z(u)$.

**9.4. Taylor expansion of $\hat{I}^{[lp^s-1]}_{p^s}(u)$.** Denote

$$\hat\Phi_{p^s}(v,u) := \tilde\Phi_{p^s}(v,z(u)) = \left(\prod_{i=1}^{n-1}\Big(v-\prod_{j=1}^{i}u_j\Big)\right)^{(p^s-1)/2} v^{(p^s-1)/2}, \tag{9.10}$$

$$\hat P_{p^s}(v,u) := \Big(\frac{\hat\Phi_{p^s}(v,u)}{v-u_1},\dots,\frac{\hat\Phi_{p^s}(v,u)}{v-u_1\cdots u_{n-1}},\frac{\hat\Phi_{p^s}(v,u)}{v}\Big) = \sum_i \hat P^i_{p^s}(u)\,v^i, \tag{9.11}$$

where $\hat P^i_{p^s}(u)$ are $n$-vectors of polynomials in $u$ with coefficients in $\mathbb{Z}$. For a positive integer $l$, we have

$$\hat P^{lp^s-1}_{p^s}(u) \;=\; \hat I^{[lp^s-1]}_{p^s}(u)\,, \tag{9.12}$$

where $\hat I^{[lp^s-1]}_{p^s}(u)$ is defined in (9.8).

For $l = 1,\dots,g$, denote

$$u^{l,s} = (-1)^{\delta_l}(u_1\cdots u_{n-2l})^{-l}\prod_{i=1}^{n-2l}(u_1\cdots u_i)^{\frac{p^s-1}{2}}, \tag{9.13}$$

or

$$u^{g,s} = (-1)^{\delta_g}u_1^{\frac{p^s-1}{2}-g},\quad u^{g-1,s} = (-1)^{\delta_{g-1}}u_1^{3\frac{p^s-1}{2}-g+1}u_2^{2\frac{p^s-1}{2}-g+1}u_3^{\frac{p^s-1}{2}-g+1},\dots$$
$$u^{1,s} = (-1)^{\delta_1}u_1^{(n-2)\frac{p^s-1}{2}-1}u_2^{(n-3)\frac{p^s-1}{2}-1}\cdots u_{n-2}^{\frac{p^s-1}{2}-1},$$

see $(-1)^{\delta_l}$ in (7.1). Denote

$$C^{l,s} = \binom{\frac{p^s-1}{2}}{l}\Big(0,\dots,0,\frac{2l}{p^s-1},1,\dots,1\Big), \tag{9.14}$$

where 0 is repeated $2g-2l$ times and 1 is repeated $2l$ times, cf. formula (7.3).

THEOREM 9.3. *For $l = 1,\dots,g$, the polynomial $\hat I^{[lp^s-1]}_{p^s}(u)$ has the following form,*

$$\begin{aligned}\hat I^{[lp^s-1]}_{p^s}(u) &= u^{l,s}T^{l,s}(u)\,,\\ T^{l,s}(u) &= (T^{l,s}_1(u),\dots,T^{l,s}_n(u)),\end{aligned} \tag{9.15}$$

*with coordinates $T^{l,s}_j$ defined as follows. If $j = 1,\dots,n-1$, then*

$$\begin{aligned}T^{l,s}_j &= u_{j+1}\cdots u_{n-2l}\sum\nolimits^{l,s,j}\binom{\frac{p^s-3}{2}}{a_j}\prod_{i=1,\,i\neq j}^{n-1}\binom{\frac{p^s-1}{2}}{a_i}\\ &\times\prod_{i=1}^{n-2l-1}(u_{i+1}\cdots u_{n-2l})^{a_i}\prod_{i=1}^{2l-1}(u_{n-2l+1}\cdots u_{n-2l+i})^{a_{n-2l+i}},\end{aligned} \tag{9.16}$$

*where the summation $\sum^{l,s,j}$ is over all $a_1,\dots,a_{n-1}\in\mathbb{Z}$, $0\leqslant a_i\leqslant\frac{p^s-1}{2}$, such that $a_1+\dots+a_{n-2l} = a_{n-2l+1}+\dots+a_{n-1}+l-1$, if $j\leqslant n-2l$; and such that*

$a_1+\dots+a_{n-2l}=a_{n-2l+1}+\dots+a_{n-1}+l$, *if* $n-2l<j\leqslant n-1$;

(9.17)

$$T_n^{l,s}=\sum\nolimits^{l,s,n}\prod_{i=1}^{n-1}\binom{\frac{p^s-1}{2}}{a_i}\prod_{i=1}^{n-2l-1}(u_{i+1}\cdots u_{n-2l})^{a_i}\prod_{i=1}^{2l-1}(u_{n-2l+1}\cdots u_{n-1})^{a_{n-2l+i}},$$

*where the summation* $\sum^{l,s,n}$ *is over all* $a_1,\dots,a_{n-1}\in\mathbb{Z}$, $0\leqslant a_i\leqslant\frac{p^s-1}{2}$, *such that* $a_1+\dots+a_{n-2l}=a_{n-2l+1}+\dots+a_{n-1}+l$.

*The constant term of* $T^{l,s}(u)$ *equals* $C^{l,s}$.

Notice that the factor $u_{j+1}\cdots u_{n-2l}$ in (9.16) equals 1 if $j\geqslant n-2l$.

PROOF. Make the change of variables $v=wu_1\cdots u_{n-2l}$ in (9.11),

$$P^{\circ}_{p^s}(w,u):=\hat P_{p^s}(wu_1\cdots u_{n-2l},u)=\sum_i\hat P^i_{p^s}(u)\,(wu_1\cdots u_{n-2l})^i=:\sum_i P^{\circ,i}_{p^s}(u)\,w^i\,.$$

Hence

$$\hat I^{[lp^s-1]}_{p^s}(u)\;=\;\big(u_1\cdots u_{n-2l}\big)^{-(lp^s-1)}\,P^{\circ,lp^s-1}_{p^s}(u)\,.$$

We transform the factors in the polynomial $P^{\circ}_{p^s}(w,u)$ as follows. For any positive integers $k$ and $i\leqslant n-2l$ we write

(9.18)

$$\begin{aligned}(wu_1\cdots u_{n-2l}-u_1\cdots u_i)^k&=(u_1\cdots u_i)^k(wu_{i+1}\cdots u_{n-2l}-1)^k\\&=(u_1\cdots u_i)^k\sum_{a=0}^{k}(-1)^{k-a}\binom{k}{a}(wu_{i+1}\cdots u_{n-2l})^a,\end{aligned}$$

and if $i>n-2l$, we write

(9.19)

$$\begin{aligned}(wu_1\cdots u_{n-2l}-u_1\cdots u_i)^k&=(wu_1\cdots u_{n-2})^k(1-u_{n-2l+1}\cdots u_i/w)^k\\&=(wu_1\cdots u_{n-2})^k\sum_{a=0}^{k}(-1)^a\binom{k}{a}(u_{n-2l+1}\cdots u_i/w)^a.\end{aligned}$$

Notice that for factors in (9.11), we have $k=\frac{p^s-1}{2}$ or $k=\frac{p^s-1}{2}-1$. This explains the binomial coefficients in (9.16) and (9.17).

We prove formula (9.16) for $j=1$, the proof for other values of $j$ is similar.

Our goal is to calculate the first coordinate of the vector

$$\big(u_1\cdots u_{n-2}\big)^{-(lp^s-1)}\,P^{\circ,lp^s-1}_{p^s}(u).$$

That is we need to calculate the coefficient of $w^{lp^s-1}$ in

$$\begin{aligned}&\big(u_1\cdots u_{n-2}\big)^{-(lp^s-1)}u_1^{\frac{p^s-1}{2}-1}(wu_2\cdots u_{n-2l}-1)^{\frac{p^s-1}{2}-1}(u_1u_2)^{\frac{p^s-1}{2}}\\&\qquad(wu_3\cdots u_{n-2l}-1)^{\frac{p^s-1}{2}}\cdots\\&\cdots(u_1\cdots u_{n-2l})^{\frac{p^s-1}{2}}(w-1)^{\frac{p^s-1}{2}}(wu_1\cdots u_{n-2l})^{\frac{p^s-1}{2}}(1-u_{n-2l+1}/w)^{\frac{p^s-1}{2}}\cdots\\&\cdots(wu_1\cdots u_{n-2l})^{\frac{p^s-1}{2}}(1-u_{n-2l+1}\cdots u_{n-1}/w)^{\frac{p^s-1}{2}}(wu_1\cdots u_{n-2l})^{\frac{p^s-1}{2}},\end{aligned}$$

which is the same as the coefficient of $w^{l-1}$ in

$$\begin{aligned}(9.20)\quad&u_2\cdots u_{n-2l}\,u^{l,s}\,(wu_2\cdots u_{n-2l}-1)^{\frac{p^s-1}{2}-1}(wu_3\cdots u_{n-2l}-1)^{\frac{p^s-1}{2}}\cdots\\&\cdots(w-1)^{\frac{p^s-1}{2}}(1-u_{n-2l+1}/w)^{\frac{p^s-1}{2}}\cdots(1-u_{n-2l+1}\cdots u_{n-1}/w)^{\frac{p^s-1}{2}}\,.\end{aligned}$$

Expanding the binomials we obtain formula (9.16) for $j = 1$.

The constant term of $T^{l,s}(u)$ is given by the summands in (9.16) and (9.17), corresponding to $a_1 = \dots = a_{n-2l-1} = a_{n-2l+1} = \dots = a_{n-1} = 0$ and $j = n-2l, \dots, n$. Theorem 9.3 is proved. □

**9.5. Taylor expansion of holomorphic solutions.** Recall the multivalued holomorphic solutions of system (2.1) described in Section 3,

$$I^{(\gamma)}(z) = \int_\gamma \Big(\frac{\Phi(x,z)}{x-z_1}, \dots, \frac{\Phi(x,z)}{x-z_n}\Big) dx\,.$$

We make the same changes of variables in the integrals $I^{(\gamma)}(z)$ as we did in the previous sections. Namely, first we change the integration variable $x$ and set $x = v + z_n$, then we make the change of variables $z$ and set $z = z(u)$. The resulting integral is

$$\hat I^{(\gamma)}(u) = \int_\gamma \Big(\frac{\hat\Phi(v,u)}{v-u_1}, \dots, \frac{\hat\Phi(v,u)}{v}\Big) dv\,,$$

where $\hat\Phi(v,u) = \Big(\prod_{i=1}^{n-1}\big(v - \prod_{j=1}^{i} u_j\big)\Big)^{-1/2} v^{-1/2}$.

For $l = 1, \dots, g$, we change the integration variable $v$ and set $v = wu_1\dots u_{n-2l}$. Then

$$\begin{aligned}\hat\Phi(wu_1\dots u_{n-2l}, u) &= e^{(l-n/2)\pi i}\,(u_1\cdots u_{n-2l})^{-l}\prod_{i=1}^{n-2l}(u_1\cdots u_i)^{-1/2}\\ &\times \Big((1-wu_2\cdots u_{n-2l})(1-wu_3\cdots u_{n-2l})\cdots(1-w)\\ &\quad\times (1-u_{n-2l+1}/w)\cdots(1-u_{n-2l+1}\cdots u_{n-1}/w)\Big)^{-1/2} w^{-l}\,.\end{aligned}$$

Choose the integration cycle $\gamma = \gamma_l$ to be the circle $|w| = 1/2$ oriented counterclockwise. We assume that all the variables $u_2, \dots, u_{n-1}$ lie inside the circle. We fix the branch of the function

$$\begin{aligned}\Big((1-wu_2\cdots u_{n-2l})(1-wu_3\cdots u_{n-2l})\cdots(1-w)\\ \times (1-u_{n-2l+1}/w)\cdots(1-u_{n-2l+1}\cdots u_{n-1}/w)\Big)^{-1/2}\end{aligned} \tag{9.21}$$

over the circle by choosing the argument of the function in (9.21) at $w = 1/2$, $u_2 = \dots = u_{n-1} = 0$ to be 0. We multiply the circle with the chosen branch of the integrand by $\frac{e^{n\pi i/2}}{2\pi i}$. This finishes the description of $\gamma_1$.

The resulting integral is

$$\begin{aligned}&\hat I^{(\gamma_l)}(u)\\ &= \frac{e^{n\pi i/2}}{2\pi i}\int_{|w|=1/2}\Big(\frac{\hat\Phi(wu_1\dots u_{n-2l},u)}{wu_1\dots u_{n-2l}-u_1}, \dots, \frac{\hat\Phi(wu_1\dots u_{n-2l},u)}{wu_1\dots u_{n-2l}}\Big) u_1\dots u_{n-2l}\,dw.\end{aligned}$$

Denote

$$u^l := (u_1\cdots u_{n-2l})^{-l}\prod_{i=1}^{n-2l}(u_1\cdots u_i)^{-1/2}\,. \tag{9.22}$$

THEOREM 9.4. *For $l = 1, \dots, g$, the function $\hat I^{(\gamma_l)}(u)$ has the following form,*

$$\hat I^{(\gamma_l)}(u) = u^l\, T^l(u)\,, \qquad T^l(u) = (T^l_1(u), \dots, T^l_n(u)), \tag{9.23}$$

*with coordinates $T^l_j$ defined as follows. If $j = 1, \dots, n-1$, then*

$$\begin{aligned} T^l_j \;&=\; u_{j+1}\cdots u_{n-2l} \sum\nolimits^{l,j} \binom{\frac{-3}{2}}{a_j} \prod_{i=1,\, i\neq j}^{n-1} \binom{\frac{-1}{2}}{a_i} \\ &\times \prod_{i=1}^{n-2l-1} (u_{i+1}\cdots u_{n-2l})^{a_i} \prod_{i=1}^{2l-1} (u_{n-2l+1}\cdots u_{n-2l+i})^{a_{n-2l+i}}\,, \end{aligned} \tag{9.24}$$

*where the summation $\sum^{l,j}$ is over all $a_1, \dots, a_{n-1} \in \mathbb{Z}_{\geqslant 0}$ such that $a_1+\dots+a_{n-2l} = a_{n-2l+1} + \dots + a_{n-1} + l - 1$, if $j \leqslant n - 2l$; and such that $a_1 + \dots + a_{n-2l} = a_{n-2l+1} + \dots + a_{n-1} + l$, if $n - 2l < j \leqslant n - 1$;*

$$T^l_n = \sum\nolimits^{l,n} \prod_{i=1}^{n-1} \binom{\frac{-1}{2}}{a_i} \prod_{i=1}^{n-2l-1} (u_{i+1}\cdots u_{n-2l})^{a_i} \prod_{i=1}^{2l-1} (u_{n-2l+1}\cdots u_{n-1})^{a_{n-2l+i}}\,, \tag{9.25}$$

*where the summation $\sum^{l,n}$ is over all $a_1, \dots, a_{n-1} \in \mathbb{Z}_{\geqslant 0}$ such that $a_1+\dots+a_{n-2l} = a_{n-2l+1} + \dots + a_{n-1} + l$.*

*The power series $T^l(u)$ converges in the polydisc $\{(u_2, \dots, u_{n-1}) \in \mathbb{C}^{n-1} \mid |u_i| < 1,\ i = 2, \dots, n-1\}$.*

PROOF. We prove formula (9.16) for $j = 1$, the proof for other values of $j$ is similar.

The function $T^l_1(u)$ equals

$$\begin{aligned} &\tfrac{e^{n\pi i/2}}{2\pi i} \textstyle\int_{|w|=1/2} u_1^{\frac{-3}{2}} (wu_2\cdots u_{n-2l} - 1)^{\frac{-3}{2}} (u_1u_2)^{\frac{-1}{2}} (wu_3\cdots u_{n-2l} - 1)^{\frac{-1}{2}} \cdots \\ &\quad \cdots (u_1\cdots u_{n-2l})^{\frac{-1}{2}} (w-1)^{\frac{-1}{2}} (wu_1\cdots u_{n-2l})^{\frac{-1}{2}} (1 - u_{n-2l+1}/w)^{\frac{-1}{2}} \cdots \\ &\quad \cdots (wu_1\cdots u_{n-2l})^{\frac{-1}{2}} (1 - u_{n-2l+1}\cdots u_{n-1}/w)^{\frac{-1}{2}} \\ &\quad \times (wu_1\cdots u_{n-2l})^{\frac{-1}{2}}\, u_1\cdots u_{n-2l}\, dw \end{aligned}$$

$$\begin{aligned} &= \frac{(-1)^{l-1}}{2\pi i}\, u_1^{-1} (u_1\cdots u_{n-2l})^{-l+1} \prod_{i=1}^{n-2l} (u_1\cdots u_i)^{\frac{-1}{2}} \\ &\times \int_{|w|=1/2} (1 - wu_2\cdots u_{n-2l})^{\frac{-3}{2}} \Big( (1 - wu_3\cdots u_{n-2l}) \cdots (1-w) \\ &\times (1 - u_{n-2l+1}/w) \cdots (1 - u_{n-2l+1}\cdots u_{n-1}/w) \Big)^{\frac{-1}{2}} \frac{dw}{w^l}\,. \end{aligned}$$

Expanding the binomials we obtain formula (9.24) for $j = 1$.

The convergence property is clear. □

**9.6. Formal solutions over $\mathbb{Q}_p$ and truncation.** For $l = 1, \dots, g$, the formal series $\hat I^{(\gamma_l)}(u)$ is a formal solution of system (2.1), in which the change of variables $z = z(u)$ is performed and which is considered over the field $\mathbb{Q}_p$ of $p$-adic numbers.

LEMMA 9.5. *The formal series $\hat I^{(\gamma_l)}(u)$, $l = 1, \dots, g$, are linear independent over the field $\mathbb{Q}_p$.*

PROOF. The proof follows from the fact that the monomials $u^l$, $l = 1, \dots, n$, are linear independent over $\mathbb{Q}_p$. □

**9.7. Example** $n = 3$. In this case we have $g = 1$, $u_1 = z_1 - z_3$, $u_2 = \frac{z_2 - z_3}{z_1 - z_3}$,

(9.26)

$$\hat{I}^{(\gamma_1)}(u_1, u_2) = u_1^{-3/2} \sum_{a=0}^{\infty} \left( \binom{-\frac{3}{2}}{a}\binom{-\frac{1}{2}}{a}, \binom{-\frac{1}{2}}{a+1}\binom{-\frac{3}{2}}{a}, \binom{-\frac{1}{2}}{a+1}\binom{-\frac{1}{2}}{a} \right) u_2^a$$
$$= u_1^{-3/2} \sum_{a=0}^{\infty} \binom{-\frac{1}{2}}{a+1}\binom{-\frac{3}{2}}{a} \Big( \frac{a+1}{-1/2 - a}, 1, \frac{-1/2}{-1/2 - a} \Big) u_2^a$$

(9.27)

$$\hat{I}^{[p^s-1]}(u_1, u_2) =$$
$$= (-1)^{\frac{p^s-3}{2}} u_1^{\frac{p^s-3}{2}} \sum_{a=0}^{\frac{p^s-1}{2}} \left( \binom{\frac{p^s-3}{2}}{a}\binom{\frac{p^s-1}{2}}{a}, \binom{\frac{p^s-1}{2}}{a+1}\binom{\frac{p^s-3}{2}}{a}, \binom{\frac{p^s-1}{2}}{a+1}\binom{\frac{p^s-1}{2}}{a} \right) u_2^a$$
$$= (-1)^{\frac{p^s-3}{2}} u_1^{\frac{p^s-3}{2}} \sum_{a=0}^{\frac{p^s-1}{2}} \binom{\frac{p^s-1}{2}}{a+1}\binom{\frac{p^s-3}{2}}{a} \Big( \frac{a+1}{(p^s-1)/2 - a}, 1, \frac{(p^s-1)/2}{(p^s-1)/2 - a} \Big) u_2^a$$

## 10. $p$-Adic convergence

Consider the field $\mathbb{Q}_p$ with the standard $p$-adic norm $|t|_p$, $t \in \mathbb{Q}_p$. In this section we consider the polynomial solutions $((-1)^{\delta_l} T^{l,s}(u))_{s=1}^{\infty}$ and the formal solutions $\hat{I}^{(\gamma_l)}(u)$ as functions on $\mathbb{Q}_p^{n-1}$.

Recall that $\mathbb{Z}_p \subset \mathbb{Q}_p$ denotes the ring of $p$-adic integers.

**10.1. Teichmuller representatives.** For $t \in \mathbb{Z}_p$ there exists the unique solution $\omega(t) \in \mathbb{Z}_p$ of the equation $\omega(t)^p = \omega(t)$ that is congruent to $t$ modulo $p$. The element $\omega(t)$ is called the *Teichmuller representative.* It also can be defined by $\omega(t) = \lim_{s\to\infty} t^{p^s}$. The *Teichmuller character* is the homomorphism

$$\mathbb{F}_p^{\times} \to \mathbb{Z}_p^{\times}, \qquad \alpha \mapsto \omega(\alpha), \qquad \alpha \in \mathbb{F}_p^{\times}.$$

For $\alpha \in \mathbb{F}_p$, $r > 0$, define the disc

$$D_{\alpha,r} = \{t \in \mathbb{Z}_p \mid |t - \omega(\alpha)|_p < r\}.$$

The space $\mathbb{Z}_p$ is the disjoint union of the discs $D_{\alpha,1}$, $\alpha \in \mathbb{F}_p$. The function $\omega : \mathbb{Z}_p \to \mathbb{Z}_p$, $t \mapsto \omega(t)$, is a locally constant function equal to $\omega(\alpha)$ on the disc $D_{\alpha,1}$.

For a subset $S \subset \mathbb{Z}_p$ and a function $f : S \to \mathbb{Z}_p$ define the norm

$$\| f \| = \sup_{t\in S} |f(t)|_p\,.$$

LEMMA 10.1. *For any $\alpha \in \mathbb{F}_p$ the sequence of polynomial functions $(x^{p^s})_{s=1}^{\infty}$ uniformly converges on $D_{\alpha,1}$ to the constant function $\omega(\alpha)$.*

PROOF. We use the "fundamental inequality" from [**Ro**, II.4.3]: if $|t|_p \leqslant 1$, then $|(1+t)^{p^s} - 1|_p \leqslant |t|_p \cdot \max(|t|_p, 1/p)^s$. Now let $t \in D_{\alpha,1}$. Then $t^{p-1} = 1 + t_1$, $|t_1|_p \leqslant 1/p$. We have $|t^{p^{s+1}} - t^{p^s}|_p = |t^{p^s}|_p |t^{(p-1)p^s} - 1|_p \leqslant |(1+t_1)^{p^s} - 1|_p \leqslant 1/p^{s+1}$.

For positive integers $s_1, s_2$ and $t \in D_{\alpha,1}$ we have $|t^{p^{s_1+s_2}} - t^{p^{s_1}}|_p = |t^{p^{s_1+s_2}} - t^{p^{s_1+s_2-1}} + t^{p^{s_1+s_2-1}} + \dots + t^{p^{s_1+1}} - t^{p^{s_1}}|_p \leqslant 1/p^{s_1+1}$. Hence the sequence $(x^{p^s})_{s=1}^{\infty}$ is a Cauchy sequence.

For $t \in D_{0,1}$, we have $|t^{p^s}|_p \leqslant 1/p^s$. For $t_1, t_2 \in D_{\alpha,1}$, $\alpha \neq 0$, we have $t_1/t_2 = 1+t$ with $|t|_p \leqslant 1/p$ and $|t_1^{p^s} - t_2^{p^s}|_p = |(1+t)^{p^s} - 1|_p \leqslant 1/p^{s+1}$. The lemma is proved. □

For $\alpha \in \mathbb{F}_p$ consider the sequence of polynomial functions $(x^{(p^s-1)/2})_{s=1}^{\infty}$ on $D_{\alpha,1}$. This sequence uniformly converges to 0 on the disc $D_{0,1}$.

Let $\alpha^{(p-1)/2} = 1$. Let $\beta \in \mathbb{F}_p$ be such that $\beta^2 = \alpha$. The function $D_{\beta,1} \to D_{\alpha,1}$, $t \mapsto t^2$, is an analytic diffeomorphism. The inverse function $D_{\alpha,1} \to D_{\beta,1}$ will be denoted by $x^{1/2}$. There are two square roots $\pm x^{1/2}$. The root $x^{1/2}$ corresponds to the chosen $\beta \in \mathbb{F}_p$ and the root $-x^{1/2}$ corresponds to $-\beta \in \mathbb{F}_p$.

We change the variable $x$, set $x = y^2$, and lift the sequence $(x^{(p^s-1)/2})_{s=1}^{\infty}$ to the sequence $(y^{p^s-1})_{s=1}^{\infty}$ of polynomial functions on $D_{\beta,1}$.

LEMMA 10.2. *The sequence of polynomial functions $(y^{p^s-1})_{s=1}^{\infty}$ uniformly converges on $D_{\beta,1}$ to the function $\omega(\beta)/y$. In other words, the sequence of polynomial functions $(x^{(p^s-1)/2})_{s=1}^{\infty}$ uniformly converges on $D_{\alpha,1}$ to the function $\omega(\beta)x^{-1/2}$.*

PROOF. The lemma follows from Lemma 10.1. □

Let $\alpha^{(p-1)/2} = -1$. Then $\omega(\alpha)^{(p^s-1)/2} = (-1)^{1+p+\dots+p^{s-1}} = (-1)^s$ and the sequence $(x^{(p^s-1)/2})_{s=1}^{\infty}$ has no limit on $D_{\alpha,1}$.

**10.2. Approximation of binomial coefficients.** It is known that $\binom{-\frac12}{a}, \binom{-\frac32}{a} \in \mathbb{Z}_p$ for $a \in \mathbb{Z}_{\geqslant 0}$.

LEMMA 10.3. *Let $l_1 \geqslant 0$, $l_2 \geqslant 0$ be integers. Then there exists an integer $s_0 \geqslant 0$, such that for any integer $s \geqslant s_0$ and any integer $a$ with $\frac{p^s-1}{2} - l_1 \geqslant l_2 + a \geqslant 0$ we have*

(10.1)
$$\left|\binom{-\frac12 - l_1}{l_2 + a} - \binom{\frac{p^s-1}{2} - l_1}{l_2 + a}\right|_p \leqslant 1/p^{s-d-a}, \qquad \text{where} \quad d = l_1 + l_2 - 1/2.$$

PROOF. We have
$$\binom{-\frac12 - l_1}{l_2 + a} = \frac{1}{(-2)^{l_2+a}(l_2+a)!}\prod_{k=1}^{l_2+a}(2(l_1+k)-1),$$
$$\binom{\frac{p^s-1}{2} - l_1}{l_2 + a} = \frac{1}{(-2)^{l_2+a}(l_2+a)!}\prod_{k=1}^{l_2+a}(2(l_1+k)-1-p^s).$$

The $p$-adic norm of $\binom{-\frac12-l_1}{l_2+a}$ is $\leqslant 1$. The difference $\binom{-\frac12-l_1}{l_2+a} - \binom{\frac{p^s-1}{2}-l_1}{l_2+a}$ is the sum of products $\frac{1}{(-2)^{l_2+a}(l_2+a)!}\prod_{k=1}^{l_2+a}(2(l_1+k)-1)$ in each of which at least one of the factors $2(l_1+k)-1$ is replaced with $-p^s$. We prove that even one such replacement implies that this summand of the difference has $p$-adic norm $\leqslant 1/p^{s+1/2-l_1-l_2-a}$.

Indeed, let $2(l_1+k)-1 = bp^c$, $p \nmid b$. We have $2(l_1+k)-1 \leqslant 2(l_1+l_2+a)-1 \leqslant 2(l_1 + \frac{p^s-1}{2}) - 1$. Hence $bp^c \leqslant 2l_1 - 2 + p^s$. Hence for any $s$, large enough, we have $c \leqslant s$, and the replacement of $2(l_1+k)-1$ with $-p^s$ makes the norm of that summand $\leqslant 1/p^{s-c}$.

We also have $bp^c \leqslant 2(l_1+l_2+a)-1$. Hence $c \leqslant p^{c-1} < \frac{p^c}{2} \leqslant \frac{bp^c}{2} \leqslant a+l_1+l_2-\frac12$. This shows that each summand has the $p$-adic norm $1/ \leqslant p^{s-c} \leqslant 1/p^{s+1/2-l_1-l_2-a}$. The lemma is proved. □

**10.3. Example $n = 3$, continuation.** Consider the formal power series

$$(10.2)\qquad T^1(x) = \sum_{a=0}^{\infty} \left( \binom{-\frac{3}{2}}{a}\binom{-\frac{1}{2}}{a}, \binom{-\frac{1}{2}}{a+1}\binom{-\frac{3}{2}}{a}, \binom{-\frac{1}{2}}{a+1}\binom{-\frac{1}{2}}{a} \right) x^a$$

in (9.26) and the sequence of polynomials

$$(10.3)\quad T^{1,s}(x) = \sum_{a=0}^{\frac{p^s-1}{2}} \left( \binom{\frac{p^s-3}{2}}{a}\binom{\frac{p^s-1}{2}}{a}, \binom{\frac{p^s-1}{2}}{a+1}\binom{\frac{p^s-3}{2}}{a}, \binom{\frac{p^s-1}{2}}{a+1}\binom{\frac{p^s-1}{2}}{a} \right) x^a$$

in (9.27) as functions on $\mathbb{Z}_p$.

PROPOSITION 10.4. *The power series $T^1(x)$ uniformly convergence on $D_{0,1}$. The sequence of polynomial functions $(T^{1,s}(x))_{s=1}^{\infty}$ uniformly converges on $D_{0,1}$ to the function $T^1(x)$.*

PROOF. The fact that the binomials $\binom{-\frac{1}{2}}{a}, \binom{-\frac{3}{2}}{a}$ are $p$-adic integers implies the uniform convergence of the power series $T^1(x)$ on $D_{0,1}$.

Let us write $T^1(x) = \sum_{a=0}^{\infty} T_a^1 x^a$ and $T^{1,s}(x) = \sum_{a=0}^{\frac{p^s-1}{2}} T_a^{1,s} x^a$, where $T_a^1, T_a^{1,s} \in \mathbb{Z}_p^3$. Then $T^1(x) - T^{1,s}(x) = \sum_{a=\frac{p^s+1}{2}}^{\infty} T_a^1\, x^a + \sum_{a=0}^{\frac{p^s-1}{2}} (T_a^1 - T_a^{1,s})\, x^a$. Clearly the $p$-adic norm of the first sum is $\leqslant 1/p^{\frac{p^s-1}{2}}$. By Lemma 10.3 if s is big enough, then for each summand of the second sum and $t \in D_{0,1}$ we have $|(T_a^1 - T_a^{1,s})\, t^a|_p \leqslant 1/p^{s-d}$ for some $d$ independent of $s$ and of the summand. This proves the proposition. □

Consider the formal series

(10.4)

$$\hat{I}^{(\gamma_1)}(u_1, u_2) = u_1^{-3/2} \sum_{a=0}^{\infty} \left( \binom{-\frac{3}{2}}{a}\binom{-\frac{1}{2}}{a}, \binom{-\frac{1}{2}}{a+1}\binom{-\frac{3}{2}}{a}, \binom{-\frac{1}{2}}{a+1}\binom{-\frac{1}{2}}{a} \right) u_2^a$$

and the sequence of polynomials

$$(10.5)\quad (-1)^{\frac{p^s-3}{2}}\, \hat{I}^{[p^s-1]}(u_1, u_2) =$$

$$= u_1^{\frac{p^s-3}{2}} \sum_{a=0}^{\frac{p^s-1}{2}} \left( \binom{\frac{p^s-3}{2}}{a}\binom{\frac{p^s-1}{2}}{a}, \binom{\frac{p^s-1}{2}}{a+1}\binom{\frac{p^s-3}{2}}{a}, \binom{\frac{p^s-1}{2}}{a+1}\binom{\frac{p^s-1}{2}}{a} \right) u_2^a$$

as functions on $D_{\alpha,1} \times D_{0,1}$, where $\alpha = \beta^2$ for some $\beta \in \mathbb{F}_p$. Then the function $u_1^{1/2} : D_{\alpha,1} \to D_{\beta,1}$ is well-defined and the series $\hat{I}^{(\gamma_1)}(u_1, u_2)$ is a well-defined function on $D_{\alpha,1} \times D_{0,1}$.

THEOREM 10.5. *The sequence of polynomial functions $\left((-1)^{\frac{p^s-3}{2}} \hat{I}^{[p^s-1]}(u_1, u_2)\right)_{s=1}^{\infty}$ uniformly converges on $D_{\alpha,1} \times D_{0,1}$ to the function $\omega(\beta)\hat{I}^{(\gamma_1)}(u_1, u_2)$.*

PROOF. The theorem follows from Lemma 10.2 and Proposition 10.4 . □

**10.4. $p$-Adic convergence for arbitrary $n$.** Given $l$, $1 \leqslant l \leqslant g$, consider the sequence of polynomials $(-1)^{\delta_l} \hat{I}_{p^s}^{[lp^s-1]}(u)$ and the series $\hat{I}^{(\gamma_l)}(u)$. Here $u = (u_1, \dots, u_{n-1})$.

We multiply the polynomials and the series by the same factor $(u_1 \cdots u_{n-2l})^l = (z_{n-2l} - z_n)^l$ and study the convergence of the sequence of polynomials $J^{l,s} :=$

$(u_1\cdots u_{n-2l})^l(-1)^{\delta_l}\hat I^{[lp^s-1]}_{p^s}(u)$ to the series $J^l := (u_1\cdots u_{n-2l})^l\hat I^{(\gamma_l)}(u)$. Introduce new variables:

$$x_1=\prod_{i=1}^{n-2l}\prod_{j=1}^{i}u_j=\prod_{i=1}^{n-2l}(z_i-z_n), \qquad x_2=u_2\cdots u_{n-2l}=\frac{z_{n-2l}-z_n}{z_1-z_n},$$
$$x_3=u_3\cdots u_{n-2l}=\frac{z_{n-2l}-z_n}{z_2-z_n}, \quad \dots \quad x_{n-2l}=u_{n-2l}=\frac{z_{n-2l}-z_n}{z_{n-2l-1}-z_n},$$
$$x_{n-2l+1}=u_{n-2l+1}=\frac{z_{n-2l+1}-z_n}{z_{n-2l}-z_n}, \quad x_{n-2l+2}=u_{n-2l+1}u_{n-2l+2}=\frac{z_{n-2l+2}-z_n}{z_{n-2l}-z_n},$$
$$\dots \quad x_{n-1}=u_{n-2l+1}\cdots u_{n-1}=\frac{z_{n-1}-z_n}{z_{n-2l}-z_n}.$$

Let $x=(x_1,\dots,x_{n-1})$. Then

$$J^l(x)=x_1^{-\frac12}\,Q^l(x_2,\dots,x_{n-1})=x_1^{-\frac12}(Q^l_1(x_2,\dots,x_{n-1}),\dots,Q^l_n(x_2,\dots,x_{n-1})),$$

with coordinates $Q^l_j$ defined as follows. If $j=1,\dots,n-1$, then

(10.6)

$$Q^l_j = x_{j+1}\sum\nolimits^{l,j}\binom{\frac{-3}{2}}{a_j}\prod_{i=1,\,i\neq j}^{n-1}\binom{\frac{-1}{2}}{a_i}\prod_{i=2}^{n-2l}x_i^{a_{i-1}}\prod_{i=n-2l+1}^{n-1}x_i^{a_i}, \quad j=1,\dots,n-2l-1,$$
$$Q^l_j = \sum\nolimits^{l,j}\binom{\frac{-3}{2}}{a_j}\prod_{i=1,\,i\neq j}^{n-1}\binom{\frac{-1}{2}}{a_i}\prod_{i=2}^{n-2l}x_i^{a_{i-1}}\prod_{i=n-2l+1}^{n-1}x_i^{a_i}, \qquad j=n-2l,\dots,n-1,$$

where the summation $\sum^{l,j}$ is over all $a_1,\dots,a_{n-1}\in\mathbb Z_{\geqslant 0}$ such that $a_1+\dots+a_{n-2l}=a_{n-2l+1}+\dots+a_{n-1}+l-1$, if $j\leqslant n-2l$; and such that $a_1+\dots+a_{n-2l}=a_{n-2l+1}+\dots+a_{n-1}+l$, if $n-2l<j\leqslant n-1$;

$$Q^l_n=\sum\nolimits^{l,n}\prod_{i=1}^{n-1}\binom{\frac{-1}{2}}{a_i}\prod_{i=2}^{n-2l}x_i^{a_{i-1}}\prod_{i=n-2l+1}^{n-1}x_i^{a_i}, \tag{10.7}$$

where the summation $\sum^{l,n}$ is over all $a_1,\dots,a_{n-1}\in\mathbb Z_{\geqslant 0}$ such that $a_1+\dots+a_{n-2l}=a_{n-2l+1}+\dots+a_{n-1}+l$.

We also have

$$J^{l,s}(x)=x_1^{\frac{p^s-1}{2}}\,Q^{l,s}(x_2,\dots,x_{n-2})$$
$$=x_1^{\frac{p^s-1}{2}}(Q^{l,s}_1(x_2,\dots,x_{n-2}),\dots,Q^{l,s}_n(x_2,\dots,x_{n-2})),$$

with coordinates $Q^{l,s}_j(x)$ defined as follows. If $j=1,\dots,n-1$, then

(10.8)

$$Q^{l,s}_j = x_{j+1}\sum\nolimits^{l,j}\binom{\frac{p^s-3}{2}}{a_j}\prod_{i=1,\,i\neq j}^{n-1}\binom{\frac{p^s-1}{2}}{a_i}\prod_{i=2}^{n-2l}x_i^{a_{i-1}}\prod_{i=n-2l+1}^{n-1}x_i^{a_i}, \quad j=1,\dots,n-2l-1,$$
$$Q^{l,s}_j = \sum\nolimits^{l,j}\binom{\frac{p^s-3}{2}}{a_j}\prod_{i=1,\,i\neq j}^{n-1}\binom{\frac{p^s-1}{2}}{a_i}\prod_{i=2}^{n-2l}x_i^{a_{i-1}}\prod_{i=n-2l+1}^{n-1}x_i^{a_i}, \qquad j=n-2l,\dots,n-1,$$

where the summation $\sum^{l,s,j}$ is over all $a_1,\dots,a_{n-1}\in\mathbb Z$, $0\leqslant a_i\leqslant\frac{p^s-1}{2}$, such that $a_1+\dots+a_{n-2l}=a_{n-2l+1}+\dots+a_{n-1}+l-1$, if $j\leqslant n-2l$; and such that

$a_1 + \cdots + a_{n-2l} = a_{n-2l+1} + \cdots + a_{n-1} + l$, if $n - 2l < j \leqslant n - 1$;

$$Q_n^{l,s} = \sum\nolimits^{l,n} \prod_{i=1}^{n-1} \binom{\frac{p^s-1}{2}}{a_i} \prod_{i=2}^{n-2l} x_i^{a_{i-1}} \prod_{i=n-2l+1}^{n-1} x_i^{a_i}, \tag{10.9}$$

where the summation $\sum^{l,s,n}$ is over all $a_1, \dots, a_{n-1} \in \mathbb{Z}$, $0 \leqslant a_i \leqslant \frac{p^s-1}{2}$, such that $a_1 + \cdots + a_{n-2l} = a_{n-2l+1} + \cdots + a_{n-1} + l$.

PROPOSITION 10.6. *The power series $Q^l(x_2, \dots, x_{n-1})$ uniformly convergens on $D_{0,1}^{n-2}$. The sequence of polynomial functions $(Q^{l,s}(x_2, \dots, x_{n-1}))_{s=1}^{\infty}$ uniformly converges on $D_{0,1}^{n-2}$ to the function $Q^l(x_2, \dots, x_{n-1})$.*

PROOF. The fact that the binomials $\binom{-\frac{1}{2}}{a}, \binom{-\frac{3}{2}}{a}$ are $p$-adic integers implies the uniform convergence of the power series $Q^l(x_2, \dots, x_{n-1})$ on $D_{0,1}^{n-2}$. The proof of the uniform convergence of $(Q^{l,s}(x_2, \dots, x_{n-1}))_{s=1}^{\infty}$ to $Q^l(x_2, \dots, x_{n-1})$ follows from Lemma 10.3 in the same way as the uniform convergence in the proof of Proposition 10.4. □

Consider the formal series $J^l(x) = x_1^{-\frac{1}{2}} Q^l(x_2, \dots, x_{n-1})$ and the sequence of polynomials $J^{l,s}(x) = x_1^{\frac{p^s-1}{2}} Q^{l,s}(x_2, \dots, x_{n-2})$ as functions on $D_{\alpha,1} \times D_{0,1}^{n-2}$, where $\alpha = \beta^2$ for some $\beta \in \mathbb{F}_p$. Then the function $x_1^{1/2} : D_{\alpha,1} \to D_{\beta,1}$ is well-defined and the series $J^l(x)$ is a well-defined function on $D_{\alpha,1} \times D_{0,1}^{n-2}$.

THEOREM 10.7. *The sequence of polynomial functions $\left(J^{l,s}(x)\right)_{s=1}^{\infty}$ uniformly converges on $D_{\alpha,1} \times D_{0,1}^{n-2}$ to the function $\omega(\beta) J^l(x)$.*

PROOF. The theorem follows from Lemma 10.2 and Proposition 10.6. □

## Appendix A. The case $n = 3$ and Dwork's theory

by Steven Sperber and Alexander Varchenko

In this appendix we consider only the special case $n = 3$ of previous considerations and show how this special case is related to Dwork's theory in the classical paper [**Dw**].

### A.1. Dwork on Legendre family.

A.1.1. Consider the Legendre family of elliptic curves $E(\lambda)$ defined by the affine equation

$$y^2 = x(x-1)(x-\lambda).$$

Let $\gamma = \gamma(\lambda)$ be a family of 1-cycles on the curves $E(\lambda)$ flat under the Gauss-Manin connection. Then the function

$$h^{(\gamma)}(\lambda) = \int_\gamma \frac{dx}{y} \tag{A.1}$$

satisfies the hypergeometric differential equation

$$\lambda(1-\lambda)h'' + (1-2\lambda)h' - (1/4)h = 0. \tag{A.2}$$

All solutions of this equation are obtained in this way. One of the solutions $h^{(\delta_1)}(\lambda)$, for a suitable $\delta_1$, equals

$$F(\lambda) = {}_2F_1\Big(\frac{1}{2},\frac{1}{2};1;\lambda\Big) = \sum_{k=0}^{\infty}\binom{-1/2}{k}^2\lambda^k, \tag{A.3}$$

where ${}_2F_1$ is the classical hypergeometric function.

A.1.2. If $h(\lambda)$ is the elliptic integral in (A.1), then its derivative is

$$h' = \frac{1}{2}\int_\gamma \frac{dx}{(x-\lambda)y}. \tag{A.4}$$

A.1.3. Equation (A.2) can be written as a system of first order linear differential equations for column 2-vectors $I = (h, h')$,

$$\frac{dI}{d\lambda} = B(\lambda)I, \qquad B(\lambda) = \begin{pmatrix} 0 & 1 \\ \frac{1}{4\lambda(1-\lambda)} & \frac{2\lambda-1}{\lambda(1-\lambda)} \end{pmatrix}. \tag{A.5}$$

A.1.4. In [**Dw**] Dwork considers equation (A.2) over the field $\mathbb{Q}_p$ of $p$-adic numbers and studies analytic properties of the solution $F(\lambda)$. We remind these properties below.

A.1.5. Let $D \subset \mathbb{Q}_p$ be an open subset. A function $f : D \to \mathbb{Q}_p$ is called analytic at $\alpha \in D$ if $f$ can be presented by a power series $\sum_{k=0}^{\infty} c_k(\lambda-\alpha)^k$ with nonzero radius of convergence.

A function $f : D \to \mathbb{Q}_p$ is called analytic on $D$ if $f$ is the uniform limit on $D$ of a sequence of rational function regular on $D$. In that case $f$ is analytic at every point of $D$.

For $i = 1, \dots, N$ let $f_i : D_i \to \mathbb{Q}_p$ be an analytic function on some domain $D_i$. Assume that $D_i \cap D_{i+1} \neq \emptyset$ for $i = 1, \dots, N-1$, and $f_i = f_{i+1}$ on $D_i \cap D_{i+1}$, then we say that the collection of functions $f_i$ defines an analytic element on $\cup_{i=1}^N D_i$. Cf. [**Dw**, Section 0].

A.1.6. Igusa noted in [**Ig**] that modulo $p$ the polynomial

$$g(\lambda) = \sum_{j=0}^{(p-1)/2}\binom{-1/2}{j}^2\lambda^j \tag{A.6}$$

is the unique polynomial solution of equation (A.2) of degree less than $p$ up to multiplication by a constant. Define

$$\mathfrak{D}_1 = \{\lambda \in \mathbb{Z}_p \mid |g(\lambda)|_p = 1\}, \qquad \mathfrak{D}_2 = \{\lambda \mid \lambda^{-1} \in \mathfrak{D}_1\}, \qquad \mathfrak{D} = \mathfrak{D}_1 \cup \mathfrak{D}_2.$$

Notice that $\mathfrak{D}_1$, $\mathfrak{D}_2$ are open and $\mathfrak{D}_1 \cap \mathfrak{D}_2 \neq \emptyset$. More precisely,

$$\mathfrak{D}_1 \cap \mathfrak{D}_2 = \{\lambda \in \mathbb{Z}_p \mid |g(\lambda)|_p = 1,\ |\lambda|_p = 1\}.$$

A.1.7. Dwork considers the functions

$$f(\lambda) = \frac{F(\lambda)}{F(\lambda^p)}, \qquad \eta(\lambda) = \frac{F'(\lambda)}{F(\lambda)}, \tag{A.7}$$

defined in a neighborhood of $0 \in \mathfrak{D}_1$ as ratios of the corresponding convergent power series expansions.

Dwork proves that $f(\lambda)$ can be analytically continued to the domain $\mathfrak{D}_1$. For that, he indicates a sequence of regular rational functions on $\mathfrak{D}_1$, that sequence uniformly converges on $\mathfrak{D}_1$, and its limit equals $F(\lambda)/F(\lambda^p)$ in a neighborhood of 0, see [**Dw**, Lemma 3.4].

From that Dwork deduces that $\eta(\lambda)$ has analytic continuation to the domain $\mathfrak{D}_1$ in the same sense, see [**Dw**, Lemma 3.1].

Since $\eta(\lambda)$ is analytic on $\mathfrak{D}_1$, the function $\eta(1/\lambda)$ is analytic on $\mathfrak{D}_2$.

Using the properties of equation (A.2) Dwork shows that $\eta(1-\lambda) = -\eta(\lambda)$ on $\mathfrak{D}_1$ and shows that $\eta(\lambda) = -\eta(1/\lambda)/\lambda^2 - 1/(2\lambda)$ on $\mathfrak{D}_1 \cap \mathfrak{D}_2$. Hence the function $\eta(\lambda)$ on $\mathfrak{D}_1$ and the function $-\eta(1/\lambda)/\lambda^2 - 1/(2\lambda)$ on $\mathfrak{D}_2$ define an analytic element on $\mathfrak{D}$.

We will use the formulas

$$\eta(1-\lambda) = -\eta(\lambda), \qquad \eta(1/\lambda) = -\lambda^2\eta(\lambda) - \lambda/2\,, \tag{A.8}$$

in Section A.5.

A.1.8. For $\alpha \in \mathbb{Q}_p$ let $V_\alpha$ be the space of germs at $\alpha$ of holomorphic solutions of equation (A.2). For $\alpha \neq 0, 1$ we have $\dim V_\alpha = 2$ and for $\alpha = 0, 1$ we have $\dim V_\alpha = 1$.

For $\alpha \in \mathfrak{D}_1$ let $U_\alpha$ be the space of germs at $\alpha$ of analytic functions defined by the equation

$$\frac{du}{d\lambda} = \eta(\lambda)u\,. \tag{A.9}$$

By [**Dw**, Lemma 3.2], $U_\alpha$ is a subspace of $V_\alpha$. We have $U_\alpha = V_\alpha$ for $\alpha = 0, 1$.

For $\alpha \in \mathfrak{D}_2$ let $U'_\alpha$ be the space of germs at $\alpha$ of analytic functions defined by the equation

$$\frac{du}{d\lambda} = (-\eta(1/\lambda)/\lambda^2 - 1/(2\lambda))u\,. \tag{A.10}$$

By [**Dw**, Lemma 3.2], $U'_\alpha$ is a subspace of $V_\alpha$. For $\alpha \in \mathfrak{D}_1 \cap \mathfrak{D}_2$ we have $U_\alpha = U'_\alpha$.

A.1.9. By [**Dw**, Lemma 4.2], for $\alpha \in \mathfrak{D}$ the subspace $U_\alpha \subset V_\alpha$ also can be characterized as the subspace of germs at $\alpha$ of holomorphic functions bounded in their disc of convergence.

More precisely, let $\mathbb{C}_p$ be the metric completion of the algebraic closure $\bar{\mathbb{Q}}_p$ of the field $\mathbb{Q}_p$. Let $\alpha \in \mathfrak{D}$. Let $u(\lambda) = \sum_{k=0}^\infty c_k(\lambda - \alpha)^k$ be an element of $V_\alpha$. Consider $u(\lambda)$ as a germ at $\alpha \in \mathbb{Q}_p \subset \mathbb{C}_p$ of an analytic function on $\mathbb{C}_p$. The germ $u(\lambda)$ is called bounded on its disc of convergence if $u(\lambda)$ is bounded on its disc of convergence in $\mathbb{C}_p$.

Let $r$ be the radius of convergence of $u(\lambda) = \sum_{k=0}^\infty c_k(\lambda-\alpha)^k$. Define $|u(\lambda)|_0 = \sup_k |c_k| r^k$. Then $u(\lambda)$ is bounded in its disc of convergence if and only if $|u(\lambda)|_0 < \infty$.

A.1.10. The function $F(\lambda)$ is a holomorphic solution of equation (A.2) on the disc $D_{0,1}$. A second solution of (A.5) is of the form $G(\lambda) = F(\lambda)\log\lambda + H(\lambda)$, where the function $H(\lambda)$ is holomorphic on $D_{0,1}$. Dwork specifies $H(\lambda)$ by [**Dw**, Equation (4.19)]. Then

$$\mathcal{F}(\lambda) = \begin{pmatrix} F & G \\ F' & G' \end{pmatrix} \tag{A.11}$$

is a fundamental matrix of solutions of equation (A.5).

In [**Dw**] Dwork introduces a $2\times 2$-matrix function $\mathcal{A}(\lambda)$, and then proves the formula

$$\mathcal{A}(\lambda) = \mathcal{F}(\lambda)\, M\, \mathcal{F}(\lambda^p)^{-1}, \qquad M = \begin{pmatrix} (-1)^{(p-1)/2} & b \\ 0 & (-1)^{(p-1)/2}p \end{pmatrix}, \tag{A.12}$$

where $b$ is a suitable number, see [**Dw**, Lemma 6.2] and formulas on page 72 in [**Dw**]. Dwork shows that $\mathcal{A}(\lambda)$ extends to an analytic function on the domain $\mathfrak{D}_3 \cup \mathfrak{D}_4$, where

$$\text{(A.13)}\quad \mathfrak{D}_3 = \{\lambda \in \mathbb{Z}_p \mid |\lambda|_p = 1, |\lambda - 1|_p = 1\}, \qquad \mathfrak{D}_4 = \{\lambda \in \mathbb{Z}_p \mid \varepsilon < |\lambda|_p < 1\},$$
$$\mathfrak{D}_3 \cup \mathfrak{D}_4 = \{\lambda \in \mathbb{Z}_p \mid \varepsilon < |\lambda|_p \leqslant 1,\ |\lambda - 1|_p = 1\}.$$

Here $\varepsilon$ is some explicit number, $0 < \varepsilon < 1$. See the bottom of page 62 in [**Dw**] and the first sentence of the proof of Theorem 6 in [**Dw**].

Formula (A.12) immediately implies that

$$\text{(A.14)}\qquad \mathcal{A}(\lambda)\mathcal{F}(\lambda^p) = \mathcal{F}(\lambda)M$$

on $\mathfrak{D}_4$. The matrix $\mathcal{A}(\lambda)$ is called the matrix of the Frobenius transformation of solutions of equation (A.5) relative to the fundamental matrix $\mathcal{F}(\lambda)$.

It follows from formula (A.14) that

$$\text{(A.15)}\qquad \mathcal{A}(\lambda)\begin{pmatrix} F(\lambda^p) \\ F'(\lambda^p) \end{pmatrix} = (-1)^{(p-1)/2}\begin{pmatrix} F(\lambda) \\ F'(\lambda) \end{pmatrix}$$

on $\mathfrak{D}_4$. This can be reformulated as the relation

$$\text{(A.16)}\qquad \mathcal{A}(\lambda)\begin{pmatrix} 1 \\ \eta(\lambda^p) \end{pmatrix} = (-1)^{(p-1)/2} f(\lambda)\begin{pmatrix} 1 \\ \eta(\lambda) \end{pmatrix}$$

on $\mathfrak{D}_4$. By the already formulated analytic properties of $\eta(\lambda)$ and $\mathcal{A}(\lambda)$, relation (A.16) can be analytically continued to the domain $\mathfrak{D}_1 \cap \mathfrak{D}_3$.

Equation (A.16) implies that for any $\alpha \in \mathbb{F}_p^\times - \{1\}$ such that $\omega(\alpha) \in \mathfrak{D}_1$, the vector $(1, \eta(\omega(\alpha)))$ is an eigenvector of the Frobenius matrix $\mathcal{A}(\omega(\alpha))$ with eigenvalue $(-1)^{(p-1)/2} f(\omega(\alpha))$,

$$\text{(A.17)}\qquad \mathcal{A}(\lambda)\begin{pmatrix} 1 \\ \eta(\omega(\alpha)) \end{pmatrix} = (-1)^{(p-1)/2} f(\omega(\alpha))\begin{pmatrix} 1 \\ \eta(\omega(\alpha)) \end{pmatrix}.$$

It is known that the zeta function of the elliptic curve defined over $\mathbb{F}_p$ by the equation $y^2 = x(x-1)(x-\alpha)$ has two zeros, which are $1/((-1)^{(p-1)/2} f(\omega(\alpha)))$, $(-1)^{(p-1)/2} f(\omega(\alpha))/p$. It is also known that $|f(\omega(\alpha))|_p = 1$. The number $(-1)^{(p-1)/2} f(\omega(\alpha))$ is called the unit root. See [**Dw**] and also (A.53).

**A.2. KZ equations.** The KZ equations (2.1) for $n = 3$ is the following system of differential and algebraic equations for a column 3-vector $I = (I_1, I_2, I_3)$ depending on variables $z = (z_1, z_2, z_3)$:

$$\text{(A.18)}\qquad \frac{\partial I}{\partial z_1} = \frac{1}{2}\Big(\frac{\Omega_{12}}{z_1 - z_2} + \frac{\Omega_{13}}{z_1 - z_3}\Big)I, \qquad \frac{\partial I}{\partial z_2} = \frac{1}{2}\Big(\frac{\Omega_{21}}{z_2 - z_1} + \frac{\Omega_{23}}{z_2 - z_3}\Big)I,$$
$$\frac{\partial I}{\partial z_3} = \frac{1}{2}\Big(\frac{\Omega_{31}}{z_3 - z_1} + \frac{\Omega_{32}}{z_3 - z_2}\Big)I, \qquad 0 = I_1 + I_2 + I_3,$$

where $\Omega_{ij} = \Omega_{ji}$ and

$$\Omega_{12} = \begin{pmatrix} -1 & 1 & 0 \\ 1 & -1 & 0 \\ 0 & 0 & 0 \end{pmatrix}, \quad \Omega_{13} = \begin{pmatrix} -1 & 0 & 1 \\ 0 & 0 & 0 \\ 1 & 0 & -1 \end{pmatrix}, \quad \Omega_{23} = \begin{pmatrix} 0 & 0 & 0 \\ 0 & -1 & 1 \\ 0 & 1 & -1 \end{pmatrix}.$$

We introduce new variables

$$\text{(A.19)}\qquad u_1 = z_1 - z_3, \qquad u_2 = \frac{z_2 - z_3}{z_1 - z_3}, \qquad u_3 = z_1 + z_2 + z_3,$$

see (9.7). Then system (A.18) takes the form

$$\text{(A.20)}\qquad \frac{\partial I}{\partial u_1} = \frac{1}{2}\frac{\Omega_{12}+\Omega_{13}+\Omega_{23}}{u_1} I\,, \qquad \frac{\partial I}{\partial u_2} = \frac{1}{2}\Big(\frac{\Omega_{12}}{u_2-1}+\frac{\Omega_{23}}{u_2}\Big) I,$$
$$\frac{\partial I}{\partial u_3} = 0, \qquad 0 = I_1+I_2+I_3.$$

The variables in system (A.20) are separated, cf. (9.9).

Denote $\tilde W = \{(I_1,I_2,I_3)\mid I_1+I_2+I_3=0\}$. Then

$$\text{(A.21)}\qquad (\Omega_{12}+\Omega_{13}+\Omega_{23})|_{\tilde W} = -3\,\mathrm{Id}\,.$$

Hence all solutions of system (A.20) have the form

$$\text{(A.22)}\qquad I = u_1^{-3/2}(J_1(u_2),J_2(u_2),J_3(u_2)), \qquad J_1+J_2+J_3=0,$$

where the column vector $J(u_2)$ is a solution of the differential equation

$$\text{(A.23)}\qquad \frac{\partial J}{\partial u_2} = \frac{1}{2}\Big(\frac{\Omega_{12}}{u_2-1}+\frac{\Omega_{23}}{u_2}\Big) J\,.$$

**A.3. Solutions over $\mathbb{C}$.** Any solution of system (A.18) has the form

$$\text{(A.24)}\qquad I^{(\gamma)}(z) = \int_\gamma \Big(\frac{1}{x-z_1},\frac{1}{x-z_2},\frac{1}{x-z_3}\Big)\frac{dx}{\sqrt{(x-z_1)(x-z_2)(x-z_3)}}\,.$$

where $\gamma$ is a flat family of 1-cycles on the elliptic curves of our family of curves.

We change $x$ and $z$ in this integral by setting $x=(z_1-z_3)u+z_3$ and $z=z(u)$ as in (A.19). Then integral (A.24) takes the form

$$\text{(A.25)}\qquad I^{(\gamma)}(u_1,u_2) = u_1^{-3/2}\int_\gamma \Big(\frac{1}{w-1},\frac{1}{w-u_2},\frac{1}{w}\Big)\frac{dw}{\sqrt{(w-1)(w-u_2)\,w}}\,.$$

We take $\gamma=\gamma_1$ to be the circle $|w|=1/2$ oriented counter-clockwise. We assume that $u_2$ lies in this circle. We fix the branch of $\sqrt{(w-1)(w-u_2)\,w}$ over the circle by choosing the argument of $\sqrt{(w-1)(w-u_2)\,w}$ at $w=1/2$, $u_2=0$ to be $\pi/2$. We multiply the circle with the chosen branch of the integrand by $-\frac{1}{2\pi}$. This finishes the description of $\gamma_1$. See the definition of cycles $\gamma_l$ in Section 9.5.

We expand the integral $I^{(\gamma_1)}(u_1,u_2)$ as a power series in $u_2$ and obtain

(A.26)

$$I^{(\gamma_1)}(u_1,u_2) = u_1^{-3/2}\sum_{a=0}^\infty \left(\binom{-\frac32}{a}\binom{-\frac12}{a},\binom{-\frac12}{a+1}\binom{-\frac32}{a},\binom{-\frac12}{a+1}\binom{-\frac12}{a}\right)u_2^a$$
$$= u_1^{-3/2}\sum_{a=0}^\infty \binom{-\frac12}{a+1}\binom{-\frac32}{a}\Big(\frac{a+1}{-1/2-a},1,\frac{-1/2}{-1/2-a}\Big)u_2^a\,,$$

see Theorem (9.4). Denote

(A.27)

$$I := I^{(\gamma_1)} = (z_1-z_3)^{-3/2}\sum_{a=0}^\infty \binom{-\frac12}{a+1}\binom{-\frac32}{a}\Big(\frac{a+1}{-1/2-a},1,\frac{-1/2}{-1/2-a}\Big)\Big(\frac{z_2-z_3}{z_1-z_3}\Big)^a$$

$$\text{(A.28)}\qquad = u_1^{-3/2}\sum_{a=0}^\infty \binom{-\frac12}{a+1}\binom{-\frac32}{a}\Big(\frac{a+1}{-1/2-a},1,\frac{-1/2}{-1/2-a}\Big)u_2^a\,.$$

This series is a solution of system (A.18).

REMARK. Formulas (A.25) and (A.26) imply that

(A.29)

$$I(u_1,u_2)=u_1^{-3/2}\Big({}_2F_1\Big(\frac12,\frac32;1;u_2\Big),\ -\frac12\,{}_2F_1\Big(\frac32,\frac32;2;u_2\Big),-\frac12\,{}_2F_1\Big(\frac12,\frac32;2;u_2\Big)\Big),$$

where ${}_2F_1(a,b;c;\lambda)$ is the classical hypergeometric function.

**A.4. Solutions as vectors of first derivatives.** Introduce the function

$$\ell^{(\gamma)}(z)=\int_\gamma\frac{dx}{\sqrt{(x-z_1)(x-z_2)(x-z_3)}}. \tag{A.30}$$

Then

$$I^{(\gamma)}(z)=2\Big(\frac{\partial\ell^{(\gamma)}}{\partial z_1},\frac{\partial\ell^{(\gamma)}}{\partial z_2},\frac{\partial\ell^{(\gamma)}}{\partial z_3}\Big). \tag{A.31}$$

Changing the variable $x=w(z_1-z_3)+z_3$ we write

$$\begin{aligned}\ell^{(\gamma)}(z)&=(z_1-z_3)^{-1/2}\int_\gamma\frac{dw}{\sqrt{(w-1)(w-\frac{z_2-z_3}{z_1-z_3})w}},\\ &=(z_1-z_3)^{-1/2}h^{(\gamma)}\Big(\frac{z_2-z_3}{z_1-z_3}\Big)\\ &=u_1^{-1/2}h^{(\gamma)}(u_2),\end{aligned} \tag{A.32}$$

where $h^{(\gamma)}(\lambda)$ is the elliptic integral in (A.1) and $h^{(\gamma)}(\lambda)$ is a solution of equation (A.2). Denote $h(\lambda):=h^{(\gamma)}(\lambda)$. Then

(A.33)

$$I^{(\gamma)}=(z_1-z_3)^{-3/2}\Big(-h\big(\frac{z_2-z_3}{z_1-z_3}\big)-2h'\big(\frac{z_2-z_3}{z_1-z_3}\big)\frac{z_2-z_3}{z_1-z_3},2h'\big(\frac{z_2-z_3}{z_1-z_3}\big),h\big(\frac{z_2-z_3}{z_1-z_3}\big)+2h'\big(\frac{z_2-z_3}{z_1-z_3}\big)\frac{z_2-z_1}{z_1-z_3}\Big)$$

(A.34)

$$=u_1^{-3/2}(-h(u_2)-2h'(u_2)u_2,\ 2h'(u_2),\ h(u_2)+2h'(u_2)(u_2-1)).$$

Formula (A.34) relates solutions of system (A.5) and solutions of system (A.20). If $(h,h')$ is a solution of system (A.5), then equation

$$\begin{pmatrix}I_1(u_1,u_2)\\ I_2(u_1,u_2)\\ I_3(u_1,u_2)\end{pmatrix}=u_1^{-3/2}\begin{pmatrix}-1&-2u_2\\ 0&2\\ 1&2u_2-2\end{pmatrix}\begin{pmatrix}h(u_2)\\ h'(u_2)\end{pmatrix} \tag{A.35}$$

gives a solution $(I_1,I_2,I_3)$ of system (A.20). Conversely, if $(I_1,I_2,I_3)$ is a solution of system (A.20) then formula

$$\begin{pmatrix}h(u_2)\\ h'(u_2)\end{pmatrix}=u_1^{3/2}\begin{pmatrix}-1&-u_2&0\\ 0&1/2&0\end{pmatrix}\begin{pmatrix}I_1(u_1,u_2)\\ I_2(u_1,u_2)\\ I_3(u_1,u_2)\end{pmatrix} \tag{A.36}$$

gives a solution $(h,h')$ of system (A.5).

Using the cycle $\gamma_1$ we can evaluate

$$\begin{aligned}\ell^{(\gamma_1)}&=(z_1-z_3)^{-1/2}F\Big(\frac{z_2-z_3}{z_1-z_3}\Big)\\ &=u_1^{-1/2}F(u_2),\qquad\text{where}\quad F(\lambda)={}_2F_1\Big(\frac12,\frac12;1;\lambda\Big).\end{aligned} \tag{A.37}$$

Denote

$$\ell := \ell^{(\gamma_1)}, \qquad \mathcal{I} := \frac{I}{\ell} = \frac{2}{\ell}\Big(\frac{\partial \ell}{\partial z_1}, \frac{\partial \ell}{\partial z_2}, \frac{\partial \ell}{\partial z_3}\Big), \tag{A.38}$$

where $I$ is defined in (A.27). Formulas (A.33) and (A.34) imply

$$\mathcal{I} = \tfrac{1}{z_1-z_3}\big(-1 - 2\eta\big(\tfrac{z_2-z_3}{z_1-z_3}\big)\tfrac{z_2-z_3}{z_1-z_3},\ 2\eta\big(\tfrac{z_2-z_3}{z_1-z_3}\big),\ 1+2\eta\big(\tfrac{z_2-z_3}{z_1-z_3}\big)\tfrac{z_2-z_1}{z_1-z_3}\big), \tag{A.39}$$

$$= \tfrac{1}{u_1}\big(-1-2\eta(u_2)u_2,\ 2\eta(u_2),\ 1+2\eta(u_2)(u_2-1)\big), \tag{A.40}$$

where the function $\eta(\lambda)$ is defined in (A.7).

**A.5. Six coordinate systems.** System (A.18) of KZ equations has 6 distinguished coordinate systems (asymptotic zones). They are labeled by permutations $\sigma = (i,j,k) \in S_3$. The coordinate system $u^\sigma = (u_1^\sigma, u_2^\sigma, u_3^\sigma)$ is defined by the formulas

$$u_1^\sigma = z_i - z_k, \qquad u_2^\sigma = \frac{z_j - z_k}{z_i - z_k}, \qquad u_3^\sigma = z_1 + z_2 + z_3. \tag{A.41}$$

For the identity element id $= (1,2,3)$ the corresponding coordinate system is defined in (A.19).

Having one of these coordinate systems we repeat the constructions of Sections A.3-A.4 and construct a scalar function $\ell^\sigma(z)$ and vector-valued functions $I^\sigma(z)$, $\mathcal{I}^\sigma(z)$, such that $\mathcal{I}^\sigma(z) = I^\sigma(z)/\ell^\sigma(z)$. For the identity element id $= (1,2,3)$ these functions are $\ell(z)$, $I(z)$, $\mathcal{I}(z)$ in (A.37), (A.27), (A.39). Notice that the functions $\ell(z)$ and $I(z)$ are defined as integrals over $\gamma_1$, and that $\gamma_1$ is defined with the help of coordinates $u_1, u_2, u_3$.

For any $\sigma$ the function $I^\sigma(z)$ is a power series solution of system (A.18) in the chart with coordinates $u^\sigma$, see (A.26) and (A.27).

Below we list the functions $\mathcal{I}^\sigma$ :

(A.42)

$$\begin{aligned}
\mathcal{I}^{123} &= \tfrac{1}{z_1-z_3}\big(-1-2\eta\big(\tfrac{z_2-z_3}{z_1-z_3}\big)\tfrac{z_2-z_3}{z_1-z_3},\ 2\eta\big(\tfrac{z_2-z_3}{z_1-z_3}\big),\ 1+\eta\big(\tfrac{z_2-z_3}{z_2-z_3}\big)\tfrac{z_2-z_1}{z_1-z_3}\big),\\
\mathcal{I}^{321} &= \tfrac{1}{z_3-z_1}\big(1+2\eta\big(\tfrac{z_2-z_1}{z_3-z_1}\big)\tfrac{z_2-z_3}{z_3-z_1},\ 2\eta\big(\tfrac{z_2-z_1}{z_3-z_1}\big),\ -1-2\eta\big(\tfrac{z_2-z_1}{z_3-z_1}\big)\tfrac{z_2-z_1}{z_3-z_1}\big),\\
\mathcal{I}^{213} &= \tfrac{1}{z_2-z_3}\big(2\eta\big(\tfrac{z_1-z_3}{z_2-z_3}\big),\ -1-2\eta\big(\tfrac{z_1-z_3}{z_2-z_3}\big)\tfrac{z_1-z_3}{z_2-z_3},\ 1+2\eta\big(\tfrac{z_1-z_3}{z_2-z_3}\big)\tfrac{z_1-z_2}{z_2-z_3}\big),\\
\mathcal{I}^{132} &= \tfrac{1}{z_1-z_2}\big(-1-\eta\big(\tfrac{z_3-z_2}{z_1-z_2}\big)\tfrac{z_3-z_2}{z_1-z_2},\ 1+2\eta\big(\tfrac{z_3-z_2}{z_1-z_2}\big)\tfrac{z_3-z_1}{z_2-z_2},\ 2\eta\big(\tfrac{z_3-z_2}{z_1-z_2}\big)\big),\\
\mathcal{I}^{231} &= \tfrac{1}{z_2-z_1}\big(1+2\eta\big(\tfrac{z_3-z_1}{z_2-z_1}\big)\tfrac{z_3-z_2}{z_2-z_1},\ -1-2\eta\big(\tfrac{z_3-z_1}{z_2-z_1}\big)\tfrac{z_3-z_1}{z_2-z_1},\ 2\eta\big(\tfrac{z_3-z_1}{z_2-z_1}\big)\big),\\
\mathcal{I}^{312} &= \tfrac{1}{z_3-z_2}\big(2\eta\big(\tfrac{z_1-z_2}{z_3-z_2}\big),\ 1+2\eta\big(\tfrac{z_1-z_2}{z_3-z_2}\big)\tfrac{z_1-z_3}{z_3-z_2},\ -1-2\eta\big(\tfrac{z_1-z_2}{z_3-z_2}\big)\tfrac{z_1-z_2}{z_3-z_2)^2}\big).
\end{aligned}$$

THEOREM A.1. *For $(i,j,k) \in S_3$ consider the three functions $\mathcal{I}^{ijk}$, $\mathcal{I}^{kji}$, $\mathcal{I}^{jik}$. Then $\mathcal{I}^{kji}$ is transformed to $\mathcal{I}^{ijk}$ by application of formula $\eta(1-\lambda) = -\eta(\lambda)$ and $\mathcal{I}^{jik}$ is transformed to $\mathcal{I}^{ijk}$ by application of formula $\eta(1/\lambda) = -\lambda^2\eta(\lambda) - \lambda/2$.*

PROOF. The proof is straightforward. For example we check the statement for $(i,j,k) = (1,2,3)$. In this case the functions $\mathcal{I}^{123}$, $\mathcal{I}^{321}$, $\mathcal{I}^{132}$ are

$$\begin{gathered}
\tfrac{1}{u_1}\big(-1-2\eta(u_2)u_2,\ 2\eta(u_2),\ 1+2\eta(u_2)(u_2-1)\big),\\
\tfrac{1}{u_1}\big(-1+2\eta(1-u_2)u_2,\ -2\eta(1-u_2),\ 1+2\eta(1-u_2)(1-u_2)\big),\\
\tfrac{1}{u_1u_2}\big(2\eta\big(\tfrac{1}{u_2}\big),\ -1-2\eta\big(\tfrac{1}{u_2}\big)\tfrac{1}{u_2},\ 1+2\eta\big(\tfrac{1}{u_2}\big)\tfrac{1-u_2}{u_2}\big),
\end{gathered}$$

where $u_1, u_2$ are defined in (A.19). Then formula $\eta(1-u_2) = -\eta(u_2)$ transforms the second function to the first and the formula $\eta(1/u_2) = -u_2^2\eta(u_2) - u_2/2$ transforms the third function to the first. □

Define

$$\tilde{\mathfrak{D}}_0 = \{(z_1, z_2, z_3) \in \mathbb{Q}_p^3 \mid z_i \neq z_j \ \forall i \neq j\}.$$

For any $\sigma = (i, j, k) \in S_3$ define

$$\tilde{\mathfrak{D}}_1^\sigma = \Big\{(z_1, z_2, z_3) \in \tilde{\mathfrak{D}}_0 \ \Big|\ \frac{z_j - z_k}{z_i - z_k} \in \mathbb{Z}_p,\ \Big|g\Big(\frac{z_j - z_k}{z_i - z_k}\Big)\Big|_p = 1\Big\},$$

$$\tilde{\mathfrak{D}}_2^\sigma = \Big\{(z_1, z_2, z_3) \in \tilde{\mathfrak{D}}_0 \ \Big|\ \frac{z_i - z_k}{z_j - z_k} \in \tilde{\mathfrak{D}}_1^\sigma\Big\}, \qquad \tilde{\mathfrak{D}}^\sigma = \tilde{\mathfrak{D}}_1^\sigma \cup \tilde{\mathfrak{D}}_2^\sigma, \qquad \tilde{\mathfrak{D}} = \sum_{\sigma \in S_3} \tilde{\mathfrak{D}}^\sigma,$$

where the function $g$ is defined in (A.6).

For any any $(i, j, k) \in S_3$ the functions $\mathcal{I}^{ijk}$, $\mathcal{I}^{kji}$, $\mathcal{I}^{jik}$ define an analytic element on $\tilde{\mathfrak{D}}^\sigma$, see Section A.1.7 and [**Dw**]. Theorem A.1 implies the following corollary.

COROLLARY A.2. *The functions* $(\mathcal{I}^{ijk})_{(i,j,k)\in S_3}$ *define an analytic element on* $\tilde{\mathfrak{D}}$. □

REMARK. Dwork's formulas (A.8) present the $S_3$-symmetries of the analytic element $(\eta(\lambda),\ -\eta(1-\lambda),\ -\eta(1/\lambda)/\lambda^2 - 1/(2\lambda))$. Dwork's $S_3$-symmetries reformulated as $S_3$-symmetries of the analytic element $(\mathcal{I}^{ijk})_{(i,j,k)\in S_3}$ look even more well-rounded.

**A.6. Subbundle.** Denote $\tilde{W} = \{(I_1, I_2, I_3) \in \mathbb{Q}_p^3 \mid I_1 + I_2 + I_3 = 0\}$. System (A.18) of KZ equations defines a flat connection on the trivial bundle $\tilde{W} \times \tilde{\mathfrak{D}}_0 \to \tilde{\mathfrak{D}}_0$. The flat sections of that bundle are solutions of system (A.18) of KZ equations.

For any $\alpha \in \tilde{\mathfrak{D}}$ such that $\alpha \in \tilde{\mathfrak{D}}^\sigma$ the vector $\mathcal{I}^\sigma(\alpha)$ spans a one-dimensional subspace $\tilde{U}_\alpha \subset \tilde{W}$. That subspace does not depend on $\sigma$ such that $\alpha \in \tilde{\mathfrak{D}}^\sigma$. The union of these subspaces defines a one-dimensional subbundle $\tilde{\mathcal{U}} \to \tilde{\mathfrak{D}}$ of the trivial bundle $\tilde{W} \times \tilde{\mathfrak{D}} \to \tilde{\mathfrak{D}}$.

THEOREM A.3. *The subbundle* $\tilde{\mathcal{U}} \to \tilde{\mathfrak{D}}$ *is invariant with respect to the KZ connection on* $\tilde{W} \times \tilde{\mathfrak{D}} \to \tilde{\mathfrak{D}}$.

PROOF. For any $\sigma \in S_3$ the subbundle $\tilde{\mathcal{U}} \to \tilde{\mathfrak{D}}$ is generated by the flat section $I^\sigma$ near the points where $u_2^\sigma = 0$. Hence the subbundle $\tilde{\mathcal{U}} \to \tilde{\mathfrak{D}}$ is generated by a flat section near any point of $\tilde{\mathfrak{D}}$, see Section A.1.8 and [**Dw**, Lemma 3.1]. □

REMARK. For any $\sigma \in S_3$ the flat section $I^\sigma$ generates the subbundle $\tilde{\mathcal{U}} \to \tilde{\mathfrak{D}}$ near the points where $u_2^\sigma = 0$. The power series $I^\sigma$ considered over $\mathbb{C}$ is the expansion of an integral over a cycle vanishing at the points where $u_2^\sigma = 0$. The analytic continuation over $\mathbb{C}$ of that integral over that vanishing cycle could not generate a one-dimensional subbundle of the trivial bundle $\tilde{W} \times \tilde{\mathfrak{D}} \to \tilde{\mathfrak{D}}$ since the monodromy representation of the complex KZ equations in this case is irreducible. In contrast with this fact over $\mathbb{C}$, the $p$-adic power series solutions $I^\sigma$, $\sigma \in S_3$, defined at different points glue together into a single line bundle $\tilde{\mathcal{U}} \to \tilde{\mathfrak{D}}$. This line bundle is what Dwork calls a *p-adic cycle.* This $p$-adic phenomenon was stressed by Dwork in [**Dw**] who titled his paper *P-adic Cycles.*

REMARK. The invariant subbundles of the KZ connection over $\mathbb{C}$ usually are related to some additional conformal block constructions, see [**FSV1**, **FSV2**, **SV2**, **V7**]. Apparently the subbundle $\tilde{\mathcal{U}} \to \tilde{\mathfrak{D}}$ is of a different $p$-adic nature, cf. [**V7**].

**A.7. Boundedness.** Let $\sigma \in S_3$ and $\alpha \in \tilde{\mathfrak{D}}^\sigma$. For $w \in W$ let $I(z; w)$ be the germ at $\alpha$ of the solution of the KZ equations with initial condition $I(\alpha, w) = w$. By formula (A.34), the coordinates of $I(z; w)$ have the form

$$(u_1^\sigma)^{-3/2}(-h(u_2^\sigma) - 2h'(u_2^\sigma)u_2^\sigma), \qquad (u_1^\sigma)^{-3/2}2h'(u_2^\sigma),$$
$$(u_1^\sigma)^{-3/2}(h(u_2^\sigma) + 2h'(u_2^\sigma)(u_2^\sigma - 1)),$$

where $h$ is the germ at the point $u_2^\sigma = u_2^\sigma(\alpha)$ of a solution of equation (A.2). We say that the germ $I(z; w)$ is bounded if each of the germs $-h(u_2^\sigma) - 2h'(u_2^\sigma)u_2^\sigma$, $2h'(u_2^\sigma)$, $h(u_2^\sigma) + 2h'(u_2^\sigma)(u_2^\sigma - 1)$ is bounded in its disc of convergence.

THEOREM A.4. *The germ $I(z; w)$ is bounded if and only if $w \in \tilde{U}_\alpha$.*

PROOF. Let $w \in \tilde{U}_\alpha$. Then the germ $h$ belongs to the corresponding subspace $U_{u_2^\sigma(\alpha)}$ defined in Section A.1.8. By [**Dw**, Lemma 4.2] the germ $h$ is bounded in its disc of convergence, see Section A.1.9. Hence each of the three germs $-h(u_2^\sigma) - 2h'(u_2^\sigma)u_2^\sigma$, $2h'(u_2^\sigma)$, $h(u_2^\sigma) + 2h'(u_2^\sigma)(u_2^\sigma - 1)$ is bounded in its disc of convergence.

If $w \notin \tilde{U}_\alpha$, then $h \notin U_{u_2^\sigma(\alpha)}$. By [**Dw**, Lemma 4.2] the germ $h$ is unbounded in its disc of convergence. Then at least one of the three germs $-h(u_2^\sigma) - 2h'(u_2^\sigma)u_2^\sigma$, $2h'(u_2^\sigma)$, $h(u_2^\sigma) + 2h'(u_2^\sigma)(u_2^\sigma - 1)$ is unbounded in its disc of convergence. □

**A.8. More domains.** Denote

(A.43)
$$\tilde{\mathfrak{D}}_3 = \Big\{(z_1, z_2, z_3) \in \tilde{\mathfrak{D}}_0 \;\Big|\; \frac{z_2 - z_3}{z_1 - z_3} \in \mathbb{Z}_p,\; \Big|\frac{z_2 - z_3}{z_1 - z_3}\Big|_p = 1,\; \Big|\frac{z_2 - z_3}{z_1 - z_3} - 1\Big|_p = 1\Big\},$$
$$\tilde{\mathfrak{D}}_4 = \Big\{(z_1, z_2, z_3) \in \tilde{\mathfrak{D}}_0 \;\Big|\; \frac{z_2 - z_3}{z_1 - z_3} \in \mathbb{Z}_p,\; \varepsilon < \Big|\frac{z_2 - z_3}{z_1 - z_3}\Big|_p < 1\Big\},$$

where $\varepsilon$ is the same number as in (A.13).

**A.9. Frobenius map on solutions of KZ equations.** Formula (A.14) describes the Frobenius map on solutions of equation (A.5). Solutions of equation (A.5) are identified with solutions of the KZ system (A.20) by formulas (A.35) and (A.36). That allows us to define the Frobenius map on solutions of the KZ system (A.20).

Denote

$$\text{(A.44)}\quad \mathcal{B}(u_1, u_2) = u_1^{3/2}\begin{pmatrix} -1 & -u_2 & 0 \\ 0 & 1/2 & 0 \end{pmatrix}, \quad \mathcal{C}(u_1, u_2) = u_1^{-3/2}\begin{pmatrix} -1 & -2u_2 \\ 0 & 2 \\ 1 & 2u_2 - 2 \end{pmatrix}.$$

we have $\mathcal{B}(u_1, u_2)\mathcal{C}(u_1, u_2) = \begin{pmatrix} 1 & 0 \\ 0 & 1 \end{pmatrix}$, $\mathcal{C}(u_1, u_2)\mathcal{B}(u_1, u_2) = \begin{pmatrix} 1 & 0 & 0 \\ 0 & 1 & 0 \\ -1 & -1 & 0 \end{pmatrix}$. The second matrix defines the identity operator on the space $\tilde{W} = \{(I_1, I_2, I_3) \in \mathbb{Q}_p^3 \mid I_1 + I_2 + I_3 = 0\}$.

Recall the matrix $\mathcal{F}(u_2)$ defined in (A.11). By formula (A.35) the matrix

$$\tilde{\mathcal{F}}(u_1, u_2) = \mathcal{C}(u_1, u_2)\mathcal{F}(u_2)$$

is a fundamental matrix of solutions of system (A.20). Recall the matrices $\mathcal{A}(\lambda)$, $M$ in (A.12). Denote

$$\tilde{\mathcal{A}}(u_1,u_2) = \mathcal{C}(u_1,u_2)\,\mathcal{A}(u_2)\,\mathcal{B}((u_1)^p,(u_2)^p). \tag{A.45}$$

This is a $3\times 3$ matrix valued function, whose values preserve the subspace $\tilde{W}\subset\mathbb{Q}_p^3$.

THEOREM A.5. *We have*

$$\tilde{\mathcal{A}}(u_1,u_2)\,\tilde{\mathcal{F}}((u_1)^p,(u_2)^p) \;=\; \tilde{\mathcal{F}}(u_1,u_2)\,M. \tag{A.46}$$

*The matrix $\tilde{\mathcal{A}}(u_1,u_2)$ extends to an analytic function on the domain $\tilde{\mathfrak{D}}_3\cup\tilde{\mathfrak{D}}_4$.*

PROOF. The theorem is a corollary of formula (A.14) and Dwork's statements listed in Section A.1.10. □

We call $\tilde{\mathcal{A}}(u_1,u_2)$ the matrix of the Frobenius transformation of solutions of system (A.20) relative to the fundamental matrix $\tilde{\mathcal{F}}(u_1,u_2)$ on the domain $\tilde{\mathfrak{D}}_3\cup\tilde{\mathfrak{D}}_4$.

Recall the distinguished solution

$$I(u_1,u_2) = u_1^{-3/2}(-F(u_2)-2F'(u_2)u_2,\ 2F'(u_2),\ F(u_2)+2F'(u_2)(u_2-1)) \tag{A.47}$$

of system (A.20) defined near the points where $u_2=0$, see (A.34). By (A.15) we have

$$\tilde{\mathcal{A}}(u_1,u_2)\,I((u_1)^p,(u_2)^p) \;=\; (-1)^{(p-1)/2}\,I(u_1,u_2) \tag{A.48}$$

on $\tilde{\mathfrak{D}}_4$. Recall $\ell(u_1,u_2)=u_1^{-1/2}F(u_2)$ in (A.37). Dividing both sides in (A.47) by $\ell((u_1)^p,(u_2)^p)$ we can reformulate (A.48) as

$$\tilde{\mathcal{A}}(u_1,u_2)\,\mathcal{I}((u_1)^p,(u_2)^p) \;=\; (-1)^{(p-1)/2}u_1^{(p-1)/2}f(u_2)\,\mathcal{I}(u_1,u_2) \tag{A.49}$$

on $\tilde{\mathfrak{D}}_4$, see $\mathcal{I}(u_1,u_2)$ in (A.40) and $f(u_2)$ in (A.7). As in Section A.1.10 we conclude with Dwork that relation (A.49) can be analytically continued to the domain $\tilde{\mathfrak{D}}_1^{(1,2,3)}\cap\tilde{\mathfrak{D}}_3$.

Equation (A.49) implies that for any $\alpha\in\mathbb{F}_p^\times-\{1\}$, $\beta\in\mathbb{F}_p^\times$ such that $\omega(\alpha)\in\mathfrak{D}_1$, the vector $\mathcal{I}(\omega(\beta),\omega(\alpha))$ is an eigenvector of the Frobenius matrix $\tilde{\mathcal{A}}(\omega(\beta),\omega(\alpha))$ with eigenvalue $\omega(\beta^{(p-1)/2})(-1)^{(p-1)/2}f(\omega(\alpha))$,

$$\tilde{\mathcal{A}}(\omega(\beta),\omega(\alpha))\,\mathcal{I}(\omega(\beta),\omega(\alpha)) \;=\; \omega(\beta^{(p-1)/2})(-1)^{(p-1)/2}f(\omega(\alpha))\,\mathcal{I}(\omega(\beta),\omega(\alpha))\,. \tag{A.50}$$

In this Section A.9 we described the matrix $\tilde{\mathcal{A}}(u_1,u_2)$ of the Frobenius transformation of solutions of system (A.18) written in coordinates $u_1,u_2,u_3$ corresponding to the chart labeled by the identify permutation $(1,2,3)\in S_3$, In the same way we may start with the chart corresponding to any permutation $\sigma\in S_3$ and describe the matrix of the Frobenius transformation of solutions of system (A.18) written in coordinates $u_1^\sigma,u_2^\sigma,u_3^\sigma$.

### A.10. Eigenvalue $\omega(\beta^{(p-1)/2})(-1)^{(p-1)/2}f(\omega(\alpha))$.

THEOREM A.6. *The number $\omega(\beta^{(p-1)/2})(-1)^{(p-1)/2}f(\omega(\alpha))$ is the unit root of the elliptic curve $E(\alpha,\beta)$ defined over $\mathbb{F}_p$ by the affine equation*

$$w^2 = \beta\, v(v-1)(v-\alpha). \tag{A.51}$$

PROOF. Assume that $\beta \in \mathbb{F}_p^\times$ is a square, $\beta = \gamma^2$ for some $\gamma \in \mathbb{F}_p$. Then on the one hand the change of the variable $\tilde{w} = w/\gamma$ makes $E(\alpha,\beta)$ isomorphic to $E(\alpha,1)$. On the other hand $\beta^{(p-1)/2} = 1$ and $\omega(\beta^{(p-1)/2})(-1)^{(p-1)/2}f(\omega(\alpha)) = (-1)^{(p-1)/2}f(\omega(\alpha))$, where the last number is the unit root of the elliptic curve $E(\alpha,1)$ by [**Dw**].

Assume that $\beta \in \mathbb{F}_p^\times$ is not a square. Denote by $N_{1,\beta}$ the number of points on $E(\alpha,\beta)$. Then

$$N_{1,1} + N_{1,\beta} = 4 + 4 + 2(p-3) = 2p+2. \tag{A.52}$$

Indeed the number 4+4 corresponds to the points $(0,0)$, $(0,1)$, $(0,\alpha)$, $\infty$ on $E(\alpha,\beta)$ and on $E(\alpha,1)$. The number $2(p-3)$ corresponds to $p-3$ elements of $\mathbb{F}_p - \{0,1,\alpha\}$. Namely if $v_0 \in \mathbb{F}_p - \{0,1,\alpha\}$, then exactly one of the two elements $\beta v_0(v_0-1)(v_0-\alpha)$, $v_0(v_0-1)(v_0-\alpha)$ is a square in $\mathbb{F}_p$ and exactly one of the two elliptic curves has two points over $v = v_0$, while the other curve does not have points over $v_0$.

It is known that the zeta function of the curve $E(\alpha,\beta)$ has the form

$$\exp\Big(\sum_{s=1}^{\infty} \frac{N_{s,\beta}}{s}T^s\Big) = \frac{(1-R_\beta T)(1-(p/R_\beta)T)}{(1-T)(1-pT)}. \tag{A.53}$$

Here $N_{s,\beta}$ is the number points on $E(\alpha,\beta)$ considered over the field $\mathbb{F}_{p^s}$, while the number $R_\beta$ has $|R_\beta|_p = 1$ and is called the unit root, for example see [**Mo**]. Equation (A.53) implies that for any $s$ we have $N_{s,\beta} = 1 + p^s - R_\beta^s - (p/R_\beta)^s$. In particular for $s = 1$ we have

$$N_{1,\beta} = 1 + p - R_\beta - p/R_\beta. \tag{A.54}$$

From (A.52) and (A.54) we obtain

$$0 = R_\beta + p/R_\beta + R_1 + p/R_1 = (R_\beta + R_1)(1 + p/R_\beta R_1).$$

Since the second factor is nonzero we conclude that $R_\beta = -R_1$. By [**Dw**] we have $R_1 = (-1)^{(p-1)/2}f(\omega(\alpha))$. Hence

$$R_\beta = -(-1)^{(p-1)/2}f(\omega(\alpha)) = \omega(\beta^{(p-1)/2})(-1)^{(p-1)/2}f(\omega(\alpha)).$$

The theorem is proved. □

The relation between the eigenvector $\mathcal{I}(\omega(\beta),\omega(\alpha))$ and the elliptic curve $E(\alpha,\beta)$, indicated in Theorem A.6, can be explained as follows. Over $\mathbb{C}$ the vector $\mathcal{I}$ is given by integrals over cycles on elliptic curves with equation $y^2 = (x-z_1)(x-z_2)(x-z_3)$. After the change of variables

$$x = (z_1 - z_3)w + z_3, \quad u_1 = z_1 - z_3, \quad u_2 = \frac{z_2 - z_3}{z_1 - z_3}, \quad y = (z_1 - z_3)v,$$

the equation takes the form

$$v^2 = u_1(w-1)(w-u_2)w. \tag{A.55}$$

The eigenvector $\mathcal{I}(\omega(\beta),\omega(\alpha))$ corresponds to the curve in (A.55) with $(u_1,u_2) = (\omega(\beta),\omega(\alpha))$, and $(\omega(\beta),\omega(\alpha)) \equiv (\beta,\alpha) \bmod p$.

It is more surprising that system (A.18) of KZ equations gives a bit more arithmetic information than the hypergeometric equation (A.5), despite the fact system (A.18) and equation (A.5) are equivalent by (A.35) and (A.36). Indeed Dwork's eigenvectors in (A.16) give unit roots of elliptic curves $E(\alpha,1)$ while the eigenvectors in (A.50) coming from the KZ equations give unit roots of more general elliptic curves $E(\alpha,\beta)$.

**A.11. Approximation of analytic element $(\mathcal{I}^{ijk})_{(i,j,k)\in S_3}$ by rational functions.** Let $s$ be a positive integer.

A.11.1. Let $M=(p^s-1)/2$, $\Phi_{p^s}(x,z)=\prod_{i=1}^3(x-z_i)^M$,

$$P_{p^s}(x,z)=\Big(\frac{\Phi_{p^s}(x,z)}{x-z_1},\frac{\Phi_{p^s}(x,z)}{x-z_2},\frac{\Phi_{p^s}(x,z)}{x-z_3}\Big)=\sum_i P^i_{p^s}(z)\,x^i\,.$$

Denote $I^{[p^s-1]}_{p^s}(z)\;=\;P^{p^s-1}_{p^s}(z)$. The functions

$$I^{[p^s-1]}_{p^s}(z),\quad pI^{[p^{s-1}-1]}_{p^{s-1}}(z),\quad \dots,\quad p^{s-2}I^{[p^2-1]}_{p^2}(z),\quad p^{s-1}I^{[p-1]}_{p}(z),$$

are solutions of system (A.18) modulo $p^s$ by Theorem 4.4.

Let $\Phi_{p^s}(x,z)=\sum_i\Phi^i_{p^s}(z)x^i$. Denote

$$\ell_{p^s}(z)=\Phi^{p^s-1}_{p^s}(z). \tag{A.56}$$

A.11.2. For $k=1,2,3$ let

$$P_{k,p^s}(v,z):=P_{p^s}(v+z_k,z)=\sum_i P^i_{k,p^s}(z)\,v^i\,.$$

Denote $I^{[p^s-1]}_{k,p^s}(z)\;=\;P^{p^s-1}_{k,p^s}(z)$. The functions

$$I^{[p^s-1]}_{k,p^s}(z),\quad pI^{[p^{s-1}-1]}_{k,p^{s-1}}(z),\quad \dots,\quad p^{s-2}I^{[p^2-1]}_{k,p^2}(z),\quad p^{s-1}I^{[p-1]}_{k,p}(z),$$

are solutions of system (A.18) modulo $p^s$ by Theorem 9.1.

Let $\Phi_{k,p^s}(v,z):=\Phi_{p^s}(v+z_k,z)=\sum_i\Phi^i_{k,p^s}(z)v^i$. Denote

$$\ell^{[p^s-1]}_{k,p^s}(z)=\Phi^{p^s-1}_{k,p^s}(z). \tag{A.57}$$

A.11.3. Recall the homomorphisms $\mathbb{Z}\to\mathbb{Z}/p^s\mathbb{Z}$, $\mathbb{Z}[z]\to(\mathbb{Z}/p^s\mathbb{Z})[z]$, $\mathbb{Z}[z]^3\to(\mathbb{Z}/p^s\mathbb{Z})[z]^3$ denoted by $\pi_s$. Recall the subring $\mathbb{Z}[z]_{p^r}\subset\mathbb{Z}[z]$ of quasi-constants modulo $p^r$.

A.11.4. Define the filtration

$$0=\mathcal{M}^0_{p^s}\subset\mathcal{M}^1_{p^s}\subset\dots\subset\mathcal{M}^{s-1}_{p^s}\subset\mathcal{M}^s_{p^s}=\mathcal{M}_{p^s}\,, \tag{A.58}$$

where $\mathcal{M}^t_{p^s}\;=\;\big\{\pi_s\big(\sum_{r=1}^t c_r(z)\,p^{s-r}I^{[p^r-1]}_{p^r}(z)\big)\mid c_r(z)\in\mathbb{Z}[z]_{p^r}\big\}$, $t=1,\dots,s$. Every element of $\mathcal{M}_{p^s}$ is a polynomial solution of system (A.18) with coefficients in $\mathbb{Z}/p^s\mathbb{Z}$.

Define the filtration

$$0=\mathcal{M}^0_{k,p^s}\subset\mathcal{M}^1_{k,p^s}\subset\dots\subset\mathcal{M}^{s-1}_{k,p^s}\subset\mathcal{M}^s_{k,p^s}=\mathcal{M}_{k,p^s}\,, \tag{A.59}$$

where $\mathcal{M}^t_{k,p^s}\;=\;\big\{\pi_s\big(\sum_{r=1}^t c_r(z)\,p^{s-r}I^{[p^r-1]}_{k,p^r}(z)\big)\mid c_r(z)\in\mathbb{Z}[z]_{p^r}\big\}$, $t=1,\dots,s$. Every element of $\mathcal{M}_{p^s}$ is a polynomial solution of system (A.18) with coefficients in $\mathbb{Z}/p^s\mathbb{Z}$ by Theorem 9.1.

By Theorem 9.1 filtrations (A.58) and (A.59) coincide, $\mathcal{M}^t_{p^s}=\mathcal{M}^t_{k,p^s}$ for any $k,t$.

A.11.5. Define the filtration

$$0=\mathcal{L}^0_{p^s}\subset\mathcal{L}^1_{p^s}\subset\dots\subset\mathcal{L}^{s-1}_{p^s}\subset\mathcal{L}^s_{p^s}=\mathcal{L}_{p^s}\,, \tag{A.60}$$

where $\mathcal{L}^t_{p^s}\;=\;\big\{\pi_s\big(\sum_{r=1}^t c_r(z)\,p^{s-r}\ell^{[p^r-1]}_{p^r}(z)\big)\mid c_r(z)\in\mathbb{Z}[z]_{p^r}\big\}$, $t=1,\dots,s$.

Define the filtration

$$0=\mathcal{L}^0_{k,p^s}\subset\mathcal{L}^1_{k,p^s}\subset\dots\subset\mathcal{L}^{s-1}_{k,p^s}\subset\mathcal{L}^s_{k,p^s}=\mathcal{L}_{k,p^s}\,, \tag{A.61}$$

where $\mathcal{L}^t_{k,p^s}\;=\;\big\{\pi_s\big(\sum_{r=1}^t c_r(z)\,p^{s-r}\ell^{p^r-1]}_{k,p^r}(z)\big)\mid c_r(z)\in\mathbb{Z}[z]_{p^r}\big\}$, $t=1,\dots,s$.

It is easy to see that filtrations (A.60) and (A.61) coincide, $\mathcal{L}^t_{p^s} = \mathcal{L}^t_{k,p^s}$ for any $k, t$.

A.11.6. Let $u = (u_1, u_2, u_3)$ be the coordinates in the chart corresponding to $(1, 2, 3) \in S_3$, see (A.19). Consider the functions $I^{[p^s-1]}_{3,p^s}(z) \in \mathcal{M}_{p^s}$, $\ell^{[p^s-1]}_{3,p^s}(z) \in \mathcal{L}_{p^s}$. Denote

$$\hat{I}^{[p^s-1]}(u) := I^{[p^s-1]}_{3,p^s}(z(u)), \qquad \hat{\ell}^{[p^s-1]}(u) := \ell^{[p^s-1]}_{3,p^s}(z(u)). \tag{A.62}$$

We have

(A.63)

$$(-1)^{\frac{p^s-3}{2}} \hat{I}^{[p^s-1]} = u_1^{\frac{p^s-3}{2}} \sum_{a=0}^{\frac{p^s-1}{2}} \left( \binom{\frac{p^s-3}{2}}{a}\binom{\frac{p^s-1}{2}}{a}, \binom{\frac{p^s-1}{2}}{a+1}\binom{\frac{p^s-3}{2}}{a}, \binom{\frac{p^s-1}{2}}{a+1}\binom{\frac{p^s-1}{2}}{a} \right) u_2^a,$$

$$(-1)^{\frac{p^s-1}{2}} \hat{\ell}^{[p^s-1]} = u_1^{\frac{p^s-1}{2}} \sum_{a=0}^{\frac{p^s-1}{2}} \binom{\frac{p^s-1}{2}}{a}^2 u_2^a, \tag{A.64}$$

see formula (A.63) and (10.5).

Notice that $\sum_{a=0}^{\frac{p^s-1}{2}} \binom{\frac{p^s-1}{2}}{a}^2 u_2^a$ is a solution of eqution (A.2) modulo $p^s$.

As $s \to \infty$ the sequence $\big((-1)^{\frac{p^s-3}{2}} \hat{I}^{[p^s-1]}\big)_{s=1}^{\infty}$ of vector-valued polynomials uniformly converges to the series $I$ in (A.28) near the points where $u_2 = 0$, see Theorem 10.5. Similarly as $s \to \infty$ the sequence $\big((-1)^{\frac{p^s-3}{2}} \hat{\ell}^{[p^s-1]}\big)_{s=1}^{\infty}$ of scalar polynomials uniformly converges to the series $\ell$ in (A.37) near the points where $u_2 = 0$.

Corollary A.7. *As $s \to \infty$ the sequence $\big(-\hat{I}^{[p^s-1]}/\hat{\ell}^{[p^s-1]}\big)_{s=1}^{\infty}$ of vector-valued rational functions uniformly converges to the function $\mathcal{I}$ in* (A.40) *near the points where $u_2 = 0$.*

A.11.7. Let $\sigma = (i, j, k) \in S_3$. Similarly to Section A.11.6 consider the coordinates $u^\sigma$ and the functions $I^{[p^s-1]}_{k,p^s}(z) \in \mathcal{M}_{p^s}$, $\ell^{[p^s-1]}_{k,p^s}(z) \in \mathcal{L}_{p^s}$. Denote

$$\hat{I}^{[\sigma,p^s-1]}(u^\sigma) := I^{[p^s-1]}_{k,p^s}(z(u^\sigma)), \qquad \hat{\ell}^{[\sigma,p^s-1]}(u^\sigma) := I^{[p^s-1]}_{k,p^s}(z(u^\sigma)). \tag{A.65}$$

Similarly to Section A.11.6 we obtain the following corollary.

Corollary A.8. *As $s \to \infty$ the sequence $\big(-\hat{I}^{[\sigma,p^s-1]}/\hat{\ell}^{[\sigma,p^s-1]}\big)_{s=1}^{\infty}$ of vector-valued rational functions uniformly converges to the function $\mathcal{I}^\sigma$ in* (A.40) *near the points where $u_2^\sigma = 0$.*

A.11.8. This Appendix A is devoted to the relation between the analytic element $(\mathcal{I}^{ijk})_{(i,j,k)\in S_3}$ and Dwork's theory in [**Dw**].

As additional information, for any $\sigma = (i, j, k) \in S_3$, Corollary A.8 indicates the sequence of polynomials $I^{[p^s-1]}_{k,p^s}(z) \in \mathcal{M}_{p^s}$, $\ell^{[p^s-1]}_{k,p^s}(z) \in \mathcal{L}_{p^s}$ whose ratio $p$-adically tends to the function $\mathcal{I}^{ijk}$ near the points $u_2^\sigma = 0$ where the function $\mathcal{I}^{ijk}$ is initially defined.

**A.12. Further directions.** In Sections 2–10 we considered system (2.1) of KZ equations with parameter $n = 2g + 1$ and constructed polynomial solutions of system (2.1) modulo $p^s$. We defined the module $\mathcal{M}_{p^s}$ of the constructed solutions and studied the limit of $\mathcal{M}_{p^s}$ as $s \to \infty$. Namely we considered a special coordinate system $u = u(z)$ in (9.7) associated with one of the asymptotic zones of the KZ equations and showed that in this coordinate system the limit of $\mathcal{M}_{p^s}$ as $s \to \infty$ produces a $g$-dimensional space of solutions of system (2.1) over $p$-adic numbers $\mathbb{Q}_p$ in the neighborhood of the point $u = 0$.

Constructions in this appendix for $g = 1$ and Dwork's theory in [**Dw**] suggest the following project. Consider all asymptotic zones of system (2.1), see their definition for example in [**V2**]. The asymptotic zones are labeled by suitable trees $T$. These trees are analogs of the elements $\sigma \in S_3$ in the appendix. Each asymptotic zone has a distinguished system of coordinates $u^T$. Probably, for every asymptotic zone the limit of $\mathcal{M}_{p^s}$ as $s \to \infty$ produces a $g$-dimensional space $V_T$ of solutions of system (2.1) considered over $\mathbb{Q}_p$ in a neighborhood of the point $u^T = 0$. Probably the spaces $V_T$ of $p$-adic solutions, defined at different places $u^T = 0$, analytically continue into a single global invariant $g$-dimensional vector subbundle of the associated KZ connection on the trivial vector bundle of rank $2g$. Following Dwork and Theorem A.4 we may expect that this subbundle is spanned at any point of the base by the germs of all solutions of the KZ equations bounded in their polydiscs of convergence. This subbundle would give a generalization of the line subbundle generated by the analytic element $(\mathcal{I}^{ijk})_{(i,j,k)\in S_3}$ constructed in this appendix. Probably, that $g$-dimensional subbundle will determine the set of unit roots of the curves with equation $y^2 = \prod_{i=1}^n (x - z_i)$ over the field $\mathbb{F}_p$ similarly to how it is done in Sections A.9 and A.10 for the elliptic curves.

## Acknowledgments

The author thanks Steven Sperber for collaboration on this project. The author thanks Pavel Etingof and Vadim Schechtman for useful discussions.

## References

[AH] J. Achter, E. Howe, *Hasse-Witt and Cartier-Manin matrices: A warning and a request*, 2017, `arXiv:1710.1072`, 1–14

[BV1] F. Beukers, M. Vlasenko, *Dwork Crystals I*, 2019, `arXiv:1903.11155`, 1–27

[BV2] F. Beukers, M. Vlasenko, *Dwork Crystals II*, 2019, `arXiv:1907.10390`, 1–12

[Cl] C. Herbert Clemens, *A scrapbook of complex curve theory*, 2nd ed., Graduate Studies in Mathematics, vol. 55, American Mathematical Society, Providence, RI, 2003, DOI 10.1090/gsm/055. MR1946768

[Dw] B. Dwork, *p-adic cycles*, Inst. Hautes Études Sci. Publ. Math. **37** (1969), 27–115. MR294346

[EFK] Pavel I. Etingof, Igor B. Frenkel, and Alexander A. Kirillov Jr., *Lectures on representation theory and Knizhnik-Zamolodchikov equations*, Mathematical Surveys and Monographs, vol. 58, American Mathematical Society, Providence, RI, 1998, DOI 10.1090/surv/058. MR1629472

[FSV1] B. Feigin, V. Schechtman, and A. Varchenko, *On algebraic equations satisfied by hypergeometric correlators in WZW models*, I, Comm. Math. Phys. **163** (1994), 173–184

[FSV2] B. Feigin, V. Schechtman, and A. Varchenko, *On algebraic equations satisfied by hypergeometric correlators in WZW models*, II, Comm. In Math. Phys. **70** (1995), 219–247

[Ig] J. Igusa, *Class number of a definite quaternion with prim discriminant*, Proc. Natl. Acad. Sci. U S A, **44(4)** (1958), 312–314

[KZ] V. Knizhnik and A. Zamolodchikov, *Current algebra and the Wess-Zumino model in two dimensions*, Nucl. Phys. **B247** (1984), 83–103

[Lu] E. Lucas, *Theorie des Fonctions Numeriques Simplement Periodiques*, American Journal of Mathematics. 1 (2) (1878) 184–196, DOI: 10.2307/2369308, JSTOR 2369308, MR1505161

[Ma] Y.I. Manin, *The Hasse-Witt matrix of an algebraic curve*, (Russian) Izv. Akad. Nauk SSSR Ser. Mat. **25** (1961), 153–172

[Mo] P. Monsky, *p-adic analysis and zeta functions*, Lectures in Mathematics, Kyoto University, Kinokuniya Book-Store Co., 1970, 123 pp

[RV1] R. Rimányi, A. Varchenko, *The $\mathbb{F}_p$-Selberg integral*, 2020, arXiv:2011.14248, 1–19

[RV2] R. Rimányi, A. Varchenko, *The $\mathbb{F}_p$-Selberg integral of type $A_n$*, 2020, arXiv:2012.01391, 1–21

[Ro] Alain M. Robert, *A Course in p-adic Analysis*, (Graduate Texts in Mathematics), Springer, 2000

[SV1] V. Schechtman, A. Varchenko, *Arrangements of Hyperplanes and Lie Algebra Homology*, Invent. Math. **106** (1991), 139–194

[SV2] V. Schechtman, A. Varchenko, *Solutions of KZ differential equations modulo p*, The Ramanujan Journal, **48** (3), 2019, 655–683, https://doi.org/10.1007/s11139-018-0068-x, arXiv:1707.02615

[SliV] A. Slinkin, A. Varchenko, *Hypergeometric Integrals Modulo p and Hasse–Witt matrices*, 2020, arXiv:2001.06869, 1–35

[V1] A. Varchenko, *Beta-Function of Euler, Vandermonde Determinant, Legendre Equation and Critical Values of Linear Functions of Configuration of Hyperplanes*, I. Izv Akademii Nauk USSR, Seriya Mat., 53:6 (1989), 1206–1235; II, Izv. Akademii Nauk USSR, Seriya Mat. 54:1 (1990), 146–158

[V2] A. Varchenko, *Asymptotic solutions to the KZ equation and crystal base*, Comm. in Math. Phys., Springer, **171** (1995) 99–137

[V3] Alexander Varchenko, *Special functions, KZ type equations, and representation theory*, CBMS Regional Conference Series in Mathematics, vol. 98, Published for the Conference Board of the Mathematical Sciences, Washington, DC; by the American Mathematical Society, Providence, RI, 2003, DOI 10.1090/cbms/098. MR2016311

[V4] A. Varchenko, *Solutions modulo p of Gauss-Manin differential equations for multidimensional hypergeometric integrals and associated Bethe ansatz*, arXiv:1709.06189, Mathematics, 2017, 5(4), 52; DOI: 10.3390/math5040052, 1–18

[V5] A. Varchenko, *Hyperelliptic integrals modulo p and Cartier-Manin matrices*, arXiv:1806.03289, Pure and Applied Math Quarterly, **16**, (2020), n. 3, 315–336

[V6] A. Varchenko, *Remarks on the Gaudin model modulo p*, arXiv:1708.06264, Journal of Singularities, **18**, (2018), 486–499

[V7] A. Varchenko, *An invariant subbundle of the KZ connection mod p and reducibility of $\widehat{\mathfrak{sl}_2}$ Verma modules mod p*, arXiv:2002.05834 , 1–14

[V8] A. Varchenko, *Determinant of $\mathbb{F}_p$-hypergeometric solutions under ample reduction*, arXiv:2010.11275, 1–22

FACULTY OF MATHEMATICS AND MECHANICS, LOMONOSOV MOSCOW STATE UNIVERSITY, LENINSKIYE GORY 1, 119991 MOSCOW GSP-1, RUSSIA; AND DEPARTMENT OF MATHEMATICS, UNIVERSITY OF NORTH CAROLINA AT CHAPEL HILL, CHAPEL HILL, NORTH CAROLINA 27599-3250

*Email address*: anv@email.unc.edu

SELECTED PUBLISHED TITLES IN THIS SERIES

780 **Erik Koelink, Stefan Kolb, Nicolai Reshetikhin, and Bart Vlaar, Editors,** Hypergeometry, Integrability and Lie Theory, 2022

779 **Samuele Anni, Valentijn Karemaker, and Elisa Lorenzo García, Editors,** Arithmetic, Geometry, Cryptography, and Coding Theory 2021, 2022

778 **Carlos Galindo, Alejandro Melle Hernández, Julio José Moyano-Fernández, and Wilson A. Zúñiga-Galindo, Editors,** $p$-Adic Analysis, Arithmetic and Singularities, 2022

777 **Bang-Yen Chen, Nicholas D. Brubaker, Takashi Sakai, Bogdan D. Suceavă, Makiko Sumi Tanaka, Hiroshi Tamaru, and Mihaela B. Vajiac, Editors,** Differential Geometry and Global Analysis, 2022

776 **Aaron Wootton, S. Allen Broughton, and Jennifer Paulhus, Editors,** Automorphisms of Riemann Surfaces, Subgroups of Mapping Class Groups and Related Topics, 2022

775 **Fernando Galaz-García, Cecilia González-Tokman, and Juan Carlos Pardo Millán, Editors,** Mexican Mathematicians in the World, 2021

774 **Randall J. Swift, Alan Krinik, Jennifer M. Switkes, and Jason H. Park, Editors,** Stochastic Processes and Functional Analysis, 2021

773 **Nicholas R. Baeth, Thiago H. Freitas, Graham J. Leuschke, and Victor H. Jorge Pérez, Editors,** Commutative Algebra, 2021

772 **Anatoly M. Vershik, Victor M. Buchstaber, and Andrey V. Malyutin, Editors,** Topology, Geometry, and Dynamics, 2021

771 **Nicolás Andruskiewitsch, Gongxiang Liu, Susan Montgomery, and Yinhuo Zhang, Editors,** Hopf Algebras, Tensor Categories and Related Topics, 2021

770 **Stéphane Ballet, Gaetan Bisson, and Irene Bouw, Editors,** Arithmetic, Geometry, Cryptography and Coding Theory, 2021

769 **Kiyoshi Igusa, Alex Martsinkovsky, and Gordana Todorov, Editors,** Representations of Algebras, Geometry and Physics, 2021

768 **Dražen Adamović, Andrej Dujella, Antun Milas, and Pavle Pandžić, Editors,** Lie Groups, Number Theory, and Vertex Algebras, 2021

767 **Moshe Jarden and Tony Shaska, Editors,** Abelian Varieties and Number Theory, 2021

766 **Paola Comparin, Eduardo Esteves, Herbert Lange, Sebastián Reyes-Carocca, and Rubí E. Rodríguez, Editors,** Geometry at the Frontier, 2021

765 **Michael Aschbacher,** Quaternion Fusion Packets, 2021

764 **Gabriel Cunningham, Mark Mixer, and Egon Schulte, Editors,** Polytopes and Discrete Geometry, 2021

763 **Tyler J. Jarvis and Nathan Priddis, Editors,** Singularities, Mirror Symmetry, and the Gauged Linear Sigma Model, 2021

762 **Atsushi Ichino and Kartik Prasanna,** Periods of Quaternionic Shimura Varieties. I., 2021

761 **Ibrahim Assem, Christof Geiß, and Sonia Trepode, Editors,** Advances in Representation Theory of Algebras, 2021

760 **Olivier Collin, Stefan Friedl, Cameron Gordon, Stephan Tillmann, and Liam Watson, Editors,** Characters in Low-Dimensional Topology, 2020

759 **Omayra Ortega, Emille Davie Lawrence, and Edray Herber Goins, Editors,** The Golden Anniversary Celebration of the National Association of Mathematicians, 2020

758 **Jan Šťovíček and Jan Trlifaj, Editors,** Representation Theory and Beyond, 2020

757 **Kaïs Ammari and Stéphane Gerbi, Editors,** Identification and Control: Some New Challenges, 2020